AA007684

9th Annual International Conference on Sciences and Engineering (9thAIC-SE 2019)

Strengthening Academic and Industrial Collaborations to Enhance Research and Innovation

IOP Conference Series: Materials Science and Engineering
Volume 796

Banda Aceh, Indonesia
18 - 19 September 2019

ISBN: 978-1-7138-3506-6
ISSN: 1757-8981

Printed from e-media with permission by:

Curran Associates, Inc.
57 Morehouse Lane
Red Hook, NY 12571

Some format issues inherent in the e-media version may also appear in this print version.

This work is licensed under a Creative Commons Attribution 3.0 International Licence.
Licence details: http://creativecommons.org/licenses/by/3.0/.

No changes have been made to the content of these proceedings. There may be changes to pagination and minor adjustments for aesthetics.

Printed with permission by Curran Associates, Inc. (2021)

For permission requests, please contact the Institute of Physics at the address below.

Institute of Physics
Dirac House, Temple Back
Bristol BS1 6BE UK

Phone: 44 1 17 929 7481
Fax: 44 1 17 920 0979

techtracking@iop.org

Additional copies of this publication are available from:

Curran Associates, Inc.
57 Morehouse Lane
Red Hook, NY 12571 USA
Phone: 845-758-0400
Fax: 845-758-2633
Email: curran@proceedings.com
Web: www.proceedings.com

TABLE OF CONTENTS

MECHANICAL AND INDUSTRIAL ENGINEERING

STATISTICS

IOP Conf. Series: Materials Science and Engineering **796** (2020) 011001 doi:10.1088/1757-899X/796/1/011001

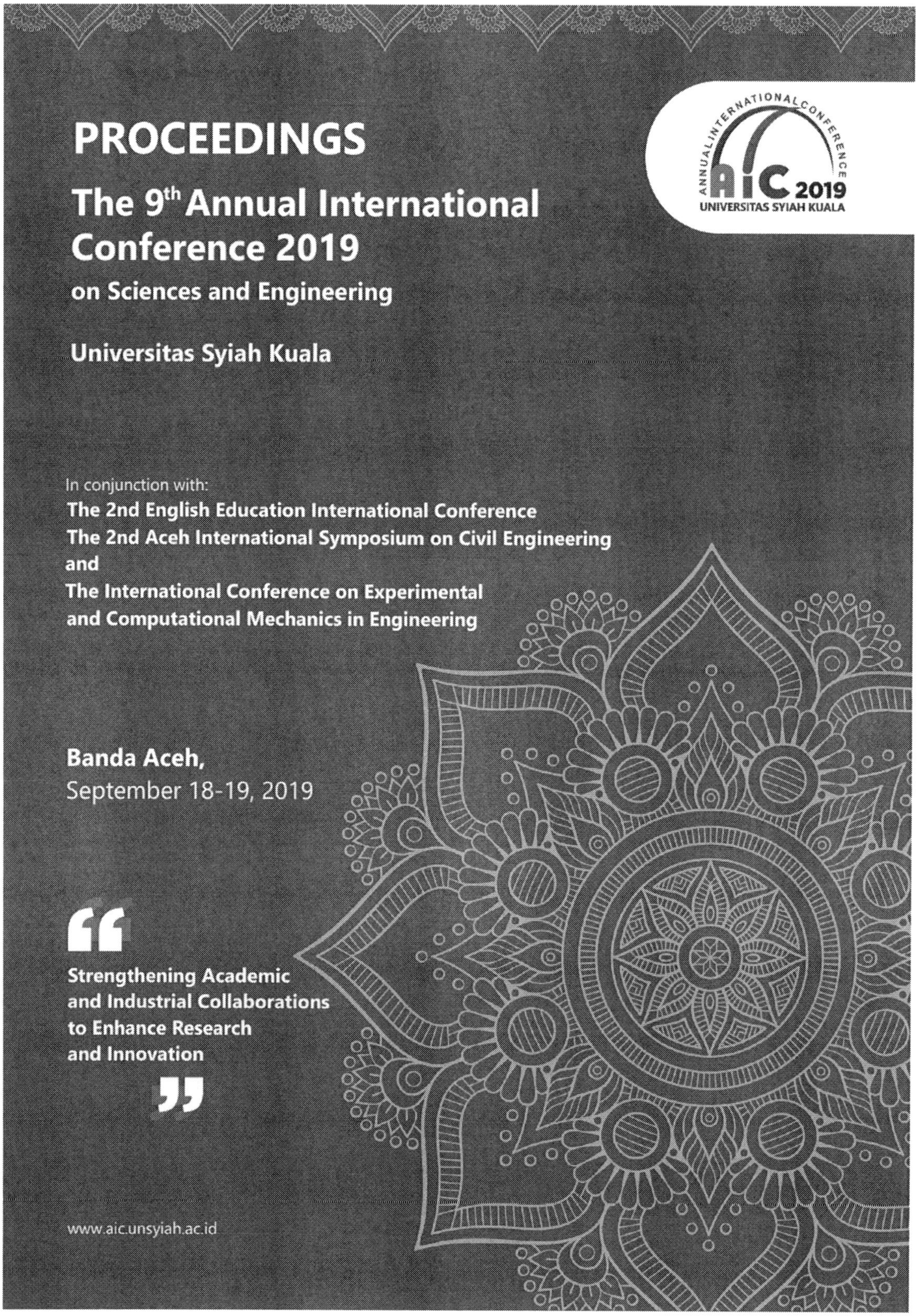

Content from this work may be used under the terms of the Creative Commons Attribution 3.0 licence. Any further distribution of this work must maintain attribution to the author(s) and the title of the work, journal citation and DOI.
Published under licence by IOP Publishing Ltd

IOP Conf. Series: Materials Science and Engineering **796** (2020) 011001 doi:10.1088/1757-899X/796/1/011001

PROCEEDINGS

The 9th Annual International Conference 2019

on Sciences and Engineering

"Strengthening Academic and Industrial Collaborations to Enhance Research and Innovation"

Universitas Syiah Kuala
Banda Aceh,
September 18-19, 2019

www.aic.unsyiah.ac.id

IOP Conf. Series: Materials Science and Engineering **796** (2020) 011001 doi:10.1088/1757-899X/796/1/011001

Conference Information

Conference Name	The 9th Annual International Conference (AIC) 2019
Conference Logo	
Host and Organizer	Universitas Syiah Kuala
Venue	Academic Activity Center (AAC) Prof. Dayan Dawood, Universitas Syiah Kuala
Official Language	English
Secretariat	Lembaga Penelitian and Pengabdian Masyarakat (LPPM), Syiah Kuala University Phone +62 651 7555262 Fax. +62 651 7555261 Website https://aic.unsyiah.ac.id Email aic@unsyiah.ac.id

IOP Conf. Series: Materials Science and Engineering **796** (2020) 011001 doi:10.1088/1757-899X/796/1/011001

Honorary Advisory Board

Rector of Universitas Syiah Kuala	Prof. Dr. Samsul Rizal
Vice Rector I	Prof. Dr. Marwan
Vice Rector II	Dr. Agussabti
Vice Rector III	Dr. Alfiansyah Yulianur
Vice Rector IV	Dr. Hizir
Head of Institute for Research and Community Services	Dr. Taufik Fuadi Abidin

Organizing Committee

General Chair	Dr.-Ing. Rudi Kurniawan
General Co-Chair	Dr. Syarizal Fonna
Treasurer and Finance Chair	Rina Indriana
Technical Program Committee Chair	Prof. Dr. Sri Aprillia

The 9th AIC on Sciences and Engineering 2019

Conference Chair	Dr. Sarwo Edhy Sofyan
Conference Co. Chair	Dr. Bambang Setiawan
Publication Chair	Prof. Dr. Nasrul Arahman
Publication Co. Chair	Dr. Saiful
Publicity and Promotion Chair	Dr. Cut Dewi
	Dr. M. Rizal
Correspondence Chair	Andriansyah, MT

IOP Conf. Series: Materials Science and Engineering **796** (2020) 011001 doi:10.1088/1757-899X/796/1/011001

International Scientific Committee

Scientific Committee Chair	Prof. Dr. Sri Aprilia., Universitas Syiah Kuala, Indonesia
Scientific Committee Member	Prof. Dr. T. M. Indra Mahlia., The University of Technology Sydney, Australia
	Prof. Dr. Dieter Schramm, Universität Duisburg-Essen, Germany
	Prof. Mochammad Noer Ilman, S.T., M.Sc., Ph.D , Univertas Gadjah Mada, Indonesia
	Prof. Dr. Husni Husin, Universitas Syiah Kuala, Indonesia
	Dr. Elwaleed Awad Khidir, Jubail University College, Kingdom Saudi Arabia
	Assoc. Prof. Dr. Azma Putra , Universiti Teknikal Malaysia, Malaysia
	Assoc. Prof. Dr. Syahrir Ridha, Universiti Teknologi Petronas, Malaysia
	Assoc. Prof. Dr.-Ing. Faizal A. Saman , Univertas Hasanuddin, Indonesia
	Dr. Thirein Myo , Military Technological College Oman, Oman
	Dr. Syifaul Huzni, Universitas Syiah Kuala, Indonesia
	Dr. Taufik Fuadi Abidin, Universitas Syiah Kuala, Indonesia
	Dr. Fitri Arnia, Universitas Syiah Kuala, Indonesia
	Dr. Zahrul Fuadi, Universitas Syiah Kuala, Indonesia
	Dr. Teuku Budi Aulia, Universitas Syiah Kuala, Indonesia
	Dr. Ashfa, Universitas Syiah Kuala, Indonesia
	Dr. Nasrullah Idris, Universitas Syiah Kuala, Indonesia
	Dr. Muliadi Ramli, Universitas Syiah Kuala, Indonesia
	Dr. Eng. Sugiarto, Universitas Syiah Kuala, Indonesia

IOP Conf. Series: Materials Science and Engineering **796** (2020) 011001 doi:10.1088/1757-899X/796/1/011001

About Unsyiah

Syiah Kuala University (Unsyiah) is the oldest public university in Aceh. Standing on September 2, 1961 by Decree of the Minister of Higher Education and Science Number 11 of 1961, dated July 21, 1961. The establishment of Unsyiah confirmed by Decree of the President of Republic Indonesia, number 161 of 1962, dated April 24, 1962 in Kopelma Darussalam, Banda Aceh. Unsyiah domiciled in the capital of Aceh province with the main campus located in the city of Students (Kopelma) Darussalam, Banda Aceh. Currently, Unsyiah has more than 30,000 students studying in 12 faculties and the Graduate Program.

About AIC

The Annual International Conference (AIC) is not only to share the scientific knowledge but also an important to open the network among the participant and also an open-door event to promote Unsyiah for the researchers in national and International level.

The 9th AIC 2019 provide an excellent forum for sharing knowledge and research findings among academicians, professionals, and governments. Topics of interests cover all theoretical and practical aspects of Sciences and Engineering, Health and Life Sciences, and Social Sciences. The primary goal AIC 2019 is organized to promote multidisciplinary understanding of the latest research and innovations through discussion, exchange, and sharing among researcher, practitioners, academicians and students.

IOP Conf. Series: Materials Science and Engineering **796** (2020) 011001 doi:10.1088/1757-899X/796/1/011001

WELCOME SPEECH FROM RECTOR OF UNIVERSITAS SYIAH KUALA

Assalamualaikum Wa Rahmatullahi Wa Barakatuh,

In the name of Allah, the Most Beneficent and the Most Merciful. May peace, mercy, and blessings of Allah be upon you.

Dear colleagues, professors, lecturers, researchers, ladies and gentlemen. On behalf of Universitas Syiah Kuala, I would like to express my sincere gratitude and welcome you to the 9th Annual International Conference (AIC) 2019. The AIC 2019 is held in conjunction with the 2nd English Education International Conference (EEIC), the 2nd Aceh International Symposium on Civil Engineering (AISCE), and the 1st International Conference on Experimental and Computational Mechanic in Engineering (ICECME).

Moreover, I honorably welcome our keynote speakers Prof. Moses Tade from Curtin University, Australia, Dr.-Ing. Joewono Prasetijo from Universiti Tun Hussein Onn, Malaysia, Prof. Diana Elizabeth Waturangi from Atma Jaya Catholic University, Indonesia, and Assoc. Prof. Dr. Rob Amery from University of Adelaide, Australia. We also have with us invited speakers Dr. Luciana from Universitas Katolik Indonesia Atma Jaya, Indonesia, and Dr. Is Helianti from Pusat Teknologi Bioindustri, Indonesia.

It is my hope that the 9th AIC 2019 would be able to achieve its objective in providing an effective forum for academician, researchers, and practitioners to strengthen academics and industrial collaborations to enhance research and innovation. We believe that academics and industry must not work alone but must collaborate. Because with collaboration, the goals become easier, more effective, and efficient to achieve. And, we can change to a better environment and life.

We also believe knowledge, research and technology are for all people, for minorities, people with disabilities, and to help anyone. No matter how much we can accomplish by ourselves, whether it be research or development, it is never sufficient in this world of knowledge. Therefore, the focal drive of this conference is to exchange ideas, and by participating in this exchange, it is hoped that all parties who may benefit from the conference can apply it in managing activities in their areas. It is pleasing to note that the agenda of this conference covers a wide range of interesting topics related to all theoretical and practical aspects. There are Sciences and Engineering, Social Sciences, and Health and Life Sciences.

Last but not least, my deepest gratitude goes to the Advisory Board, Organizing Committee, International Scientific Committee, institutions, companies, and volunteers who have directly and indirectly supported the success of this seminar. The committee has organized a vibrant scientific program and is working hard to present highly respected and internationally notorious speakers to lead it. Although we try our finest to be professional, on behalf of Syiah Kuala University, please accept our sincere apologies should there be inconveniences that occur before, during, or after the event. I wish you a very productive conference with exciting and encouraging discussions and exchange of knowledge so that together we can anticipate a future by strengthening academics and industrial collaborations to enhance research and innovation.

May God bless us all with good health to make this event a successful and enjoyable one!

Best Regards,

PROF. DR. IR. SAMSUL RIZAL, M.ENG
Rector of Universitas Syiah Kuala

IOP Conf. Series: Materials Science and Engineering **796** (2020) 011001 doi:10.1088/1757-899X/796/1/011001

WELCOME SPEECH FROM THE GENERAL CHAIR OF THE 9TH AIC 2019

Assalamualaikum Wa Rahmatullahi Wa Barakatuh,

On behalf of the organizing committees, I would like to welcome all of you to Banda Aceh, Indonesia for the 9th Annual International Conference (AIC) 2019, Universitas Syiah Kuala. The AIC is an annual international conference carried out regularly by the university since 2011.

This year, we organize the AIC in conjunction with the 2nd English Education International Conference, the 2nd Aceh International Symposium on Civil Engineering 2019 and the 1st International Conference on Experimental and Computational Mechanic in Engineering.

Under the theme of the conference "**strengthening academics and industrial collaborations to enhance research and innovation**", this conference features rich programs, including Universitas Syiah Kuala Innovation Expo 2019, four keynote speeches delivered by Prof. Moses Tade from Curtin University, Australia, Dr.-Ing. Joewono Prasetijo from Universiti Tun Hussein Onn, Malaysia, Prof. Diana Elizabeth Waturangi from Atma Jaya Catholic University, Indonesia, and Assoc. Prof. Dr. Rob Amery from University of Adelaide, Australia. Our invited speakers this year are Dr. Luciana from Universitas Katolik Indonesia Atma Jaya, Indonesia, and Dr. Is Helianti from Pusat Teknologi Bioindustri, Indonesia. The 9th AIC 2019 offers a special opportunity to bring together professors, researchers and scholars around the globe, and serves as a platform to deliver innovative research results and latest trends and development in the fields of sciences and engineering, social sciences, and health and life sciences.

All accepted papers in this conference have the opportunities to be published in various reputable publications such as IOP Material Science and Engineering Proceedings, Key Engineering Materials, Journal of Serangga, Microbiology Indonesia, Humanities and Social Sciences Reviews, International Journal of Language Studies, among others.

The conference has received more than 400 submitted papers, whereby 330 papers have been accepted by the committees for presentation and to be included in the proceedings or selected for journals. To all members of the organizing committees, the international scientific committee, the reviewers, and the collaboration partners, we would like to thank all of them for their tremendous efforts to organize this conference successfully.

We look forward to having a successful conference, and we hope that all the attendees enjoy and benefit from this conference.

Best Regards,

DR.-ING. RUDI KURNIAWAN
General Chair of the 9th AIC 2019

IOP Conf. Series: Materials Science and Engineering **796** (2020) 011001 doi:10.1088/1757-899X/796/1/011001

KEYNOTE SPEAKERS

Professor Moses O. Tadé

John Curtin Distinguished Professor, Faculty of Science & Engineering

Curtin University, Perth, Western Australia

Brief Biography of Professor Moses O. Tadé

Professor Moses Tadé received his Bachelor degree in Chemical Engineering with first class honours from University of Ife (now Obafemi Awolowo University), Ile-Ife, Nigeria. He won the prestigious Commonwealth Scholarship for his Masters and PhD in Chemical Engineering at Queen's University, Kingston, Ontario, Canada. Professor Tadéhas been an academic at Curtin University for over 3 decades. He was awarded a John Curtin Distinguished Professor at Curtin University in 2011 for his significant contributions to Chemical Engineering research at the highest level in the field of Process Systems Engineering.He has received substantial funding from both industry and the Australian Research Council (ARC) for his projects. He has successfully supervised over 50 PhD students, several Masters' students and research fellows. He has published 5 books on various aspects of his work as well as over 400 research papers in refereed international journals and conference proceedings with an H-index of 55and over 9,500 citations (excluding self-citations). He is the founding Editor-in-Chief of Asia-Pacific Journal of Chemical Engineering (www.apjChemEng.com), a Fellow of IChemE and an Honorary Fellow of Engineers Australia. He was listed in the Top 100 Australia's most Influential Engineers in 2008 and was the Chair of IChemE in Australia in 2012 and 2013. He was the President of the Australian Council of Engineering Deans (in 2015 and 2016) and a member of the ARC College of Experts (from 2014 to 2017) where he chaired the ARC Panel for Mathematics, Physics, Chemistry and Earth Sciences (MPCE) for the Discovery and DECRA grants in 2016. Professor Tade has been a plenary and keynote speaker at several conferences worldwide. He continues to evolve his leadership and management skills to harness the capabilities of the new generation of leaders.

IOP Conf. Series: Materials Science and Engineering **796** (2020) 011001 doi:10.1088/1757-899X/796/1/011001

ABSTRACT

Applications of Data Mining in Chemical Engineering

The ever-increasing size and complexity of chemical and process industry introduces several difficulties in their monitoring and optimization. On the other hand, the decreasing cost and widespread availability of the measurement instrumentation has resulted in heavily instrumented and digitalized industrial processes. This has resulted in the abundant availability of generated process data for profitably used on process improvement. In addition, industries are constrained to minimize operating costs due to economic pressures and increased competition. Traditional methods for process improvement are time and cost intensive. The availability of relatively cheap data offers the possibility of utilizing it for a broad range of purposes resulting in economic and efficiency gains. The digitalization of the chemical industry is the new revolution that is currently ongoing and it is expected to improve their productivity substantially. In this talk, the applications of data mining methods to the chemical process industry for realizing a wide range of objectives will be addressed. The applications of data mining in the chemical industry includes predictive analytics, data reconciliation, process monitoring, fault detection and diagnosis, root cause analysis, process improvement and optimization, controller performance monitoring and improvement, abnormal situation management and alarm management. This is made possible by the availability of several mature and established methods ranging from simple regression analysis to deep learning using neural networks.

IOP Conf. Series: Materials Science and Engineering **796** (2020) 011001 doi:10.1088/1757-899X/796/1/011001

Dr.-Ing. Joewono Prasetijo

Head of Department of Transportation Engineering Technology
Faculty of Engineering Technology

Universiti Tun Hussein Onn Malaysia

Pagoh Higher Education Hub

KM 1, Jalan Panchor 84600 Panchor, JOHOR

Tel: Cell: +6 011-1774 3911, Office: +6 06 974 2043, Fax: +6 06 974-2193 E-mail: joewono@uthm.edu.my, Joewono.Prasetijo@alumni.ruhr-uni-bochum.de

Brief Biography of Dr.-Ing. Joewono Prasetijo

Dr.-Ing. Joewono Prasetijo is the Head of Department of Rail Transportation Engineering Technology, Faculty of Engineering Technology, Universiti Tun Hussein Onn Malaysia. He is also the Head of Focus Group of Sustainable Transport and Safety Stusies (STSS) and Chairman of Tactical Committee of Road Safety Strategic Plan (NBOS Cluster) in Batu Pahat (MySafe Road @ Batu Pahat), Road Safety Department, Ministry of Transport Malaysia. He has contribute several research papers in traffic engineering; capacity and conflict analysis and road safety engineering; development of accident prediction models.

Dr.-Ing. Joewono Prasetijo is Dr.-Ing/PhD graduated from Ruhr Universität Bochum, Germany. He is member of American Society of Civil Engineers (ASCE), U.S.A. and World Conference on Transport Research (WCTRS), U.K. He have just awarded by the Minister of Transport Malaysia for the Best Project ASEAN NCAP (New Car Assessment Program) in 2018.

IOP Conf. Series: Materials Science and Engineering **796** (2020) 011001 doi:10.1088/1757-899X/796/1/011001

ABSTRACT

The Most Relevant Contributory Factors for Road Accidents: Experiences From Johon Federal Roads (F001, F023 & F050)

Road accidents are one of the most relevant issues in today's society. It causes hundreds of accidents every year from over the world. Every year 1.2 million of people are killed and between 20 and 50 million people are offended due to the road accidents. Three main types of accidents in Malaysia is collision with passenger cars, collisions with other motorcycles and single-motorcycle accidents. F0050 is ranked the district with the highest road fatalities in Johor for five consecutive years. Motorcyclists and their pillion riders were the highest contributors - 60% or 648 fatalities, followed by car drivers and passengers numbering 266 fatalities. One reason of accident occurrence can be lack of road design consistency which most drivers make fewer errors in the vicinity of geometric features. Geometric design consistency is emerging as an important component in highway design relate to the safety performance. This study explore the use of Poisson regression with excess zero outcomes on the response variable. A generalized linear modelling technique such as Poisson regression model and Negative Binomial model was found to be insignificant in explaining and handle over dispersion which due to high amount of zeros thus Zero Inflated model was introduced to overcome the problem. The application work on the number of road accidents on F001 Jalan Johor Bahru - Segamat, F023 Segamat - Tangkak - Muar and F050 Kluan - Ayer Hitam - Batu Pahat. Data on road accident were collected for five-year period from 2010 through 2014. The result from analysis show that ZINB model performed best, in terms of the comparative criteria based on the P value less than 0.05.

IOP Conf. Series: Materials Science and Engineering **796** (2020) 011002 doi:10.1088/1757-899X/796/1/011002

Peer review statement

All papers published in this volume of *IOP Conference Series: Materials Science and Engineering* have been peer reviewed through processes administered by the proceedings Editors. Reviews were conducted by expert referees to the professional and scientific standards expected of a proceedings journal published by IOP Publishing.

Content from this work may be used under the terms of the Creative Commons Attribution 3.0 licence. Any further distribution of this work must maintain attribution to the author(s) and the title of the work, journal citation and DOI.
Published under licence by IOP Publishing Ltd

IOP Conf. Series: Materials Science and Engineering **796** (2020) 012001 doi:10.1088/1757-899X/796/1/012001

Effect of vegetable oils as cutting fluid on wear of carbide cutting tool insert in a milling process

Arya R. Nasution[1], Z. Fuadi[2]*, I. Hasanuddin[2], R. Kurniawan[2]

[1]University of Muhammadiyah Sumatera Utara, Jalan Kapten Mukhtar Basri No.3 Medan, Indonesia
[2]Graduated Study Program, Department of Mechanical Engineering, Universitas Syiah Kuala, Jl. Tgk. Syech Abdurrauf No. 7 Darussalam, Banda Aceh, Indonesia

E-mail: zahrul.fuadi@unsyiah.ac.id

Abstract. This research focusses on the effect of vegetable oils, i.e. crude palm oil (CPO) and coconut oil (CO), used as the cutting fluid on the wear of carbide cutting tool insert in a face milling process. The performances of the tool, in term of wear and the surface roughness of the workpiece, were investigated and compared to those resulting from a similar milling process but using conventional cutting fluid, which is a commercial soluble oil emulsion (SOE). The results show that at the spindle speed of 360 rpm and the feed rate of 80 mm/min, the tool wear was smaller in the case of CO than that in the case of CPO cutting fluid, which is 0.16 mm^2 compared to 0.40 mm^2, respectively. The tool worn area in these cases are still larger than that in the case of SOE, which is 0.09 mm^2. However, at higher spindle speed of 490 rpm and feed rate of 80 mm/min, the smallest tool worn area occurred for the case of CO cutting fluid, which is 0.04 mm^2, compared to 0.1 mm^2 and 0.11 mm^2 for the case of CPO and SOE cutting fluids, respectively. As for the workpiece, the achieved surface roughness, Ra, were relatively similar for all the evaluated cases.

1. Introduction

Cutting tools and cutting fluids are the most important factors in a machining process because they determine the final product quality. The cutting fluid prevents the tool from overheating, reduces friction, keeps a cool working condition, provides a tolerable surface finish, acts as cleaner, increases chip removal, and prevents corrosion [1,2,3]. Therefore, it is desirable for a cutting fluid to have properties such as high thermal conductivity, good lubricity, high flash point, stable oxidation, and corrosion resistance. Cutting fluids also affect the tool wear, influencing the finished quality of the workpiece. Less tool wear will result in longer tool's life and improves the dimensional accuracy of the workpiece [4,5].

Straight oils and soluble oils are often used as cutting fluids in a machining process. Soluble oils, also refer to as emulsions, comprise of water, mineral oil, emulsifiers and additives. Soluble oils improves cooling capabilities and lubrication but tends to promote corrosion due to water content. Straight oils are excellent in lubricity, rust protection, and easily maintained but have poor heat

---

* To whom any correspondence should be addressed.

Content from this work may be used under the terms of the Creative Commons Attribution 3.0 licence. Any further distribution of this work must maintain attribution to the author(s) and the title of the work, journal citation and DOI.
Published under licence by IOP Publishing Ltd

dissipation, smoking and misting problem, oil film production on the finished workpiece, and poor cutting operation [5,6].

The use of conventional cutting fluids in industries has a negative impact on environment. Particularly, if they are not disposed properly, they will cause harm to both environment and public health [7,8,9]. The wasted fluids contain lubricating oils, additives, metal particles, and surfactant [10]. Additives in mineral oils contain sulphur, chlorinated paraffin, phosphorous compound, fatty material, and metal chlorides [11,12]. High concentration of oils and additives containing corrosion inhibitors, stabilizers, and antibacterial agents also prevents the waste from direct disposal to the treatment plants [13]. Due to this environment concerns, it is important to consider a human safe and environment friendly cutting fluids. Vegetable oils are excellent candidate for this application because they are sustainable, eco-friendly, less toxic, and most importantly, bio-degradable [14].

Vegetable based cutting fluids has advantages such as high viscosity index [15], higher flash point [16], and better lubricity due to larger molecular weight [17]. Vegetable based cutting fluids are also reported to have a better thermal conductivity, specific weight, storage stability, corrosion prevention and provide improved performance in terms of cutting force, feed force, and workpiece quality [17].
The performances of several vegetable oils have been investigated, such as coconut oil, palm oil, olive oil, sesame oil [14], aloe vera oil, cotton seed oil [18], canola based oil [19], and groundnut oil [20]. Coconut oil is reported to have a good performance to be used as cutting fluid for stainless steel because it produces uniform thickness and low toll wear compared to palm oil, olive oil, sesame oil [14] or groundnut oil [20]. Aloe vera oil provides a better surface finish and lower tool wear compared to conventional fluid. Canola oil has a good performance due to its lubricity. It is also reported that cotton seed oil gives a better performance compared to conventional SAE40 oil although the friction coefficient was relatively similar [19]. Those results indicated that vegetable oils have a great potential as the replacement of conventional cutting fluids to save the environment.

Figure. 1 EMCO F3 machine tool

In this research, the tool wear and the quality of the finished workpiece in term of surface roughness, Ra, in a milling process were investigated using three kinds of cutting fluid, i.e. crude palm oil (CPO), coconut oil (CO), and commercial emulsion fluid, i.e. soluble oil emulsion (SOE). Although the performance of palm oil as cutting fluid has been reported before, in this case, it was used in the form of non-refined crude palm oil, which is cheaper than refined palm oil. The performance is compare to those of coconut oil and soluble oil emulsion fluids.

The objective of this research is to find out the effect of vegetable oils used as cutting fluid on cutting tool's wear and the surface finish quality of the workpiece. The investigation was conducted using the insert carbide cutting as the tool and grey cast iron as the workpiece in a face milling process.

2. Methodology

A machine tool was used in this research (Figure 1). This machine tool can perform several milling process; peripheral milling, face milling and end milling. In this research, the face milling process was used in investigating the effect of vegetable oil used as cutting fluid on the wear of insert carbide cutting tool.

Schematically, a process of face milling is shown in Figure 2. The spindle rotates at a certain revolution speed, rpm, in a fixed position while the workpiece moves in the direction of motion at certain velocity; feed rate, mm/min. Normal milling process uses three cutting tool, as indicated in Figure 2. However, in this investigation, only one cutting tool was used. This is done so that contact occurs only at one cutting tool and the wear can be directly related to the particular surface interaction. The operation parameters used in this research are given in Table 1. There are total of four tests conducted at various spindle revolution speed and feed rate.

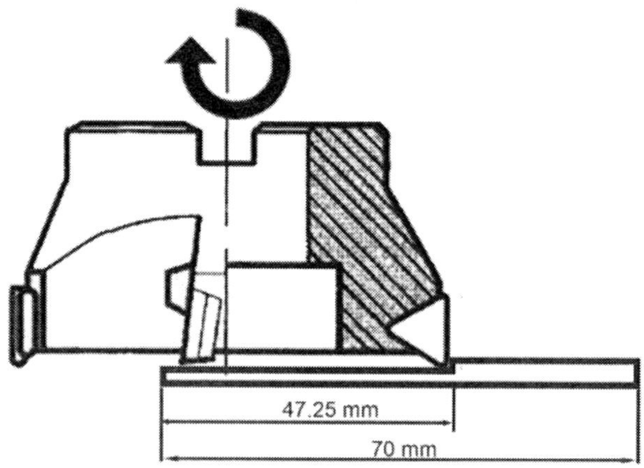

Figure 2. Schematic diagram of the face milling process

The cutting tool used in this research is the insert carbide cutting tool (*solid carbide*) series TPKN 22 VC2. The material has a hardness of 1300 - 1800 (Vickers) and heat resistant of 1000° C. The geometry and dimension of the cutting tool is given in Figure 3. The workpiece used in this study is the grey cast iron. The mechanical properties of grey cast iron is given in Table 2.

Table 1. Parameters of spindle speed and feed rate

Spindle rev., (rpm)	**Feed rate, mm/min**
360	60
360	80
490	60
490	80

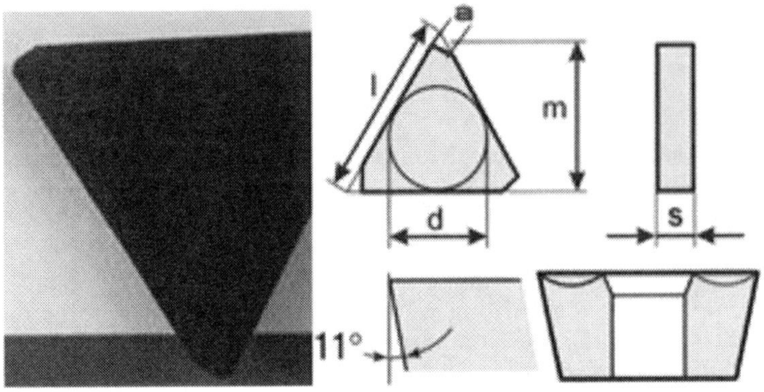

Figure 3. Cutting tool geometry and dimension

Table 2. Material properties of grey cast iron

Density	**7.34 gr/cm³**
Elasticity Modulus	**162 GPa**
Shear Strength	**610 MPa**
Poisson ratio	**0.33**
Hardness, Vickers	**321**

Table 3. Properties of cutting fluid

Cutting fluid	**Viscosity, Ns/m² 27°C**	**Viscosity, Ns/m² 80°C**
Crude Palm Oil (CPO)	**48.7**	**3.5**
Coconut Oil (CO)	**3.19**	**2.2**
Soluble Oil Emulsion (SOE) composed of dromus : water (1:5)	**0.13**	**0.09**

The cutting fluid is the most important factor in this investigation. The fluids used are bio-oils, which are crude palm oil (CPO), coconut oil (CO), and conventional soluble oil emulsion (SOE) composed of commercial dromus and water with a ratio of 1:5. The properties of the fluids are given in Table 3.

The wear evaluation is done by examining the area lost at the cutting tool after the cutting process. An optical microscope, GX-71, was employed for the evaluation. An example of worn cutting tool is given in Figure 4. Figure 4(a) indicates the wear of the cutting tool at two different locations; A and B. The wear of the tool is calculated as the accumulation of all area losses; in this case the sum of area A and B

3. Results and Discussion

3.1 *Effect of spindle speed and feed rate on the tool wear*

Figure 5 shows the microscopic image of the tool cutting before (Figure 5(a)) and after (Figure 5(b)) the cutting process. In this case, the spindle speed was 360 rpm and the feed rate was 60 mm/min. The test was conducted using SOE cutting fluid. From the figure it can be seen that wear occurred during

IOP Conf. Series: Materials Science and Engineering **796** (2020) 012001 doi:10.1088/1757-899X/796/1/012001

the process. Some areas of the tool have been removed in the process at location A and B, which are 26936.73 μm² and 34705.51 μm², respectively. Thus, the total area loss in this case is 0.06 mm². It seems that the tip of the cutting tool was chipped over during the cutting process. It also seems that some wear have occurred along the tip of the cutting tool from A to B. However, this wear could not be evaluated due to limitation in the analysis equipment.

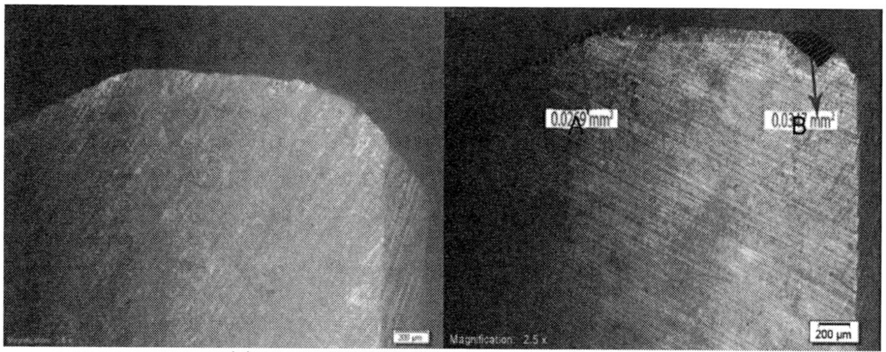

(a) (b)

Figure 4. Tool wear showing worn area; before machining process (a), after machining process (b)

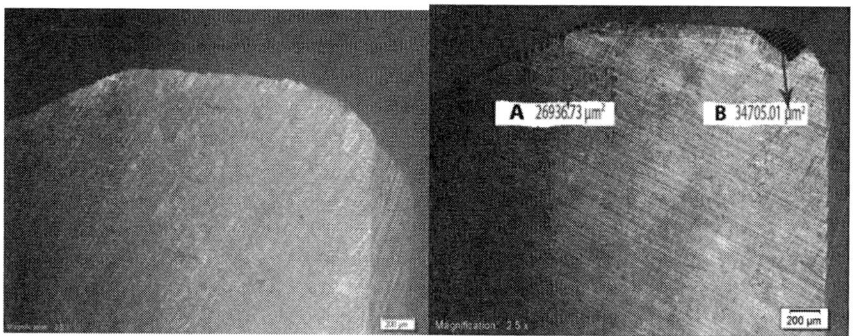

(a) (b)

Figure 5. Tool wear of the tool at 360 rpm spindle speed, before the machining process (a), after the machining process (b)

The total area losses of the cutting tool for other cutting conditions at spindle speed of 360 rpm are given in Figure 6. At the feed rate of 60 mm/min, the highest tool wear occurred for the test condition with CO used as the cutting fluid, with a total area loss of 0.4 mm². This is significantly higher than those using SOE and CPO, which are 0.06 mm² and 0.08 mm², respectively. Here, the wear of the tool was relatively similar for the condition in which SOE and CPO was used as the cutting fluid.

A similar trend occurred for the feed rate of 80 mm/min for the same spindle speed of 360 rpm, as also shown in Figure 6. At this condition, the total area losses for the SOE, CPO, and CO were 0.09 mm², 0.16 mm², and 0.4 mm², respectively. The total worn area of the cutting tool increased by half for the case of SOE and doubled for the case of CPO as the feed rate increases. However, the total area loss of the tool was the same for the case of CO cutting fluid.

Figure 7 shows the effect of feed rate on the total area loss for the spindle speed of 490 rpm at two different feed rates; 60 mm/min and 80 mm/min. At 60 mm/min, the largest area loss occurred for the case of CPO, which is 0.22 mm², followed by that of CO, which is 0.17 mm². The smallest area loss for this condition is for the SOE cutting fluid, which is 0.12 mm². At 80 mm/min feed rate, as also shown in Figure 7, the smallest area loss occurred for the case of CO, which is 0.04 mm². Whereas for the case of SOE and CPO, the area losses are 0.11 mm² and 0.1 mm², respectively.

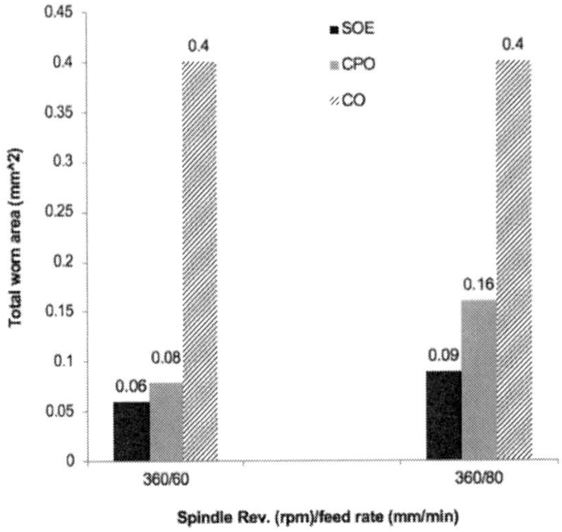

Figure 6. Effect of feed rate on total worn area of the tool at spindle speed of 360 rpm

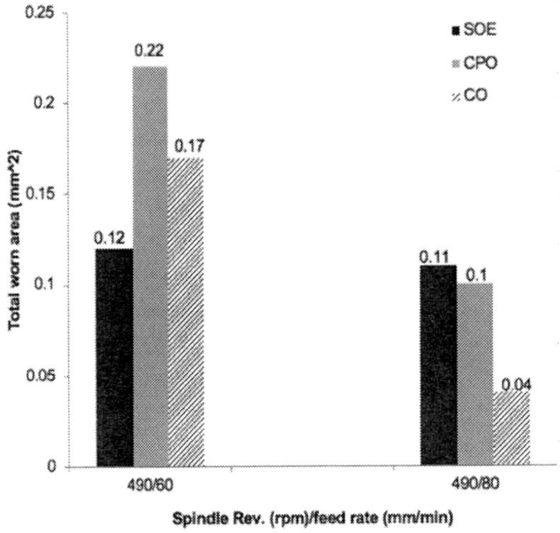

Figure 7. Effect of feed rate on total worn area of the tool at spindle speed of 490 rpm

From Figure 6 and Figure 7, it can be seen that total area loss of the cutting tool insert is affected by both the spindle speed and the feed rate. This is obvious because higher spindle speed means more interaction between the tool and the workpiece. Figure 6 implies that CO is not quite suitable for cutting fluid at the particular condition because it caused heavy wear on the cutting tool. However, the CO worked well at spindle speed of 490 rpm especially at the feed rate of 80 mm/min. Particularly, at this spindle speed and feed rate, CO has a better performance as the cutting fluid compare to CPO or SOE.

Up to this point, the particular reason of better tool wear properties in in the case of CO at higher spindle speed is still not clear thus requires further analysis and investigation. However, one possible explanation is the fluid's viscosity. Particularly at higher spindle speed, the temperature at contact interface may increase significantly and change the viscosity of the oil. At lower spindle speed, CO and CPO have a lower performance compared to commercial SOE possibly because their viscosity are comparatively higher, as shown in Table 3. However, due to temperature increase at higher spindle speed, the viscosity of CO or CPO reduces, resulting in improved cooling effect performance thus improving the tool wear properties. As it is shown in Figure 7, at spindle speed of 360 rpm, the vegetable oils, particularly CO, has a relatively lower performance compared to that of the SOE.

However, these performances improved at the spindle speed of 490 rpm. At this speed, both CO and CPO have a better performances compare to commercial SOE fluid.

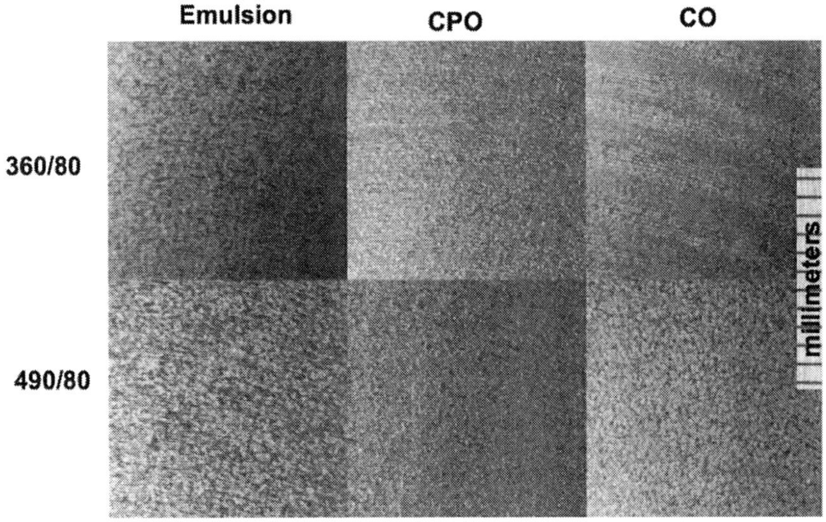

Figure 8. Effect of various parameters on surface quality of the workpieces

From this analysis, it can be argued that CPO and CO fluids can be applicable as the cutting fluids for milling process, in term of the tool wear properties. This is particularly true for the case of carbide cutting tool insert working on grey cast iron workpiece. Since both fluids are renewable vegetable oils, their application can be beneficial for environment and sustainability.

In this analysis, it was observed that wear of the carbide cutting tool insert are in types of flank wear and crater wear. Thus, a more suitable parameter for wear evaluation is term of volume loss. However, this has not been conducted in this analysis due to limitation. Nevertheless, the results shown in this analysis have clearly indicated the potential of bio-oils, coconut oil and crude palm oil, as the cutting fluids.

3.2. Effect of spindle speed and feed rate on the surface roughness of the workpiece

One of the most important factor determining the final quality of a milling process is the surface roughness of the product. Here, it is the surface roughness of the workpiece. Commonly in a machining process, the surface roughness is affected by various factors such as the spindle speed, the cutting rate, the cutting depth, the properties of both the tool and workpiece, and the lubrication. In this case, only two parameters were taken into account for the evaluation; the spindle speed and cutting rate. Figure 8 shows the images of the finished workpieces as the result of milling process at various spindle speeds, feed rates, and cutting fluids. By physical observation, the finished quality of the workpiece in the case of CO cutting fluid matches the quality of that in the case of SOE. In order to better compare the results, the roughness of each workpiece was evaluated by using a surface profilometer taking the parameter of arithmetic average value of the roughness profile, Ra, μm.

The results are given in Figure 9 and Figure 10. It is obvious from the figures that smoother surface roughness were achieved at higher spindle speed and/or cutting rate. The effect of cutting fluid on the surface roughness of the workpiece is more obvious at the feed rate of 60 mm/min. For the spindle speed of 360 rpm, the smoothest surface was achieved for the case of SOE fluid, which is Ra=0.89 μm. The highest roughness occurred for the case of CPO fluid, which is Ra=1.99 μm, followed by that of CO fluid, which is Ra=1.56 μm. The surface roughness decreases at the spindle speed of 490 rpm, as shown in Figure 10. At the feed rate of 60 mm/min, it seems that CPO is the least favourable fluid because it produced a relatively rough surface compare to the other fluids.

On the other hand, the difference among the surface roughness of the workpiece is less observable for the case of feed rate of 80 mm/min for both spindle speed of 360 rpm and 490 rpm. At spindle speed of 80 mm/min, the roughness of the workpiece for the case of SOE fluid and that of CO fluid are Ra=0.6 μm and Ra=0.64 μm, respectively. For the case of CPO, the roughness is Ra=0.7 μm. The roughness of the workpiece for the three condition decreased as the spindle speed increases, as shown Figure 10. The largest decrease occurred for the case of SOE fluid, which is 50%. This is obvious since the commercial SOE is the cutting fluid that has been optimized for the machining process.

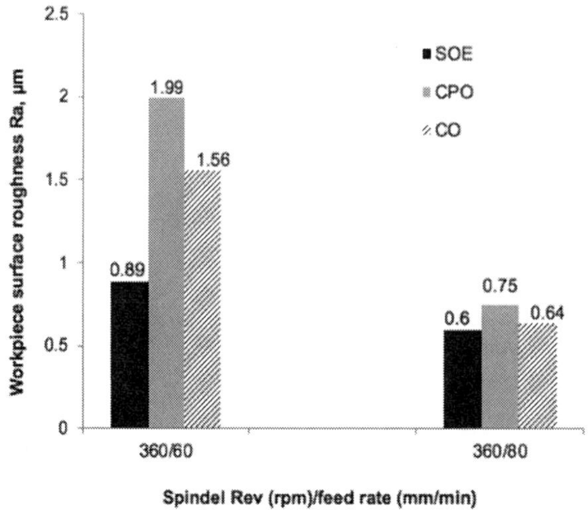

Figure 9. Effect of feed rate on the surface roughness of the workpiece at spindle speed of 360 rpm

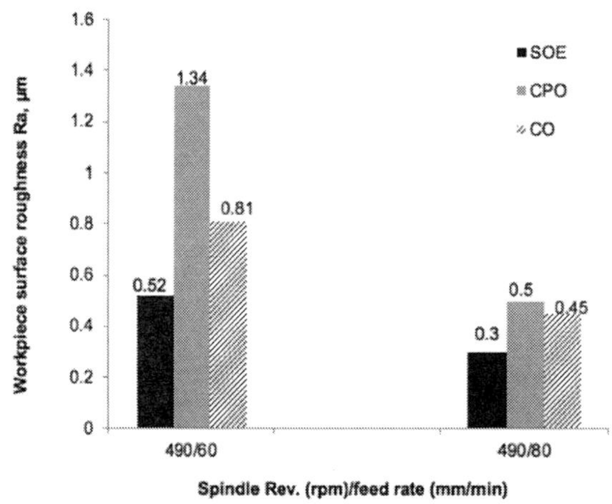

Figure 10. Effect of feed rate on the surface roughness of the workpiece at spindle speed of 490 rpm

The results of this investigation is comparable to previous results reported in various literatures. It was reported that improvement of wear up to 67% was achieved when using vegetable oil in the case of minimum quantity lubrication condition, as well as improvement of the workpiece surface roughness up to 31% [21]. Here in this research, the improvement of tool wear was achieved particularly at higher spindle speed. However, improvement in surface roughness of the workpiece was not clearly observed in this research, thus requires further analysis and investigations.

4. Conclusion
The investigation on the effect of vegetable oils as cutting fluid on the wear of the carbide cutting tool insert and the surface roughness of the grey cast iron workpiece in a face milling process has been conducted. The results indicates that coconut oil has a better performance as a cutting fluid compared

to crude palm oil or commercial soluble oil emulsion, particularly at higher spindle speed and feed rate since it gave the smallest wear of the tool. However, their effects on the surface roughness of the workpiece are less observable. The results imply that coconut oil is highly potential to be used as cutting fluid in machining processes, especially at higher spindle speed.

Acknowledgment

The authors would like to thanks the Production Process Laboratory, Faculty of Engineering, University of Muhammadiyah Sumatera Utara, Laboratory of Materials and Metrology Laboratory, Universitas Syiah Kuala for supports in this research.

References

[1] VM Mamidi, AM Xavior, 2014. A review on election of cutting fluids. Journal of Research in Science and Technology, 1 (5), pp. 1-19
[2] WL Feng, Y Yin, MDL Mendoza, LD Wang, P Chen, YD Liu, LK Cai, LH Zhang, 2018. Oil recovery from waste cutting fluid via the combination of suspension crystallization and freeze-thaw processes. 172 pp. 481–487
[3] S Debnath, MM Reddy, QS Yi, 2014. Environmental friendly cutting fluids and cooling techniques in machining: a review. J. Clean. Prod. 83 pp. 33–47
[4] JA Ghani, M. Rizal, CH Che Haron, 2014. Performance of green machining: a comparative study of turning ductile cast iron FCD700. Journal of cleaner production 85
[5] EP De Garmo, JT Black, RA Kosher, 1984. Materials and processes in manufacturing. Maxwell MacMillan Publication (7th Edition), New York, USA pp. 578-581
[6] RF Avila, AM Abrao, 2001. The effect of cutting fluids on the machining of hardened AISI 4340 steel. Journal of Materials Processing Technology 119 pp. 21-26
[7] WL Feng, Y Yin, MDL Mendoza, LD Wang, P Chen, YD Liu, LK Cai, LH Zhang, 2018. Oil recovery from waste cutting fluid via the combination of suspension crystallization and freeze-thaw processes. J. Clean. Prod. 172 (2018) pp. 481–487
[8] CJ Van der Gast, AS Whiteley, M Starkey, CJ Knowles, IP Thompson, 2003. Bio-augmentation strategies for remediating mixed chemical effluents. Biotechnol. Progr. 19 pp. 1156–1161
[9] ZG Jiang, F Zhou, H Zhang, Y Wang, JW Sutherland, 2015. Optimization of machining parameters considering minimum cutting fluid consumption. J. Clean. Prod. 108 pp. 183–191.
[10] C Griffith, H Daigle, 2018. Manipulation of pickering emulsion rheology using hydrophilically modified silica nanoparticles in brine, J. Colloid Interface Sci. 509 pp. 132–139.
[11] US Dixit, DK Sarma, JP Davim, 2012. Environmentally friendly machining. Springer Science & Business Media, NewYork, USA.
[12] EM Trent, 2000. Metal Cutting (2nd Edition). Butterworths & Co., London, U.K.
[13] I Rodriguez-Verde, L Regueiro, R Pena, JA Álvarez, JM Lema, M Carballa, 2014. Feasibility of spent metalworking fluids as co-substrate for anaerobic co-digestion. Bioresour. Technol., vol. 155 pp. 281–288.
[14] KK Gajrani, PS Suvin, SV Kailas, MR Sankar, 2019. Hard machining performance of indigenously developed green cutting fluid using flood cooling and minimum quantity cutting fluid. J. Clean. Prod., 206, pp. 108-123
[15] U Krahenbuhl, 2002. Vegetable oil-based coolants improve cutting performance (cutting fluids). Tooling and Production. Nelson Publishing.
[16] S Woods, 2005. Going green. Cut. Tool. Eng. 57 (2), pp. 48-51.
[17] MMA Khan, NR Dhar, 2006. Performance evaluation of minimum quantity lubrication by vegetable oil in terms of cutting force, cutting zone temperature, tool wear, job dimension and surface finish in turning AISI-1060 steel. J. Zhejiang Univ. - Sci. 7 (11), pp. 1790-1799.
[18] SM Agrawal, S Lahane, NG Patil, PK Brahmankar, 2014. Experimental investigations into wear characteristics of M2 steel using cotton seed oil. Procedia Engineering 97, pp. 4-14.
[19] ME Kuram, C Huseyin, O Babur & E Demirbas, 2012. Performance analysis of developed vegetable-based cutting fluids by D-optimal experimental design in turning process. International Journal of Computer Integrated Manufacturing, 25 (12).

[20] KK Sharafadeen, JK Odusote, 2013. Performance evaluation of vegetable oil-based cutting fluids in mild steel machining. Chemistry and Materials Research Vol.3 (9), pp. 35-45.
[21] SA Lawal, IA Choudhury ,Y Nukman, 2013. A critical assessment of lubrication techniques in machining processes: a case for minimum quantity lubrication using vegetable oil-based lubricant. Journal of Cleaner Production 41, pp. 210-221

IOP Conf. Series: Materials Science and Engineering **796** (2020) 012002 doi:10.1088/1757-899X/796/1/012002

Lean Manufacturing Production Management Model using the Johnson Method Approach to Reduce Delivery Delays for Printing Production Lines in the Digital Graphic Design Industry

A Gomero-Campos[1], R Mejia-Huayhua[1], C Leon-Chavarri[1], C Raymundo-Ibañez[2,4] and F Dominguez[3]

[1] Ingeniería Industrial, Universidad Peruana de Ciencias Aplicadas (UPC), Lima 15023, Perú.
[2] Dirección de Investigación, Universidad Peruana de Ciencias Aplicadas (UPC), Lima 15023, Perú.
[3]Escuela Superior de Ingeniería Informática, Universidad Rey Juan Carlos, Mostoles, 28933, España.

E-mail: carlos.raymundo@upc.edu.pe

Abstract. Several factors compel graphic design companies to improve efficiency and competitiveness in their production lines. However, these companies are not prepared to take on this challenge, as they report delays in 20% of their deliveries, caused by high setup times, low machine availability, and poor work scheduling. In this context, this study proposes a new production management model fed by the interaction of lean manufacturing tools and the Johnson scheduling method. This model has been validated by direct application at the SISSA. The results obtained were the reduction of the setup time to 15 minutes, increased machine availability up to 24%, and an efficient scheduling of its tasks. All of these reduced the percentage of delivery delays from 20% to 6%.

1. Introduction

The graphic design industry worldwide has been recovering from the double impact it has suffered as a result of the global financial crisis and digital revolution. Nonetheless, among its different printing technologies, digital technology companies record the largest growth in both income and printing volumes. However, this growth has stagnated in regions such as Africa and South America (Andigraf Magazine). In South America, the country that has distinguished itself as one of the countries with the greatest economic dynamism is Peru wherein the graphic design industry is expected to recover, which compels companies to adapt and face new challenges that the market brings (Interempresas Magazine).

The Peruvian printing industry is one of the most emerging industries in the country owing to the quality of its work and competitive prices. According to the AGUDI Magazine, more than 9800 companies comprise the graphic design industry, which include printing activities (86%), publishing

[4] To whom any correspondence should be addressed.

Content from this work may be used under the terms of the Creative Commons Attribution 3.0 licence. Any further distribution of this work must maintain attribution to the author(s) and the title of the work, journal citation and DOI.
Published under licence by IOP Publishing Ltd

activities (8%), paper manufacturing and production activities (5%), and others (1%). The Ministry of Production's report on printing activity productivity reveals that they experienced a 6.9% decrease in production in the first quarter of 2017 compared to the first quarter of 2016, and in the following year, a 1.8% decrease in quarterly production in 2018 was observed compared to the production levels reported for 2017.

This has forced the domestic market into a highly competitive environment, characterized by the industry's need to implement more agile and efficient production systems that can meet the increasingly personalized demand in terms of lead time and quality.

This report seeks to present a solution to the critical issues facing the graphic design industry by using engineering tools that can reduce production times. To this end, the authors propose the use of a production management system focused on work and process organization. The low success rates that the Lean technique has had thus far may be attributable to the fact that the authors have not considered the low-volume production and high product variability of this industry. Additionally, according to the Comunigraf Magazine, the printing industry deals with the lack of production planning and often does not have a job scheduling system. Moreover, when implementing the Lean manufacturing technique, scheduling problems tend to precede the waste presented. To achieve this, we proposed a production management model focused on workplace organization (5S) and process organization (Lean manufacturing) supplemented by proper work scheduling (Johnson method).

2. State of the art

2.1. Lean Manufacturing

The lack of a methodology to help diagnose the current state of companies and to identify production waste factors is one of the reasons why managers cannot make improvements. Value Stream Mapping (VSM) has proven to be one of the most useful diagnostic tools for manufacturing companies, as it helps identify more than waste and provides improvement opportunities for manufacturing factory managers. After identifying improvement opportunities for production lines, it is necessary to opt for solution techniques related to the problems found [1] [2] [3] [4]. One of the Lean manufacturing techniques is the Single-Minute Exchange of Dies (SMED), which involves separation and conversion of internal setup operations into external operations and aims at achieving setup times in less than ten minutes.

This technique was first used on stamping presses, but its applicability was later extended to other machines. SMED has proven to be effective in reducing cycle times. However, the authors state that besides focusing on cycle times, workplace cooperation is also required, and the technique has been improved by additional procedures involving the 5S method, which also improves workplace environment and relationships. The implementation of the SMED technique coupled with the 5S method has resulted in great improvements, increasing SMED effectiveness [5] [6]. On the contrary, the Total Productive Maintenance (TPM) technique is an innovative approach to maintenance that optimizes equipment effectiveness, eliminates breakdowns, and promotes autonomous maintenance for operators through day-to-day activities. Researchers show that successful TPM implementation in an organization largely depends on the organization's effort to comprehensively implement TPM initiatives. In addition to commitment, support, and involvement from senior management, developing a realistic TPM implementation plan, instilling skills and knowledge related to autonomous maintenance, and improvement in equipment is required. TPM implementation not only improves the Overall Equipment Effectiveness (OEE) of large industries but also OEE rates at small industries through an innovative strategy [6] [7] [8] [9].

2.2. Johnson Method

The abovementioned techniques (SMED and TPM) are used to improve production processes by eliminating waste. However, these "Lean" techniques are still at an early stage of implementation in printing companies, having a low "Lean" implementation success rate of 5% compared to other manufacturing industries, which report up to 46%. This may be grounded in the fact that past authors have not considered this industry to be characterized by low-volume production and high product variability. This research aims at implementing, prior to "Lean" tools, a scheduling methodology for print jobs related to the areas involved and at giving rise to the implementation of the "Lean" tools mentioned [7] [10] [11].

3. Contributions

3.1. Proposed Model

This research proposes a production management model focused on the two production dimensions shown in Figure 1:

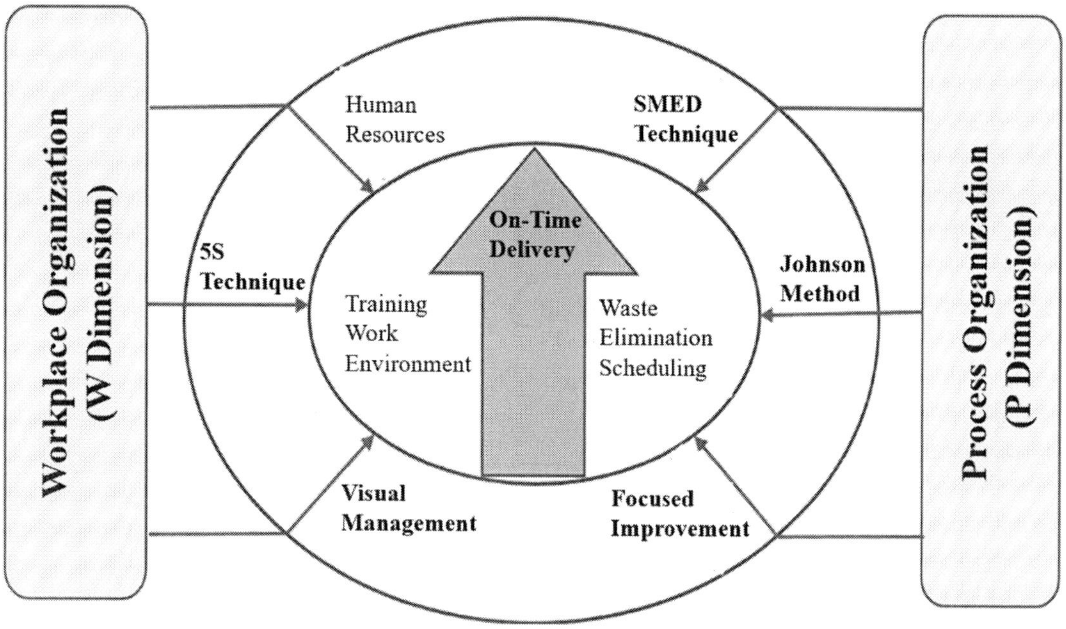

Figure 1. Proposed production management model

The first dimension to be implemented in the model is the W dimension, which is based on the following techniques:

- 5S is an important component in the context of a previous implementation of Lean manufacturing methodologies aiming at generating order and discipline and managing activities on the implementation site.
- Visual Management is a technique that aims at improving process efficiency based on graphics and colours.
- Improving human resources management is a component that focuses on employee training to increase their awareness and commitment, thereby improving workplace environment and enhancing productivity.

On the contrary, the P dimension comprises improving the production processes for which the following techniques are implemented:

- Lean Manufacturing (SMED and Focused Improvement)—a methodology to increase productivity by reducing cycle times and increasing machine availability

- Johnson method—a technique used to efficiently schedule the production jobs to fulfill product deliveries

3.2. Proposed Method
The implementation of the proposed model is described in four stages based on the development of the following main techniques to be developed:

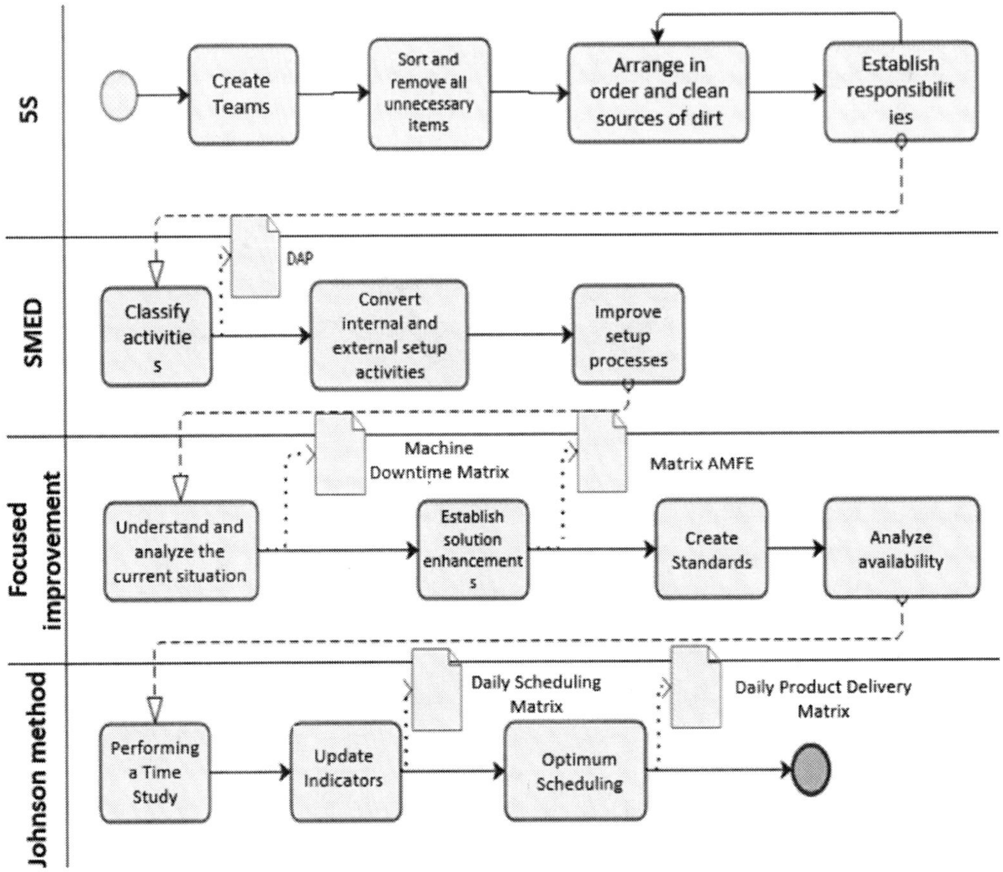

Figure 2. Implementation flowchart

3.3. Ratios
The proposed model aims at generating a more efficient and competitive production line in the digital printing industry, which requires an analysis of ratios that help meet the proposed objective.

3.3.1. On-Time Delivery. Goal: To measure compliance in the delivery of SPs on the agreed date (achieve 90% compliance).
Formula:

$$On - Time\ Delivery\ (\%) = \frac{No.of\ on - time\ delivered\ orders}{Total\ no.of\ orders\ delivered}\ 100\% \tag{1}$$

3.3.2. Machine Availability. Goal: To measure equipment availability and classify the company's competitiveness and reach 85% availability.
Formula:

$$Availability\,(\%) = \frac{(total\ hours - machine\ downtime)}{total\ hours} \tag{2}$$

3.3.3. Setup Time. Goal: To reduce the setup time caused by machine downtime to or less than 15 minutes. Formula:

$$Average\ Set\ up\ time = \frac{Cumulative\ Set\ up\ time}{Total\ downtime\ due\ to\ Set\ up} \tag{3}$$

3.3.4. Idle Time due to Scheduling. Goal: To reduce idle time due to scheduling to 0.
Formula:

$$Idle\ Time = Total\ Available\ Hours - Actual\ Hours\ Worked \tag{4}$$

4. Validation

4.1. Case Study

This section discusses the Lean manufacturing methodologies applied along with the Johnson method in a production management model at a digital printing company and the impact that they exerted as per the objectives set.

Sistema de Impresiones S.A. was established in 1994, and since its inception, the company has been dedicated to designing and printing Point-of-Sales (POS) marketing collateral items using the on-demand production system on different types of substrates to respond to customer demands. To manufacture a wide range of products and to remain flexible as per changing customer needs, the company uses different types of substrates such as banners, vinyl, cardboard, sintra, and foam among others. Furthermore, the company counts on the machinery necessary to print on any material as per customer instructions.

Figure 3. An image from the company

4.2. Situation

One of the main problems faced by Sistema de Impresiones S.A. is its continuous loss of contracts. According to the company, some of their regular clients have decided not to award them contracts because they have found other providers who can best meet their requirements. This issue is rooted in late and damaged deliveries. The on-time delivery rate averages 79.8% owing to different factors that increase the amount of time lost each month.

Subsequently, a 70% machine availability rate was achieved. The desired availability for the printing industry should exceed 85% of total hours. In fact, Sistema de Impresiones has been found to record up to 2.5 idle hours per day because of poor work scheduling and lack of clear procedure standardization.

4.3. Implementation
The following activity scheduling was taken into account for the implementation:

- Implementation Kick-off
- Initial Audit
- 5S Implementation
- SMED Implementation
- Focused Improvement Implementation
- Johnson Method Implementation

- 5S implementation focused on engaging employees and making them aware that their activities are essential for order fulfilment. Therefore, they must be able to distinguish between non-value-added activities and activities that increase productivity. Thus, both productivity and work environment are improved.
- During SMED implementation, we identified all the printing process activities and activities that can be performed by previous or following processes.

To this end, we created a Process Analysis Diagram to identify all the activities; then, using a time study, we obtained average times for each activity. Subsequently, we identified the internal setup activities that will become external. Finally, the remaining internal setup activities were improved by reducing non-productive times, as shown in Table 1 below. Activity times were reduced to an average of 40.6 minutes; setup times were reduced to 15 minutes; and non-value-added activities were eliminated.

- For focused improvement, we created an FMEA matrix that identified the most relevant potential RPNs for potential failure: tray printing. Next, arrangements were established for focused maintenance. Machine availability increased to 93%, which presents a statistical error of 0.06 in a t-student test.
- Then, the Johnson method was implemented for effective job scheduling and order fulfilment planning.

Thereafter (i.e., after applying the Johnson method), the on-time delivery trend increased in April. In fact, in the weeks following the implementation, it reached an average of 93%.

4.4. Results
The other assessed indicators included machine availability and setup time. Table 1 depicts how these indicators have changed after deploying the model proposed in the digital printing industry.

Table 1. Results achieved

	Before	**After**
Availability	70%	93%
Time Setup	25 min	15 min
Idle Hours	2.5 h/day	0 h/day
On-Time Delivery	79%	93%

The post-model implementation results obtained were positive, as according to the above traffic light indicator, these results are within the industry expectations in terms of a more efficient production line.

5. Conclusions

This research aimed at decreasing lead times for a printing production line that resulted in late deliveries to customers. After the application of a set of Lean tools in conjunction with the Johnson Scheduling Method, lead times experienced a 12% reduction. The implementation of this model will allow companies to meet their production plans with the least possible number of downtimes and machine breakdowns. Thus, the sector may focus its efforts to continue growing under constant optimization strategies, offering quality products to customers, and, above all, fulfilling their orders on time. Hence, companies operating within the sector may better satisfy customer demands and reduce their percentage of contract losses. In spite of the positive results obtained from applying the model in the printing production line, an approximate loss of 50 hours per month remains because of non-value-added activities. Therefore, the development of a Process Management System can be considered to help reduce these activities and to increase on-time delivery rates.

6. References

[1] Ben Fredj-Ben Alaya, L 2016 VSM a powerful diagnostic and planning tool for a successful Lean implementation: a Tunisian case study of an auto parts manufacturing firm *Prod Plan Control* **27** 1-16.

[2] Bevilacqua M, Ciarapica F E, De Sanctis I, Mazzuto G and Paciarotti C 2015 A Changeover Time Reduction through an integration of lean practices: a case study from pharmaceutical sector *Assem Autom* **35**(1) 22-34.

[3] Mohd Rojani J and Zahraee S M Production Line Analysis via Value Stream Mapping: A Lean Manufacturing Process of Color Industry *Procedia Manuf* **2** 6-10.

[4] Sahoo S and Sudhir Y 2018 Lean production practices and bundles: a comparative analysis International *J Lean Six Sigma* **9**(3) 374-398.

[5] Vipan K and Amit B 2015 The Implementation of Single Minute Exchange of Die with 5'S in Machining Processes for reduction of Setup Time *Int J Recent Technol Eng* **2** (2) 32-39.

[6] Vinod Y, Rakesh J, Murari L M, Avinash P and Andrew L 2018 The impact of lean practices on the operational performance of SMEs in India *Ind Manag Data Syst.*

[7] Jain A, Bhatti R S and Singh H 2015 OEE enhancement in SMEs through mobile maintenance: a TPM concept *Int J Q Int J Qual Reliab Manag* **32** (5) 503-516.

[8] Sachin M and Sharma S 2016 Total productive maintenance, total quality management and operational performance: An empirical study of Indian pharmaceutical industry J Qual Maintenance Eng.

[9] Kanti Angustiady T and Cudney E A 2018 Total productive maintenance *Total Qual Manag Bus.*

IOP Conf. Series: Materials Science and Engineering **796** (2020) 012002 doi:10.1088/1757-899X/796/1/012002

[10] Lastra F, Meneses N, Altamirano E, Raymundo C, Moguerza J M 2019 Production management model based on lean manufacturing for cost reduction in the timber sector in Peru", *Adv Intel Syst Comput* **971** 467-476.
[11] Mandariaga Neto F 2013 Lean manufacturing: exhibition adapted to the repetitive manufacturing of product families through discrete processes (*Madrid: Bubok*).

IOP Conf. Series: Materials Science and Engineering **796** (2020) 012003 doi:10.1088/1757-899X/796/1/012003

Design, prototyping, and experiment of the ocean current power generation system using Bach Savoniusrotor

Yusiran[1], Erwin[1], Lazuardi[1]
[1]Department of Physic, Faculty of Mathematical and Science, Riau University, Pekanbaru, Indonesia

E-mail: yusiran123@gmail.com

Abstract. Design, prototyping, and experiment of ocean current power generation system have been carried out using the Bach Savonius rotor. The introduced ocean current power generation prototype is validated on real measurement obtained in the Rupat Strait, Indonesia. In this research, the influence of the velocity of ocean currents, the number of turbine blades, and the blade arc angle of the generator output power are studied. The results showed that the turbine output power is strongly influenced by the velocity of ocean currents where the velocity values of ocean currents varied in the range 0,63-1,98 m/sec. The maximum electrical power of the turbine occurs at a current velocity of 1,98 m/sec of 26,88 Watts. The number of turbine blades has a significant effect on turbine output power. The turbine reaches maximum power is found in the rotor with a number of 3 blades with a power coefficient of 0,1176 on the tip speed ratio of 0,359. The blade arc angle is varied at angles of 90°, 135° and 165°. The blade arc angle 135° gives the best performance with a power coefficient of 0,102 on the tip speed ratio of 0,298.

1. Introduction

Ocean current energy is one of the interesting renewable energy resources to develop because it has several advantages, which are predictable, scalable and have no visual impact [1]. Ocean's current exploration near the coast is not a new idea. The application of ocean current energy resources has been started since 1966 in France [2]. The process of converting the kinetic energy of ocean currents into electrical energy can be done using vertical axis ocean current turbines and horizontal axis ocean current turbines. Although the energy source of ocean currents is more promising, the problem that arises in the process of applying ocean current power generation technology is the high cost of making and installing tools. The cause of the increase in the cost of making and installing tools is limited accessibility and underwater work. To solve these problems, researchers have proposed a variety of concepts. Gorlov designed a helical turbine that was able to harness ocean current energy at low velocity without dam construction. This concept is offered in order to reduce environmental damage due to dam construction which damages ecosystems in the region and reduces the cost of power plants [3]. Another type of turbine developed by Marine Current Turbine Ltd is a kobold turbine mounted under afloat [4]. The research mentioned above is some practical effort to reduce investment costs for ocean currents. However, the cost constraints have not been completed with the development of both types of gorlov and kobold turbines, the reason being that the turbine manufacturing process is

[1]Corresponding author : yusiran123@gmail.com

Content from this work may be used under the terms of the Creative Commons Attribution 3.0 licence. Any further distribution of this work must maintain attribution to the author(s) and the title of the work, journal citation and DOI.
Published under licence by IOP Publishing Ltd

IOP Conf. Series: Materials Science and Engineering **796** (2020) 012003 doi:10.1088/1757-899X/796/1/012003

complicated and still costs quite a lot. There is one type of turbine that is easy to make, cheaper and can work at low fluid flow speeds, called the Savonius turbine, Savonius turbine are usually composed of two curved blades with a flow passage around the center. In the original design, each blade shows a semicircular shape [5]. Savonius turbine still has a disadvantage, the turbine efficiency is still very low. The researchers sought to correct this weakness by optimizing turbine parameters. Only a few researchers have used a type of conventional Savonius turbine to be applied as a water flow turbine. Hassanzadeh conducted an experimental study to create a new configuration of ocean current turbines using conventional Savonius turbines [6].

To obtain a significant performance improvement, the author worked to design and conduct an experimental study of Savonius turbine power generation using the Bach Savonius rotor and study the influence number of turbine blades and blade arc angle of power generation system performance. The experiment was designed to allow measurements of ocean current velocity and operational variables to determine the electrical power output, coefficients of performance and dynamic torque.

2. Theory

2.1. Rotor Geometry

Bach is one of the second common profiles used in the Savonius rotor which was submitted by Bach six years after the conventional profile. Bach replaces a semicircular profile with a Bach profile consisting of a straight line and an arc. Figure 1 shows the geometry of the Savonius Bach rotor.

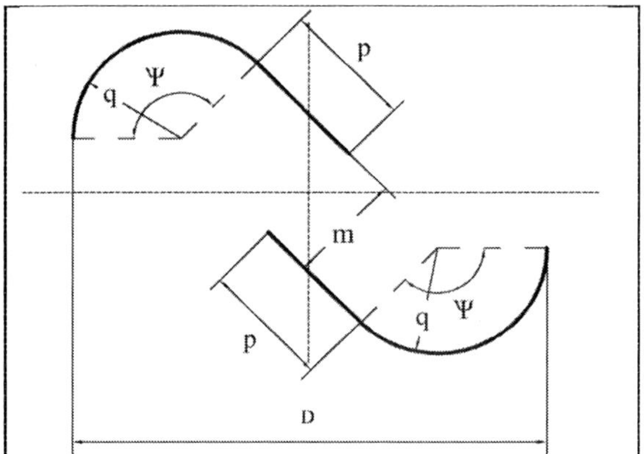

Figure 1. Bach blade profile

In wind turbine research, a comparative study of conventional Savonius Bach and Savonius turbine rotor has been carried out. The results showed that Bach Rotor achieved better performance than Conventional Savonius turbines [6].

2.2. Mathematical Formulation

The power generated by ocean currents is calculated using equation 1 [7];

$$P_T = \frac{1}{2}\rho A V_T^3 \tag{1}$$

P_T is the power produced by ocean currents, ρ is the density of seawater (1.025 kg/m³), A is sweep area and V_T is the velocity of ocean currents (m/sec).

The power extracted by the Savonius turbine from the free flow of the seawater can be calculated by using equation 2 [8];

$$P_M = T.\omega \tag{2}$$

Where T is torque (N-m) generated by the turbine and ω is the rotational speed (rad/sec) of the turbine. In order to evaluate how much energy is used by the turbine from the free stream flow, the term TSR is defined. It relates free stream velocity with the speed at the blade tip and is expressed as [10]

$$TSR = \frac{\omega . R}{V_T} \tag{3}$$

The performance of the turbine is expressed non-dimensionally by using the term coefficient of power Cp. It is defined as the ratio of power extracted by the turbine, P_M (W) with respect to the maximum energy, P_T (W) of the free stream flow and is expressed as [11];

$$Cp = \frac{P_M}{P_T} \tag{4}$$

3. Experimental Methodology

An experimental study was conducted to test real ocean current power system models under real operating conditions. The purpose of the experimental approach is to determine the performance of Bach-rotor.

3.1. Design Concept of the Model

The model of ocean current power generation system using the Bach Savonius turbine is shown in figure 2. As shown in figure 3 the ocean current power system is equipped with a float, gearbox, rotor, gearbox, bearing and shaft. The spesification of the prototype showns in table 1.

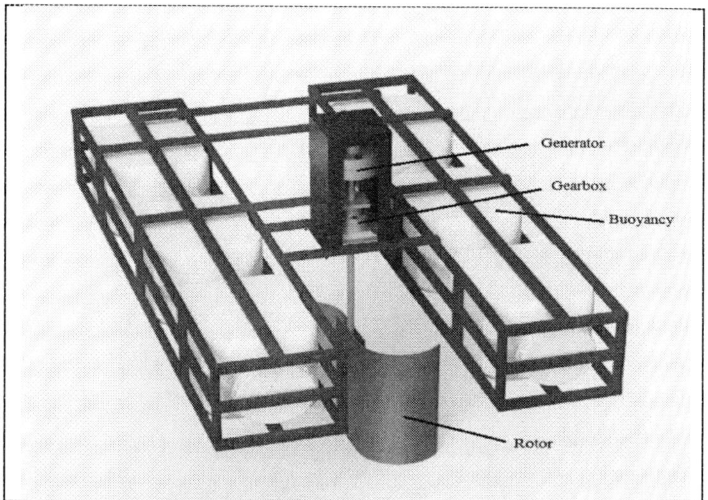

Figure 1. The model of ocean current power generation system.

Table 1. Spesification of prototype ocean current power generation system using Bach-rotor.

Rated power	65 Watt
Floating support dimensions	
Lenght	2,2 m
Widht	1,2 m
High	0,5 m
Generator rated power	200 Watt

Gearbox ratio	1:12
Turbine	
Materials	Alumunium
Thickness	2 mm
Diameters	0,6 m
High	0,5 m
Aspect Ratio	0,8

3.2. Fabrication
The blades of the Bach rotor models were fabricated from 2 mm aluminum sheets due to good manufacturability. To bend the aluminum sheets to design profiles using a bending machine. The experiment was used Bach Savonius rotor with two, three and four-blades. The blade arc angle for investigating the performance of Bach-rotor with the various number of the blade is 135°. Bach Savonius rotor shown in figure 2 and Blade arc angle of the Bach-rotor shown in figure 3. The blade arc angle varied in angles 90°, 135° and 165° with the number of blades keeps in two blades.

Figure 2. Bach Savonius rotor

Figure 3. Blade arc angle at Bach Savonius rotor

3.3. Experimental Set-Up
The experiment setup, as shown in figure 3, was used to test the ocean power generation system models. The free mainstream of ocean current velocity was measured using the current meter. The rotating speed of the rotors at specific ocean velocity was measured using a digital laser tachometer. A DC Generator was used for electrical power output and dynamic torque. To measure the electric current, a 20-ohm resistor was connected to the maximum output current. In this study, the Rupat Strait was chosen as a potential site for the turbine. The mean velocity of Rupat strait is 1,42 m/s.

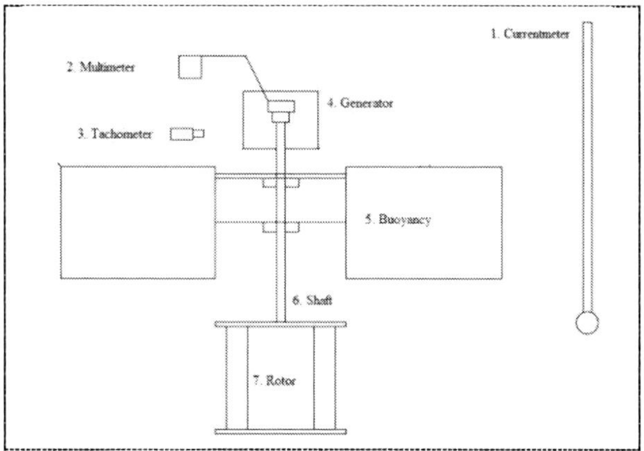

Figure 4. Schematic setup for the experiment of ocean power generation using Bach-rotor in the sea current.

4. Result and Discussion

Figure 5 shows the prototype. The prototype was tested in Rupat Strait, Riau Province, Indonesia. The real values of ocean current velocity in the Rupat Strait shown in figure 6.

Figure 5. Prototype testing in Rupat Strait, Indonesia

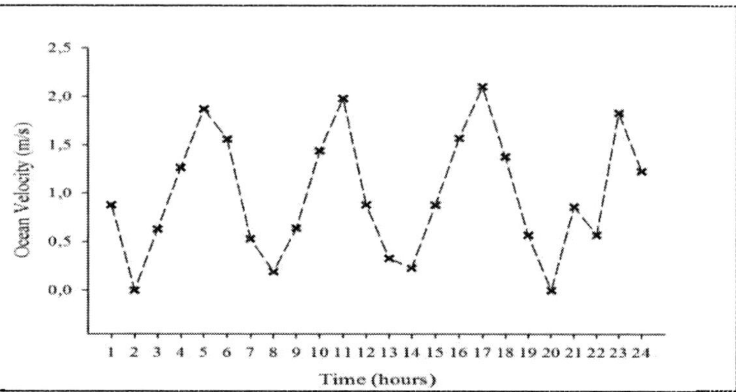

Figure 6. Real data ocean current velocity in Rupat Strait

As shown in figure 6 it can be seen that ocean current velocity in Rupat Strait has a lunar semidiurnal tidal constituent. The maximum velocity of ocean current in the Strait is 2,02 m/s and the minimum velocity is 0,1 m/s. This appears to be rather a slow velocity, but it is sufficient for generating electricity because the density of water is 830 times higher than the density of air. The present work focuses on studying the electrical power output at various ocean current velocity, the number of blade and blade arc angle of Bach Savonius rotor.

4.1. Effect of Ocean Current Velocity
To investigate the efficiency of the turbine, three blades Bach rotor was chosen in this discussion. Figure 6 shows the electrical power.

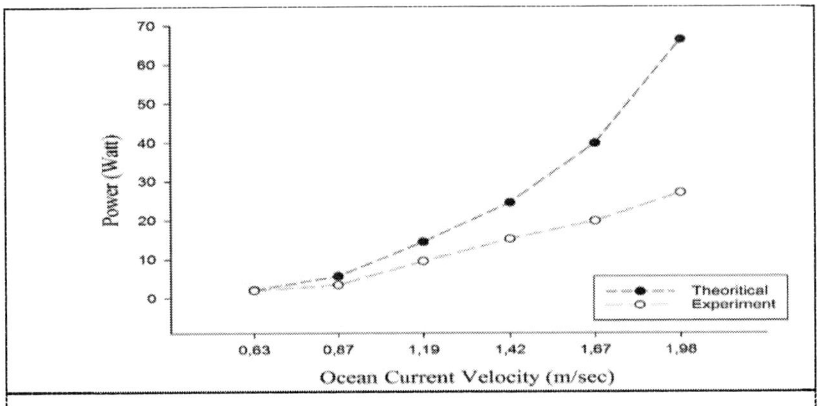

Figure 6. The electrical power output of Bach-rotor

The power extracted by Bach-rotor increases with *the* increase of ocean current velocity. Turbine power is a function of the ocean current velocity, it is suitable with equation 1. In this study, *the* author was designed the rated electrical power approach at 65 Watts, but in the experiment result *in* only 26,88 Watts. The variations of turbine rotational speed versus ocean current velocity for three blades Bach- rotor presented in figure 7. As *can be seen in figure 7,* the rotational speed of *the* turbine increases with increases *in* ocean current velocity.

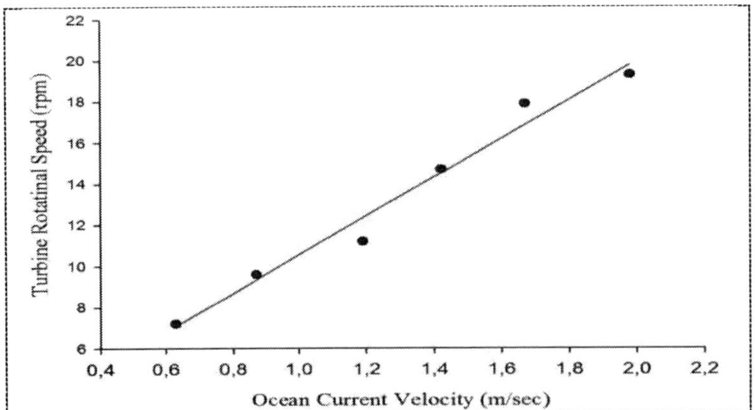

Figure 7. Measured turbine rotational speed for three blades turbine

4.2. Effect of Number of Blades

Figure 7 shows variation in electrical power with ocean current velocity for investigated rotors performance. The three blades gives high electrical output power compared to two and four blades rotors. The maximum can be reaches of three blades Bach-rotor is 26,88 Watt at ocean current velocity 1,98 m/s.

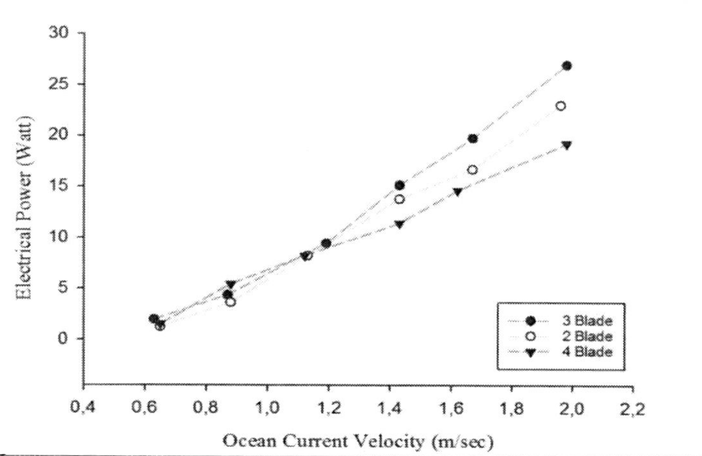

Figure 7. The relation between electrical power and ocean current velocity for Bach-rotor with two, three and four blades

The number of blade has an impact on the hydrodinamic performance of the Bach-rotor, which can observed from Table 2. It showns that a low speed ratio, where ocean current velocity 1,42 m/s and the swept area 0,3 m^2, the three blade has good performance because the Cpon this turbine is higher than other. According to table 2, rotor with three blade have coefficient of performance 0,1176 on TSR0,359.

Table 2. The results of the calculation of dynamic torque, TSR, and the coefficient of turbine power with the variation in the number of blades.

n	Ocean Velocity (m/s)	ω (rad/s)	Torque (N-m)	*TSR*	*Cp*
2	1,42	1,421	33,25	0,325	0,1073
3	1,42	1,539	33,68	0,359	0,1176
4	1,42	1,331	29,43	0,281	0,0668

4.3. Effect of Blade Arc Angle

Effect of blade arc angles is studied here for 90°, 135° and 165° angles. Figure 8 shows the relation between electrical power and ocean current velocity for 90°, 135° and 165° blade arc angles arrangement with the number of two blades. The electrical power for 135° is higher than electrical power obtained for 90° and 165° blade arc angles. This may be due to the 90° blade arc angle is much shorter and straighter created high negative torque in returning blades. The increase of blade arc angle up to 135° makes an effect of obstruction come into play. Thus the Bach-rotor design is optimized at 135° blade arc angles.

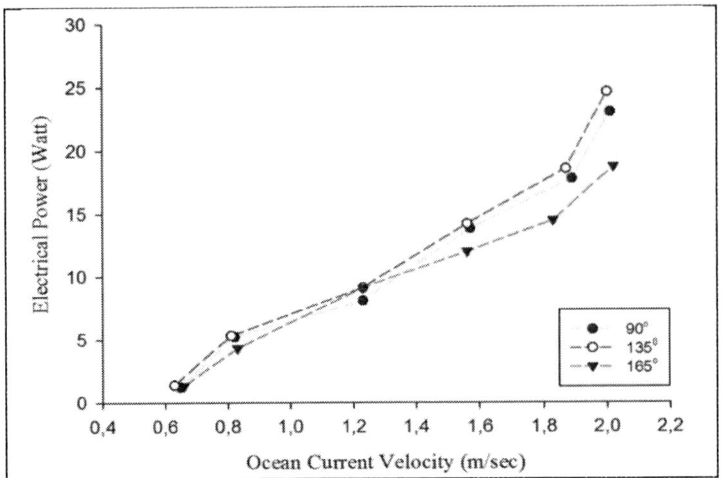

Figure 8. Electrical power of Bach-rotor ocean power system of various blade arc angle

5. Conclusions

In the present study, the performance of the ocean current power generation system using Bach-rotor was evaluated experimentally. The performance of the ocean current power system was also experimentally compared to the Bach-rotor with a varied number of blade and blade arc angles. From the study, the following conclusions can be summarized; The electrical power and rotational speed of the turbine increase with increases ocean current velocity, which is maximum for velocity 1,98 m/s. Three blades of Bach Savonius turbine give the best performance with the coefficient of performance 0,11 at a tip speed ratio 0, 359. Blade arc angle influences turbine performance. The angle of the arc blades giving the best performance is at a value of 135° which can reach a maximum power of 23.06 Watt and a maximum power coefficient value of 0.102 at a tip speed ratio value of 0.298.

Aknowledgments

The authors gratefully acknowledge the financial support of Astra Green Energy Community 2018.

References

[1] R. Vennell, S.W. Funke, S. Draper, C. Stevens, and T. Divett 2015 Designing large arrays of tidal turbines: a synthesis and review *Renew Sustain. Energy Rev***41** 454-472

[2] Kamoji, M. Kedare, S, Prabhu, S. (2009) Performance Test on Helical Savonius Roto *Renewable Energy*, 34(3), 521-529

[3] Gorlov AM 1998 Helical turbines fot the Gulf Stream: Conceptual approach design of a large-scale floating power farm Marine Technology 35175

[4] Mauller G, Kauppert K 2004 Performance characteristic of water wheels J of hydraulic research 425

[5] Akimoto H, Tanaka K, and Uzawa J 2013 A conceptual study of floating axis water current turbine for low-cost energy capturing from river, tide and ocean current Renewable energy57 283

[6] Hassanzadeh R, Yaakob U, Gholami A, Pourmahdavi M, Hoissenzadeh M, Taheri MM, Ahmed Y 2018 Experimental investigation of an innovative configuration for new marine current turbine Renewable energy181

[7] Khan, M. J., Iqbal, M. T. & Quiacoe, J. E.(2008). River current energy conversion system: Progress, prospects and challenges Renewable Sustain Energy Reviews,8(3), 21772179.

[8] Wu H N, Chen L J, Ming H Y, Li W Y and Chen B F 2012 On design and performance prediction of the horizontal-axis water turbine *Ocean Eng***50**23-30

IOP Conf. Series: Materials Science and Engineering **796** (2020) 012004 doi:10.1088/1757-899X/796/1/012004

LEAN Production Management Model based on Organizational Culture to Improve Cutting Process Efficiency in a Textile and Clothing SME in Peru

R Cespedes-Pino[1], J Hurtado-Laguna[1], I Macassi-Jaurequi[1], C Raymundo-Ibañez[2,4] and F Dominguez[3]

[1] Ingeniería Industrial, Universidad Peruana de Ciencias Aplicadas (UPC), Lima 15023, Perú.
[2] Dirección de Investigación, Universidad Peruana de Ciencias Aplicadas (UPC), Lima 15023, Perú.
[3]Escuela Superior de Ingeniería Informática, Universidad Rey Juan Carlos, Mostoles, 28933, España.

E-mail: carlos.raymundo@upc.edu.pe

Abstract. In recent years, homegrown SMEs have had low production levels when compared with Chinese garment imports, losing their competitive advantage in the domestic market. SMEs represent 96% of garment companies in Peru and have a positive impact on the creation of jobs. The search for a technique to improve SME output was conducted in various studies; however, the efforts did not bear fruit over time. Thus, this article seeks to improve the low production efficiency in textile and clothing SMEs. Therefore, we proposed a model and validated it in the production area of a denim clothing manufacturing company in Peru. We conducted business diagnostics and found a production efficiency problem. Later, we adapted the Lean production management model to the prevailing organizational culture. The main result was that the company's production efficiency increased from 68% to 71%. Finally, employee commitment, along with the combination of the Lean model and organizational culture, allowed the improvements to stand the test of time after their implementation.

1. Introduction

In recent years, the main problems in the textile and clothing industry are the lack of productivity and competitiveness. We analyzed these problems to identify their causes, including high logistics costs, labor and tax cost overruns, low international prices, very low exchange rates, and the closure of markets because of the international recession [1]. This industry counts on a large number of micro and small enterprises, which represent 96% of the organizational structure of the clothing industry in Peru, which is the most important manufacturing activity [2]. According to the Global Competitiveness Report 2017–2018, which evaluates the factors that drive productivity and growth in 137 countries, Peru ranks 72nd with a GDP per capita of US$6,198.6 and 2,496 hours worked per year, which depicts a setback of five positions when compared with the 2016–2017 Report [3]. Nowadays, low production efficiency harrows SMEs, with most of them pointing to the lack of

[4] To whom any correspondence should be addressed.

Content from this work may be used under the terms of the Creative Commons Attribution 3.0 licence. Any further distribution of this work must maintain attribution to the author(s) and the title of the work, journal citation and DOI.
Published under licence by IOP Publishing Ltd

IOP Conf. Series: Materials Science and Engineering **796** (2020) 012004 doi:10.1088/1757-899X/796/1/012004

qualified personnel as well as escalating costs as impediments to improving productivity. However, this is also a matter of entrepreneurial culture with respect to innovation [4].

It is necessary to study the low production efficiency in the textile and clothing industry because it accounts for 7.2% of manufacturing GDP, occupies an important place in nontraditional exports (11% of the total), and has a positive impact on Peru's economic performance because it creates both direct and indirect employment [5].

This research study is necessary because it seeks to integrate Lean tools with organizational culture in textile and clothing SMEs—something that previous studies have not attempted. The expected result is the improvement of production efficiency in textile SMEs through the alignment of organizational culture with Lean process improvement methodologies. Thus, we hope that the improvements implemented in most cases prevail over time with the appropriate personnel guidelines.

2. State of The Art

2.1. Production Management Models

SMEs deal with functional barriers, which consist of human resources, production, and financial issues that show that the difference between areas represents obstacles to SMEs that affect their exports [6]. This problem generated doubt with regard to determining the productive management model to implement in industrial SMEs. Therefore, the companies decided to follow a traditional management system based on mass production. They aim to achieve significant results in costs by operating in large, highly standardized batches and producing at maximum capacity [7].

2.2. Lean

The cases related to Lean tools aim at combining techniques to improve SME productivity by eliminating waste. The SMEs in question apply improvement programs to identify problems and propose solutions by verifying the manner in which their results affect their production process [8]. SMEs establish the best possible way for companies' improvement and profitability. Therefore, environment assessment is important for establishments in that it enables them to determine whether the tools have been adjusted with respect to their dimensions [9][10].

2.3. Organizational Culture

Organizational culture shows the importance of creating a commitment and involvement identity in a company, which refers to the improvement of organizational processes and performance [11]. With regard to this, family businesses are an impediment to organizational culture because their practices are transmitted through generations, growing with no option to improve. Therefore, a study evaluated both cases through the organizational culture assessment instrument (OCAI), which showed that there are no outstanding differences and explains how organizational culture, through various types of family businesses, contributes as a first step toward adopting cultural patterns based on the resources that are needed to compete [12].

3. Contribution

3.1. Proposed Model

The general model is based on the organizational mission with three Lean stages (5S methodology, Poka–yoke, and standardized work) and is complemented by organizational culture through its dimensions (adaptability, involvement, and standardized work). The model structure is presented below (Figure 1):

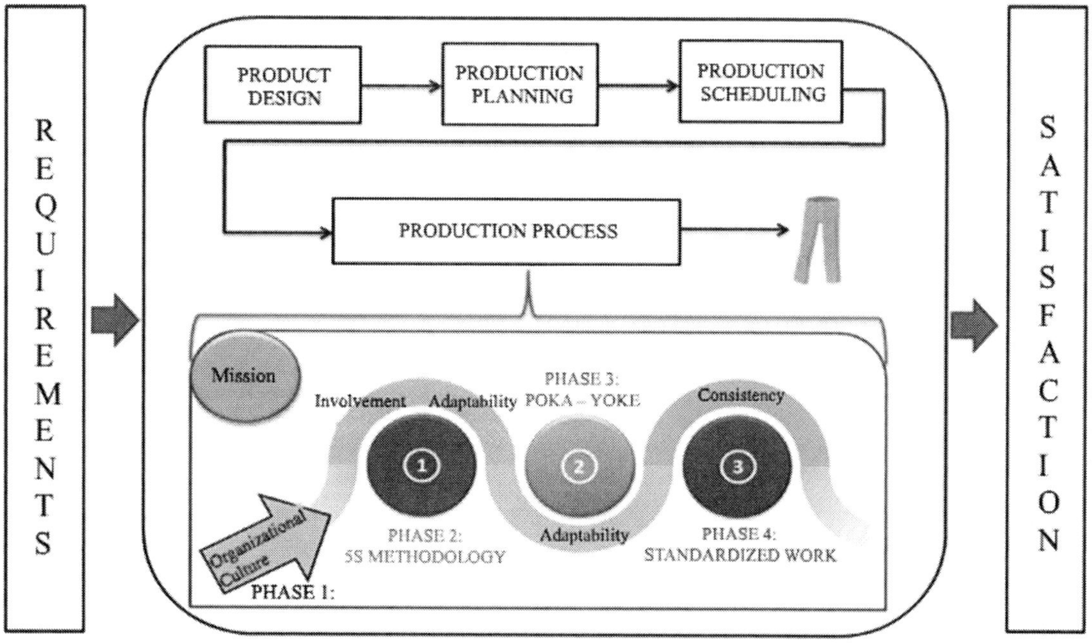

Figure 1. Lean production management model based on organizational culture

3.1.1. PHASE 1. Organizational mission and culture: To align the company's objectives with organizational culture dimensions to form the Denison Model audits.

3.1.2. PHASE 2. Stage 1 – 5s Methodology: Improve the company's working environment and product quality through five steps. Dimension 1: Adaptability and involvement: Involve the personnel in workshops to support the implementation of improvements such as work in the company.

3.1.3. PHASE 3. Stage 2 – Poka–Yoke – Mistake proofing: Prevent errors through a visual warning to a work method that requires it. Dimension 2: Involvement: Involve the employee in the implementation and the form of prevention applicable to their work.

3.1.4. PHASE 4. Stage 3 – Standardized work: Record the standard process in a flow that is within the reach of everyone. Dimension 3: Consistency: Maintain improvements with constant evaluations.

3.2. Proposed Method
The following depicts the entire sequence of activities involved in the working of the proposed model as a tool to improve production efficiency in a textile and clothing SME (Figure 2).

IOP Conf. Series: Materials Science and Engineering **796** (2020) 012004 doi:10.1088/1757-899X/796/1/012004

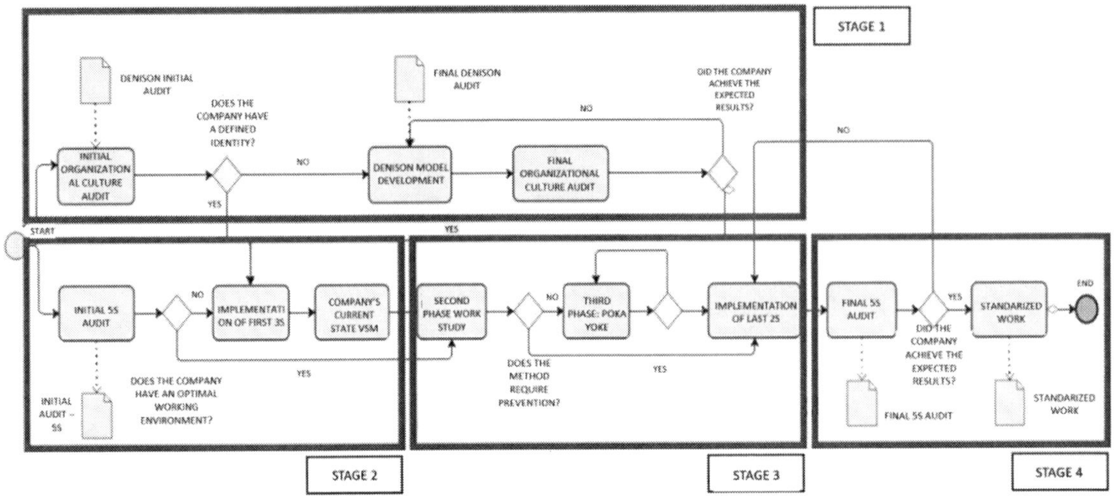

Figure 2. Process flowchart

3.3. Indicators

3.3.1. Inventory Shrinkage (IS) Ratio:

$$\text{IS} = \frac{Kg.of\ shrinkage}{Kg.of\ denim\, fabric}\ 100\% \quad (1)$$

3.3.2. Error Occurrence (EO) Ratio:

$$\text{ER} = \frac{No.of\ erros\ (jeans)}{No.of\ cutting\ operations\ per\, piece}\ 100\% \quad (2)$$

3.3.3. Production Efficiency (PE) Ratio:

$$\text{PE} = \frac{Actual\ output}{Effective\ capacity}\ 100\% \quad (3)$$

4. Validation

4.1. Case Study:
From the textile and clothing industry, we selected the company Bonice Jeans S.A.C. for this case study. This company is located in the Gamarra Commercial Center and is dedicated to the manufacture of denim clothing for women.

4.2. Business Diagnostics
We analyzed the supply chain and found the main problem to be the low cutting process efficiency. The initial diagnostic presented the following results: 17% shrinkage for each denim fabric roll; 15% errors in operations of each production batch; labor productivity of 7.9 and production process efficiency of 68%.

4.3. Application of The Model in The Case Study

4.3.1. First Phase – Denison Organizational Culture Model. We used the Denison Model, which consists of a questionnaire. The scoring is based on the following scale: (1) Strongly disagree; (2) Disagree; (3) Neutral; (4) Agree; and (5) Strongly agree. By applying the improvement, the results of the study positioned the company within the normal degree with a score of 17.9, surpassing the initial 12.9.

Table 1. Denison final matrix

Levels/ Dimension types	Agree	Neutral	Disagree	Total	Degree of compliance with organizational culture
Strategic direction		3		3	Normal
Goals		3		3	Normal
Vision		3		3	Normal
Values		3		3	Normal
Agreement	4			4	Good
Integration		3		3	Normal
Teamwork			2	2	Poor
Capability development		3		3	Normal
Empowerment			2	2	Poor
Creating change		3		3	Normal
Customer focus		3		3	Normal
Learning	4			4	Good

4.3.2. Second Phase – Workplace Organization Using the Lean 5S Method. We carried out the initial audit, by means of which we assessed the current development of each of the denim clothing manufacturing process operations. To this end, we considered a rating scale of 0–50. Then, we created a matrix to evaluate each step of the method according to the objectives by considering the following score: (1) Strongly disagree; (2) Disagree; (3) Neutral; (4) Agree; and (5) Strongly agree. The final 5S matrix showed a result of 26.6, exceeding the score of 9.6 achieved in the initial audit, which indicates that the company is in a state that is considered to be normal.

Table 2. Final 5S matrix

Goals/ Steps	Optimal work environment	Ease of search	Pleasant workplace	Cultural awareness	Total
Set in order	3	2	2	4	11
Sort	2	2	4	5	13
Shine	3	3	2	3	11
Standardize	2	2	3	2	9
Sustain	3	2	2	5	12

4.3.3. Third Phase – Critical process control through the Poka – Yoke Lean tool. We created a matrix to evaluate the alternatives according to the criteria and selected a tape measure to attach it to the cutting machine as a reference to indicate the proper height at which to cut.

4.3.4. Fourth Phase – Production Process Consolidation by Means of a Standardized Work Lean Tool. The manufacturing time was reduced from 30.28 minutes to 28.31 minutes to manufacture a pair of jeans, which was recorded on a standardized worksheet.

Finally, we present the final metrics used to evaluate the results of each phase.

Table 3. Summary of ratios

Phases	Ratios	Previous	Current
Phase 1: Culture Management	Denison Audit	12.9	17.8
Phase 2: Workplace Organization	Production within the Normal Range (PNR)	16%	11%
Phase 3: Critical Process Analysis	Shrinkage Ratio (SR)	17%	12%
Phase 4: Production Process Control	Error Occurrence (ER) Ratio	15%	10%
Project: Production Efficiency Improvement	Production Efficiency (PE)	68%	71%

As shown above, in the first phase, the culture level increased by 5 units; moreover, the last three model phases achieved improvements with a reduction of 5% each, and the main production efficiency ratio increased by 3%.

5. Conclusions

The implementation of the proposal resulted in a 3% increase in production efficiency over the previous year (2018), reaching 71%, which led to the generation of an additional 30 trousers per month. The involvement of operators is essential in evaluation to ensure the continuity and efficiency of the proposed model. The model contributes to citizenship by shaping employee behavior for them to identify with the company's values and put them into practice in society.

6. References

[1] Arrarte R, Bortesi L and Michue E 2018 Productivity and Competitiveness article in the Peruvian Textile and Clothing Industry 2015 - 2018 *Quipukamayoc* **25** 113-121
[2] Klaus S, Xavier S I M, World Economic Forum 2017 Global Competitiveness Report 2017 -2018 (*World Economic Forum*)
[3] ComexPerú *ComexPerú*. Retrieved from https://www.comexperu.org.pe/articulo/el-repunte-de -las-exportaciones-textiles.
[4] PRODUCE 2015 Textile and Clothing Industry (*Lima*).
[5] Loayza N 2016 Productivity as the key to growth and development in Peru and the world *J Econ Stud* **31** 9-28
[6] G. Triveño 2015 About the textile sector.
[7] IEES Sector Report 2018 Textile and clothing industry (*Lima*).
[8] Andrade Y, Cardenas L, Viacava G, Raymundo C and Dominguez F 2019 Lean manufacturing model for the reduction of production times and reduction of the returns of defective items in textile industry *Adv. Int. Sys. Comp* **954** 387-398
[9] Aurora I. and Cavazos J. L. M. 2017 Transfer of knowledge within the company: analysis of precursor variables in a Lean-Kaizen environment *Nova Scientia*
[10] Kumar S, A. Dhingra, Singh B 2018 Process improvement through Lean-Kaizen using value stream map: a case study in India *Int. J. Adv. Manuf. Tech.* **96** 2687–2698
[11] Ahmady G A, Aghdas N and Mehrpour M 2016 Effect of Organizational Culture on knowledge Management Based on Denison Model *Procedia Soc Behav Sci* **230** 387-395
[12] Urban W 2015 The Lean Management Maturity Self-Assessment Tool Based on Organizational Culture Diagnosis *Procedia Soc Behav Sci* **213** 728-733

IOP Conf. Series: Materials Science and Engineering **796** (2020) 012005 doi:10.1088/1757-899X/796/1/012005

The experimental study of identification trigger components in ash deposits at the boiler pipe of the electric steam power plant in Nagan Raya, Aceh province

A F Syah[1], A Gani[2] and Khairil[1, 3]

[1] Department of Mechanical Engineeering, Faculty of Engineering
Universitas Syiah Kuala 23374, Aceh, Indonesia

[2] Department of Chemical Engineeering, Faculty of Engineering
Universitas Syiah Kuala 23374, Aceh, Indonesia

E-mail: khairil@unsyiah.ac.id

Abstract. Coal energy is used as the greatest main source of power plant in this day and age. The consequence of the coal energy exertion is the ash deposits are cumulated in the used boiler. In this experiment, the fly ash and ash deposit are analyzed by using SEM/EDX (scanning electron microscope/ energy dispersive x-ray) to obtain the tendency of the ash deposits with a base to acid ratio. the sample of ash deposit was selected at three layers of observation to discover the slagging index of the combustion process. The result of the study indicates that the tendency of the ash deposits is 0.959736 and the slagging index on three layers observation are 0.4987, 0.264183, and 0.140248. The result of the slagging index on three layers observation is categorized as low. The trigger compound of cumulation is K_2O which is found more in the ash deposits around the pipe and constitutes in the lowest melting point compounds as compared to the other compounds contained in the slagging ash.

1. Introduction

The cumulation of deposits in the boiler's pipe is one of the effects of combustion process commonly happened in the high temperature furnace in the boiler. Most of the formed ash deposits is caused by the inorganic materials which fuse in the high temperature furnace and adhere to the boiler wall. Those problems affect the overall work of the boiler. The deposits in the boiler's pipe forms several linings with various structures and colors. Nevertheless, the exertion of various coal energy causes various deposits effects. The deposits formation occurs in three mechanisms, they are first the particle displacement in the cause of gradient concentration (diffusion), second the displacement of small particles in the cause of temperature (thermophoresis), and third the ash precipitation in the cause of large inertia quantity in the impacted deposits (inertial impact). CaO, Al_2O_3, and SiO_2 are the primary sources of deposits formation in the boiler's pipe as the initial lining of deposits formation (the trigger). Furthermore, the other metal elements (Mg, Al, and Fe) are bound to form the deposit compounds. On the other hand, the deposits formation occurred in *the water wall* with a high temperature boiler has the initial lining (the trigger) formed as a thin white layer with 10% of S, Na, Si, and Ca. The condensed $CaSO_4$ in the combustion process leads to the formation of a strong initial deposits compounds due to the displacement of small particle as the result of the temperature gradient (thermophoresis). Moreover, molten aluminosilicate formed in the farther layer of deposits detains solid particle elements as in CaO, Al_2O_3, etc., to form the outer layer of deposits. [1-4].

In this study, the fly ash and the ash deposits from the combustion process in the boiler of the electric steam power plant in Nagan Raya are observed by *Scanning Electron Microscope / Energy Dispersive X-Ray Analysis (SEM/EDX)* to discover the trigger compounds of slagging formation in boiler's pipes

[3] To whom any correspondence should be addressed.

Content from this work may be used under the terms of the Creative Commons Attribution 3.0 licence. Any further distribution of this work must maintain attribution to the author(s) and the title of the work, journal citation and DOI.
Published under licence by IOP Publishing Ltd

in the furnace. Furthermore, the aim of study is to identify the compounds found in the ash deposits to obtain the tendency of the fly ash and the slagging index of the combustion process in the boiler.

2. Experimental

The samples of ash deposit generated during combustion at Nagan Raya steam power plant under boiler conditions are shown in table 1. Figure 1 shows the condition of the sample ash deposit after it was formed in a furnace at the Nagan Raya steam power plant which operated for 6 months with a loading of around 70 MW. The ash deposits formed and attached on the pipe in the furnace is obtained and observed with SEM/EDX. The observation of the ash deposits consists of three parts and has been decided beforehand (figure 2). The three parts of the ash deposits are the ash attached on the water pipe wall, the middle part ash, and the external part which connects directly to the furnace. The samples are studied on several parts of examination to observe the image in enlargement up to 2000 times using SEM, moreover, the samples as in chemical composition are presented by EDX.

Figure 1. The ash deposit formed on a pipe in the furnace during the combustion process at the Nagan Raya steam power plant.

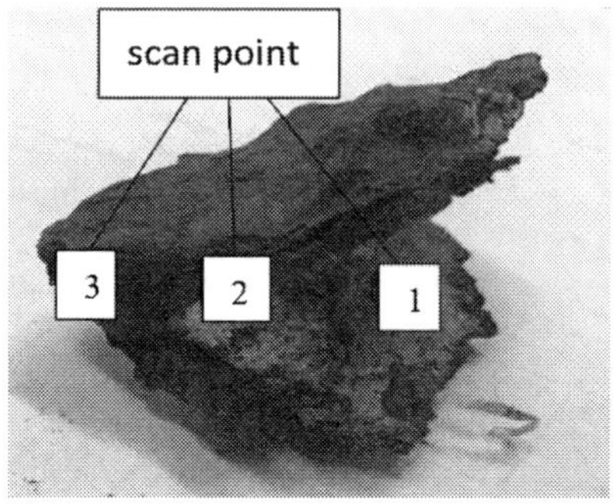

Figure 2. The examples of ash deposit to be observed at 3 scanning points

3. Result and Discussion

3.1. Amount of ash formed

Proximate analysis of coal in the Nagan Raya electric steam power plant (table 1) obtained the weight of ash content in every 1 kg of coal is 8.3194%. the amount of ash produced by normal coal combustion in the furnace is 4,3344 - 4,5007 T/h

Table 1. Fuel characteristics of proximate analysis and normal coal consumption in combustion Nagan Raya electric steam power plant

Proximate analysis (%)				Coal consumption (T/h)	
Total moisture	Volatile content	Ash content	Fixed carbon	Min	Max
35,55	43,67	8,3194	38,67	52,1	54,1

3.2. The tendency of ash deposits

The amount of ash deposits in combustion process is determined by several methods of observation. One of the methods used is observing the chemical compositions of the ash resulted from the combustion process. Equation 1 can be used if the compound content is CaO < 7.5 and Na2O < 2.5. The tendency of the ash deposits can be obtained using the equation 1

$$R\frac{b}{a}=\frac{Fe_2O_3+CaO+MgO+K_2O+Na_2O}{SiO_2+Al_2O_3+TiO_2} \quad (1)$$

The result of SEM/EDX test for base-acid ratio in the ash resulted from the combustion process is presented in Table 2.

Table 2. The value of acid-base compounds in the ash resulted from the combustion process of the electric steam power plant in Nagan Raya region.

The amount of acidic compounds			The amount of basic compounds				
SiO_2	Al_2O_3	TiO_2	Fe_2O_3	CaO	MgO	K_2O	Na_2O
5,49	4,481	0,92	6,59	2,118	1,006	0,774	0,31

The ratio accumulation above shows that the base to acid ratio contained in the ash from the combustion process in the boiler of the electric steam power plant in Nagan Raya is 0.959736.

Base-acid ratio directly reacts to the temperature point of the ash fusion. The value escalation in base-acid ratio of the ash deposits decreases the Initial Deformation Temperature (IDT), the Softening Temperature (ST), the Hemispheric Temperature (HT), and the Flow Temperature (FT). In the base-acid ratio ≤ 0.15, the fusion temperature of the ash reaches out 1873 K and decreases gradually along with the escalation of based-acid ratio that is formed in the process. In the base-acid ratio 0.75 – 2, the fusion temperature of the ash deposits reaches out 1473 K which easily causes slagging. In the base-acid ratio > 2, the dependency of the base-acid ratio towards the fusion temperature of the formed ash does not exist [5].

3.3. Slagging index

The slagging index of the cumulation of ash deposits is obtained using equation 2

$$Fs = \frac{(Fe_2O_3 + CaO + MgO + K_2O + Na_2O)S}{SiO_2 + Al_2O_3 + TiO_2} = R\frac{b}{a} \times S \qquad (2)$$

The value of the compounds can be discovered by observing the samples of ash deposits that are formed in the process. The study is undertaken on three-point samples of ash deposits resulted from the combustion process of the electric steam power plant in Nagan Raya. They are the part around the pipes, the middle part of ash deposits, and the external part of ash deposits.

Figure 3 is the portrait of the ash deposit particles on the part around the pipes and compound compositions found in the observation. The lines in the portrait shows the result of mineral attached on groups of fine ash caused by combustion on fluidized-bed boiler type. The particles do not merge as one because of inadequate temperature level in fluidized-bed combustion type. According to Raask, 1985, the intergranular cracks from the lines are caused by the high level of sulfur in each stem lines. The white dots in the portrait are chars or ashes [6].

IOP Conf. Series: Materials Science and Engineering **796** (2020) 012005 doi:10.1088/1757-899X/796/1/012005

Figure 3. The result of SEM/EDX observation of the slagging ash around the pipe. The enlargement image of ash deposits up to 2000 times 30 μm

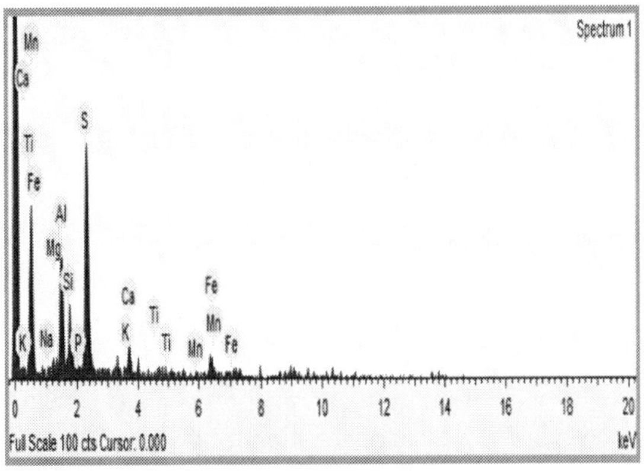

Figure 4. The result of SEM/EDX observation of the slagging ash around the pipe. The chemical compositions in the observation.

Figure 5 and 6 are the portraits of ash deposits on the middle part of ash and the compound compositions found in the observation. The portraits are similar to figure 3 and 4 because it formed intergranular lines as the result of the high level of sulfur in the lines. Moreover, the bigger formed of the line stems are caused by the increasing amount of silica in the middle part observation compared with the part around the pipe.

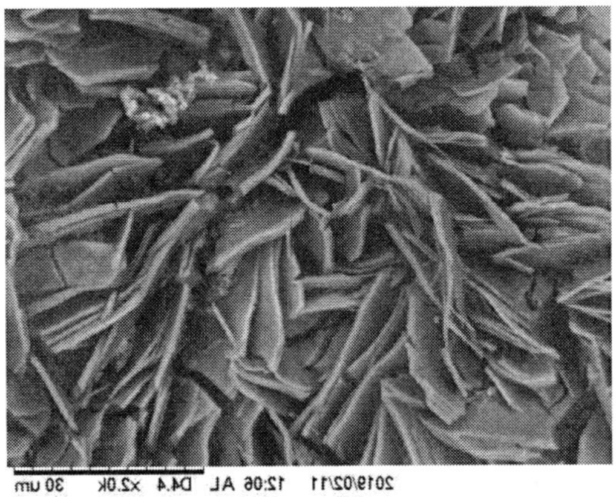

Figure 5. The result of SEM/EDX observation of the slagging ash in the middle part. The enlargement image of ash deposits up to 2000 times 30 μm

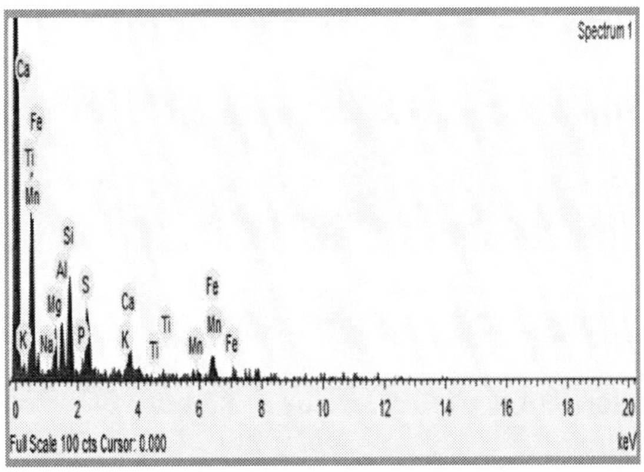

Figure 6. The result of SEM/EDX observation of the slagging ash in the middle part. The chemical compositions in the observation.

Figure 7 and 8 are the observation of the ash deposits external part of 2000 times enlargement with the size of 30 μm and the compound compositions found in the observation. The distinction for each clot in each particle portrays different characteristic of the formed ash. Moreover, the picture shows that the diameter of the ash particles is differed based on the size; large and small [7].

Figure 7. The result of SEM/EDX observation of slagging ash in the external part. The enlargement image of ash deposits up to 2000 times 30 μm

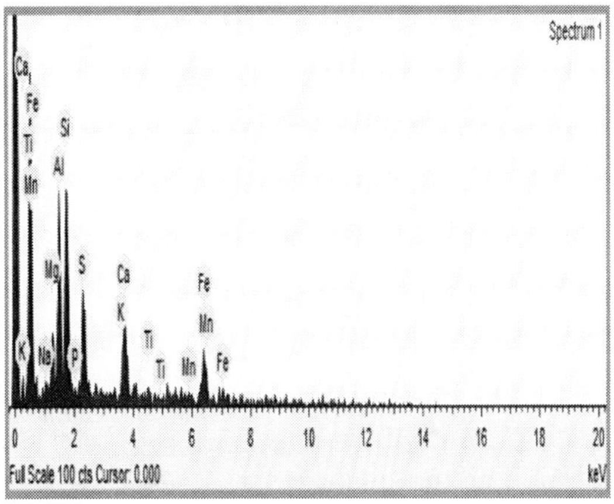

Figure 8. The result of SEM/EDX observation of slagging ash in the external part. The chemical compositions in the observation.

Using the same equation for each sample, the slagging index is discovered with the value of; the part around the pipe 0.4987, the middle part of ash deposits 0.264183, the external part of ash deposits 0.140284. According to the result of slagging index observation, the researcher concludes that the slagging index of the electric steam power plant in Nagan Raya is classified as low. It is caused by the use of sand containing big amount of silica as the heat transfer media in the boiler of the electric steam power plant in Nagan Raya. The calculation result of the slagging index describes the most accretion of slagging occurs in the part around the pipe due to the bigger amount of sulfur found in it. Furthermore, in that part occurs the large heat transfer that put an impact on the transformation of inorganic components contained in the melted coal because of the high-temperature combustion in the boiler reshaping the deposits caused by the heat transfer.

3.4. The cause of the ash formation

From the three points observation of the ash deposits found in the boiler of the electric steam power plant in Nagan Raya, the trigger components causing the ash deposits can be discovered by observing the part around the pipe as presented in Table 3.

Table 3. The chemical compositions of ash deposits around the pipe

Chemical components	SiO_2	Al_2O_3	TiO_2	Fe_2O_3	CaO	MgO	Na_2O	K_2O	P_2O_5	MnO
Wt (%)	8,723	16,912	0	14,782	6,732	0,29	1,211	3,119	0,868	0

The ash deposits are formed from the compounds that are not burned-out by the used fuel. In this case, the compounds undergo the transformation to gas phase because the temperature in the boiler gets up to 1300K, particularly in case of the low melting point compound. In line with the transfer of heat in the boiler's pipes, the compounds experiences temperature digression that causes the compounds transformed into fluid. The temperature digression is caused by heat absorption of fluids in the internal part of boiler's pipe with lower temperature than the compounds undergoing heating process. The process causes the compounds caused to clot into solid. The lowest melting point compound is the initial trigger of ash transformation in the boiler's pipe. Diphosphorus pentaoxide (P_2O_5) is the compound with the lowest temperature that is 613K, however, it has the least composition and cannot be found in all scanning. As a result, kalium oxide (K_2O) is the lowest melting point compound of 7 compounds obtained in the ash deposits around the pipe with the temperature of 1010 K (Lide, 2009), followed by iron (III) oxide (Fe_2O_3) and natrium oxide (Na_2O). The kalium oxide (K_2O) bonds with the other compounds to form the ash deposits along the path of the heat transfer pipe.

4. Conclusion

The tendency of ash deposits process occurred in the boiler's pipe of the electric steam power plant in Nagan Raya is analyzed by observing the ash resulted from the combustion-output indicates that the base-acid ratio value ($R\frac{b}{a}$) is 0.959736. It includes in high ratio level which alleviates the Initial Deformation Temperature (IDT), Softening Temperature (ST), the Hemispheric Temperature (HT), and the Flow Temperature (FT) in the ash deposits escalating the probability of the cumulation of ash deposits in the furnace. The Value of *Slagging Index* of the ash deposits observed in three points observation which is around the pipe, the middle part, and the external part respectively are 0.4987, 0.264183, 0.140248. It includes in the group of low slagging index. The most dominant compound found in the ash deposits in the boiler's pipe of the electric steam power plant in Nagan Raya are sulfur and silica, furthermore, the compositions containing in the ash deposits are SiO_2, Al_2O_3, CaO, MgO, K_2O, Na_2O, P_2O_5, TiO_2, and MnO. Kalium oxide (K_2O) is the initial trigger of ash transformation in the boiler's pipe because of that compound is the lowest melting point compound of 7 compounds obtained in the ash deposits around the pipe with the temperature of 1010 K

Acknowledgements

The author would like to greatly appreciate for research permission from PT. PLN (Persero) Unit Pelaksana Pembangkitan Nagan Raya with contract number No. 093/MUM.01.01/UPKNGR/2018

Reference

[1] Naruse I, Nakayama K and Khairil 2000 Ash Deposition Characteristics in Pulverized Coal Reaction Under High Temperature Conditions *J Chem Eng Japan* **33** (3)

[2] Wei B, Tan H and Wang Y 2017 Investigation of Characteristics and Formation Mechanisms of Deposits on Different Positions in Full-Scale Boiler Burning High Alkali Coal *Appl Therm Eng **119** 449-458.*

[3] Xianpeng Z, Dunxi Y and Fangqi Liu 2018 Scavenging of Refractory Elements (Ca, Mg, Fe) by Kaolin During Low Rank Coal Combustion *Fuel* **223** 198-210.

[4] Zhou H, Zhou B and Li L 2013 Experimental Measurement of the Effective Thermal Conductivity of Ash for High Sodium Coal (Zhun Dong Coal) in a 300 KW Test Furnace *Energ fuels* **27** 7008-7022.

[5] Marek P 2002 Evaluation of the influence of biomass co-combustion on boiler furnace slagging by means of fusibility correlations *Biomass Bioenerg* **28** 375-383.

[6] Raask E 1985 Mineral Impulrities in Coal Combustion. Behavior, Problem and Remedial Measures (*London: Hemisphere Publishing Coorporation*).

[7] Khairil 2013 Teknik Pembakaran (Bahan Bakar Padat) (*Banda Aceh: Syiah Kuala University Press*)

Maintenance Management Model under the TPM approach to Reduce Machine Breakdowns in Peruvian Giant Squid Processing SMEs

A Gallesi-Torres[1], A Velarde-Cabrera[1], C León-Chavarri[1], C Raymundo-Ibañez[2,4] and F Dominguez[3]

[1] Ingeniería Industrial, Universidad Peruana de Ciencias Aplicadas (UPC), Lima 15023, Perú.
[2] Dirección de Investigación, Universidad Peruana de Ciencias Aplicadas (UPC), Lima 15023, Perú.
[3]Escuela Superior de Ingeniería Informática, Universidad Rey Juan Carlos, Mostoles, 28933, España.

E-mail: carlos.raymundo@upc.edu.pe

Abstract. In recent years, the giant squid processing industry in Peru exhibited a 59% increase in exports with respect to 2018. According to estimates, this industry generates approximately 30,900 jobs per year. However, some SMEs experience low productivity, such as the PECEPE company, due to plant downtime. This represents 26% of the available time, which translates into the loss of 1760 tons every year. The most constraining external factor the sector faces is uncertainty in resource availability caused by changing weather conditions and informal fishing activities. Although there is a large number of research studies on the fishing industry and resource extraction, literature on processing plant operations is scarce. Within this context, this study seeks to promote a high impact sector in Peru, as well as fostering processing plant competitiveness and productivity. Hence, to address these issues, the authors propose a maintenance management model under the TPM approach. As part of the results from model implementation, a 39% decrease was reported in plant downtime, while maintenance costs also decreased by 16%, which, in turn, increased machine availability and production by 784 tons per year.

1. Introduction

Between 2008 and 2016, small-scale giant squid landings in Peru represented approximately 53% of the world's total, with an average catch of 556,000 MT, thus turning Peru into the country with the highest historical resource capture record in the world. However, even though Peru ranks first in resource extraction, it ranks fourth in exports, with just a 7% market share, which is due to the low levels of competitiveness exhibited by Peruvian hydrobiological resource processing plants [1]. According to the Peruvian Industry Chamber (SNI, for its Spanish acronym), the 127 companies currently operating in this sector reported daily production averages at 48 MT in 2016 [2]. However, the "Fish and Cephalopods of Peru" company, PECEPE, reports an average production of 30 MT/day,

[4] To whom any correspondence should be addressed.

Content from this work may be used under the terms of the Creative Commons Attribution 3.0 licence. Any further distribution of this work must maintain attribution to the author(s) and the title of the work, journal citation and DOI.
Published under licence by IOP Publishing Ltd

which denotes a gap in comparison to the sector due to low productivity. In 2018, the company reported its overall equipment effectiveness (OEE) rate at 52.15%, which is below the 85% benchmark established for the manufacturing sector at the international level [3].

According to estimates, this industry generates approximately 30,900 jobs per year. The main external factors that affect this industry is uncertainty in resource availability as giant squids migrate depending on the weather conditions. The above, in addition to port infrastructure, acts of nature and the informality of fishing [4].

In this context, this research study seeks to promote a high impact sector in Peru, wherein most fishing companies are still under development. This is evidenced by the existing gap between the number of landings and the position currently held by the country in terms of resource exports at the global level. Another critical item is the lack of literature on improving operations and asset maintenance management in hydrobiological product processing plants as most of the sector plants are semi-automated and use processing lines [5]. However, there are vast opportunities for the development of value-added products. For example, countries such as China and India, which capture and land 50% less resources than Peru, successfully triple Peruvian exports due to the greater added value offered by their products and the greater levels of competitiveness exhibited by their plants [6]. Therefore, this study seeks to contribute research to the sector which may contribute to improving productivity at the PECEPE giant squid processing plant by reducing machine breakdowns and their impact on increasing company profitability.

The proposal comprises a maintenance management model reinforced with lean manufacturing tools such as TPM, 6S, and continuous improvement. The pilot plan was implemented for 6 months in the giant squid ring production line—the company's flagship product.
The next section describes a literary review on the techniques used.

2. Literary review

2.1. TPM

Most TPM implementation case studies come from the manufacturing sector [7]. For instance, Gupta and Vardhan applied TPM to a renowned tractor manufacturing industry in India. Their study showed that, through TPM implementation, production costs were reduced, while OEE rates improved, which doubled revenue from sales and tripled profits over a three-year period [8] [9].

Different authors who applied TPM to manufacturing organizations identified several pre-TPM implementation issues, such as extremely high maintenance costs; emergency repairs being thrice as expensive as the actual cost; and extremely high equipment downtime numbers. Moreover, pre- and post-implementation OEE rate were calculated for each machine, determining a 15%–30% increase in OEE rates for the different machines [8] [9] [10] [11].

2.2. 6S

The integration of 5S and security (6S) is unique in the scheme proposed by Rodrigues. As opposed to previous studies that treat these elements independently, 6S focuses on work safety by guaranteeing a clean, organized environment without unnecessary items on the way [12].

One of the successful cases found in the literature was in an ink manufacturing company, which used 6S to improve overall productivity and efficiency rates at a corporate level. Here before and after organizational photography was used to evidence the benefits from 6S implementation. Then, a 6S Checklist audit validated that throughout the 8-week period in which it was deployed, the level of technique maturation was deemed as sustainable, thus ratifying the objective of the study. The benefits from implementing 6S in this company were an optimized use of workspace, less time spent in searching for materials and supplies, improvements in occupational safety and health, and complete elimination of accidents, hazards, and risks [13] [14].

3. Contribution

3.1. Proposed Model
Based on the articles selected from each typology, the proposed maintenance management model was developed using a TPM approach supported by 6S implementation, as depicted in Figure 1. As a pilot plan, this model is deployed in the giant squid ring production line at the PECEPE plant.

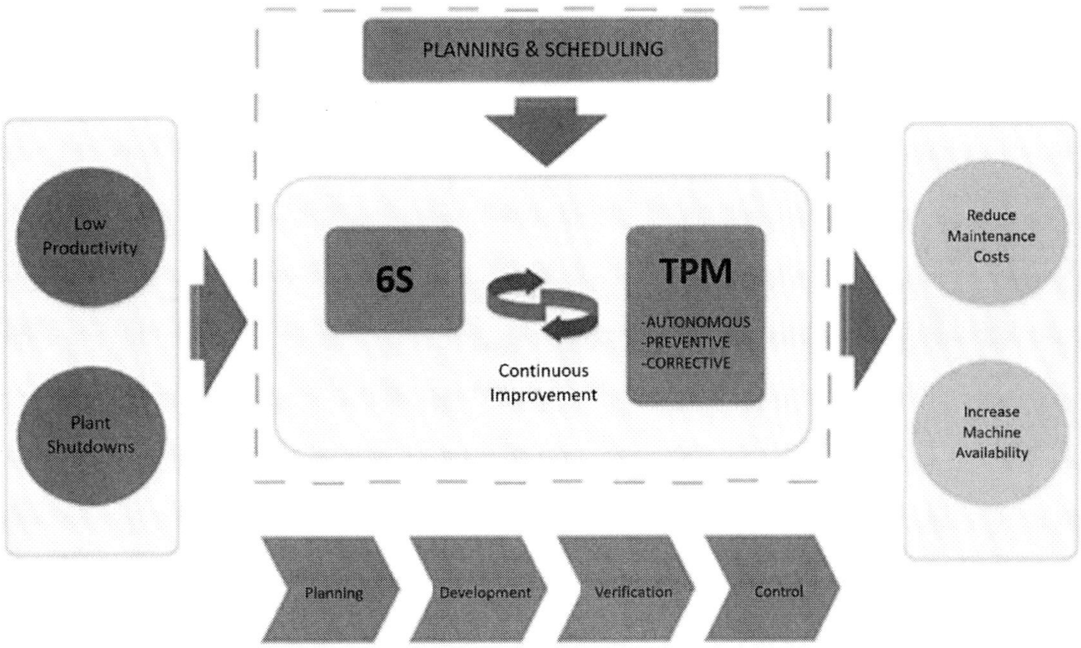

Figure 1. Maintenance management model with a TPM & 6S approach

The model comprises the following main phases: planning, development, verification, and control.

1. Planning: in this phase, the main components that will serve as the foundation for the proposed model are defined and prepared.
2. Development: the 6S, autonomous maintenance, and planned maintenance tools are implemented.
3. Verification: the techniques applied are assessed and measured through indicators that verify compliance with the objectives proposed.
4. Control: the sustainability of model techniques within the maintenance process is monitored through documentation. Opportunities for improvement are reviewed through feedback and new techniques are proposed.

3.2. Proposed Process
The improvement proposal deployed in the giant squid ring production line at PECEPE focuses on the PDCA cycle and on TPM and 6S implementation. This way, the model seeks to reduce plant downtime due to machine breakdowns, thus increasing productivity in giant squid processing.

Figure 2 denotes the model development stages.

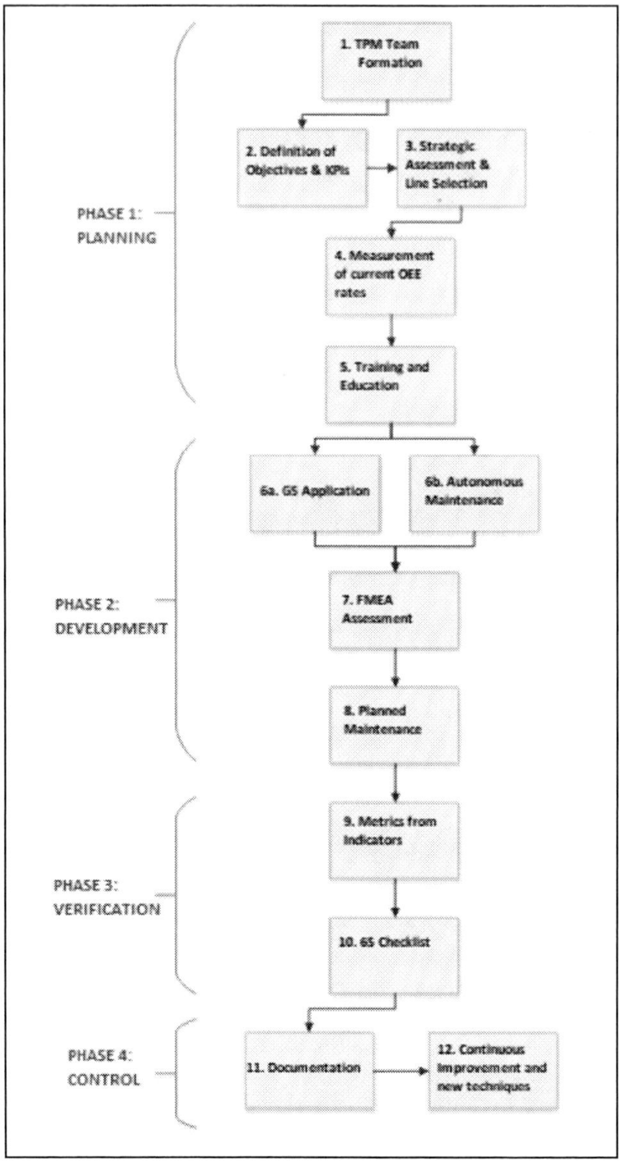

Figure 2. TPM application diagram

3.3. Indicators

Based on the proposed objectives, the following indicators were set forth to measure and control the parameters required by the company.

3.3.1. Equation 1: Overall Equipment Efficiency. This metric denotes the percentage of planned production time that is truly productive. It measures increases in availability, performance, and quality for the selected production line and is used to determine the success rate of TPM application.

$$OEE = Availability \times Performance \times Quality \tag{1}$$

IOP Conf. Series: Materials Science and Engineering **796** (2020) 012006 doi:10.1088/1757-899X/796/1/012006

Where: 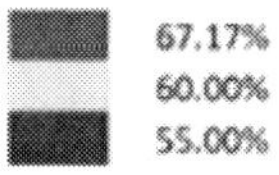

3.3.2. Equation 2: Maintenance Compliance (CM). This metric measures the percentage of compliance with preventive maintenance plans.

$$CM = (\frac{MR}{PM}) \times 100 \quad (2)$$

Where: 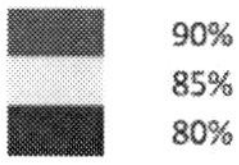

3.3.3. Equation 3: Unscheduled Plant Downtime. Quantifies unscheduled plant downtime at the giant squid ring production line level.

$$Unscheduled\ Downtime = \frac{Downtime}{Available\ Time} \quad (3)$$

Where: 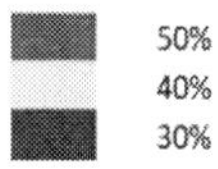

4. Validation

4.1. Successful Case

The company under study, "Fish and Cephalopods of Peru," PECEPE, has been in operation for five years in the market. It is located at 18th Km, South Pan-American Highway, within the Esmeralda Corp facilities. The pilot plan was developed in the production area, which comprises four semi-automated lines: giant squid rings, tentacles, neck, and paste. The company employs seven workers in the production area, while the rest of the plant operators are outsourced at hourly rates. An ABC analysis deemed giant squid rings as the company's flagship product as it represents more than 25% of the total sales from previous years.

Figure 3. Inside the PECEPE plant

4.2. Diagnosis
The company faces low productivity issues caused by downtimes of approximately 15.8 hours per week, which is equivalent to 26.4% of the entire production time. This generates an annual economic impact of $893,693.00as its average daily production capacity (30 MT/day) is below the average production rates reported for the sector (48 MT/day). Similarly, current OEE rates for the giant squid production line are at 52.15% in contrast to the world-class OEE standard rates at 85%.

Table 1. World-Class OEE standard rates

OEE Factors	**World Class Standard**	**OEE PECEPE**
Availability	90.00%	73.61%
Efficiency	95.00%	71.98%
Quality	99.90%	98.43%
General OEE	85.00%	52.15%

As determined, this low productivity was mainly caused by machine breakdowns and delays in repairs (64%), the plant disorganization (12%), and reprocessing (7%).

4.3. Model Application

4.3.1. Planning. Within the first phase, the following actions are performed:

- Establishment of a Maintenance Team
- Strategic Assessment and Production Line Selection
- Measuring of Current OEE Rates
- Training and Education

4.3.2. 6S Implementation
Applying the checklist 6S helps to clean and organize the workplace in order to facilitate the solution of the problem and as a basis for the implementation of the TPM. The checklist shows a score based on infractions for each item. If the score is greater than 75 then it is satisfactory.

After deploying 6S in the plant for two months, the 6S Checklist was applied for auditing purposes. In this case the score was 90, rated as optimal (the list was complete at 90%).

PHASE	ITEM						SUM
	Number of Infractions	5 4 3		2	1	0	
	Score	1	2	3	4	5	
SORT (SEIRI)	*Distinguish between what is needed and what is not*						
	Are only the required parts, materials and tools present in the work area?				X		4
	Have unnecessary items been removed from the work area?					X	5
	Are personal belongings and documents stored properly?					X	5
SET IN ORDER	*Everything must be in the right place*						
(SEITON)	Are walkways clearly marked and identified?					X	5
	Are all the parts, tools and equipment stored in their correct locations?					X	5
	Are all machines properly labeled?					X	5
SHINE (SEISO)	*Keep the workplace and surfaces clean*						
	Are cleaning supplies and equipment easily accessible?				X		4
	Are tools and equipment kept clean?				X		4
	Are the walkways free of dirt and liquid?				X		4
SUSTAIN	*Stick to the rules*						
(SHITSUKE)	Has the 6S audit been updated, made available to everyone and shared with top management?				X		4
	Have sufficient resources and time been allocated to 6S activities?					X	5
	Is the 6S score continuously reviewed and discussed?				X		4
					TOTAL		90

Figure 4. 6S Checklist. Note: Obtained from PECEPE.

As per the authors, any score above 75 is considered excellent. In this case, the final score of 90 renders the 6S implementation at the PECEPE plant as successful.

Figure 5. The Giant Squid ring production line before and after 6S application.

4.3.3. Autonomous Maintenance Application

The flowchart depicting the autonomous maintenance procedure was developed. Pre- and post-application results are presented below:

IOP Conf. Series: Materials Science and Engineering **796** (2020) 012006 doi:10.1088/1757-899X/796/1/012006

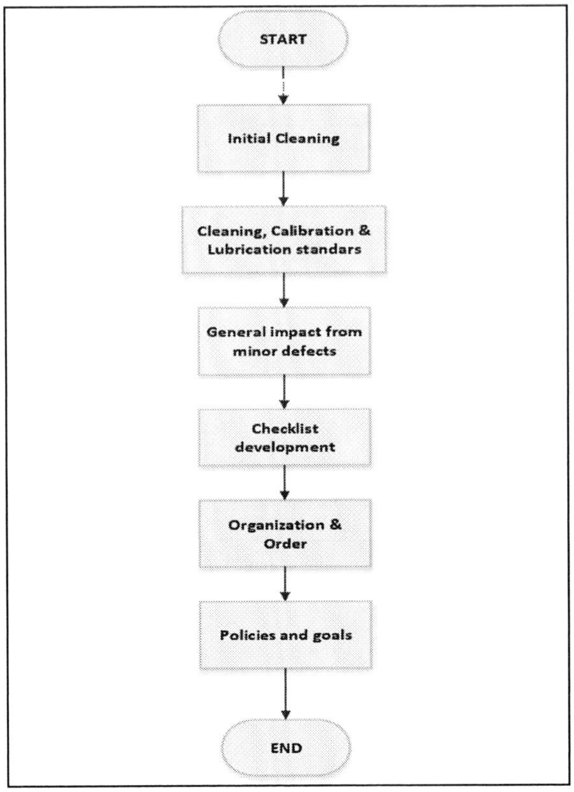

Figure 6. Autonomous maintenance flowchart

GIANT SQUID RINGS PRODUCTION LINE							
BEFORE				AFTER			
Activity	Maintenance Type	Frequency	Initial Time per Month (Min)	Activity	Maintenance Type	Frequency	Final Time per Month (Min)
Bearings Greasing	Preventive	Weekly	220	Clean Work Area	Autonomous	Daily	200
Megging of Electric Engine	Preventive	Weekly	360	Clean Conveyor Belt	Autonomous	Daily	360
Check Electrical System	Preventive	Weekly	720	Filter Cleaning	Autonomous	Daily	200
Change Weighing Card	Preventive	Weekly	400	Bearings Lubrication	Autonomous	Daily	300
Band Tape Check	Preventive	Weekly	320	Band Check	Autonomous	Daily	200
Blade Changes	Corrective	Weekly	260	Chain Adjustments	Autonomous	Daily	200
Gear Change	Corrective	Weekly	800	Blade Changes	Preventive	Monthly	20
Blade Sharpening	Corrective	Weekly	120	Gear Change	Preventive	Monthly	20
Suction Joint Adjustment	Corrective	Weekly	140	Check Electrical System	Preventive	2 months	45
Packing Changes	Corrective	Weekly	280	Electric Engine Maintenance	Preventive	2 months	70
Filter Cleaning	Corrective	Weekly	180	Wheel Bearing Changes	Preventive	3 months	15
			380				**1630**

Figure 7. Maintenance Times Before and After Autonomous Maintenance Application.

The model proposed the application of autonomous and preventive maintenance tools, which reduced downtimes by 36.17 hours.

4.3.4. Preventative Maintenance

The preventive maintenance process was redesigned as per the flowchart described in Figure 7:

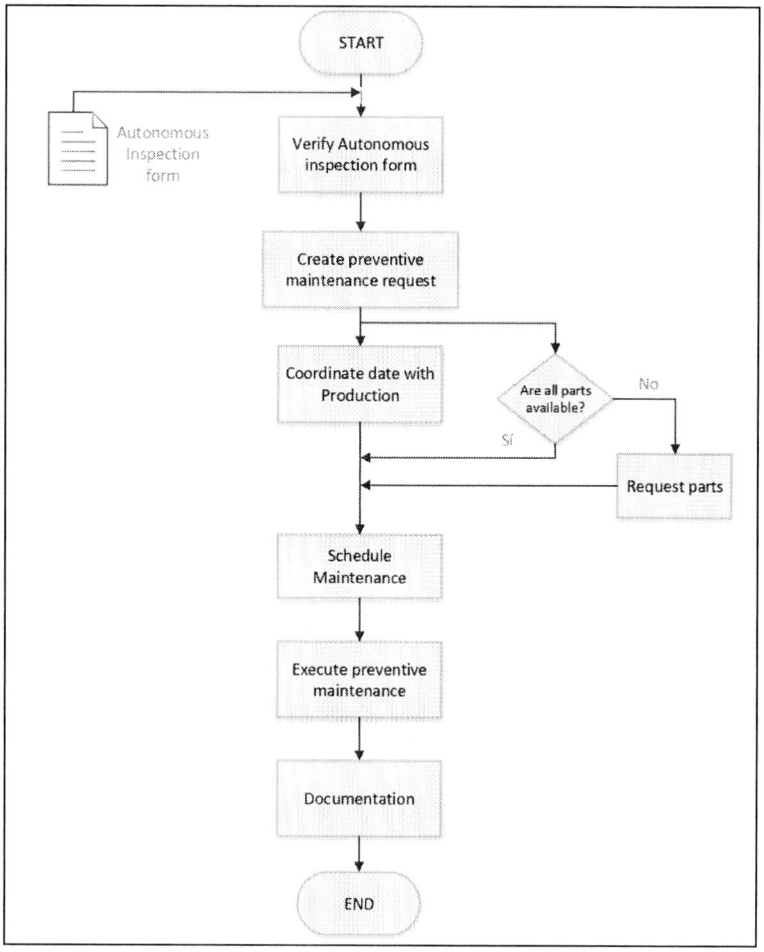

Figure 8. Autonomous Maintenance Flowchart

4.3.5. Verification: Metrics from Indicators

Table 2. Results from initial and final indicator data

	Initial data	**Final data**
MTBF	4.45 hrs	6.30 hrs
OEE	52.15%	68.66%
Quality	98.43%	98.68%
Performance	71.98%	82.80%
Availability	73.61%	84.03%
CM	67.00%	91.00%
Machine Breakdowns	26.39%	15.97%
Maintenance cost per unit	19.57%	16.44%

The data and times were taken before and after implementing the pilot plan to validate the achievement of objectives and to measure their progress.

5. Conclusions

- Model implementation reported a 39% reduction in downtimes, which increases production by 748 tons per year. This represents revenue of $405,770 obtained during the 6-month period in which the model was implemented.
- Further, overall machine availability experienced a 14% increase, which improves average times between breakdowns by 40%.
- Finally, maintenance costs were reduced from 19.57% to 16.44% per ring block, thus generating a 3% reduction in total production costs.

6. References

[1] PROMPERÚ 2018 Developing Fishing and Aquaculture Foreign Trade (*Lima*).
[2] SNI 2016 "Number of Fishing Plants," (*Lima*).
[3] Vorne 2011 Lean Production. Retrieved from: www.leanproduction.com/.
[4] PRODUCE 2015 ANUARIO ESTADÍSTICO PESQUERO Y ACUÍCOLA (*Lima*).
[5] Fullerton R, Kennedy F A and Widener S K 2014 Lean manufacturing and firm performance: The incremental contribution of lean management accounting practices *J Oper Manag* Management, **32**(7-8) 414-428.
[6] MINCETUR 2016 Peruvian Giant Squid in the Spanish Market (*Lima*).
[7] Lastra F, Meneses N, Altamirano E, Raymundo C, Moguerza J M 2019 Production management model based on lean manufacturing for cost reduction in the timber sector in Peru Adv Intel Syst Comput **971** 467-476.
[8] Gupta P and Vardhan S 2016 Optimizing OEE, productivity and production cost for improving sales volume in an automobile industry through TPM: a case study *Int J Prod Res* **54** (10) 2976-2988.
[9] Ahuja I and J S K 2008 Total productive maintenance: literature review and directions," *Int J Qual Relia Manag* **25** (7) 709-756.
[10] Peimbert-García R, Limon-Robles J, Beruvides M and García-Hernández D 2012 An economic framework for total productive maintenance (TPM) 2222-2231.
[11] Wickramasinghe and Perera A 2016 Effect of total productive maintenance practices on manufacturing perfomance: Investigation of textile and apparel manufacturing firms *J Manuf Tech Manag* **27** 713-729.
[12] Rodrigues M and Hatakeyama K 2006Analysis of the fall of TPM in companies *J Mater Process Tech* **179**(1-3) 276-279.
[13] Sukdeo N 2017 The application of 6S methodology as a lean improvement tool in an ink manufacturing company *IEEE International Conference on Industrial Engineering and Engineering Management (IEEM) (Johannesburgo)*.
[14] Bataineh O, Al-Hawari T, Alshraideh H and Dalalah D 2019 A sequential TPM-based scheme for improving production effectiveness presented with a case study," *J Qual Maintenance Eng*.

IOP Conf. Series: Materials Science and Engineering **796** (2020) 012007 doi:10.1088/1757-899X/796/1/012007

Properties of nanocrystalline cellulose from pineapple crown leaf waste

Fitriani[1], N A Sri Aprilia[1, 2] and N Arahman[1]

[1] Department of Chemical Engineering, Engineering Faculty, Universitas Syiah Kuala, Jl. Tgk. Syech Abdul Rauf No. 7, Banda Aceh 4326, Indonesia

E-mail: sriaprilia@unsyiah.ac.id

Abstract. Nanocrystalline cellulose was extracted from pineapple crown leaf as agricultural waste. Nanocrystalline cellulose was successfully extracted using chemical treatments followed by acid hydrolysis using sulfuric acid. Hydrolysis of pineapple crown leaf was carried out by using different hydrolysis time while maintaining acid concentration and temperature. The properties of nanocrystalline cellulose were obtained such % yield, moisture content, crystallinity index (CrI) and functional group present were studied. The yield shown a comparable result regardless of different reaction time of hydrolysis. While Fourier transform infrared spectroscopy (FT-IR) showed the progressive removal of non-cellulose content and showed the cellulose molecular structure remains unchanged with the reaction time. X-ray diffraction (XRD) analysis revealed that the crystallinity increased linearly with increasing hydrolysis time. The result showed potential of pineapple crown leaf to be used in nanocrystalline cellulose synthesis process and optimum hydrolysis time was determined based on the properties of nanocrystalline cellulose.

1. Introduction

Pineapple is one type of fruit that is widely used as a source of food in the world. Its processing which produces 3 billion tons of by-products each year, namely pineapple leaves has caused environmental pollution and problems in agricultural land. Generally, pineapple crowns leaf (PCL) are made up of 79-83% cellulose, 19% hemicellulose and 5-15% lignin [6]. Pineapple crown leaf (PCL) has recently been used as an ingredient in the production of textiles, paper and as a reinforcement in polymers [5]. However, in large quantities, pineapple crown leaves are dry and burned in an open environment, which causes environmental pollution issues. Then it is necessary to reduce the problem of environmental pollution by utilizing pineapple crown leaf waste as a source of cellulose. Cellulose is a biodegradable and renewable polymer material and can be processed into cellulose microfibrils, microcrystalline cellulose (MCC), nanocrystalline cellulose (NCC) and many more. This material can be obtained from agricultural by-products such as wood fiber, cotton fiber, kenaf bast, corn stalk and rice husk [1,2,4,11].

Nanocrystalline cellulose (NCC) is a new type of material from cellulose which has physical and chemical properties such as biocompatibility, non-toxic, hydrophilic, has a large surface area and is characterized by increased crystallinity, dispersion ability and biodegradation. These properties make cellulose widely used in various fields of research including being used as a polymer reinforcement, additives for biodegradable products, membrane reinforcement and highly relevant for the development of renewable biomaterials in many fields of chemistry, food, pharmacy, and recently

[2] To whom any correspondence should be addressed.

Content from this work may be used under the terms of the Creative Commons Attribution 3.0 licence. Any further distribution of this work must maintain attribution to the author(s) and the title of the work, journal citation and DOI.
Published under licence by IOP Publishing Ltd

used as nanocomposites [9,12]. Based on the analysis of Santos, et al [13] the properties possessed by NCC as polysaccharides can be used in liquid media applications and improve mechanical properties in nanocomposite polymers and are used as reinforcement or fillers in thermoplastic polymer matrix and also bionanocomposite films which have uses as food packaging [4]. Synthesis of NCC from biomass can be done by various methods, including mechanical, chemical and biological methods. Chemical methods that are often used in synthesizing NCC are using the hydrolysis, organosolv, alkaline solvent, oxidation, and ionic liquid methods [3]. NCC can be produced in the form of nanocrystals by using acid hydrolysis method where the amorphous region of cellulose component is disintegrated resulting with different degrees of crystallinity index.[7]. The acid concentration and hydrolysis time are two important parameters in acid hydrolysis process. Acid hydrolysis process reduces the degree of polymerization as a function of hydrolysis time. Increasing the hydrolysis time has great effect on the cellulose because of the β-1,4 glycosidic bond to acid [2]. However, the hydrolysis time parameter impact both the efficient removal of non-cellulose component and properties of acid hydrolyzed of PCL. The results of the literature extraction of NCC from PCL fibers are still little published, so there is a need for research to extract NCC by using acid hydrolysis techniques. Therefore, this study aims to report the extraction of NCC from PCL by acid hydrolysis with different reaction times. The characterization of the NCC produced properties such as yield, functional group, crystallinity index, moisture content and potential applications were analyzed.

2. Experimental

2.1 Materials

Pineapple crown leaf waste were supplied by local markets in Banda Aceh, Indonesia. NaOH and H_2O_2 were used as mercerization and bleaching agents while H_2SO_4 were used for hydrolysis. All the chemicals were purchased from CV. Karya Graha Agung, Medan, Indonesia.

2.2 Preparation of NCC

Singular pineapple crown leaves were isolated from pineapple crown. Prior to removing non-cellulosic compounds through the mercerization and bleaching treatments, the PCL fibers were washed in running water to remove soluble sugars and impurities. Dried PCL fibers (60°C for 24 h) were cut into small pieces (1 cm) before father treatment. The mercerization and bleaching of PCL fibers was performed by using NaOH and H_2O_2 1 M at 80°C for 1 h. Upon completion of the mercerization and bleaching process, the samples were rinsed with water. Following the mercerization and bleaching process, the samples was hydrolyzed using 3 M H_2SO_4 at 45°C for 1, 2, and 3 h. Upon completion of the reaction, the samples were washed and rinsed repeatedly with distilled water until it reached pH 7. The resultant NCC were grounded into fine powder by using blender after the washing and drying process. Finally, the NCC were ultrasonicated and centrifugation for 30 min and were stored until further analysis.

2.3 Characterization of NCC

2.3.1 Percentage yield

Percentage yield was calculated using the following equation:

$$\text{Yield (\%)} = \frac{weight\ after\ hydrolysis}{weight\ initial}\ x\ 100\% \tag{1}$$

2.3.2 Functional group analysis. Functional group analysis was carried out using IRAffinity-1S FTIR spectrometer with scanning ranges of 5000 cm^{-1} to 500 cm^{-1} at room temperature.

2.3.3 Crystallinity and crystal size. Crystallinity and crystal size were determined using XRD Shimadzu XDR 7000 at an operating voltage 40 kV and applied current was 30 mA. The crystallinity of the samples was calculated with Segal formula using equation:

$$\text{Crl (\%)} = \frac{I_{200} - I_{AM}}{I_{200}} \ x\ 100\% \quad (2)$$

where I_{200} symbolizes the amorphous and crystalline fractions and I_{AM} symbolizes the amorphous region [2]. The crystal size of sample was determined based on Scherrer formula using equation:

$$D_{(hkl)} = \frac{K\lambda}{\beta \cos\theta} \ x\ 100\% \quad (3)$$

where $D_{(hkl)}$ is the crystal size (nm), K is the Scherrer constant, λ is the wavelength of XRD β is the full width of reflection and θ is the Bragg angle.

2.3.4 Moisture content
The moisture content was determined in triplicate, according to the standard method of AOAC. Sample of 1,4 g each were dried at 80°C and weighted after 30 min of drying. This procedure was repeated until constant mass was reached and calculated using equation:

$$\text{Yield (\%)} = \frac{mi - md}{mi} \ x\ 100\% \quad (4)$$

where m_i symbolizes the initial mass of the sample before drying and m_d symbolizes the final mass after drying [12].

3. Result and Discussion

3.1 Percentage yield
The yield of NCC sample was determined and the data was summarized in Table 1. Different hydrolysis time (1, 2 and 3 h) were used to extract NCC from PCL while maintaining a constant H_2SO_4 concentration and temperature. The yield for 1, 2 and 3 h hydrolysis time of PCL were found to be 79.36, 78.10 and 76,23% respectively.

Table 1. Percentage yield of NCC

Sample	Yield (%)
1 h	79.36
2 h	78.10
3 h	76.23

Increasing reaction time of hydrolysis gave a lower NCC yield. This is probably due to the break-down of the more β-1,4 glycosidic bond at higher reaction time and easier removal of short chains during hydrolysis [2]. The percentage yield of NCC varied from sample to sample and depending on the type of raw material, geographical conditions, sample preparation and method of analysis which could be the reason for this variation.

3.2 Fourier Transform Infrared Spectroscopy (FTIR)
Figure 1 shows the FTIR spectra recorded for NCC treated with different hydrolysis time (1, 2, and 3 h) and concentration H_2SO_4 3 M. All samples display similar spectra, which indicates similarity in chemical compositions and all samples display absorbance regions at 2900-3500 cm^{-1} and 450-1750 in

line with Aprilia et. al. [2] corresponding to O-H bending, and CH_2 groups, C-O-C stretching and C-H. The absorption at region 1654 cm^{-1} in all sample corresponding to O-H bending vibration of absorbed water [8,12].

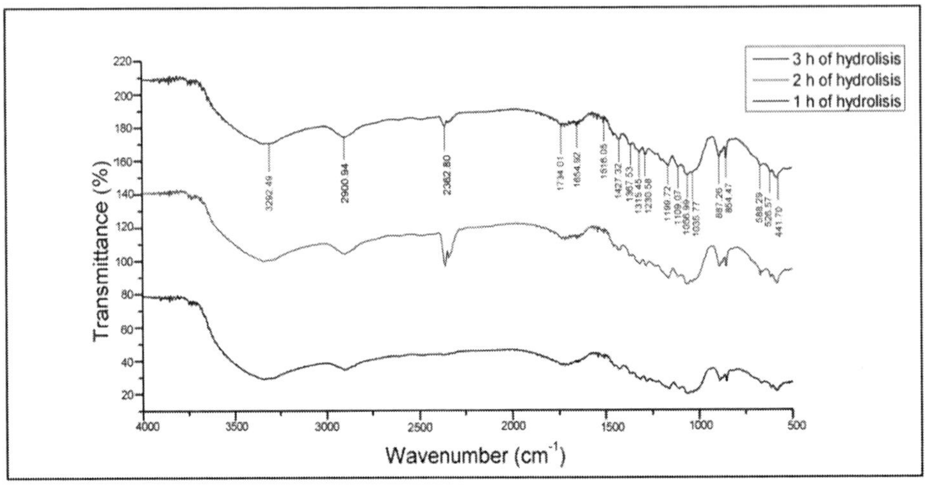

Figure 1. FTIR spectra of NCC from pineapple crown leaf waste after 1, 2 and 3 h of hydrolysis.

It is evident from Figure 1 that by increasing the hydrolysis time, the intensity of the hydrogen bonds increased. This might have been because of the degradation of the amorphous part, which enhanced the crystallinity of the NCC as time progressed. Furthermore, the OH bond shifted to lower wavenumbers with increasing hydrolysis time, which can be described to the increased hydrogen bonding and crystallinity of the NCC. Ahmad et al. [1] reached a similar conclusion by increasing the hydrolysis time, the hydrogen bond of rice husk fibers would shift to a lower wave number because of the increase in the OH bond in the rice husk chains.

The peak intensity obtained in the region of 1427 cm^{-1} shows the -CH_2 bond which is an aromatic characteristic of polysaccharides. Kumar et al. [11] explain that shifts that occur at peaks of 1488-1455cm^{-1} and 997-891 cm^{-1} indicate vibrational deformation of -CH_2 and C-O-C glycosidic bonds in the synthesis of NCC which indicate crystalline levels in the samples. This shows that the morphological structure of cellulose nanocrystals is intact and the increase in peak intensity is due to increased crystallinity of cellulose. Based on FTIR spectra, lignin and hemicelluloses have been removed via mercerization and bleaching treatments. It is evident in Figure 1, where the peaks at 1500, 1605 and 1700 cm^{-1} are absent in NCC spectra [8,12]. In addition, no significance difference was observed in all samples at the cellulose region, which indicates the cellulose molecular structure remains unchanged from the treatment.

3.3 X-ray diffraction (XRD)

Cellulose is the important component of PCL contains a crystalline structure resulting from placement of the chain by acid hydrolysis. Figure 2 shows the diffraction patterns obtained for NCC with different hydrolysis time. The peak at $2\theta = 16°$, 22.5° and 36° is correspond to cellulose I region for all samples. Additionally, the crystallinity percentage and crystal size are summarized in Table 2.

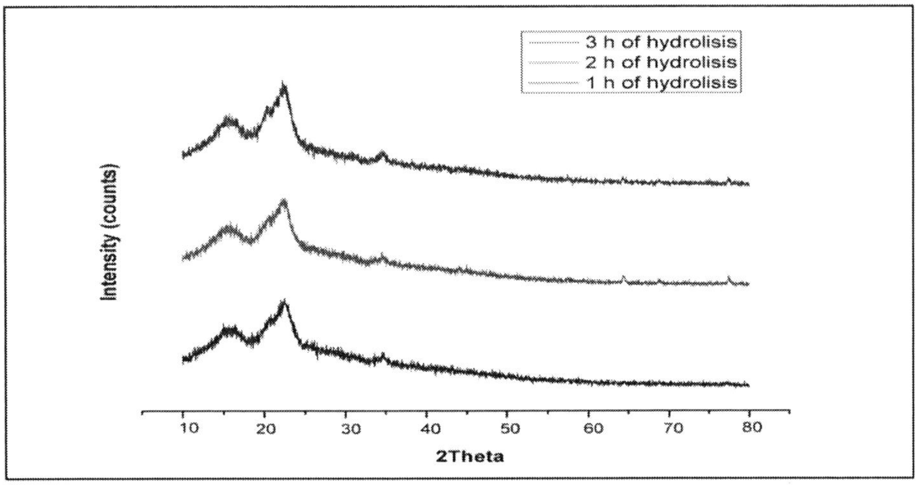

Figure 2. X-ray diffraction patterns of NCC from pineapple crown leaf waste after 1, 2 and 3 h of hydrolysis.

The crystallinity value of the samples with different hydrolysis time were 78.21, 22.94 and 77.69 % respectively. Furthermore, the crystallinity values of NCC with 2 h hydrolysis is higher compared to others. It is due to the removal of amorphous regions of cellulose by acid hydrolysis, which prompts the hydrolytic cleavage of glycosidic bonds, and finally released of more individual crystals [8,10].

Table 2. Crystallinity percentage and crystal size of NCC

Sample	Crystallinity (%)	Crystal size (nm)
1 h	78,21	16,80
2 h	83,16	22,94
3 h	77,69	19,38

In addition, during the acid hydrolysis process the amorphous part dissolve and single crystals are released and thus can improve the cellulose crystallinity [10]. An increase in crystallinity is related to increases in the rigidity of the cellulose structure, which can lead to higher tensile strength to fibers. This increase would be expected to enhance the mechanical properties of composites. A further increase in hydrolysis time at 3 h caused a reduction in the crystallinity value, probably because of damage to some of NCC crystalline region because of the high of acid concentration. Aprilia et al. [2] reached a similar data by increasing hydrolysis time from 2 h to 3 h, the crystallinity of MCC from kenaf bast decreased from 82.7% to 82%. The crystal size was found to be 16.80, 22.94, and 19.38 nm respectively. By increasing the hydrolysis time, with removal of no-cellulosic components and leaving behind the dense cellulose and leading to increased crystal size [2].

3.4 Moisture content

The moisture content of each sample is listed in Table 3. Cellulose has three hydroxyl groups that can interact with water molecules, which confirm NCC as hydrophilic. Higher moisture content of NCC has influence on potential applications as disperse phase in polymeric composites [12].

Table 3. Moisture content of NCC

Sample	Moisture content (%)
1 h	10.28
2 h	10.92
3 h	11,42

The moisture content of NCC with different hydrolysis time were 10.28, 10.92 and 11,42 % respectively. Higher moisture content can be attributed to the higher surface area of nanoparticles. As seen in FTIR result show band related to O-H bonding, which confers hydrophilic behavior for NCC. This trend is in agreement with previously reported results [12]. By increasing the hydrolysis time, the intensity of the hydrogen bonds increased and might have been because of the degradation of the amorphous part and also the moisture content may depend on several other factors as climatic and storage conditions [14].

4. Conclusions

NCC has been successfully extracted from PCL using chemical treatments involving mercerization, bleaching and acid hydrolysis treatments. The results obtained from FTIR analysis confirmed the removal of the non-cellulosic compounds of PCL. The crystallinity analysis shows that NCC give high crystallinity providing evidence that acid hydrolysis does not change crystal structure from cellulose I of PCL. The proposed of different hydrolysis time gives comparable % yield and % crystallinity making it a potential for condition process hydrolysis. The NCC isolated from PCL show interesting properties that can be used in several applications with high added- value. Their high hydrophilicity allows the use of NCC in liquid media applications and as reinforcement in polymeric nanocomposites. The diverse potential applications of the obtained NCC motivates the use of PCL residues as a source of NCC extraction since they are abundant and inexpensive by-product. Using PCL wastes for NCC extraction arises as an important contribution to solve the agroindustrial disposal problem.

Acknowledgement

This work was supported by the Universitas Syiah Kuala, Aceh, Indonesia under Research, Program *Riset Unggulan Unsyiah Percepatan Doktor (PRUU-P*D) Grant no. 860/SPK/PNPB/2019.

References

[1] Ahmad Z, Roziazian N N, Rahman R., Mohamad A F and Ismail W I 2016 Isolation and characterization of Microcrystalline Cellulose from Rice Husk *MATEC Web of Conf.* **1** 1-6

[2] Aprilia N A S, Davoudpour Y, Zulqarnain W, Abdul khalil HPS, Hazwan C I, Hossain M S, Dungani R, Fizree H M, Zaidon A, and Haafiz M K M 2016 Physicochemical characterization of microcrystalline extracted from kenaf bast *Bioresour* **11** (2) 3875-3889

[3] Aprilia N A S, Ambarita A C, Karmila, Armando M A, and Guswara F Y 2017 Isolation and characterization of cellulose nanofiber (CNF) from sugarcane bagasse by acid hydrolysis with addition of ferric chloride catalyst ($FeCl_3$) *Orient J. Phys. Sciences* **2** (2) 103-108

[4] Azeredo H M, Rosa M F and Mattoso L H C 2017 Nanocellulose in bio-based food Packaging applications *Ind Crops Prod* **97** 664–671

[5] Braga R M, Queiroga T S, Calixto G Q, Almeida H N, Melo D M A, Melo M A F, Freitas M A F, and Curbelo F D S 2015 The energetic characterization of pineapple crown leaves *Environ Sci Pollut R* **22** 18987–18993

[6] Brinchi L, Cotana F, Fortunati E, and Kenny J M 2013 Production of nanocrystalline cellulose from lignocellulosic biomass: technology and applications *Carbohydr. Polym.* **94** 154–169

[7] Cherian B M, Leao A L, de Souza S F, Costa L M M, de Olyveira G M, Kottaisamy M, Nagarajan E R, and Thomas S 2010 Cellulose nanocomposites with nanofibres isolated from pineapple leaf fibers for medical applications *Carbohydr. Polym.* **86** 1790–1798

[8] Haafiz M K M, Hassan A, Zakaria Z, and Inuwa I M 2014 Isolation and characterization of cellulose nanowhiskers from oilpalm biomass microcrystalline cellulose *Carbohydr. Polym.* **103** 119-125

[9] Habibi Y, Lucia LA, and Rojas O J 2010 Cellulose nanocrystals: chemistry, self-assembling and application *Chem Rev* **110** 479–3500

[10] Johar N A, Dufresne and Ahmad I 2012 Extraction, preparation and characterization of cellulose fibers and nanocrystals from rice husk *Ind Crops Prod* **37** 93-99.

[11] Kumar R, Hu F, Hubbell C A, Ragauskas A J, and Wayman C E 2013 Comparison of laboratory delignification methods, their selectivity, and impacts on physiochemical characteristic cellulosic biomass *Bioresour* **130** 372-381

[12] Prado K S and Spinace M A S 2019 Isolation and characterization of cellulose nanocrystals from pineapple crown waste and their potential uses *Int J Biol Macromol* **122** 410–416

[13] Santos R M, Flausino N W P, Silverio H A, Martin D F, Dantas N O, and Pasquini D 2013 Cellulose nanocrystals from pineapple leaf, a new approach for the reuse of this agrowaste, *Ind Crops Prod* **50** 707–714.

IOP Conf. Series: Materials Science and Engineering **796** (2020) 012008 doi:10.1088/1757-899X/796/1/012008

TPM Maintenance Management Model Focused on Reliability that Enables the Increase of the Availability of Heavy Equipment in the Construction Sector

A Palomino-Valles[1], M Tokumori-Wong[1], P Castro-Rangel[1], C Raymundo-Ibañez[2,4] and F Dominguez[3]

[1] Ingeniería Industrial, Universidad Peruana de Ciencias Aplicadas (UPC), Lima 15023, Perú.
[2] Dirección de Investigación, Universidad Peruana de Ciencias Aplicadas (UPC), Lima 15023, Perú.
[3]Escuela Superior de Ingeniería Informática, Universidad Rey Juan Carlos, Mostoles, 28933, España.

E-mail: carlos.raymundo@upc.edu.pe

Abstract. The purpose of this paper is to present a maintenance study focused on total productive maintenance (TPM) and reliability-centered maintenance (RCM). Its approach is based on the first pillars of TPM, preventive and autonomous maintenance, as well as the FMEA analysis of RCM for maintenance analysis, which was conducted in this study. The implementation of TPM was successful in that various preventive maintenance (PM) policies assigned to the assets were implemented and it was demonstrated that TPM application in the construction industry could reduce the excessive accumulation of maintenance with the same effective optimization, and with support from RCM analysis and its heavy equipment systems analysis. Excessive corrective maintenance accounts for high investment and delay rates in work times of the assigned project. Traditional methods of availability guarantee, such as reactive or routine maintenance, are insufficient to satisfy a heavy equipment maintenance plan; therefore, what is called for is the systematic application of RCM and TMP because they allow the selection and application of effective PM tasks. An approach that develops and thoroughly analyzes the strategies of continuous corrective and PM is used with an atmosphere of uncertainty and with operational data limited by criticism. Results show a 90% improvement in availability.

1. Introduction

The role that an effective maintenance framework plays in organizations has become more important in maintaining competitiveness in the global market. Zegarra [1] details that construction companies at an international level are immersed in World Class Maintenance, which refers to having policies and good working practices in administrative fields, as well as in the technical field. Today, the construction sector represents an important section of the country's economy, and its growth is driven by the great dynamism shown by public investment. According to INEI, companies in the construction sector grew in this fourth quarter by 79.6% when compared with 2016 growth numbers in the same

[4] To whom any correspondence should be addressed.

Content from this work may be used under the terms of the Creative Commons Attribution 3.0 licence. Any further distribution of this work must maintain attribution to the author(s) and the title of the work, journal citation and DOI.
Published under licence by IOP Publishing Ltd

quarter. The projects assigned for 2019 involve infrastructure improvement and the expansion of a category that the Peruvian Chamber of Construction reckons would reach 4.5% growth when compared with the previous year's growth numbers.

The studies conducted indicate that the adoption of a good maintenance strategy allows us to keep the machines operational. However, the increase in mechanical forms and the complexity of the systems lead to an increase in unplanned failure shutdowns. Qarahasanlou, Barabadi, and Ayele [2] point to component failure as resulting in downtime; the lack of system availability; and consequently, substantial losses in production performance, high maintenance costs, and so on. Jain & Shingh [3] indicate (1) that small- and medium-sized enterprises must adopt, through their department or maintenance area, focused improvements, both preventive and autonomous, that are part of the TPM model, (2) and that the implementation of the indicated methodology increases availability, performance, quality, and efficiency. On the other hand, En-Nhaili and Bouami [4] use the system dynamics approach to model and analyze a maintenance system in order to evaluate the reliability of the equipment. Shmatkov and Shmatkova [5] indicate that the operating cost per hour can be reduced by an increase in equipment operating efficiency (EOE) in much the same way that Davies [6] pointed out years ago regarding the EOE, which is a measure of the efficiency and performance of the equipment that is directly linked to TPM. This measure focuses mainly on production losses where maintenance policy is one effect among many others. In the development within the semiconductor industry, Shin & Ahmad [7] tell us that the key to the success of TPM is the development of the practice of autonomous maintenance (AM), which refers to human capital among operators supported by engineers to perform simple daily maintenance activities, apart from the planned maintenance. Conversely, Carlson [8] tells us that the RCM approach allows us to improve this availability and thus to maximize performance through FMEA (Failure Mode and Effects Analysis) for the evaluation of the development of preventive maintenance (PM). The result is an improvement in the availability of equipment and a reduction of the maintenance cost.

2. State of the art

2.1. Maintenance Management Model

Riquero [9] asserts that the quick changes in business demand in the current economic environment has led organizations toward higher efficacy and efficiency in all aspects, and Edwards and Love [10], on the basis of their research, have determined that fixed-time (scheduled) and daily maintenance of infrastructure and machinery for construction projects is a basic requirement for productive and safe operations at the site. Among their results, courses for operator skills, onsite management training, and higher regularization stand out. Consequently, a model composed of protocols and maintenance procedures—that may be implemented throughout the organization—has to be designed.

2.2. TPM – (Total Productive Maintenance)

C-C Shen [11] explores in detail the performance of TPM development in companies, by means of planning, execution, and methodology objectives with the work team, operators, and management; this will enable the improvement of equipment and operations through failure reduction. The result is the 12-step TPM implementation and a process that may take 3–5 years to implement. While Amad, Abdul, Kamaruddin, and Min [12] agree that the key to TPM's success is developing the autonomous maintenance (MA) practice and increasing the availability and reliability of machinery; this is connected with the development of human capital among operators supported by technicians and engineers to conduct daily maintenance activities, which are easy but the key is to follow up the process on top of the planned maintenance.

2.3. RCM – Reliability-Centered Maintenance:

The maintenance response when it comes to error prevention is to set a maintenance program. Prabhakar [13] shows a maintenance model based on preventive maintenance (PM) and predictive maintenance (PdM) that gets integrated with the initial processes of reliability-centered maintenance (RCM). Sillivant [14] underlines that the proposed analysis of studies of RCM aims mainly at improving the availability of machines so as to maximize production. However, as he clears up for the operation processes, there are difficulties in adopting the RCM standardized methodology, mainly because of the complexity and the great deal of analysis that it entails (a long time is required for study and analysis), which results in an extended implementation, involving the services of an optimal number of qualified people.

2.4. RCM AND TPM

As Fah and Ekpiwhre [15] put it, maintenance is defined as a set of actions taken to guarantee that the systems or components provide the planned functions with a key target, that of preserving or restoring reliability. Rimawan and Bambang [16] state that the integration of a maintenance system for total productivity is vital to the determination of the degree of collaboration for the general efficiency of heavy equipment. The operation progress of each machine should be thoroughly monitored to measure its efficacy and efficiency. However, Holt and Edwards [17] point out that a key element of operational productivity is the operator's capacity, when faced with the need to act quickly, to keep the operability of heavy machines.

3. Contribution

3.1. Fundamentals

The existing model enables the monitoring of the need for the prevention of mistakes by means of a maintenance program based on condition and time-controlled tasks.

Unlike the initial model mentioned by author Deepak Prabhakar and Dr Jagaty Raj, who propose predictive maintenance, that is to say, "on time" maintenance practiced on equipment. Autonomous maintenance will be implemented as the equipment is of the heavy machinery type and as the functions/activities that they fulfill make it impossible to carry out any kind of maintenance during the process. Apart from that, authors Shin, Rosmaini Ahmad, Kamaruddin, and Abdul help us in integrating the autonomous methodology to adjust the activity timeframe to the four stages of MA so as to take preventive and ongoing improvement measures toward operation. After executing TPM, RCM is performed, where it is mentioned how the methodology helps to reduce the pre-fixing logistic time and to define the "how to," in accordance with the terms defined in the TPM steps and in relation to the application of the optimal maintenance system for heavy equipment, either preventive or autonomous (major pillars used). The objective is to build and totally execute an organizational culture that is centered around Kaizen and on the zero defect or breakdown principle.

3.2. Proposed Model

3.2.1. Autonomous TPM. To develop the autonomous maintenance practice (AM) and increase the availability and reliability of machines by means of daily tasks performed by operators, including checkups, lubrications, cleaning, and interventions. The advantage is the improvement of equipment performance and personnel skills.

3.2.2. Preventive TPM. This involves preserving equipment or premises by performing reviews and repairs that may guarantee working and reliability. The advantage is the prevention of failure and the retention of equipment capacity.

3.2.3. RCM. This type of RCM is a methodology that is based on the analysis of failure in a productive system by directly attacking the root of the problem. The main objective is to increase availability and reduce maintenance costs.

3.3. Components of the Model Proposed

The devices made are the following records for the correct functioning and monitoring of the maintenance program:

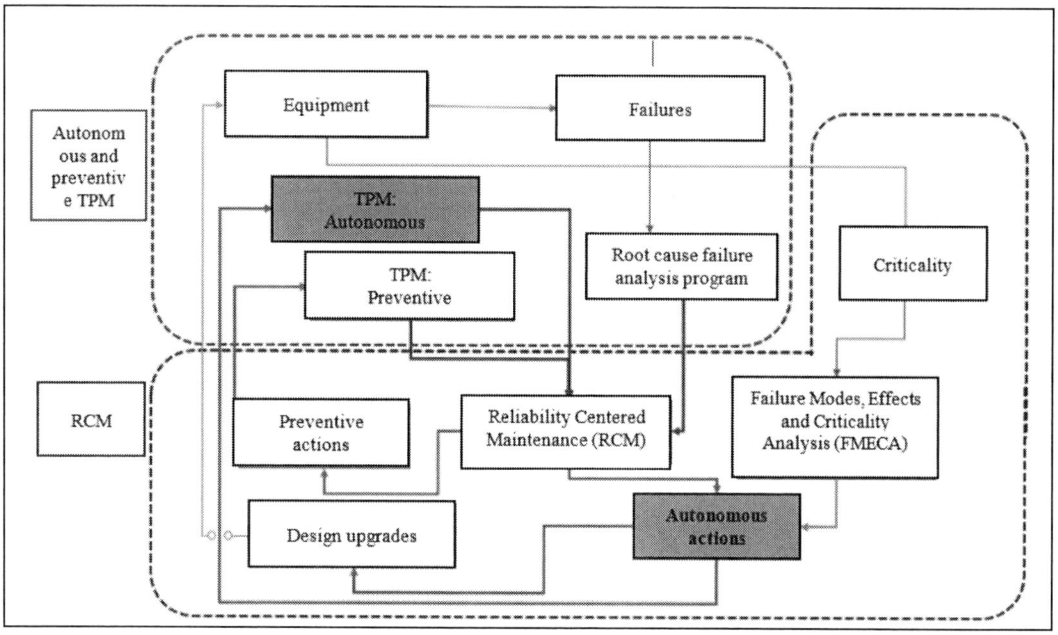

Figure 1. Maintenance model

Maintenance training program							
No.	**Training steps**	**4/03/2019**			**5/03/2019**		
		9:00 a.m.	10:00 a.m.	11:00 a.m.	12:00 p.m.	1:00 p.m.	2:00 p.m.
1	Work team building and line maintenance activity description.						
2	Skid steer and backhoe machine selection and description.						
3	Machine technical specifications.						
4	Specifications of each machine system, piece, and part.						

Figure 2. Maintenance training program

The audit program for autonomous maintenance, which takes place in the first part of the model, is intended to execute an audit on workers about the correct functionalities (Figure 3).

IOP Conf. Series: Materials Science and Engineering **796** (2020) 012008 doi:10.1088/1757-899X/796/1/012008

AUTONOMOUS MAINTENANCE AUDIT PROGRAM SHEET								
EQUIPMENT No.: MC1 AREA: Maintenance RESPONSIBLE: Engineer Perez AUDITOR: START DATE: 07/06 ENDING DATE: 08/16								
AUDIT ITEMS		EVALUATION SCALE (BAD-GOOD)					UPGRADE RECOMMENDATION	
		1	2	3	4	5	YES	NO
1	Team AM program goals understanding			X			X	
2	Team AM program goals understanding		X				X	
3	Management team AM activities planning				X			X
4	Operation team AM activities implementation			X			X	
5	Cooperation between teams				X			X
6	Communication between teams			X			X	
Decision according to evaluated score >50 = good and looking for future upgrades 40–49 = minor upgrades <39 = requires modification in AM practice	TOTAL	25						

Figure 3. Maintenance audit

This is interpreted as the steps that are required for training evaluation and that enable the project manager to measure the range of activities and the collaboration of MA teams.

A logbook of each machine records the type of maintenance and repairs made under the program proposed. This recording is conducted in the final part of our model in accordance with the improvement of the design that composes the RCM block (Figure 4).

This allows us to identify the most frequent failures and the tools necessary for its operating reliability, which are represented in the total life hours of each skid steer loader or backhoe machine.

3.4. Suggested Method

The investigation detail is adapted as a change in maintenance processes in which TPM and RCM coexist by representing the main axis for the development of maintenance activities and processes. (Figure 5)

3.5. Indicators

The indicators shown below are as follows

$$TPEF = \frac{Total\ operating\ hours\ of\ equipment}{Total\ reported\ failures} \tag{1}$$

$$TPPR = \frac{Corrective\ intervention\ total\ time}{Total\ corrective\ shutdowns} \tag{2}$$

$$\text{Availability} = \frac{\text{TPEF}}{\text{TPEF} + \text{TPPR}} \tag{3}$$

IOP Conf. Series: Materials Science and Engineering **796** (2020) 012008 doi:10.1088/1757-899X/796/1/012008

MAQUINARIAS ACCESORIOS INDUSTRIALES Y MINEROS S.A.C.					
DISTRIBUTION, IMPORT, EXPORT, TECHNICAL ASSISTANCE, MAINTENANCE, AND REPAIRING OF MINING EQUIPMENT					
EQUIPMENT	SKID STEER				**MACH 05**
TRADEMARK	CASE	**MAINTENANCE**		**YEAR**	2018
SERIES	JAFSR220CEM468870	**ENGINE**	210820	**HOURS No.**	10.173
MODEL	SR220	**SERIES**	3	**SERVICE DATE**	04/14/2018
NEXT CHANGE				**HOURS No.**	10.250
SERVICES					
DESCRIPTION				**QUANTITY**	**CODE**
DIESEL ENGINE OIL				1	MULTIGRADE 15W40 SHELL OIL
CHAINS COMPARTMENT OIL (BOTH SIDES)					
HYDRAULIC OIL					
ENGINE COOLANT					
MATERIALS					
MOBILITY FOR TASKS OUT OF GARAGE					
MAINTENANCE OF:					250 HOURS
SPARE PARTS					
PART NUMBER	**DESCRIPTION**			**QUANTITY**	**CONDITION**
87679598	ENGINE OIL FILTER			1	CHANGE
84299977	FUEL FILTER			1	CHANGE
87548612	INLINE FUEL FILTER			1	CHANGE
47456328	HYDRAULIC FILTER			--	--
84217229	PRIMARY AIR FILTER			1	CHANGE
87682999	SECONDARY AIR FILTER			1	CHANGE
COMMENT	**LOCATION**			**LIMA**	
NAME					
SIGNATURE					

Figure 4. Machinery logbook

4. Validation

4.1. Case Study

The company under review is MACISAC PERU SAC. Its capital is 100% Peruvian, and it has had a presence in the national market for 9 years. The company rents machinery and accessories for the development and implementation of specialized engineering projects during external domestic gas pipe installation, sanitation, construction, and agroindustry. Nowadays, MACISAC has a line of eight skid steer loaders and one backhoe, which bear the CASE trademark, and which are being used in the water supply networks project for the client Grupo Cobra (1 year and 8 months up to date), which is located in San Bartolo district, Lima province, Peru.

4.2. Diagnosis

The following graphic shows the effective capacity of each piece of heavy machinery in the current situation of the company under study.

A simulation using the Arena Simulation Software package was carried out, with time and activity data gathered to perform a maintenance program for heavy machinery (Figure 7) during the analysis of skid steer loaders.

- Average current maintenance hours: 13.18 hours
- Maximum current maintenance hours: 57.68 hours
- Average time between failures – Current: TPEF: 24.62
- Average time for failure repair – Current: TPPR: 15.24
- Availability rate – Current: 85%

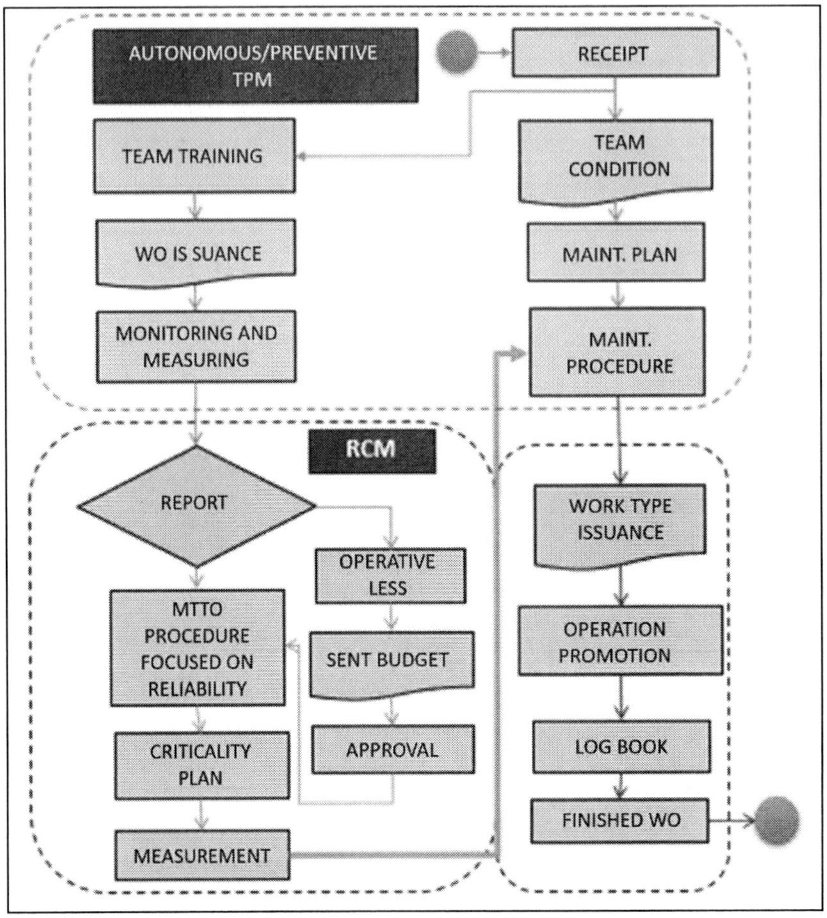

Figure 5. Process mapping (proposed)

Figure 6. SR220 Case skid steer

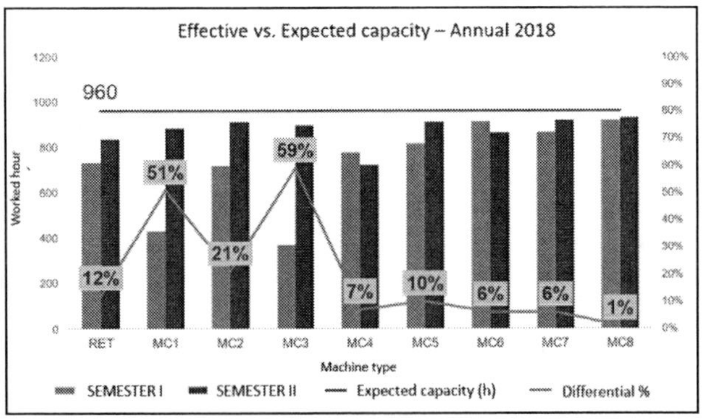

Figure 7. Effective vs. Expected capacity

4.3. *Apply Model in C.E.*

To reduce shutdown hours and improve machinery availability, an improvement proposal in the maintenance program of the company was conducted. Using these methodologies, the activities for each skid steer loader are to be standardized to the real hours that they must operate. Thus, processes will be more efficient. The proposed flow chart for a maintenance program has the following steps:

4.3.1. Machines arrival. This is the first step of the maintenance program, where a machine (e.g. a skid steer loader) has had a failure that will be evaluated, and the necessary activities carried out for it to continue working. In this phase, the machine may be revised in the garage or field.

4.3.2. Machines receipt. In this phase, a responsible operator will receive the machine and make a decision about earmarking it for preventive or corrective maintenance.

4.3.3. Machine status analysis. It is based on machine failure analysis, which has been previously recorded, with the purpose of avoiding them in the future with PM. This is because they may have serious consequences and their criticality level is high.

- **Repair:** The diagnostic tests detect a minor failure type, the likes of which can be repaired at once without detailed planning; hence, repairs may be carried out in the field.
- **Planning:** Maintenance type to be carried out has been planned. Preventive/Corrective.

4.3.4. Work order issuance. This activity is carried out before allocating resources, so as to monitor it in a better way and to manage the inventory of required parts for the maintenance task. Thus, the following operation times are reduced as pauses are shorter.

4.3.5. Available resources monitoring. If the required inventory is not available, materials, tools, and/or spare parts will be ordered.

4.3.6. Maintenance operation type assignment that will be performed on the machine. Regarding the current model of the company, this proposal includes autonomous maintenance at this stage as it is carried out to root out the causes of failure found in the heavy machine. At the same time, it is measured and monitored according to the stipulated parameters. A team trained for each required task carries it out and considers all the steps that have been taken and appear in the records. This team receives constant feedback, as well as audits.

4.3.7. Machine logbook preparation. The results and observations were recorded in writing. All the details of the maintenance activities performed for each machine are registered there, and this is part of a database for the modal analysis of failures and effects.

4.3.8. Exit of machine in good condition. The heavy machine returns to its task in the field, and the project continues to be in process without delays.

4.4. Results

On the basis of the results obtained from our proposed model, heavy machine availability greater than 85% in the company MACISAC PERU SAC was revealed. This means that on a monthly basis, shutdowns will no longer occur throughout the construction project. Similarly, the time for the proposed maintenance model was measured.

The availability increases to 81% (Table 1) compared to 62% obtained firstly, shows that our proposed model has innovative ideas gathered from different authors who have also dealt with the same problems in the sector under study. Apart from these integrated models, the best methodologies for organization performance in the water pipes installation project in San Bartolo were integrated. This allows for better control over the maintenance activities performed and helps to determine the critical components to mainly supply these resources.

With the help of the simulation model, the results showed an improvement in the saving of hours because of heavy machine maintenance. This was taken to cost per hour of work, which allowed us to see the savings in overtime and downtime per skid steer loader as shown in Table 2.

Table 1. Indicator

INDICATOR	PROPOSED
AVERAGE MAINT. HOURS	7.50
MAXIMUM MAINT. HOURS	17.80
TPEF	30.08
TPPR	7.24
AVAILABILITY RATE	81%

Table 2. Saving in overtime and downtime

REASON	AMOUNT	%
Savings in overtime	S/44,800	14 %
savings in downtime	S/266,435	86 %
TOTAL	S/311,235	100 %

Costs were allocated for each type of methodology on the basis of research on case study-related articles, manpower, consultants, and specialized personnel who will be in charge of the measurement of activities. In addition, an investment of S/18,275.00 was estimated, which contains all the necessary resources, such as records, electronic devices for real-time control, and equipment and tools for those involved in the process shown in Table 3.

Table 3. Cost of methodologies

	5`s	TPM	RCM
MO	S/1,800.00	S/1,425.00	S/1,300.00
Consultants	S/4,000.00	S/4,000.00	S/1,250.00
Specialized personnel	S/2,000.00	S/2,000.00	S/2,000.00
	S/7,800.00	**S/7,425.00**	**S/4,550.00**

Table 4. Investment

TOOL	RESULT
VAN	S/10,310.31
TIR	3.56 %
COK	1.13 %
COST-BENEFIT	2.41

The following data were considered for the validation of the economic flow.

- Annual effective rate: 20%.
- Monthly effective rate: 1.53%
- Loan: S/10,965.00
- Fee: S/1,007.21 (French Method)
- Weighted average annual cost of capital: 13.56%.

Over a period of 12 months, investment was recovered for the implementation of the methodologies and the following results were obtained (Table 4).

This means that the proposed model will be profitable for the company for the first 12 months and increasingly long-term and user-friendly for the various construction projects in which the company is going to be involved.

5. Conclusions

- With the proposal of improvement, a 5% increase is possible in the general availability of the fleet of heavy machines and skid steers, which would allow the company under study to be at competitive levels within the national machinery rental market in the construction sector.
- By simulating the proposed maintenance process, the average waiting time between failures can be reduced from 13 hours to 7 hours. This means a 15% reduction in downtime.
- The proposal was feasible; the cost-benefit ratio was greater than 1, which will generate profits and savings.

- To determine the root cause, different sampling methods were conducted in which the operators recorded the activities/incidences of the equipment. This was decisive for the hypothesis and the proposal of improvement.
- The relationship between TPM methodologies and RCM allows a ramp-up of expert operator performance and interaction with machines with the objective of reaching zero defects.

6. References

[1] Zegarra, M 2015 Modern Heavy Equipment Maintenance Management (*Alas Peruanas University*)

[2] Qarahasanlou, Barabadi and Ayele 2017 Production performance analysis during operation phase: A case study (*Proceedings of the Institution of Mechanical Engineers, Part O: Journal of Risk and Reliability*)

[3] Jain, A, Bhatti, R and Singh, H 2014 Total productive maintenance (TPM) implementacion practice: a literature review and directions *Int J Lean Six Sigma* **5**(3)

[4] En-Nhaili, A., Meddaoui, A. and Bouami, D. 2016 Effectiveness improvement approach basing on OEE and lean maintenance tools *J Clean Prod* **6**(2) 147-169

[5] Shmatkov and Shmatkvoa 2018 Economic Efficiency and Hourly Operational Cost of high -speed Equipment *Russ Eng Res* **38**(6) 469-473.

[6] Davies, C. 2003 The contribution of lean thinking to the Maintenance of Manufacturing Systems (*School of Industrial and Manufacturing Science*).

[7] Shin C., Ahmad R., Kamaruddin S. and Abdul I. 2011 Development of autonomous maintenance implementation framework for semiconductor industries *Int J Ind Syst Eng* **9**(3), 268-297.

[8] Carlson, C. S 2014 Understanding and applying the fundamentals of FMEA (*USA: Proceedings of the 2014 Annual Reliability and Maintainability Symposium*)

[9] Riquero, I., Hilario, C., Chavez, P. and Raymundo, C. 2019 Improvement proposal for the logistics process of importing SMEs in Peru through lean, inventories, and change management *Smart Innovation, Systems and Technologies* **140** 495-501.

[10] Edwars D. and Love P. 2015 A case study of machinery maintenance protocols and procedures within the UK utilities sector *Accid Anal Prev* **93** 319-329.

[11] C. – C. Shen 2015 Discussion on key successful factors of TPM in enterprises *J Appl Res Technol* **13** (3) 425-427.

[12] Ahmad R., Abdul I., Kamaruddin S. and Min C. 2011 Development of autonomous maintenance implementation framework for semiconductor industries *Int J Ind Syst Eng* **9**.

[13] Prabhakar Deepak P. and Raj Jagathy V. P. 2013 A New Model for Reliability Centered Maintenance in Petroleum Refineries *Int J Sci Technol Res* **2** (5).

[14] Silliant D 2015 Reliability Centered Maintenance cost modeling: Lost opportunity cost (*Annual Reliability and Maintainability Symposium*).

[15] Tee K F and Ekpiwhre E 2019 Reliability-based preventive maintenance strategies of road junction systems *Int J Qual Reliab Manag*.

[16] Rimawan E, Irawan A P B 2017 Analysis of Calculation Overall Equipment Effectiveness (OEE) in the Implementation of Total Productive Maintenance (TPM) PC 200-8 Excavator Grab and Magnet Type Case Study in Cakratunggal Steel Mills Company *Int J Sci Eng Res* **8** (1)

[17] Holt G D and Edwards D 2015 Analysis of interrelationships among excavator productivity modifying factors. *Int J Product Perform Manag*

IOP Conf. Series: Materials Science and Engineering **796** (2020) 012009 doi:10.1088/1757-899X/796/1/012009

A framework for developing lot-sizing and scheduling model for multiproduct

Huda Muhamad Badri[1], Nor Kamaliana Khamis and Mariyam Jameelah Ghazali
Faculty Engineering and Built Environenment, Universiti Kebangsaan Malaysia, Bangi, Malaysia.

E-mail: hudamb@siswa.ukm.edu.my

Abstract. This paper considers the issue of lot-sizing and scheduling for multi-product production facilities using flexible flow lines. Flexible flow lines are also called flow lines with parallel machines at several or all stages of production with various products. The problem encountered in the company produce machine sparepart is a scheduling that is not optimal such as work rules that are not systematic causing a long of manufacturing time, but it also does not systematically result in scheduling a plan that is not optimal lot-sizing. In solving the problem, this paper proposes adjustments to scheduling planning and lot-sizing to improve the company performance. The objective of this study is to minimize processing time and production cost so as to satisfy customers. The expectation of the conceptual framework of the lot-sizing model and scheduling model is to be able to optimize lot-sizing and scheduling to achieve multi-objective goals of production time and production cost as a result of decisions related to many machine and lot size.

1. Introduction

The company's production plan is an important step before the production process is executed. The main concern in planning is the limitations of workers and constraints in shopfloor, especially the availability and capability of the machine as well as the time and cost of production. Therefore, production planning is needed. One of the production planning is determining the optimum lot sizes to be produced which can reduce production costs.

A common problem that occurs with regard to determining production lots is the imbalance in production costs between (but not limited to) inventory costs and holding costs that affect the company's performance. This will become even more complex when determining the quantity not only for one product but also for many products. Therefore, production costs also become more complex to reduce. To solve this problem, the determination of the production lots of each product must be optimized and considered together to reduce production costs.

In addition to the problem of determining lot-sizing, scheduling is an issue considered in this study. Scheduling is defined as the process set a series of tasks to resources from time to time [1]. Effective scheduling plays a very important role in today's competitive manufacturing world. Performance criteria such as machine use, production time, due date, customer satisfaction, and product quality depend on how efficiently the work is scheduled in the system. Especially when there are many types

[1] To whom any correspondence should be addressed.

Content from this work may be used under the terms of the Creative Commons Attribution 3.0 licence. Any further distribution of this work must maintain attribution to the author(s) and the title of the work, journal citation and DOI.
Published under licence by IOP Publishing Ltd

of products in production, scheduling problem will become more complex. Therefore, good scheduling is needed. This means that multi-product scheduling needs to be optimized by sorting jobs according to the character of each different job and machine.

This problem will be more complex when combining lot-sizing and scheduling because both will provide solutions and achieve objectives simultaneously. The solution to this problem is to minimize setup cost and inventory costs by optimizing production lot-sizing decisions while minimizing production process time by optimizing job sequences that are compatible with job and machine characteristics. This paper aims to build a framework that combines two complex problems as a solution to reduce inventory costs, set-up costs, and overall production time.

2. Literature Review

The issue of lot-sizing and scheduling has been studied since the advent of mass production and the concept of corporate philosophy. In general, papers have been published to solve lot-sizing problems using a heuristic or metaheuristic approach in recent years [2, 3].

Lot-sizing and scheduling can be divided into three sub-problems. The first problem is Lot-sizing Problem (LSP), which could be the focus when there is no arrangement between the various products or arrangement is not dependent sequentially. In this case, the main objective is to minimize inventory costs and set-up costs while other goals can maximize the overall throughput of the production line.

The second problem is a scheduling problem when the aim is to determine the allocation of work and sequence of work in each machine in a multi-machine environment. The scheduling problem can be divided into two main problems i.e. flow shop scheduling and job shop scheduling. In the flow shop scheduling problem, a set of machines is arranged in series and each job is processed on several machines in the sequence of the machines that have been determined [4]. While in the job shop scheduling problem, a job can be processed through different machines in any sequences [5].

The third problem is solving lot-sizing and scheduling problems simultaneously. It is common in the literature to solve the problem of lot-sizing and scheduling simultaneously while considering the timing or costings. The main goal in this case depends on the problem. The general objective besides minimizing inventory costs and processing time is to maximize throughput by optimizing setup time. This study focuses mainly on related work using genetic algorithms (GA) to solve various types of lot-sizes and scheduling problems.

2.1. Lot-sizing Problem

Xie and Dong [6] introduced a genetic algorithm (GA) to solve General Capacitated Lot-sizing Problems (GCLSP). They introduce the representation of solutions that encode the settings for each product as binary variables, while the other decision variables come from binary variables through coding and decoding procedures. The setup time in this case is an independent sequence while the setup fee depends on the product.

Gonzalez-Ramirez et al. [7] proposed a heuristic procedure to solve the Capacitated Lot-sizing Problem (CLSP) with the aim of maximizing profits by minimizing total costs, such as holding costs, production costs, and setup costs when cost management is set for each period in which production takes place. Apart from the fixed costs, the problem is still NP-hard. The complexity of the problem increases in the case of multi-products.

Toledo et al [8] state that in addition to the lot-size for each product, the algorithm determines whether certain types of products are produced during the specified period or not, without suggesting any sequences for an independent setting time sequence. On the other hand, the setup time depends on the product and cannot be ignored so that it is included in different limits to ensure the feasibility of all solutions.

2.2. Scheduling Problem

Scheduling problems are often associated with multi-machine problems. The purpose of scheduling problem solving is to find the best machine allocation for different tasks or jobs and find the optimal sequence to process the work on each machine [9].

Scheduling problems can be divided into two main categories: identical machines and non-identical machines. Chaudhry and Drake [9] introduced a genetic algorithm to minimize the total delay in identical multi-machine problems. Identical parallel machines mean that the setup time and processing time do not depend on the machine. In addition, Chaudhry and Drake [10] assume the set-up time does not depend on the sequence and machine capacity. Machine capacity can be determined by assigning a number of different workers to different machines during the production plan. The algorithm is used with the representation of chromosomes in the form of permutation vectors where this vector includes a list of jobs to be performed, the number of workers on each machine (machine capacity) and which machines will do each job. They conclude the effectiveness of using Genetic Algorithms to solve such problems.

Balin [5] proposed a genetic algorithm to solve non-identical parallel machine problems. He proposed chromosome representation in the form of a matrix in which each row represents a machine, each column represents the type of product and each cell in these chromosomes is 0 or 1, which indicates whether a product is manufactured on a specific machine or not. Each column has only one non-zero cell, and all other cells are zero. This means that each product can only be produced on one machine while each line represents a proper schedule for each machine. Because of this coding method, Balin [5] designed a new crossover operator that works on two lines of the same chromosome (one individual) and it works by moving non-zero cells from one row to another, which means moving jobs from one machine to another. This crossover operator works based on processing time for each machine to get a new individual with a better feasible schedule. He concluded that genetic algorithms are suitable for solving parallel machine scheduling problems that are not identical and promising results.

2.3. Lot-sizing and Scheduling Problem

In many real cases, set-up depends on the sequence and has a significant effect on total costs and time. Therefore, lot-sizing and scheduling issues need to be studied simultaneously to find the best product sequence and lot-size for each product. Many algorithms have been used and many models have been proposed with different characteristics and assumptions.

Mohammadi et al.[11], introduced two rolling horizon approaches to solvew a large size capacicated lot-sizing problem with sequence dependent setups, sequence dependency rises the improtanceof solving the lot-sizing problem and scheduling problem simultaneously. The algorithm works by choosing a product and setup the machine for it, and then it starts adding other products one by one. The sequence resulting from the addition of this product may be tried to all positions. For each position, the setup costs are calculated so that the position that produces the lowest total setup cost selected and the algorithm moves to the next product. This process is repeated for all periods independently. That means that the decision variable from one period does not affect the other periods. Mohammadi et al. [12] proposed a Genetic algorithm to solve the same problem with the representation of solutions in the form of a matrix. This matrix has the decision variables needed to decide which products will be produced during each period and the sequence of products in that period.

Mohammadi and Ghomi [13] solved the same capacicated single machine multi level lot-sizing problem by combining genetic algorithm (GA) with rolling horizon approach, they designed a genetic algorithm (GA) to determine the decision variables for each period while freezing other periods based on the rolling horizon approach. Genetic Algorithms proposed by Mohammadi and Ghomi [13] have been used with the representation of chromosomes in the form of vectors which show the sequence of products for one period since the Genetic Algorithm is applied for each period separately. This algorithm was superior to the previous rolling horizon approaches of Mohammadi et. al. [11], especially for large size problems.

3. Framework Methodology

This study uses a case study in a cast iron-based spareparts company. The company is trying to maintain stability of the company considering production costs, production time and production targets so that customer satisfaction can be achieved. To achieve this goal, the company need to control the production process to meet the expected quality and quantity of products according to production targets. In accordance with this problem, the company has emphasized techniques for applying a method that is implemented into the company's system. The problems that the company is currently facing are difficulty in determining the amount of production that can result in product deficiency or excess, no accurate technique to estimate the amount of production and no technique to determine the job sequence on each machine. In this case, the determination of the amount of production will affect the performance of the company including the cost of production, the cost of purchasing materials, warehouses and production time.

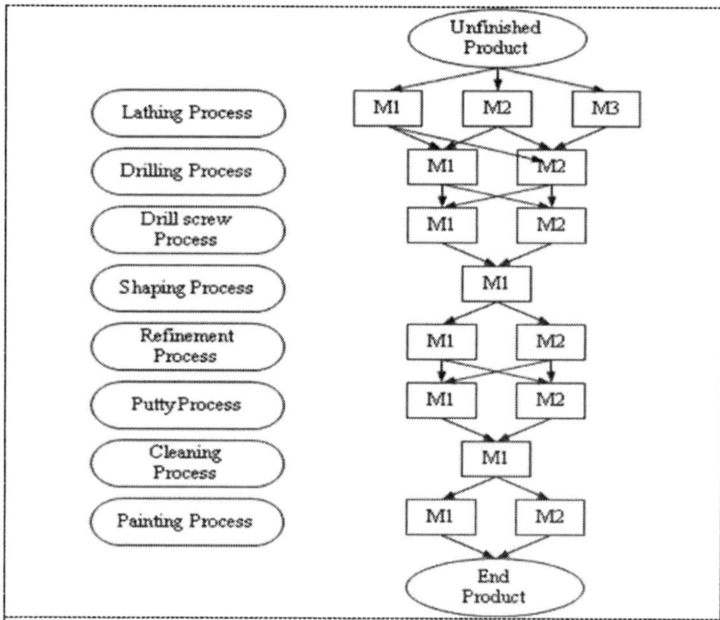

Figure 1. Production process flowchart

Based on the company's data collection in interviews with employees, there are several types of products manufactured so it is necessary to adjust the production process according to customer demand. Figure 1. shows the flow of the production process from the initial stage to the final stage. Irregular production flow makes the production process requires a long time because it can not adjust the job sequence to each machine optimally. Therefore, companies need to prepare for the production process by optimizing the lot-sizing of production and optimizing job sequences accordingly so that good scheduling is needed to optimize production in this company.

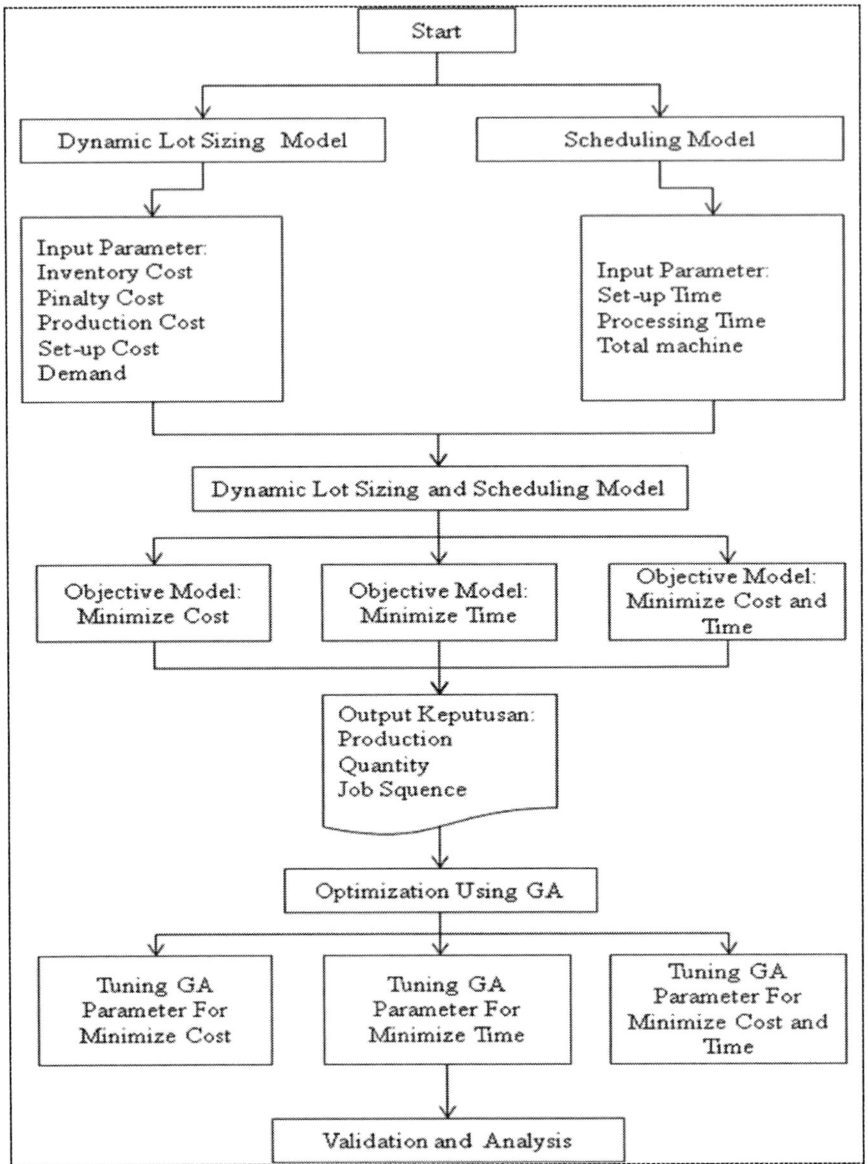

Figure 2. Reseach framework

In the diagram above, we can see the flow in making solutions to lot-sizing and scheduling problems. In this case, this study recommends some solutions that must be done by creating a conceptual model that is designed to help planners that are made to make decisions related to industrial resources. The model is a simple representation of the system or the real problem in the industry. Models can provide solutions to situations or complex problems. The methodology of the study consists of 4 phases which can be seen in Figure 3.:

IOP Conf. Series: Materials Science and Engineering **796** (2020) 012009 doi:10.1088/1757-899X/796/1/012009

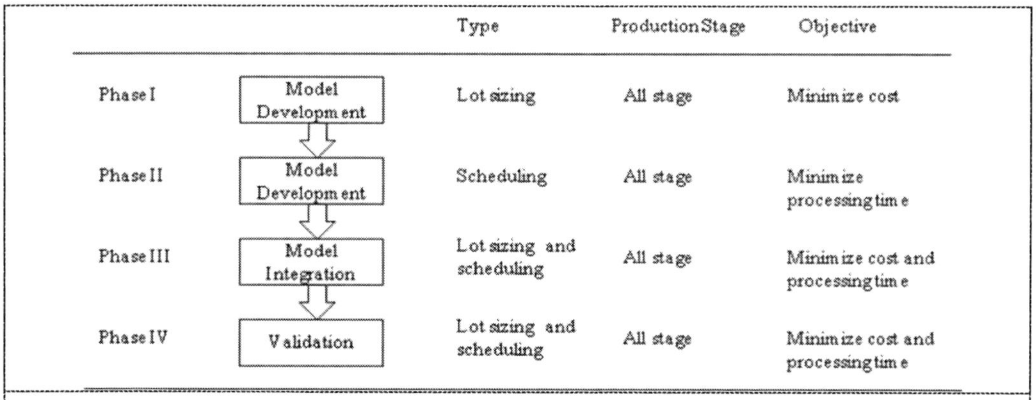

Figure 3. Phase of method solving problem

3.1 Phase 1. Build a conceptual model of lot-sizing

This study uses a dynamic lot-sizing model scenario. This model is related to the production process of products to meet customer demand that is dynamic with different amounts at each period. For example, Figure 4. shows the five production design periods.

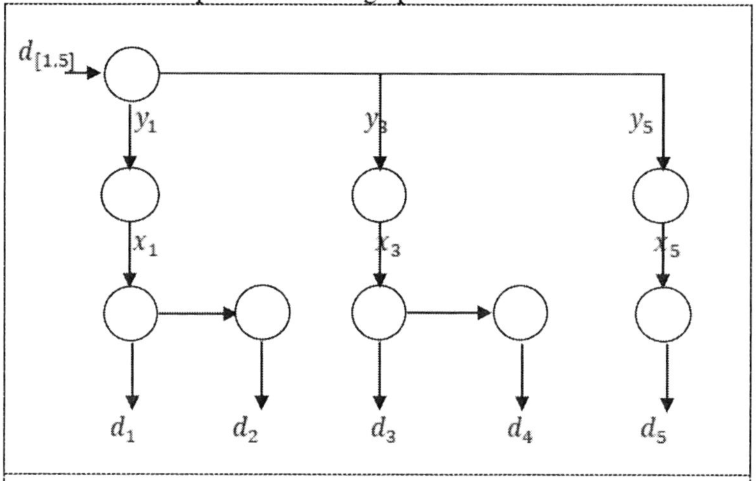

Figure 4. The network of production process with inventory

The company determines the period during which production will be carried out at a certain amount to meet demand. As shown in Figure 3, production is carried out at points y_1, y_3 and y_5. with a certain amount of x_1 , x_3 , and x_5 respectively by considering the inventory for each period. In this case, the production carried out by the company is subject to supply costs per supply in designing the production at points y_1, y_3 and y_5. In addition, the company also issued production costs per unit produced at points y_1, y_3 and y_5.

After going through the production process, the product is stored in a warehouse which is subject to inventory costs per unit. Warehouse capacity in the company's warehouse is limited. If there is excess product in the company's warehouse, the company will be charged with a holding cost penalty for each product. This is because companies have to do more maintenance or rent additional warehouses for excess products. Once the products are manufactured in accordance with the number of customer requests then the product is sent in each period.

In fact, there are transportation costs incurred by the customer for each shipment made. However, the reserved model does not consider transportation costs because the model built is only in the scope of the company which is examined as a form of simplification of the model.

Based on this reason, the model is then developed using mathematical programming techniques which include setup costs, production costs, inventory costs, and penalty costs with consideration of dynamic demand. The aim of the model is to reduce the system cost by optimizing the quantity or lot production.

3.2 Phase 2. Build a conceptual model of scheduling

Based on Figure 5, the scheduling component consists of jobs, machines, setup time and processing time. Each product has a different processing time so that it becomes one of the factors that influence scheduling. This study examines several jobs that have different characteristics. The job represents the number of products manufactured in the case study of the machinery spare parts industry. Each product or job is then processed using a machine that has certain characteristics to turn a semi-finished product into a finished product. Based on Figure 1, each stage has a different process.

In addition, jobs assigned to each machine have different setup times and processing times. Therefore, all jobs that pass several machines considering setup time and processing time will produce total time. In this case, the scheduling problem is focused on the problem of the time needed to complete the job.

To complete the whole job by considering due date, a good scheduling is needed to avoid tardiness and earliness. Therefore, the purpose of the scheduling model is to reduce production process time by optimizing the production job sequence.

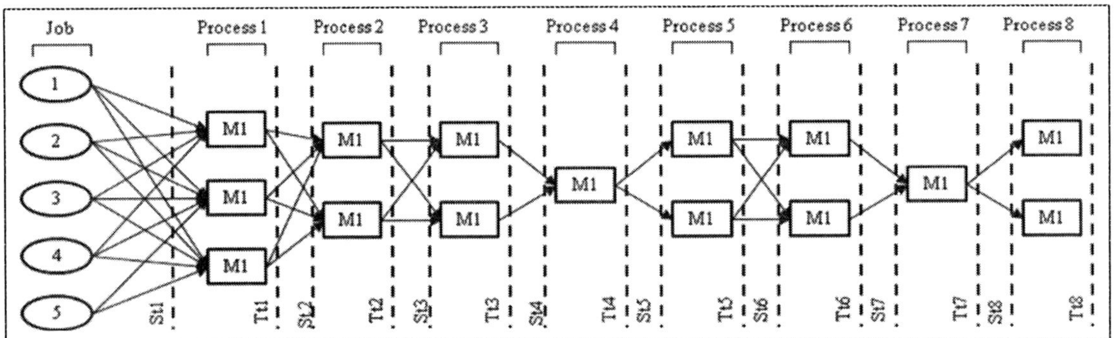

Figure 5. Schematic view of line production.

3.3 Phase 3. Combining the two models of lot-sizing and scheduling

As explained previously, lot-sizing explains some of the number of products that must be manufactured in each period while scheduling is related to job setting for production machines. The relationship of lot-sizing and scheduling is shown in Figure 6 and 7.

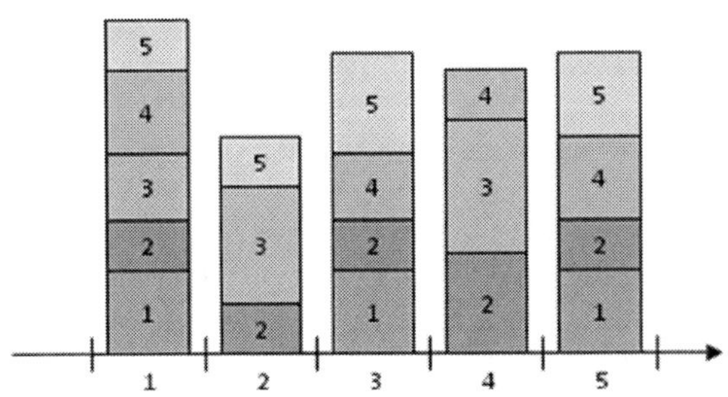

Figure 6. Lot-sizing: What quantities in which period?

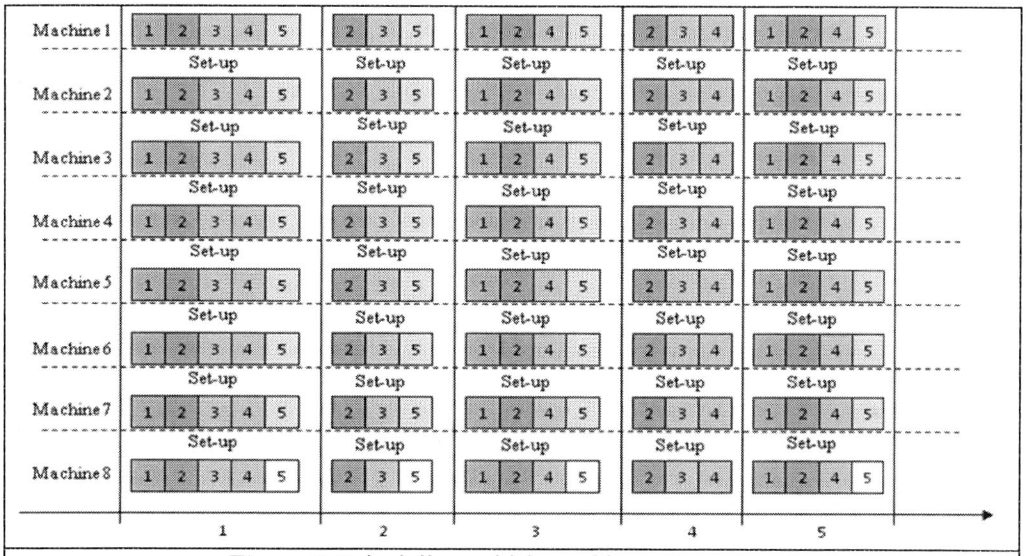

Figure 7. Scheduling: which machine and sequence?

Based on Figure 6 and 7, the combination of lot-sizing and scheduling is focused on five products i.e. 1, 2, 3, 4 and 5 which must be produced on eight machines in five periods. Lot-sizing is related to the number of products while scheduling considering machines and specific product sequences. The problem of the integration model between lot-sizing and scheduling considered is to reduce production costs and production time simultaneously. Therefore, the model reserved for this study has two important objectives. This can be achieved by optimizing production lots, especially for lot-sizing problems. Meanwhile, scheduling problems can be solved by optimizing work bookings. Both the optimization process must then be run simultaneously for 5 planning horizon that has been set.

3.4 Phase 4. Model validation

At each phase that has been described previously, this study proposes three models: a conceptual model of lot-sizing, the conceptual model of scheduling, and integration combination of lot-sizing model and scheduling model. The purpose of the first model is to reduce production costs while the purpose of the second model is to reduce production time and the purpose of the third model is to reduce production costs and production time simultaneously. These three models were developed based on the form of mathematical models coded in Microsoft excel, while optimization was done using the genetic algorithm (GA) method. Genetic algorithm (GA) is used to find production quantities and job sequences applied to 3 models. This means that although the 3 proposed models have different purposes, the optimization performed with GA is subject to lot-sizing and job sequence optimization.

To validate the three models, the performance of each model was compared. Furthermore, this raises the question of model validation:

- How capable genetic algorithm (GA) to optimize production lots and job sequences to reduce production costs.
- How does the decision affect lot-sizing and job sequence that reaches minimum production cost on production time?
- How capable genetic algorithm (GA) to optimize production lots and job sequences to reduce production time.
- How does the decision affect lot-sizing and job sequence that reaches minimum production time on production cost?

- How capable genetic algorithm (GA) to optimize production lots and job sequences to reduce production time and production cost simultanously.
- How to analyze the validity of decisions by considering lot-sizing, scheduling, production time, and production costs every minute.

4. Summary and Conclusions

Based on a case study on a spare parts company made from cast iron. This study has considered the conceptual development of lot-sizing and scheduling models to overcome the lotsizing and scheduling problems that occur in the company. The framework component consists of 4 phases where the first phase builds the concept of lot-sizing model, the second phase builds the conceptual scheduling model, the third phase combines the lot-sizing model with the scheduling model and the fourth phase validates the first three models based on the mathematical model form coded in Microsoft Excel , while the optimization is done by the genetic algorithm (GA) method. Genetic algorithm (GA) is used to find the production quantity and job sequences that are applied to three models. The aim of each model is to reduce production costs, production time, and production costs and production time simultaneously.

Acknowledgment

The authors gratefully acknowledge funding provided by the University Kebangsaan Malaysia (UKM) under grant agreement FRGD/1/2018/TK03/UKM/03/2

References

[1] Pinedo M L 2012 Scheduling: theory, algorithms, and systems (*Springer Science & Business Media*).
[2] Roshani A, Giglio D and Paolucci M 2017 A relax-and-fix heuristic approach for the capacitated dynamic lot sizing problem in integrated manufacturing/remanufacturing systems *IFAC-PapersOnLine* **50** 9008–9013.
[3] Jans R and Degraeve Z 2007 Meta-heuristics for dynamic lot sizing: A review and comparison of solution approaches *Eur. J. Oper. Res* **177** 1855–1875.
[4] Ribas I, Leisten R and Framiñan J M 2010 Review and classification of hybrid flow shop scheduling problems from a production system and a solutions procedure perspective *Comput. Oper. Res* **37** 1439–1454.
[5] Balin S 2011 Non-identical parallel machine scheduling using genetic algorithm *Expert Syst. Appl* **38** 6814–6821.
[6] Xie J D 2002 Heuristic Genetic Algorithms for General Capacitated Lot-Sizing Problems *Sci. York* **1221** 69904007.
[7] González-Ramírez R G, Smith N R and Askin R G 2011 A heuristic approach for a multi product capacitated lot-sizing problem with pricing *Int. J. Prod. Res.* **49** 1173–1196.
[8] Toledo C F M, De Oliveira R R R and Morelato França P 2013 A hybrid multi-population genetic algorithm applied to solve the multi-level capacitated lot sizing problem with backlogging *Comput. Oper. Res.* **40** 910–919.
[9] Chaudhry I A and Drake P R 2009 Minimizing total tardiness for the machine scheduling and worker assignment problems in identical parallel machines using genetic algorithms *Int. J. Adv. Manuf. Technol.* **42** 581–594.
[10] Moallemi M and Wainer G 2013 Modeling and simulation-driven development of embedded real-time systems *Simul. Model. Pract. Theory* **38** 115–131.
[11] Mohammadi M, Fatemi Ghomi S M T, Karimi B and Torabi S A 2010 Rolling-horizon and fix -and-relax heuristics for the multi-product multi-level capacitated lotsizing problem with sequence-dependent setups *J. Intell. Manuf.* **21** 501–510.
[12] Mohammadi M, Ghomi S M T F and Jafari N 2011 A genetic algorithm for simultaneous lotsizing and sequencing of the permutation flow shops with sequence-dependent setups *Int.*

J. Comput. Integr. Manuf. **24** 87–93.
[13] Mohammadi M and Fatemi Ghomi S M T 2011 Genetic algorithm-based heuristic for capacitated lotsizing problem in flow shops with sequence-dependent setups *Expert Syst. Appl.* **38** 7201–7207.

IOP Conf. Series: Materials Science and Engineering **796** (2020) 012010 doi:10.1088/1757-899X/796/1/012010

SCOR Model for a Dual-Channel Supply Chain using Drop Shipping to Reduce Overstock in Small- and Medium-Sized Retail Enterprises

G Francia-Arias[1], J Marín-Vílchez[1], I Macassi-Jauregui[1], C Raymundo-Ibañez[2, 4] and F Dominguez[3]

[1] Ingeniería Industrial, Universidad Peruana de Ciencias Aplicadas (UPC), Lima 15023, Perú.
[2] Dirección de Investigación, Universidad Peruana de Ciencias Aplicadas (UPC), Lima 15023, Perú.
[3]Escuela Superior de Ingeniería Informática, Universidad Rey Juan Carlos, Mostoles, 28933, España.

E-mail: carlos.raymundo@upc.edu.pe

Abstract. Sales are central to retail companies. One of the main problems for these companies is when products are sold later than expected, causing overstock due to lower inventory turnover, which increases inventory levels. Therefore, for many retailers, it is important to solve this problem. This is mostly applicable to companies engaged in sales; however, if we take into account the main supplier and the way they act within the supply chain, we must also consider an additional approach. Since online sales are a major innovation brought about by the new digital era, it is standard for sales strategies to focus on this new requirement of customers. In this way, the main supplier takes a leap forward on Internet sales, creating another sales channel. This is when companies under the supply chain start losing sales. According to the above, a dual supply chain model was suggested using the SCOR model and drop shipping. After the improvement proposal was implemented, the company reported a reduction of approximately S/13,000 when comparing the first quarter of 2018 to that of 2019.

1. Introduction

According to Statista, the retail industry is growing worldwide [1]. In fact, from 2013 to 2018, the industry has grown approximately $1,4 billion per year. At the national scale, this industry is fourth in the national production ranking per economic sector issued by the National Institute of Statistics and Information Technology (INEI by its Spanish acronym) [2], being surpassed only by the following sectors: manufacturing, other services (real estate and personal services), and mining and hydrocarbons. Additionally, according to the last few business demography reports published by INEI [3] [4] [5] [6], retail companies, retailers, and wholesalers report the highest number of sales drops in each quarter. Finally, as the last important piece of information provided by INEI for this economic sector [2], the retail industry also reports the highest amount of loan debt in national currency with an amount of approximately S/18,051,68, surpassing the manufacturing, real estate, and transport

[4] To whom any correspondence should be addressed.

Content from this work may be used under the terms of the Creative Commons Attribution 3.0 licence. Any further distribution of this work must maintain attribution to the author(s) and the title of the work, journal citation and DOI.
Published under licence by IOP Publishing Ltd

industries, among others. The main reason behind this is that retail companies ask for loans in order to buy several products and, thus, offer a wide variety of items on their inventories.

According to estimates from the Institute of Business Economy and Development [7], at the end of 2018, the retail industry was to report a growth of 3.2%, higher than the 3.1% estimated for Latin America. Since it is constantly growing, great importance should be placed on companies within this sector to foster their continuous growth. The retail industry has greatly progressed with time, with online product sales and promotions becoming extremely profitable for some companies. Progress has also been made in manufacturing tools that provide better inventory management and control, which is paramount for companies wishing to prevent overstock. The main characteristic of the sector is the wide variety and high quality of products offered to customers. Another characteristic of retail companies is the time spent in fulfilling and shipping customer orders, since a company often depends on these aspects to improve its sales and profitability.

It is important to solve the problems that business companies (retailers) face because, as mentioned above, the ceasing of activities is more frequent in companies in this sector than that in in companies in other economic sectors. Therefore, failure to execute demand forecasts is the second most important problem of these types of companies, according to Gestión [8]. The purpose is to provide a proposal to solve the problem of ceasing of activities of many retailers of the industry and offer retailers a completely functional alternative channel that is fully independent from the traditional channel and not limited to selling products from a single supplier.

2. State of The Art

2.1. Single- and Dual-Channel Supply Chain

According to the literature reviewed, there is a recent combination of the traditional retail (single channel) and direct online (dual channels) channels to reach a wider range of customers. Therefore, the management of the supply chain and issues therein are prioritized in the review, beginning with supply chain management.

For a single channel, we mention a success story. Nucamendi et al. [9] designed a "Newsvendor" model, which shares some characteristics with the retail sector. In addition, Zare et al. [10] concluded that a supply chain that is neutrally integrated to risk always leads to an insufficient inventory, while a negative anchorage may cause inventory excesses.

In terms of dual channels, we have Modak and Kelle [11] and Yang et al. [12]. Modak and Kelle [11] manage it according to the stochastic demand dependent on the price and lead time while Yang et al. [12] take the "Newsvendor" model as a reference pursuant to changes in consumer behavior based on stock-out to include the lead time. If we focus on retail, Saha et al. [13] use product price as a basis to suggest that the lead time is critical for buyer decision processes. This is because a longer lead time reduces acceptance and customer loyalty.

2.2. SCOR Model

Different publications address the SCOR model and its wide application in different types of industries. Some researchers like Fasika et al. [14] and Kusrini et al. [15] prove how the SCOR model can be adapted to the industry, using different approaches. Fasika et al. [14] propose an SCOR model adapted to and modified for developing countries while Kusrini et al. [15] introduce a new simultaneous approach to design the performance of the supply chain actors and the regulator on innovative products owing to a comprehensive model, which comprises the balanced scorecard (BSC) model, the SCOR model, and contributions from the regulator.

Additionally, if we consider specific success stories, we can mention Jamal et al. [16] for the modelling, assessment, and analysis of the performance of inventory management systems using the combination of the SCOR model with Petri's networks, a determinist of batches and stochasticity. Furthermore, Qing Lu et al. [17] apply the reference framework of SCOR to the context of humanitarian supply chains to develop a set of indicators for human resources organizations.

IOP Conf. Series: Materials Science and Engineering **796** (2020) 012010 doi:10.1088/1757-899X/796/1/012010

Finally, Wojciech and Richard [18] focus their study on identifying approaches and metrics of supply chain performance and its validation on different types of companies (logistics, retail, automobile, and others). As per the result of the study, the approach most used by companies is BSC.

3. Contributions

In the general model, in order to adopt drop shipping, we need to add the information flow required for the main supplier in order for the latter to send the products to the selected place in the right amount and to the correct customer (Figure 1).

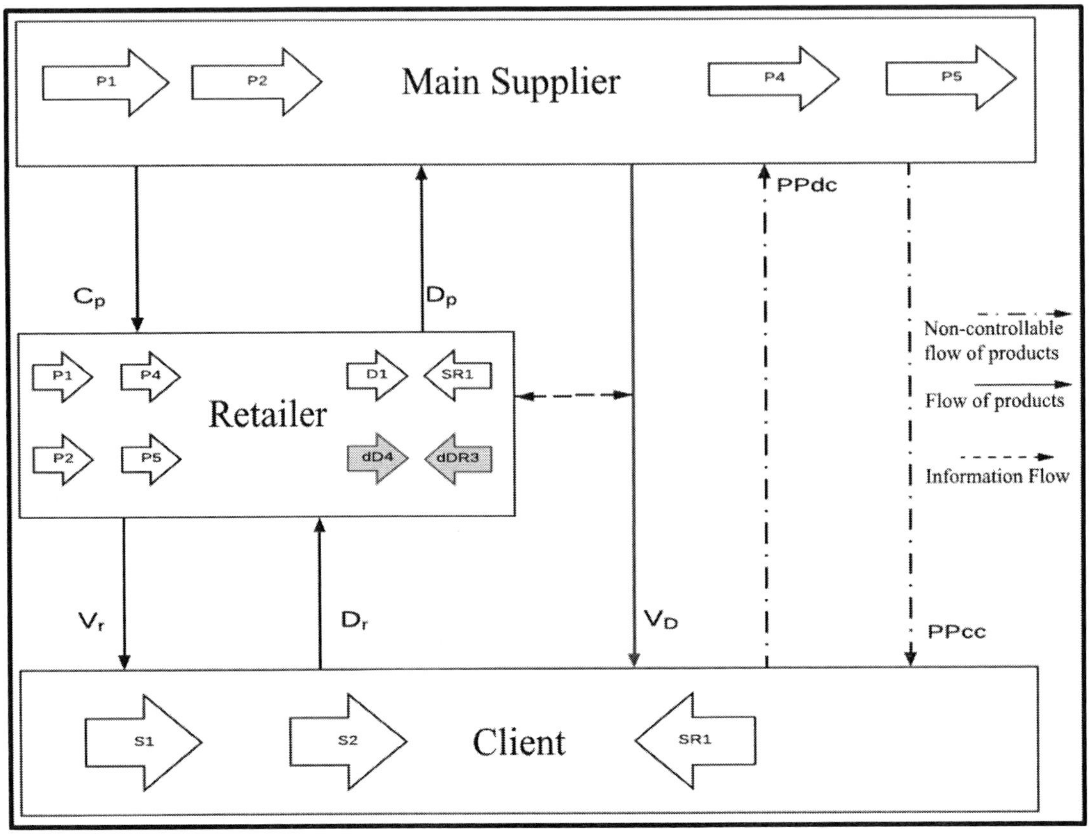

Figure 1. Proposed model for the dual channel supply chain

As a basis to apply the SCOR model in an organization, the following steps need to be followed for its correct application:

Awareness: Here is where the need for a change with the input of the initial diagnosis is evidenced, with all staff members understanding that there is a problem that needs to be fixed. Staff members also understand how the SCOR model works and the management tools it uses.

Discoveries: In this step, a model with the awareness of the current supply chain level is created, which can be viewed according to the following supply chain levels:

- First: What is the level of performance of supply chain operations?
- Second: Is the proper strategy in place and are the information and resource flows available to support and reach the desired level of performance?

Analysis: In this step, the value of the proposal is defined in the terms required by the new financial management of the company: cash to cash time cycle, order fulfilment, other channels and/or niches to cover, and other factors to be considered by the organization.

Design: Resource flows and workflows (process information) are the two key components to define flows by discovering disconnections in their processes and defining optimal flows that help eliminate disconnections.

Development and application: The execution of the model is carried out as per the deadlines established in the work plan.

3.1. Proposed Method

The flowchart below (Figure 2) depicts activities to be carried out in each application phase, in addition to inputs and outputs within the phases. These artefacts from the model's components provide information for the correct development of the said model.

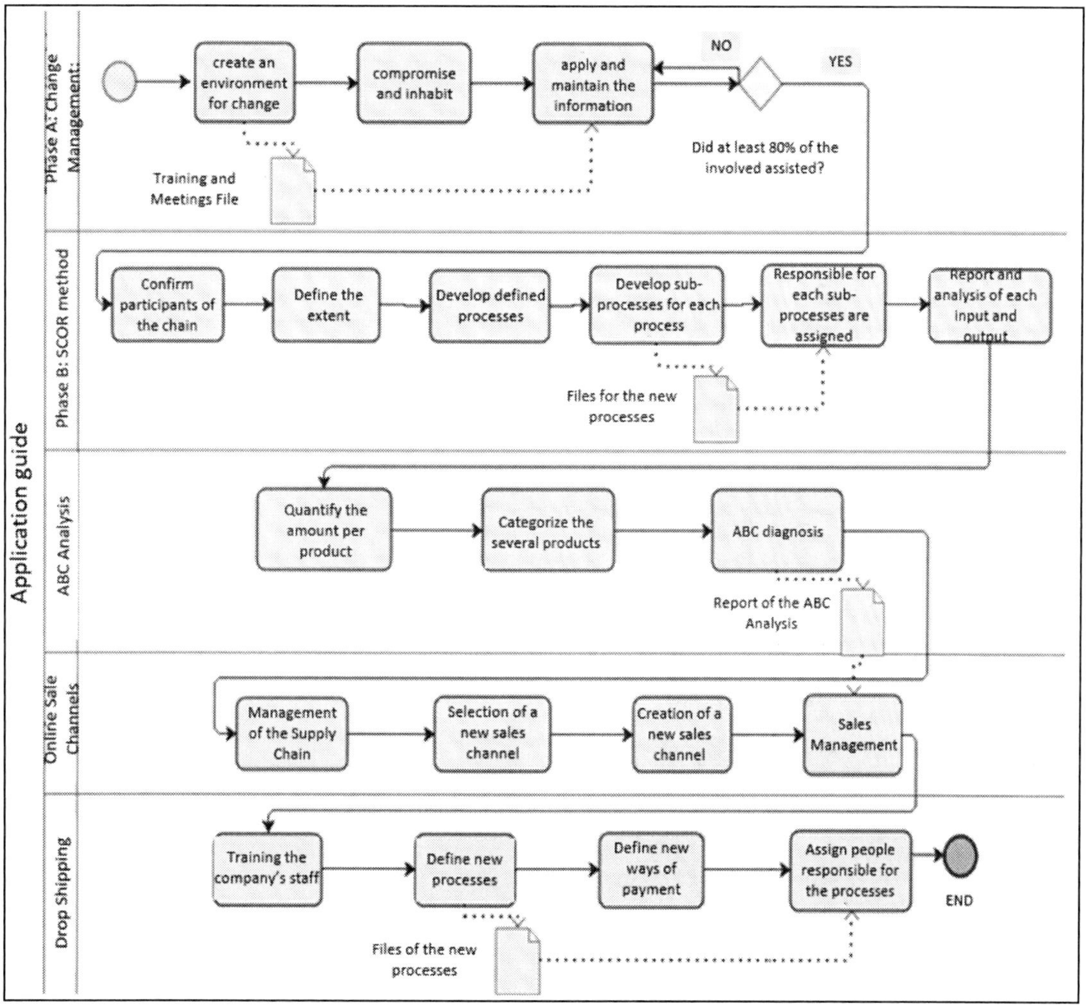

Figure 2. Proposed method

3.2. Indicators

3.2.1. Unsold Stock Ratio

$$USR = \frac{Stock\ of\ products\ at\ the\ end\ of\ the\ period\ i}{Purchases\ in\ the\ i\ period\ of\ the\ product} \quad (1)$$

3.2.2. Average Inventory for the First Trimester of the Year

$$Av.Inv. = \frac{Total\ Inventory\ in\ the\ 1st\ Trimester}{Total\ of\ months\ to\ estimate} \quad (2)$$

3.2.3. Percentage of Online Sales

$$\%O.S = \frac{Sales\ through\ online\ channel}{Total\ of\ Sales\ of\ the\ company} \quad (3)$$

4. Validation

4.1. Case Study

Kmavi E.I.RL. is a small retail company that started operations on March 19, 2013, under Business ID number 20552087357 and is located at Av. Los Portales Mz. A Lt. 03b–Los Olivos - Lima. This retailer is engaged in buying and selling five different types of products (collectibles, books and magazines, leisure goods, accessories, and home goods), using Grupo El Comercio as its sole provider.

Figure 3. Warehouse of products of Kmavi E.I.R.L.

Figure 4. Types of products

4.2. Initial Diagnosis

Taking into account the annual levels of inventory, the following data were obtained: the inventory levels registered values of S/216,670, S/282,510, and S/337,320 in 2016, 2017, and 2018, respectively. Thus, comparing 2017 inventories with 2018 inventories, the level increased 19.40%.

IOP Conf. Series: Materials Science and Engineering **796** (2020) 012010 doi:10.1088/1757-899X/796/1/012010

In 2018, the company presented an unsold stock ratio of 41%, which is considered an unsuitable value in the sector. In addition, the average inventory in the first quarter of 2018 was S/27,720.

4.3. Model Application

4.3.1. Stage 1: Change Management. This stage starts with raising the awareness of the company manager and other staff members regarding the need to perform changes in the company.

4.3.2. Stage 2: SCOR Model. At this stage, the assessment of the year is made (in this case, 2018) to validate the existence of accumulated products.

Table 1. Inventory categorization.

Inventory category (in units)							
Type of product	TC	D	SO	RT	RM	NA	NA (%)
Aggregate	17812	8509	431	0	1509	7363	41%

It can be seen that overstock reaches 41%.

4.3.3. Stage 3: ABC Analysis

Table 2. Categorization of groups of products.

Products Group	**Sales Quantity (S/)**	**% of Total Sales**	**% Accumulated Sales**	**Category of importance for the company**
Recreational	S/.37,621	39.60%	39.60%	
Collection	S/.25,414	26.75%	66.34%	A
Books and Magazines	S/.14,263	15.01%	81.35%	
Home	S/.12,128	12.76%	94.12%	B
Accessories	S/.5,588	5.88%	**100%**	C
TOTAL	**S/.95,014**	**100%**		

In this way, the most important group of products for the company can be identified.

4.3.4. Stage 4: Online Sales Channel. Once the main product groups that the company sells are identified, focus must be placed on these product groups for advertisement purposes through the new online sales channel.

4.3.5. Stage 5: Drop Shipping. Before applying drop shipping, the company informed the supplier regarding the shipping costs of its products, stating that the cost for every two clients is S/50. Thus, customers were notified about the new shipping method for their products, and nine of them selected this shipping method.

Table 3. Comparison of costs

Cost comparison for the execution of Drop Shipping			
Warehouse Cost		Drop Shipping Cost	
January	S/.1,500.00	January	S/.250.00
February	S/.1,500.00	February	S/.250.00
March	S/.1,500.00	March	S/.250.00
April	S/.1,500.00	April	S/.250.00
May	S/.1,500.00	May	S/.250.00
June	S/.1,500.00	June	S/.250.00
TOTAL	**S/.9,000**	**TOTAL**	**S/.1,500**

According to the cost comparisons, if the company continued its activities in the same manner, the costs for renting the warehouse would have been S/9,000 during the first six months of 2019. By applying drop shipping, the facilities would no longer be used; thus, only the product shipment costs will have to be considered.

4.4. Results

Table 4. Comparison of indicator's results

Indicator	**Traffic Light**			**Base Line**	**Improvement**
Unsold Stock Ratio	> 30%	21% - 30%	12% - 20%	41%	28%
Average Inventory I Quater	> S/ 20K	S/13K - S/20K	S/5K - 12K	S/ 27.7 K	S/ 14.7 K
Percentage of Online Sales	< 14%	14% - 21.3%	> 21.3%	0%	18.6%

5. Conclusions

- The execution of the improvement proposal was correctly carried out and reduced to approximately S/13,000 in the average inventory. This means a 47% reduction in contrast to the first quarter of 2018.
- To reduce the overstock and finally eliminate it, many efforts are required, one of these is to adapt to the new trends of acquisition by consumers. These seek non-physical channels for the acquisition of products and are not indifferent to the payment of the transport associated with it.
- Drop shipping is considered an additional sales channel instead of a replacement for the current channel; this is a medium-term decision, since what is sought is to completely eliminate the physical inventory and that the new internet sales channel is preferred by customers.

- The addition of the online sales channel by the main supplier entails a major risk for the rest of the chain participants.

6. References

[1] Statista 2019 Ventas minoristas totales a nivel mundial desde 2013 hasta 2018 (en billones de USD) - (Total retail sales worldwide from 2013 to 2018 – in billions of dollars) Retrieved from: https://es.statista.com/estadisticas/637363/ventas-minoristas-globales-de--a-2018/

[2] INEI 2019 Informe Técnico - Producción Nacional - N° 02 – febrero 2019 (*Lima: Technical Report – National Production – No. 2 – February, 2019*).

[3] INEI 2018 Informe de Demografía Empresarial en el Perú – I Trimestre 2018 (*Lima: Company's Demographic Report in Peru – First Trimester 2018*).

[4] INEI 2019 Informe de Demografía Empresarial en el Perú – II Trimestre 2018 (*Lima: Company's Demographic Report in Peru – Second Trimester 2018*).

[5] INEI 2019 Informe de Demografía Empresarial en el Perú – III Trimestre 2018 (*Lima: Company's Demographic Report in Peru – Third Trimester 2018*).

[6] INEI 2019 Informe de Demografía Empresarial en el Perú – IV Trimestre 2018 (*Lima: Company's Demographic Report in Peru – Fourth Trimester 2018*).

[7] El comercio 2018 CCL: Sector comercio se expandirá 3,2% este 2018 (*El Comercio - August 28, 2018. Chamber of Commerce of Lima: The Trading Industry will expand 3.2% in 2018*) Retrieved from: https://elcomercio.pe/economia/peru/sector-comercio-expandira-3-2-2018-estima-ccl-noticia-nndc-551366

[8] Gestión 2016 INEI: Estos son los cuatro problemas que limitan el crecimiento de las empresas (These are the four problems limiting growth in companies) Retrieved from: https://gestion.pe/economia/inei-son-cuatro-problemas-limitan-crecimiento-empresas-147918

[9] Abraham M, Miguel A M and Samuel N 2018 *Int J Retail Distrib Manag* **46** 726–743

[10] R. Zare, P. Chavez, C. Raymundo, J. Rojas, "Collaborative Culture Management Model to Improve the Performence in the Inventory Management of a Supply Chain", 2018 Congreso Internacional de Innovacion y Tendencias en Ingenieria, CONIITI 2018 - Proceedings 8587073.

[11] Kelle P and Modak N M 2018 *Eur J Oper Res* **272** 147–161

[12] Fu HY, Liu C, Yang J Q and Zhang X M 2017 *Appl Math Model* **42** 675–692

[13] Modak N M, Panda S, Saha S and Sana S S 2018 *J Model Manage* **13** 351–374

[14] Georgise F B, Thoben K and Wuest T 2017 *Prod Plan Control* **28** 17–32

[15] Bagyo S, Elisa K and Nur M 2016 *Int. J. Eng. Bus. Manage.* **17** 1–29.

[16] Abdeslam L, Haj E, Jamal F and Latifa E 2016 *Int. J. Eng. Bus. Manage.* **8** 1–11

[17] Mark G, Qing L and Robert D 2016 *J Humanitarian Logist Supply Chain Manage* **6** 222–239

[18] Richard W P 2015 *Int J Product Perform Manag* **64** 1068–1091.

IOP Conf. Series: Materials Science and Engineering **796** (2020) 012011 doi:10.1088/1757-899X/796/1/012011

Tribological properties of self-mated SUS304 lubricated by palm methyl ester mixed lubricant at boundary lubrication

Z Fuadi[1]*, M Faisal[2], M Dirhamsyah[1], M Tadjuddin[1], R Kurniawan[1]
[1]Department of Mechanical Engineering, Universitas Syiah Kuala, Jl. Tgk. Syech Abdurrauf No. 7 Darussalam, Banda Aceh, Indonesia
[2]Department of Chemical Engineering, Universitas Syiah Kuala, Jl. Tgk. Syech Abdurrauf No. 7 Darussalam, Banda Aceh, Indonesia

E-mail: zahrul.fuadi@unsyiah.ac.id

Abstract. Due to environment and sustainability issues, it is important to reduce the use of petroleum based products in automotive and industrial application. One of the efforts is to use bio oils as base oil for lubricants. This paper presents the experimental results on the effects of palm methyl ester mixing in a conventional lubricant on the friction and wear characteristics of stainless steel SUS304. The friction tests were conducted using a ball-on-disk tribometer at boundary lubrication condition. Three kind of lubricants were prepared, i.e. commercial SAE40, and the mixture of 0.5% and 1% palm methyl ester into SAE40 lubricant. The prepared lubricants were designated as PME0, PME5, and PME10, in reference to 0%, 0.5%, and 1% palm methyl ester mixture in the SAE40 oil. Results shows that the addition of palm methyl ester in the lubricant has reduced the value of coefficient of friction from 0.7 to 0.55 with the addition of 1% palm methyl ester in the base lubricant. As for the wear, 10% reduction of the wear scar was achieved at the condition of PME10 compared to that at PME0. One of possible explanation for this reduction is the formation of oxidative layer on the contact interface due to the existence of the methyl ester in the lubricant. This result indicates the positive impact of the palm methyl ester as additive for enhancing the performance of conventional lubricant thus requiring further analysis and investigation.

1. Introduction

Automotive and industrial machineries have been heavily relying on petroleum based products in their operation, causing pollution and environment health hazards. Lubricants, among others, vital to those machines in maintaining good and reliable operation, have also caused serious problems to environments due enormous amount of waste products, especially when not properly disposed. Therefore, it becomes very important to reduce the use of petroleum based products and replace them with a more environmentally friendly ones. Here, bio-based oils have received significant attentions, particularly due to their exhibiting good lubricity comparable to mineral and synthetic oils suitable for various engineering applications [1].

Bio oils have many advantages to be developed for base lubricants due to their low toxicity and biodegradability [2], ability to provide low wear and friction characteristics [3] and improved surface

* To whom any correspondence should be addressed.

Content from this work may be used under the terms of the Creative Commons Attribution 3.0 licence. Any further distribution of this work must maintain attribution to the author(s) and the title of the work, journal citation and DOI.
Published under licence by IOP Publishing Ltd

IOP Conf. Series: Materials Science and Engineering **796** (2020) 012011 doi:10.1088/1757-899X/796/1/012011

finished [4]. Although they have low thermal and oxidative stability, their properties can be improved chemically by transesterification [4]. Natural triglycerides in bio oils could promote the formation of protective layer on the contact interface and provide anti wear characteristic to the mineral based oils [5]. Inclusion of 5% modified soybean oil in hexadecane-based oil has improved the wear up to 50% [6]. The viscosity index of mineral-based lubricant also increased by addition of 5% palm methyl esters [7]. Castor oil methyl ester can also enhance the lubricity due to its unique fatty acid content [8].

Significant reduction in wear and friction can be achieved by mixing bio oils derived ester with mineral based oils without additional conventional additives [9]. Trimethylolpropane and pentaerythritol ester derived from palm oil match the properties of a full formulated lubricants in terms of friction coefficient [10]. Addition of 3% palm oil based trimethylolpropane in an ordinary lubricant resulted in wear scar diameter reduction up to 30% in boundary lubrication condition [11].

In this research, palm methyl ester were mixed with mineral lubricant to find out its effect on the friction and wear characteristics of stainless steel SUS304. Previously, it is reported that palm methyl ester has an ability to form a protective tribolayer on the contact interface of SUS304 [12]. Thus, in this investigation, the palm methyl ester has been applied as an additive to the base lubricant to find out its effect on the lubricant properties.

2. Methodology

Severe lubrication condition in a mechanical system occurs during heavy loading condition at low sliding speed. In this condition, the interaction between contact interfaces mainly occur among the contacting asperities in which viscosity of lubricant has less important role. Therefore, in order to simulate the severe contact condition between the contact interface, a ball on disk tribometer was used in this investigation. A ball on disk contact interaction provides a high pressure contact in a concentrated location and enables the evaluation of the material wear characteristic in a severe contact condition.

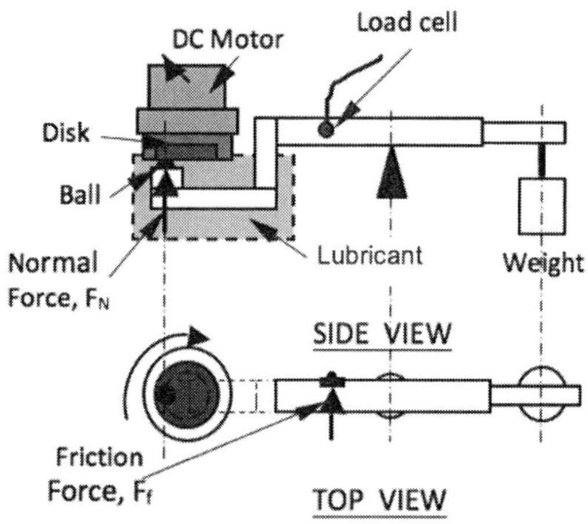

Figure. 1 Ball on disk tribometer

The schematic diagram of the ball-on-disk tribometer is given in Figure 1. The tribometer consists of a ball and disk holder, a DC motor, gravitational normal loading cantilever beam, and the motor control system. The ball specimen has a diameter of 8 mm and the thickness of 4 mm. A load cell is installed to record the friction force as the friction took place, which is connected to a strain amplifier and a personal computer installed with a data acquisition software. The lubricant chamber was installed to soak the contact between the disk and the ball. A heater system was installed in the lubricant chamber to control the temperature of the lubricant. The heater system is controlled independently.

The operating parameter of the friction tests conducted in this investigation is given in Table 1. The friction tests were conducted with the spindle revolution of 100 rpm. During the friction tests, the sliding track has a diameter of 21 mm, resulting in the sliding velocity of 0.0628 m/s. The friction tests were conducted for a sliding distance of 700 m. The friction tests were conducted in ambient room temperature of 27oC. Dead weight of 15 N were given during the test, which resulted in a 15 N normal load.

The friction tests were conducted in three lubricant conditions, as given in Table 2. A commercial lubricant, which has a specification of SAE40, was mixed with various concentration of palm methyl ester, i.e. 0%, 0.5%, and 1%. The mixture were prepared by hand shaking using a bottle container. The prepared lubricants are assigned as PME0, PME5, and PME10 for respective concentration of 0%, 0.5%, and 1% palm methyl ester. The material of the ball and disk specimens are stainless steel SUS304. The properties of SUS304 is given in Table 3. To obtain a relatively similar initial contact roughness at friction test, the contact surface of all specimens were hand rubbed using sandpaper grid #1000.

Table 1. Operating conditions

Parameter, unit	**Value**
Spindle rev, rpm	200
Sliding distance, m	1700
Normal load, N	15
Temperature, °C	Ambient, 27

Table 2. Test conditions

Ball	Disk	Lubricant + %PME
SUS304	SUS304	PME0 - SAE40 + 0%
SUS304	SUS304	PME5 - SAE40 + 0.5%
SUS304	SUS304	PME10 - SAE40 + 1%

Table 3. Properties of SUS304

Young's Modulus, MPa	200 GPa
Hardness, Vickers	200
Chemical composition, %	<0,08% C, 17,5-20% Cr, 8-11% Ni, <2% Mn, <1% Si, <0,045% P, 0,03% S

The properties of the palm methyl ester is given in Figure 2. Palm methyl ester has a viscosity of 4.415 mm2/s at 40 oC which is much lower than that of the commercial lubricant used. Since the volume fraction of the methyl ester added is very small, it is expected to have a negligible effect to the viscosity of the base lubricant. Thus, the properties modification of the mixed lubricant is expected to be caused by the incorporation of elements and component of the methyl ester, given in Figure 2, i.e. methyl palmitate, oleic acid, and methyl stearate. Palm methyl ester also contain oxygen and glycerol as side product.

3. Results and Discussion

3.1. Effect of palm methyl ester addition in lubricant on the coefficient of friction (CoF)

Figure 3 shows the coefficient of friction for various test conditions with different concentration of palm methyl ester in the lubricant. As it is mentioned previously, the contact condition is severe indicated by high contact pressure on the contact interface. In this tests, the diameter of the ball of 8 mm and the normal load of 15 N have resulted in initial contact pressure of 1.2 GPa (Hertzian).

It is shown in Figure 3, high contact pressure has resulted in a relatively high coefficient of friction for all lubrication conditions. The average CoF for the three contact conditions are 0.7, 0.65, and 0.55 for PME0, PME5, and PME10, respectively. This CoF value were achieve for a sliding distance of 600 m. This relatively high value of coefficient of friction indicated that the friction condition was in boundary regime for all test condition. In boundary regime, the contact is dominated by solid friction in which contact interaction occurs among the contact asperities and the viscosity of the lubricant is less important. However, the main role of the lubricant in this boundary regime condition is to form a protective layer (tribofilm) that protects the material from severe wear.

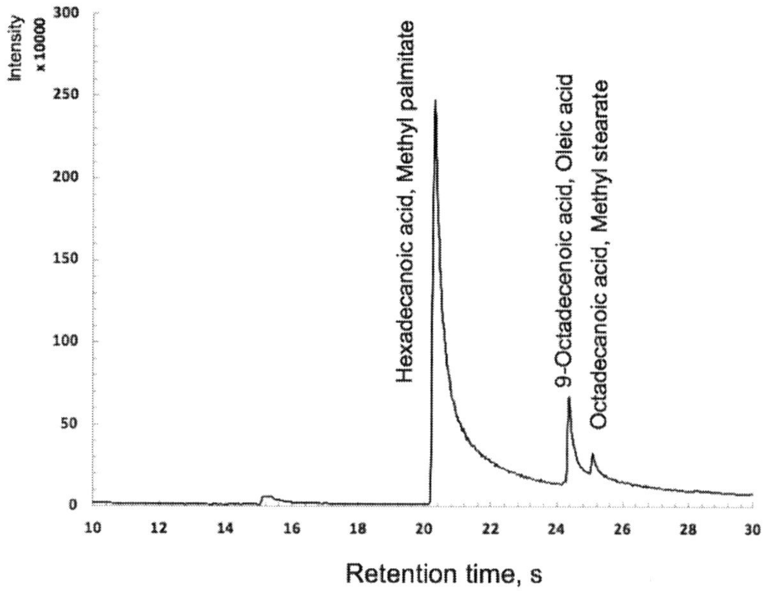

Figure 2. Infrared spectrogram of the palm methyl ester

Figure 3 also indicated that the addition of palm methyl ester in the lubricant has reduced the value of CoF although not significantly. The coefficient of friction reduced from 0.7 to 0.55 with addition of 1% palm methyl ester in the base lubricant, i.e about 10% reduction. This indicates the positive impact of the palm methyl ester on the contact interface, i.e. the palm methyl ester had form some protective layer that reduces the friction. Theoretically, this could occur due to increased lubricity as a result of protective layer formation [13, 14]. Palm methyl ester has an ability to form such a protective layer particularly on stainless steel material SUS304 [12].

This finding is also comparable to that reported by [11] in which the CoF reduced from 0.13 to 0.115 with the addition of 1% palm methyl ester to base oil in the case of using four ball tribometer test and cast iron as the friction material. In that case, the lubricant used has a viscosity of 101.86 cSt, which was relatively similar to the viscosity of the lubricant used in current investigation. The difference of the friction coefficients achieved here compared to those in ref. [11] could be attributed to the difference in the material and contact pressure used.

3.2. Effect of palm methyl addition in base lubricant on material wear

Figure 4 shows the wear of the ball and the corresponding disk for the case of PME0 condition. Figure 4(a) is the 3D micrograph the word track on the disk shown in Figure 4(b) while Figure 4(c) is the 3D micrograph the word track on the ball shown in Figure 4(d). The diameter of worn cross section width of the ball is about 890 μm, which is well correspond to that on the disk side. As it is shown in the figure, severe wear of the material occurred both on the disk (Fig. 4(a) and (b)) and the ball (Fig. 4(c) and (d)). Surface morphology of the disk (Fig 4(a)) indicated plastic deformation on the disk side and transferred material on the ball side (Fig. 4(c)). Plastic deformation and transferred material occurred presumably due to relatively soft material of SUS304.

The contact interface of the disk is quite clear although some small noticeable black spot area existed. Commonly, blackish area on the contact interface occurs due to formation of graphite layer consist of carbon or due to oxidation. In the case of SUS304 material, a relatively small amount of carbon in the material makes it not preferable for graphite formation. Thus, the formation of graphite is not likely in this case. On the other hand, oxidation is also not likely occurring in this case due to the absence of oxygen. Therefore, a relatively clear contact interface in this case could be resulted by the contact condition with small amount of carbon and the absence of oxygen. High contact pressure applied to the contact asperities caused deformation and adherence to each other. Continues sliding of the ball on the disk caused some material of the disk adhered to the ball. Extreme pressure continuously occurring on the ball could also cause high temperature and plastic deformation on the ball contact interface.

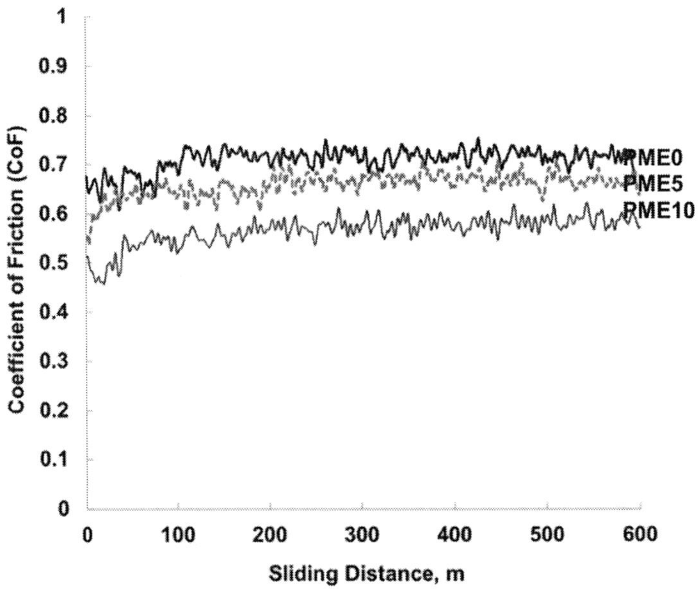

Figure 3. Effect of PME content in base lubricant on the CoF

The worn morphology of the material for the case of PME5 is shown in Figure 5. With the addition of 0.5% of PME in the base lubricant, the worn cross section area of the ball is about 850 μm, which is also well corresponded to the width of the worn track on the disk. Severe wear condition was also observed at this condition but with less transferred material on the ball contact interface. Ploughing and material removal seemed occurring on both the ball and disc contact interface. Only small effect of 0.5% palm methyl ester addition on the worn area on the ball and the disc compared to the case without palm methyl ester addition. The worn cross section width of the ball decreased from 880 μm to 850 μm. As it is shown for the coefficient of friction, the decrease was also not quite significant, which is from 0.7 to 0.65 in the average.

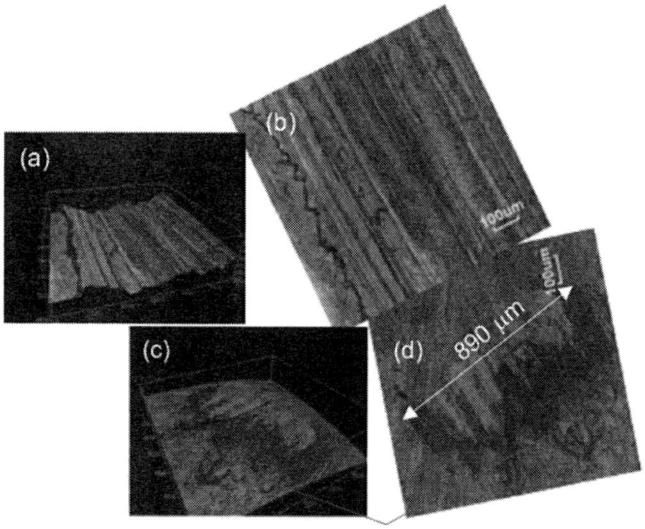

Figure 4. Contact interfaces of the disk; (a) and (b), and the ball; (c) and (d) for PME0 condition

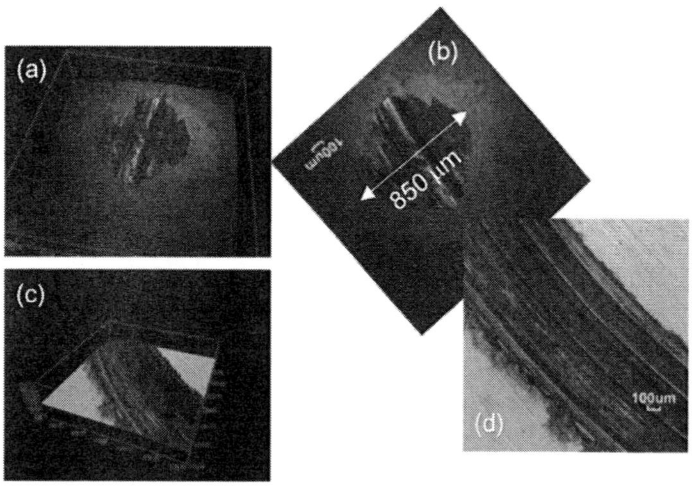

Figure 5. Contact interfaces of the ball; (a) and (b), and the disk; (c) and (d) for PME5 condition

However, the wear scar of the disk appears darker with the presence of 0.5% palm methyl ester on the base lubricant. This could be caused by the oxidation assisted by the oxygen content in the palm methyl ester, which has a chemical formulae of R3-COOCH3. The oxidation process could also be promoted by local high temperature at contact asperity during the friction due to high contact pressure. Darker contact interface could also indicate the formation of tribofilm on the contact interface due to a tribo-chemical reaction. However, this has not been clarified yet and further analysis using surface characterization equipment is necessary to investigate the tribofilm formation.

Figure 6 shows the topography of contact interface of friction condition of PME10, i.e. 1% palm methyl ester mixed in base lubricant. In this case, the worn cross sectional width of the ball is about

IOP Conf. Series: Materials Science and Engineering **796** (2020) 012011 doi:10.1088/1757-899X/796/1/012011

800 μm, which is well correspond to the width of the worn track on the disk. The wear condition on the ball in this case is quite similar to that of PME0. Material removal on the ball occurred as well as plastic deformation on the disk side and the wear properties still indicate severe wear of the material. However, the width of the worn cross section of the ball is 10% smaller than that in the case of PME0. This also well correspond to 10% reduction of the CoF in the case of PME0 compared to PME10, which is 0.7 compared to 0.55. This, somewhat, shows the positive effect of palm methyl addition to the base lubricant on the wear of stainless steel material, particularly SUS304.

The contact interface of the disk indicates darker area across the worn track. Similar to the case of PME5, this indicates the presence of oxidative layer on the contact interface promoted by the palm methyl ester. It has been reported in ref [11] that palm methyl ester added lubricant produced a black-coloured contact interface, especially at high temperature. This is due to oxidation of the lubricant at higher temperature, producing corrosive acids, thus promoting corrosive wear [15].

The formation of this blackish layer has some positive effects on the tribological performance of the contact pair because the layer can act as a sacrificial layer which is removed at friction continues [16]. The formation of lubricating film on the contact interface can also be accelerated with the presence of inorganic acid such as Fe3O4 and Fe3O4 form due to oxygen existence [17].

Therefore, the reduction of CoF and the improvement of wear on the material in this investigation could be related to the formation of some oxidative layer, indicated by the presence of dark layer on the contact interface. In this case, the source of oxygen is the palm methyl ester. Thus, a hypothesis that can be proposed is that the addition of small amount of palm methyl ester to the base lubricant could accelerate the formation of oxidative layer acting as a lubrication film on the contact interface. Therefore, the properties of based lubricant could be enhanced by adding a small amount of palm methyl ester, in this case 1%. Nevertheless, the existence of such oxidative layer must be clarified further using surface characterization method.

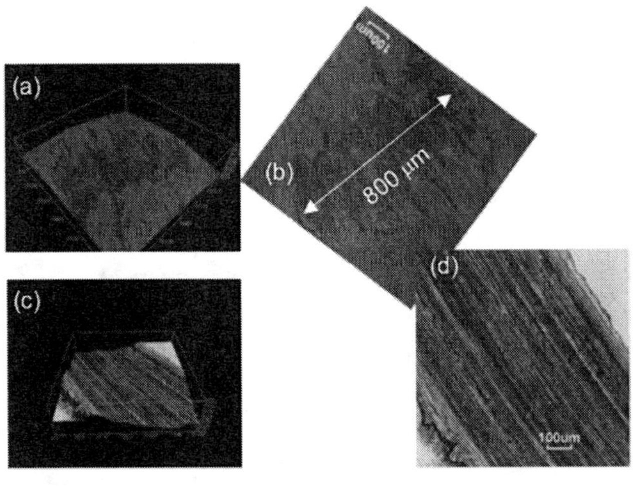

Figure 6. Contact interfaces of the ball; (a) and (b), and the disk; (c) and (d) for PME10 condition

In this investigation, the maximum addition of palm methyl ester used is 1%. Further addition of palm methyl ester on the base lubricant could either give better or worse effect on the lubricity at the contact interface. As reported in [11], an optimum number of palm methyl ester is 3% to 5%. However, at 3% addition, surface delamination of the contact interface occurred.

This investigation has not taken into account the effect of temperature in the friction test. As it has been mentioned before, temperature effect is important because it may affect the oxidation rate at contact interface thus affecting the properties and behaviour of the oxidative layer formation. This effect needs to be clarified in the future.

4. Conclusion

The investigation on the effect palm methyl ester addition on a base lubricant on friction and wear behaviour of stainless steel SUS304 has been conducted. The results shows that palm methyl ester addition to the base lubricant has reduced both the friction and wear scar dimension of the contact material by 10% with 1% concentration of the methyl ester in the base lubricant. Although the real cause of this friction and wear reduction is still not clarified, it is hypothesized that the reduction is attributed to the existence of oxidative layer on the contact interface indicated by darker contact interface in the case of palm methyl added lubricant. While the existence and the properties of this oxidative layer needs to be investigated further, it has been shown that palm methyl ester is potential to be applied as bio-additive to enhance the performance of a mineral based lubricant, thus requires further investigation and analysis.

Acknowledgment

The authors would like to thank LPPM Universitas Syiah Kuala for funding of this research through PLK research scheme ref. no.: 58/UN11.2/PP/PNBP/SP3/2018 in 2018, which is highly acknowledged. The authors also thank Mr. TM Damirsyah and Mr. Mursalin for their valuable supports during experiment and data analysis, which are highly appreciated.

References

[1] Syahir A Z, Zulkifli N W M, Masjuki H H, Kalam M A, Alabdulkarem A, Gulzar M, Khuong L S and Harith M H 2017 J Clean Prod 68 997
[2] Lathi P S and Mattiasson B 2007 Appl Catal B: Environ 69 207
[3] Asadauskas S, Perez J M and Duda J L 1997 Lubr Eng 53 35
[4] Lawal S A, Choudhury I A and Nukman A 2013 J Clean Prod 41 210
[5] Maleque M, Masjuki H H and Sapuan S 2003 Ind Lubr Tribol 55 137
[6] Erhan S Z, Sharma K, Liu Z and Adhvaryu A 2008 J Agric Food Chem 56 8919
[7] Maleque M A, Masjuki H H and Haseeb A S M A 2000 Wear 239 117
[8] Goodrum J W and Geller D P 2005 Bioresour Technol 96 851
[9] Rajendiran A, Sumathi A, Krishnasamy K, Kabilan S and Ganguli D 2016 Tribology International 99 47
[10] Zulkifli N W M, Azman S S N, Kalam M A, Masjuki H H, Yunus R and Gulzar M 2016 Tribology International 93B 555
[11] Zulkifli N W M, Kalam M A, Masjuki H H, Shahabuddin M and Yunus R 2013 Energy 54 167
[12] Fuadi Z, Adachi K, Muhammad T 2018 Tribology Letters 66 (3) 88
[13] Masjuki H H and Maleque M A 1996 Wear 198 293
[14] Choi U S, Ahn B G, Kwon O K and Chun Y J PII: S0301–679X(97)00039-X.
[15] Briscoe B J, Thomas P S and Williams D R 1992 Wear 153 (1) 263
[16] Quinn T J F, Sullivan J L and Rowson D M 1984 Wear 94 (2) 175
[17] Lu X, Cotter J and Eadie D T 2005 Wear 259 (7–12) 1262

IOP Conf. Series: Materials Science and Engineering **796** (2020) 012012 doi:10.1088/1757-899X/796/1/012012

Warehouse management model using FEFO, 5s, and chaotic storage to improve product loading times in small- and medium-sized non-metallic mining companies

P Espinoza-Camino[1], I Macassi-Jaurequi[1], C Raymundo-Ibañez[2,4] and F Dominguez[3]

[1] Ingeniería Industrial, Universidad Peruana de Ciencias Aplicadas (UPC), Lima 15023, Perú.
[2] Dirección de Investigación, Universidad Peruana de Ciencias Aplicadas (UPC), Lima 15023, Perú.
[3]Escuela Superior de Ingeniería Informática, Universidad Rey Juan Carlos, Mostoles, 28933, España.

E-mail: carlos.raymundo@upc.edu.pe

Abstract. This article addresses one of the main problems faced by small- and medium-sized business in the non-metallic mining sector in Peru. These companies own warehouses and face the major problem of failing to deliver orders correctly and in a timely manner. This problem usually occurs when the business grows from a small- to medium-sized company in a short span of time; this situation leads to new processes within warehouses that are mostly not standardized. Besides, facilities are no longer optimal in space and the workers are not properly trained. The case study shows that the orders were not delivered on time due to factors such as lack of product identification, although the products have an expiration date and a warehouse without signaling and surrounded by traffic. To tackle this situation, a labeling process has been designed for the products, an adequate distribution technique is used in the warehouse through a newly designed warehouse layout, and a First Expired, First Out system has been implemented. Similarly, the design is accompanied by the 5s tool to provide a basis for order and continuous improvement. The results show that deliveries with delays were reduced from 38% to 10%. These results show that companies can grow rapidly and maintain quality of service through orderly management.

1. Introduction

Over the last 20 years, Peru has experienced significant economic growth. In 2018, growth was 4%, which allowed many companies to develop and thrive during these years [1]. The non-metallic mining sector has grown by more than 150% in the last decade (2008–2018) [2]. Currently, there is significant growth specifically among small- and medium-sized enterprises; however, operational efficiency is quite low [3] because they face many difficulties in developing their businesses due to factors such as non-standardized processes, lack of operating space, and lack of technology used by larger companies. Hence, every one out of four medium-sized companies in Peru shows a 50% drop in sales due to poor management [3] because the abovementioned situations create operational problems. In the specific

[4] To whom any correspondence should be addressed.

Content from this work may be used under the terms of the Creative Commons Attribution 3.0 licence. Any further distribution of this work must maintain attribution to the author(s) and the title of the work, journal citation and DOI.
Published under licence by IOP Publishing Ltd

case of non-metallic mining, delays occur in delivering products to the client, most of which are construction and mining companies [4], which could lead to losing these clients. According to the Peruvian government, the construction sector in Peru will continue to grow at a rate of 7% in 2019 [5]. This will continue to boost the demand for non-metallic minerals, and small- and medium-sized companies will continue to face the aforementioned difficulties.

The main motivation for this research lies in the concern that Peru's economic growth is based on the construction and metal mining sectors. This will cause products from non-metallic mining companies to become essential as they are the input for the construction and metal mining sector, and this is normally the standard behavior in developed countries [6]. Therefore, the small- and medium-sized sectors of the non-metallic mining industry must have efficient processes for loading and unloading products so as not to affect operations and, consequently, avoid harming the end customer. The aim of this research is to propose improvements in warehouse management, which are economically viable and can reduce or mitigate delays in shipments to the customer. The proposed model is based on implementing the 5s philosophy as a lean manufacturing tool, a warehouse design with First Expired, First Out (FEFO) logic, and chaotic development to face the technological limitations that ensure the products to be sent are not expired and reach the customer within the expected time framework, thus meeting the final quality standard.

2. State of The Art

2.1. Lean Manufacturing

This work philosophy allows for the reduction of waste in the production system. Simulations show that it is possible to reduce up to 30% of the transit time inside the warehouse. There are also studies that indicate improvements of 82% and 79%, respectively, for the process of loading and unloading products in large companies when using the 5s tool [7] [8] [9] [10].

Most articles and case studies on implementing the lean manufacturing philosophy are developed and focused on large companies whose problems and operations generally differ from those of small- and medium-sized companies. However, it is important to consider that the commitment of business leaders is the most important factor in preventing failure during implementation [11] [12].

2.2. FEFO

Simulations with perishable and non-perishable products have been performed to determine the performance of the different types of methods for managing product entry and exit in both First-In, First-Out (FIFO) and FEFO storage systems. In the non-perishable scenario, FIFO inventory management required 32% of the time used by the FEFO system. In the perishable products scenario, FEFO inventory management reduced waste (expired products) by 50% with respect to FIFO [13] [14].

The inventory system that performs best in managing product storage, supply, and handling with limited shelf life is FEFO. Many companies use the FIFO inventory system for their products regardless of whether the products are perishable. However, these companies show declines because the product's storage time exceeds its shelf life. Nevertheless, a study shows that when implementing the FEFO system in warehouses with perishable products, declines were reduced by 50% due to correctly applying the inventory system for the type of product being handled [14] [15] [16] [17].

2.3. Chaotic Storage

The best-known storage systems are on-demand storage, dedicated storage, and chaotic storage. Warehouses that randomly place their products inside warehouses are called chaotic warehouses. They require less storage space than dedicated warehouses, where products have pre-determined spaces within a warehouse. However, dedicated or on-demand warehouses have less transit time within a warehouse. This can be verified through some studies showing that chaotic storage reduces transit time within warehouses [18] [19]. For example, in one case study, chaotic storage reduced transit time by 32.8%, but on-demand storage reduces it by 52.5% [18]. Chaotic warehouses need information systems to help identify, track, and manage products within warehouses, and an inventory management model is needed to allocate products in certain spaces [20] [21].

3. Contribution

3.1. Proposed Model

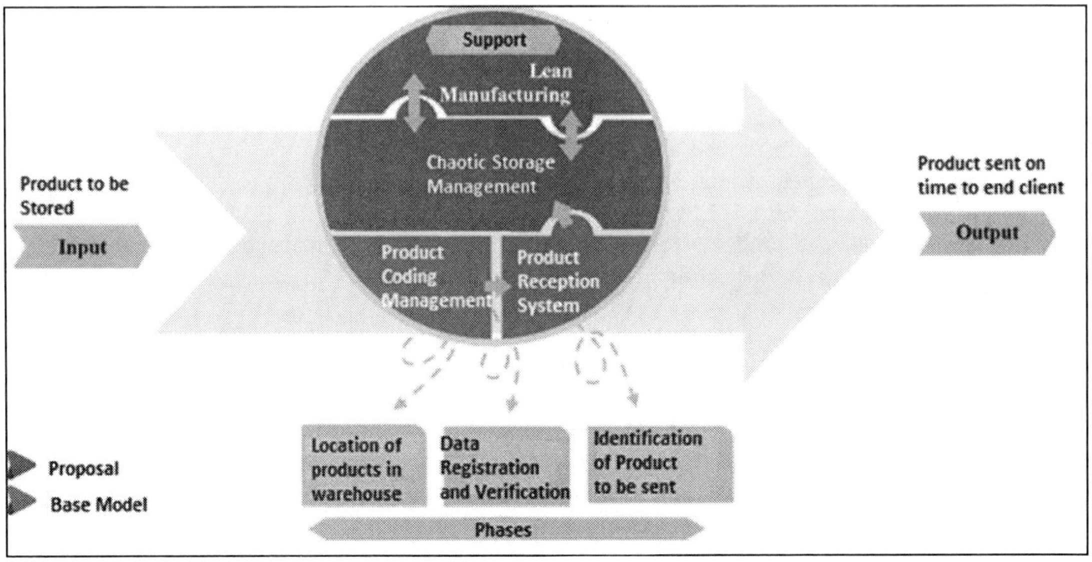

Figure 1. Layout of the model proposed

3.1.1. Managing Product Coding and Receiving. These dimensions use tools designed to create low-cost labels. The goal is to quickly identify shelf life and allow for the management of locations within the warehouse. Similarly, product receiving confirms that the product complies with security measures, has a visible expiration date, and that the order planning is followed. For this dimension, tools such as checklists were used to control label use.

3.1.2. Chaotic Storage Management. This dimension allows for the implementation of a storage system where the chaotic concept is combined with FEFO logic. The main purpose is to know warehouse's stock at all times without having to move the products and manage the future input or output of products for a given period of time by moving identifiers according to time. To guarantee the application of the FEFO technique, traffic light rules are used for inputs and outputs. The main tools used are a stock map, which provides visual instructions to operators who load and unload products in the warehouse, and a matrix where a record of warehouse movements is consolidated daily and saved so that the responsible parties can later make decisions on accepting orders within the warehouse or evaluate its capacity according to fluctuations in demand.

3.1.3. Lean Manufacturing Strategy. This dimension aims to build the foundation for the warehouse to constantly review compliance with its new procedures. The strategy consists of planned audits and analyzing information from the records to make decisions or develop action plans.

3.2. Proposed Method

The process of applying the methodology is sequential and Figure 2 shows the tools that have been applied during implementation.

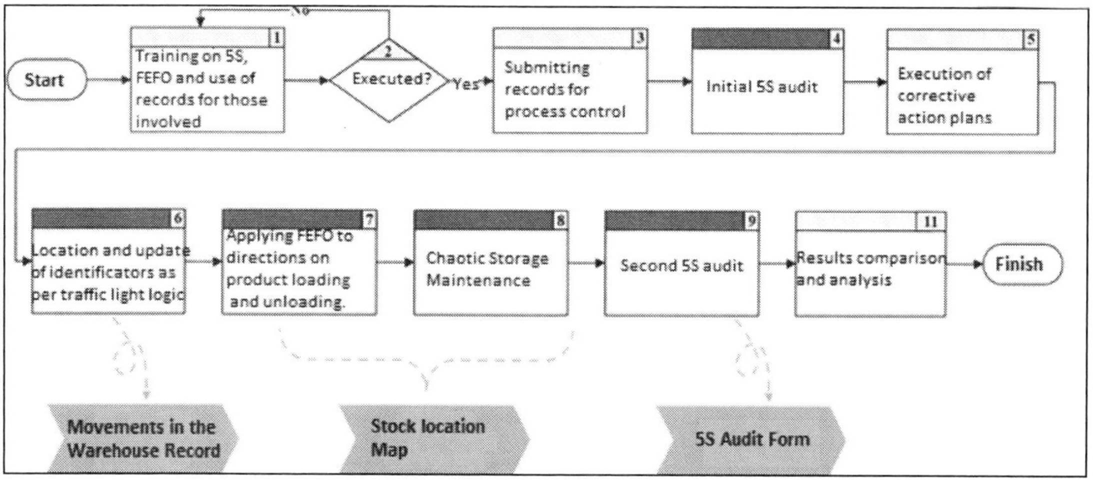

Figure 2. Method to execute the proposed model

3.3. Indicators

- Percentage of monthly shipment delays. This indicator allows for monitoring the problem of delays during shipments to clients.

$$Metric = \frac{(Total\ delayed\ shipments\ in\ the\ month)}{(Total\ shipments\ in\ the\ month)} x\ 100 \quad (1)$$

If the result is 0%, it indicates that the model is serving its main purpose and 100% indicates all shipments are delayed and the techniques applied are not resulting in any improvement.

- Deviation of monthly handling process time with respect to ideal time. This indicator shows the shipment delay time compared to the time the process should ideally take. The metric to calculate this indicator is as follows:

$$M = \frac{(Real\ process\ time - Ideal\ time)}{(Ideal\ time)} x\ 100 \quad (2)$$

When the result is above 0%, this should be interpreted as the process needing to be revised.

- Monthly expired product stock quantity indicator. This indicator tracks the effectiveness of the FEFO system. The metric to calculate this indicator is as follows:

$$Metric = Sum\ of\ expired\ products\ in\ the\ motnh \quad (3)$$

- Results of 5s audits. This indicator makes it possible to identify whether the 5s philosophy is being applied correctly within the company and also allows action plans to be monitored. This indicator is calculated as follows:

$$Metric = Sum\ of\ the\ scores\ for\ each\ \text{“}S\text{”} - 50 \quad (4)$$

These scores are determined with specialized questions for each "S." Possible values range from 0 to 50 points. A score equal to or less than 25 points indicates that the company is not applying the 5s philosophy, a score greater than 25 indicates that the company is applying the tool but has yet to generate action plans for correction or improvement, and finally, obtaining a score of 50 indicates that the tool works successfully.

4. Validation

4.1. Case Study

The company produces, stores, and markets quicklime, carbonate, and hydrated lime to construction and mining companies in Peru and has problems with the quicklime warehouse's shipping process. The company's main objective is to reduce these expensive delays. The scope of the project was to apply the proposed method.

4.2. Initial Results

The initial results found as part of the case study assessment are described below using the defined indicators. Table 1 shows a summary of the initial results.

Table 1. Initial results of the case study

% Delays during shipping	% Deviation from process time compared to ideal time	Amount of expired products (MT)	Audit Result (points)
38%	11.6%	16	10

Initial records showed delays for 217 orders in 2017 and 180 in 2018. However, for the purpose of assessing equivalent periods, data from the first quarter are considered where the percentage of monthly delays during shipping was initially 27% for 2017 and 23% for 2018. With respect to monthly process time, a time study was conducted, where a standard time of 4.87 hours was obtained and an additional margin of 2.6% was considered to find that the ideal time for the process should be equal to or less than 5 hours and less than 6 hours. Thus, initially the process took 5.57 hours, with times of even more than 6 hours. Additionally, within the warehouse, 16 metric tons of expired products were initially found per month, which could be identified only by performing individual quality tests. Finally, with respect to the 5s audit results indicator, the initial score was 10 points out of the 50 required to find that the 5s management system is successfully implemented.

4.3. Applying the Model to the Case Study

The proposed model comprises four dimensions which were implemented following the methodology proposed above. The model's application is detailed below:

4.3.1. Training and handing out forms. More than 20 people involved in the process were trained and evaluated on 5s tools, FEFO, and the use and importance of recording data. 100% of staff attended and the result of written evaluations was also 100%. The frequency for future training was determined to ensure constant discipline and training on this work philosophy. In addition, 22 forms were provided to record data and subsequently calculate indicators.

4.3.2. Performing the Initial 5s Audit. In this task, an initial audit is carried out to find the state of the warehouse and determine the following actions, which in turn allows the classification, elimination, relocation, or disposal of any unnecessary materials within the warehouse. The cleanliness and transport options for vehicles and operators were also reviewed. Finally, the results were used to develop action plans to correct observations. As a result, more than 50 MT of material including expired products, obsolete tools, empty big bags, etc. were eliminated, reprocessed, relocated, or sold. Moreover, 10 documents were approved, including policies, procedure manuals, and instructions.

4.3.3. Locating and Updating Identifiers as per the Traffic Light Logic. In this task, the products were relocated according to lot and with identifiers in each zone with material. Each identifier follows the rules of the traffic light based on the product's age.

Color	Months	Action
Green	4–6	Ship first
Yellow	1–4	Ship only if there are no green products
Red	<1	Ship only if there are no green or yellow products

Figure 3. Traffic light rules

They are updated daily and before the next day's following shifts using a movement record in the warehouse where the respective control is performed and determines the warehouse's capacity to continue receiving the product or the urgency with which each zone of products must be shipped.

4.3.4. Applying the FEFO logic to loading, unloading and maintaining chaotic storage. In this task, the previous task record is taken as input, and orders for the product's shipping and arrival are transferred. This creates a stock location map where the person in charge of the morning shift specifies the areas where the product must be taken, in the case of shipping, and the areas where the product should be placed when it arrives at the warehouse. This can be seen with the tool defined and shown in Figure 4.

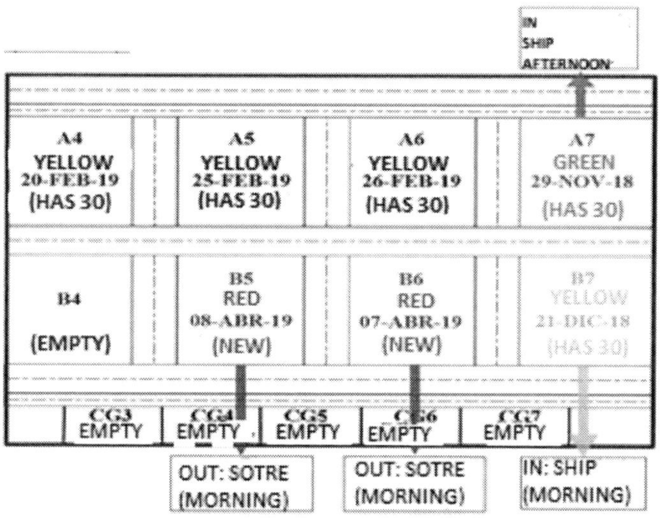

Figure 4. Map of warehouse stock

After the information is delivered to the operators in charge, the manager must update the identifier according to whether it has changed, been separated because the area was empty or if it must be included among the population of areas and again updates the record of the previous activity.

4.3.5. Performing the Second 5s Audit. In this task, an audit was carried out again where the input was the plans released as part of the first audit, which must be implemented or at least in progress. In the last audit, the score was 46 out of 50 points.

4.4. Summary of Results

During the first quarter of the year, the proposed model was implemented. Table 2 shows the results after implementation, using defined achievement indicators:

Table 2. Indicators for validation

Detail / Indicator	% Delays during shipping	% Deviation from process time compared to ideal time	Amount of expired products	Audit result
Initial Result	38%	10%	16	10
Final Result	8%	2%	0	46
Improved?	YES	YES	YES	YES

The results show a dramatic drop in customer delivery delays. Also, the total time of the shipping process is closer to the ideal time of 5 hours, largely because expired products were removed from the warehouse and there is a defined process as well as controls. Finally, the latest audit reveals that the initial action plans have been followed and that some points need to be improved to be fully successful. It is important to mention that results are expected to improve over time as improvements are consolidated and the system continues to be enhanced.

5. Conclusions

The proposed model reduces delays in shipments to the customer by up to 19% and reduces the number of expired products in the warehouse by up to 100%. The proposed model reduced the amount of time in the case study shipping process from 5.55 hours to 5.10 hours. The proposed model is capable of improving warehouse management with a small investment and high effectiveness in reducing shipping delays. This is very useful for small- or medium-sized companies that do not have large investments for improvement.

6. References

[1] Semana Económica (SE). Vizcarra: We have completed 20 years of continuous growth, but the rates are insufficient. Retrieved from: http://semanaeconomica.com/article/legal-y-politica/politica/328318-vizcarra-hemos-cumplido-20-anos-de-crecimiento-continuo-pero-las-tasas-son-insuficientes/

[2] Gestión (GE). Non-metallic mining production grew more than 150% in the last decade. Retrieved from: https://gestion.pe/economia/produccion-minera-metalica-crecio-150-ultima-decada-262811

[3] Gestión (GE). One in four medium-sized companies have 50% drops in sales due to poor management. Retrieved from: https://gestion.pe/economia/empresas/cuatro-medianas-empresas-registran-caidas-50-ventas-mala-gestion-115183

[4] Supervisory Body for Investment in Energy and Mining (Osinergmin) 2016 Sectoral economic analysis report. Mining Sector. The non-metallic mining sector in Peru. Retrieved from: http://www.osinergmin.gob.pe/seccion/centro_documental/Institucional/Estudios_Economicos/RAES/RAES-Mineria-Agosto-2016-GPAE-OS.pdf

[5] América Económica (AE). Construction sector in Peru will grow 7% in 2019 due to mining and project unlocking, says deputy minister. Retrieved from: https://www.americaeconomia.com/negocios-industrias/sector-construccion-del-peru-crecera-7-el-2019-por-mineria-y-destrabe-de

[6] Supervisory Body for Investment in Energy and Mining (Osinergmin) 2016 Sectoral economic analysis report. Mining Sector. The non-metallic mining sector in Peru. Retrieved from: http://www.osinergmin.gob.pe/seccion/centro_documental/Institucional/Estudios_Economicos/RAES/RAES-Mineria-Diciembre-2016-GPAE-OS.pdf

[7] Atieh A, Kaylani H, Al-abdallat Y., Qaderi A, Ghoul L, Jaradat L, Hdairis I 2016 Performance improvement of inventory management system processes by an automated warehouse management system *Procedia CIRP* **41** 568-572.

[8] Riquero I, Hilario C, Chavez P, Raymundo C 2019 Improvement proposal for the logistics process of importing SMEs in Peru through lean, inventories, and change management *Smart Innov Syst Tech* **140** 495-501.

[9] Zahraee S, Hashemi A, Abdi A, Shahpanah A and Rohani J 2014 Lean Manufacturing Implementation Through Value Stream Mapping: A Case Study *J Teknol* **68** 119-124.

[10] Dharmapriya U and Kulatunga A 2011 New Strategy for Warehouse Optimization–Lean warehousing *Proc Int Conf on Ind Eng Oper Manag (Kuala Lumpur)* 513-519.

[11] AlManei M and Salonitis K 2017 Lean implementation frameworks: the challenges for SMEs *Procedia CIRP* **63** 750-755.

[12] Mostafa S, Dumrak J and Soltan H 2013 A framework for lean manufacturing." *Prod Manuf Res* **1** 44-64.

[13] Wagenaar J 2010 Consumption order formulations for lot-sizing problems (*Roterdão, Holland: Erasmus School of Economics*).

[14] Ilic A, Staake T and Fleisch E 2009 Using sensor information to reduce the 1773 carbon footprint of perishable goods *IEEE Pervasive Comput* **8** 22–29.

[15] van der Vorst J, van Kooten O, Marcelis W, Luning P and Beulens A Quality controlled logistics in food suppy chain networks: Integrated decision-making on quality and logitics to meet advanced customer demands *The 14 Th International Annual Euroma Conference, 17-20 June* (*Ankara, Turkey*).

[16] Hertog M, Uysal I, McCarthy U, Verlinden B and Nicolaï B 2014 Shelf life modelling for first -expired-first-out warehouse management *Philos T R Soc A* **372** 20130306.

[17] Bertolinia M, Bottania E, Rizzia A, Volpia A and Renzi P 2013 Shrinkage reduction in perishable food supply chain by means of an RFID-based FIFO management policy *Int J RF Tech* **5** 123-136.

[18] Graves S and Hausman W 1978 Scheduling Policies for Automatic Warehousing Systems: Simulation Results *AIIE Trans* **10** 260-270.

[19] Quintanilla S, Pérez A, Ballestín F and Lino P 2015 Heuristic algorithms for a storage location assignment problem in a chaotic warehouse. Retrieved from https://www.tandfonline.com/doi/abs/10.1080/0305215X.2014.969727

[20] Hsieh S and Tsai K.-C 2001 A BOM Oriented Class-Based Storage Assignment in an Automated Storage/Retrieval System *Int J Adv Manuf Tech* **17** 683-691.

[21] Ulbrich A, Galka S and Günthner W 2008 Investigation of the Influences of Article and Order Structure on the Dimensioning of Zone-Picking-Systems *Proceedings of the 22nd EUROPEAN Conference on Modelling and Simulation ECMS2008*.

The Effect of Stray Current from Sacrificial Anode Cathodic Protection (SACP) System on the Adjacent Unprotected Reinforcing Steel

S Fonna[1], **R Kurniawan, Israr, R Fachri, S Huzni**
Department of Mechanical and Industrial Engineering, Universitas Syiah Kuala, Darussalam, Banda Aceh, Indonesia

E-mail: syarizal.fonna@unsyiah.ac.id

Abstract. Concrete protects reinforcing steel from corrosion, but the steel may be corroded if exposed to coastal environment. Cathodic protection is a popular method to protect steel structures from corrosion. Previous researchers have conducted studies on sacrificial anode cathodic protection (SACP). However, the effect of stray current on the unprotected reinforcing steel in concrete was not reported. This study aims to analyze the effect of the stray current phenomenon on reinforcing steel which are not protected by SACP. Three types of specimens were used in the study: (1) unconnected anode-cathode, (2) connected anode-cathode, and (3) partially connected anode-cathode. The specimens were treated with 30 weeks of wet-dry cycle. Half-cell potential technique was used to measure corrosion potential. The results show that the first specimen had an average potential value of -505.44 mV, meanwhile the second specimen had average potential value -883.66 mV. According to ASTM C876 criteria, the first specimen is categorized into high corrosion risk. While, the second specimen was protected from corrosion based on NACE standard. However, the result on the third specimen shows that only the first reinforcing steel was protected from corrosion, while the other steels were categorized as severe corrosion risk level. The result on third specimen indicates that the stray current produced by SACP system could cause negative effect on the reinforcing steel by increasing susceptibility to corrosion, instead of protecting them.

Keywords: Corrosion, Reinforced Concrete, Cathodic Protection, Zn Anode

1. Introduction

Steel and concrete are the most common building materials over the past hundred years along composite materials. Reinforced concrete with rebar steel is the staple of civil construction. Standards for concrete and steel were determined by the limits of their strength and compositions [1].

Banda Aceh is an aggressive area for corrosion because it is located on the coast. The air in the coastal area contains evaporated salt from sea. When the wind blew from the sea, it brings salt vapor that penetrates into the buildings. Banda Aceh is also an area that has been affected by tsunami in

[1] syarizal.fonna@unsyiah.ac.id

Content from this work may be used under the terms of the Creative Commons Attribution 3.0 licence. Any further distribution of this work must maintain attribution to the author(s) and the title of the work, journal citation and DOI.
Published under licence by IOP Publishing Ltd

2004. The tsunami washed and immersed the buildings with seawater and the reinforcing steel in concrete have become more corrosive [2]

In addition to aggressive environment, stray current is also the cause of corrosion. Stray current is an electric current that flows through another path into reinforced concrete. The current will move from one metal conductor to another metal conductor through ionic medium when the metals are not connected. Stray current generally occurs in pipes. This is due to a number of external DC currents. Pipes with stray current get cathodic protection from stray current at the point of entry, but the effect of amends will appear at the point of exit. Stray current on a structure is generally marked with a rapid corrosion rate [3].

Research related to reinforced concrete with cathodic protection systems has been widely carried out by researchers. However, the effect of stray current from sacrificial anode cathodic protection on reinforced concrete has not been reported. This study analyses the impact of the stray current phenomenon on reinforced steel which are not protected by cathodic protection.

2. Materials and Methods

2.1. Material

The sacrificial anode used in this study was Zn alloy anode which was manufactured according to ASTM B418 [4]. The Zn alloy sacrificial anode was 25 mm in length and 50 mm in width. The Zn alloy used in this study can be seen in figure 1.

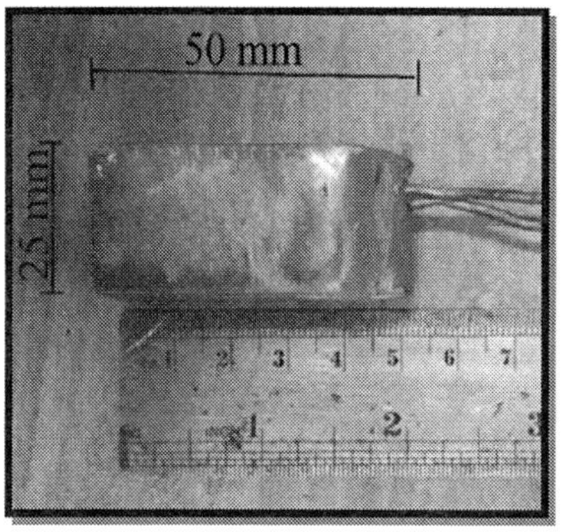

Figure 1. Sacrificial anode Zn alloy

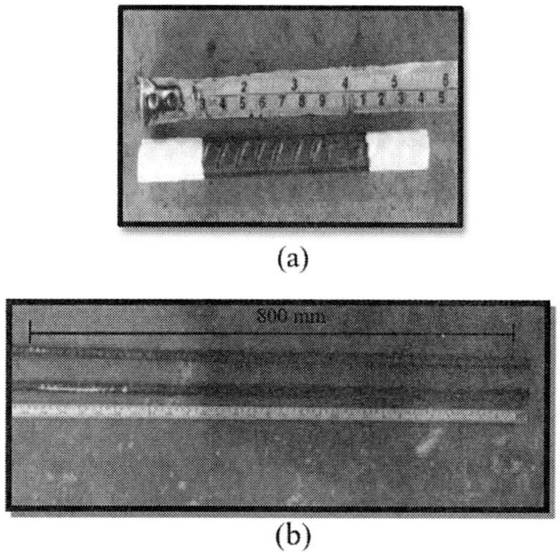

Figure 2. Dimensions (a) stirrup rebar steel (b) longitudinal rebar steel

Reinforcing steel that acts as a cathode material was carbon steel, according to SNI 07-02-2052-2002 [5]. The reinforcing steel was 13 mm in diameter, with stirrup rebar steel 140 mm in length and longitudinal rebar steel 800 mm in length. The reinforced steel can be seen in figure 2.

2.2 Specimen preparation

2.2.1 Mix proportion. The concrete specimens for this study were prepared with mix proportions K-250, this is due to ACI 211.1-91 [6]. The mix proportions of concrete are shown in Table 1.

IOP Conf. Series: Materials Science and Engineering **796** (2020) 012013 doi:10.1088/1757-899X/796/1/012013

Table 1. The proportion of mixed concrete in 1 m^3

No.	Material	Unit (Kg)	Information
1	Portland cement	373.8	-
2	Coarse Aggregate	923	Passes 9.52 and Detained 2.38
3	Fine Aggregate I	335.6	Passes 4.76 - Remain
4	Fine Aggregate II	419.6	Passes 1.20 - Remain
5	Water	228	-

2.2.2 Specimen size. Three types of specimens were used: connected anode-cathode, non-connected anode, and anode-cathode connected on first reinforcement. The geometry of the specimen is based on previous study [7]. The specimens were 100 cm in length, 10 cm in width, and 10 cm in height as shown in figure 4.

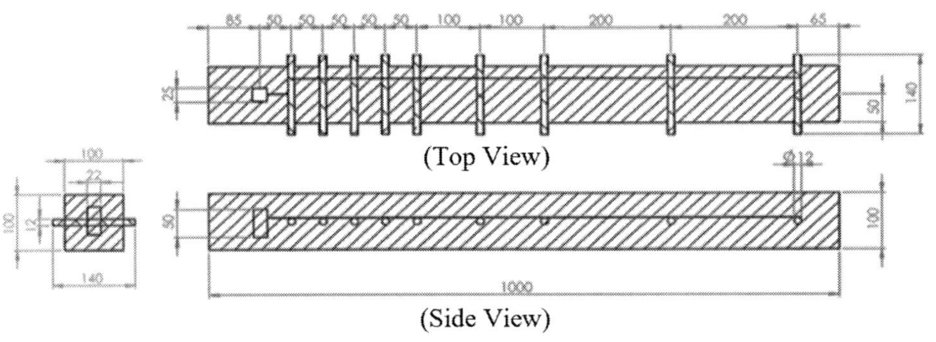

Figure 3. Dimensions of specimen connected anode-cathode (mm)

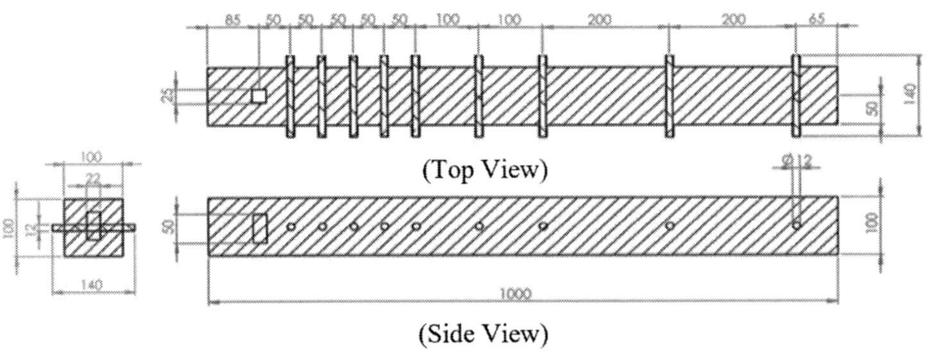

Figure 4. Dimensions of specimen unconnected anode-cathode (mm)

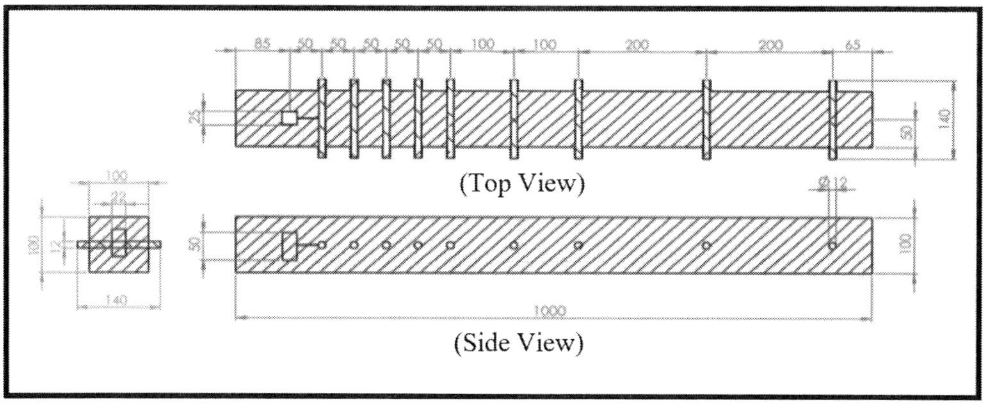

Figure 5. Dimensions of specimen connected anode-cathode on first reinforced steel (mm)

2.3 Testing

2.3.1 Determine the location of rebar steel. In this study, Profometer5+ was used to locate the reinforced steel without damaging the surface reinforced concrete. Figure 6 shows the profometer 5+ used in this study.

2.3.2 Half-cell potential measurement. The measurement of the corrosion potential value by using a Digital Half-Cell Potential Meter SCRIBE DHC (PC1018). This measurement process refers to ASTM C876 [8]. The measurement of half-cell potential on reinforced concrete can be seen in figure 7. To measure the potential values, the specimen was immersed and removed from soaking medium of 3.5% NaCl. Then the surface of specimen with exposed reinforced steel was cleaned from the remnants of corrosion products. This reinforced steel was used for electrical connection between specimens and half-cell potential meter.

Figure 6. Profometer 5+

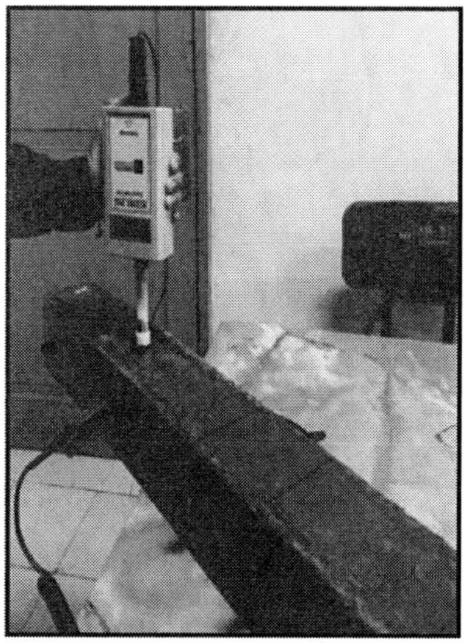

Figure 7. Half-cell potential measurement

The reference electrode was then connected to the cleaned reinforcing steel. Digital Half-Cell Meter was used to measure the corrosion potential and the measured number was taken at the point of stabilization (when the displayed number is close to zero). A thin foam was placed at the surface point of the specimen. This foam was used as a connector to the tip of the reference electrode (RE).

3. Results and Discussion

Corrosion potential value that measured on the surface of reinforced concrete was used to determine the level of corrosion risk which occurs in reinforced steel. ASTM C876 is a standard to interpret corrosion potential value data. An excerpt of ASTM C876 is shown in Table 2. Sacrificial anode cathodic protection acceptance criteria to protect reinforcing steel from corrosion standard refers to NACE RP-0169,2002 [9].

Table 2. The ASTM C876 criteria for corrosion risk of reinforced concrete.

Reference Electrode ($Cu/CuSO_4$)	Reference Electrode (Ag/AgCl)	Corrosion Risk
≥ -200 mV	≥ -106 mV	Low (10% risk of corrosion)
-200 to -350 mV	-106 to -256 mV	Intermediate corrosion risk
≤ -350 mV	≤ -256 mV	High (<90% risk of corrosion)
≤ -500 mV	≤ -406 mV	Severe corrosion

The measurement results on all specimens starting from week 0 show that corrosion did not occur at all, this is due to the corrosion potential above -200 mV (ASTM C876-91 16). As the wet-dry cycle was carried out on all specimens for 30 weeks, the corrosion potential was getting negative slightly every week.

Figure 8 illustrates the line graph of the distribution of corrosion potential in unconnected anode-cathode specimens. The distribution of corrosion potential was getting negative every week of measurement. In week 24th, the corrosion potential value does not tend to increase. The results of measurements for 36 weeks on the unconnected anode-cathode specimens show that these specimens categorized into high-risk corrosion rates as can be seen in Table 2.1 ASTM C876-91 [8]. This happens because the specimen is not connected to the anode.

Figure 9 shows the line graph of the distribution of corrosion potential values in the connected anode-cathode specimen. Corrosion potential become more negative gradually every week. On the 30th week, the corrosion potential of all test points had entered the protected criteria according to NACE RP-0169, 2002 [9].

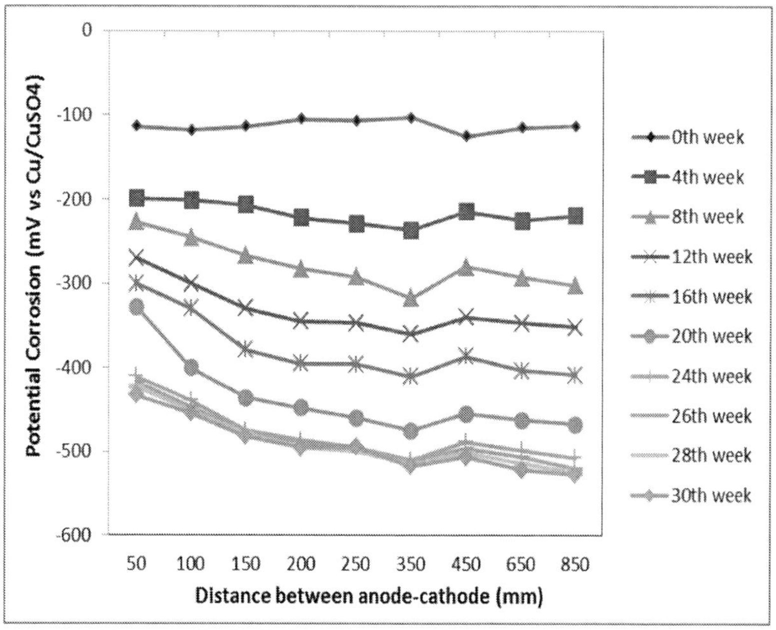

Figure 8. corrosion potential distribution (unconnected anode-cathode)

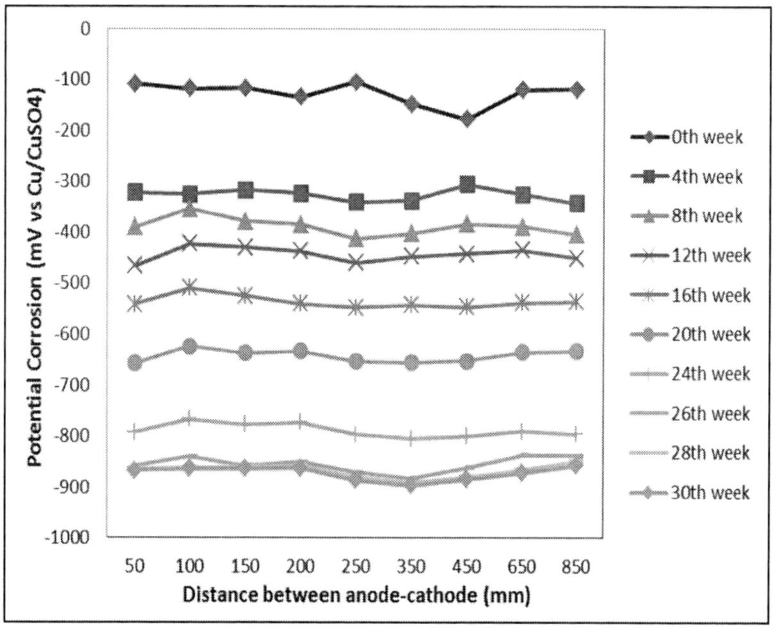

Figure 9. corrosion potential distribution (connected anode-cathode)

Figure 10 presents the line graph distribution of corrosion potential in anode-cathode connected on first reinforced steel. In this figure, the corrosion potential was uneven. This is due to partial connection where only the first test point (50mm from anode) is connected to the anode. Test point 1 of this specimen falls into the protected criteria. Test point 2 of this specimen (100 mm from anode) also had corrosion potential in the protected criteria. However, this test point had corrosion caused by stray current.

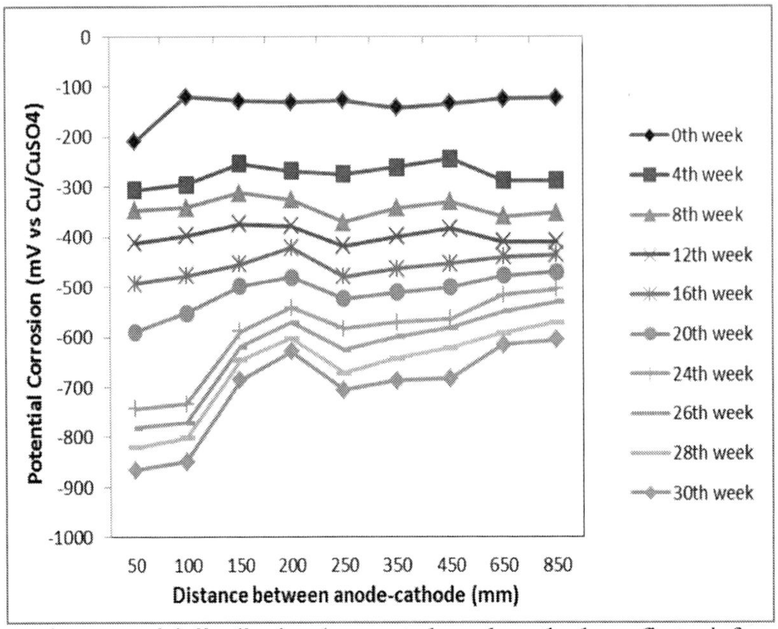

Figure 10. corrosion potential distribution (connected anode-cathode on first reinforced steel)

Due to stray current on test point 2, the current from anode went to the third test point until ninth, but the current that originated from anode did not create a loop of current for sacrificial anode cathodic protection. As a result, third test point up to ninth test points had severe corrosion rates. The phenomenon of stray current causes corrosion potential of the specimen to be more negative than the unconnected anode-cathode specimens. The uneven distributions of corrosion potential at test point 2 to test point 9 is thought to be caused by permeability of the concrete.

4. Conclusion

The phenomenon of stray current occurred on reinforced concrete that was not connected to the anode. The stray current causes the unconnected steel anode to be at severe corrosion risk [8]. The steel with stray current has a more negative corrosion potential than unconnected anode-cathode. At the 30th week, the unconnected anode-cathode specimen already showed corrosion potential at a high level (< -500 mV), according to ASTM C876-91. On 30th week, the corrosion potential on connected anode-cathode specimen had reached the protection criteria value according to NACE RP-0169. The use of Zn anode connected to the cathode may protect reinforced concrete from corrosion as expected and prevents the stray current phenomenon.

References

[1] Chess P M 1998 Cathodic Protection of Steel in Concrete (*Copenhagen: Cathodic Protection International*).

[2] Fonna S, Ridha M, Huzni S, Walid W, Mulya T and Arifin A 2017 Corrosion Risk of RC Buildings after Ten Years the 2004 Tsunami in Banda Aceh – Indonesia *Proc Eng.*

[3] Bertolini L, Carsana M, Pedeferri P 2007 Corrosion behavior of steel in concrete in the presence of stray current *Corrosion Science* **49** 1056-1068.

[4] ASTM B418 2006 Standard specification for cast and wrought galvanic zinc anodes *Annual Book of ASTM Standard.*

[5] SNI 07-2052 2002 Baja Tulangan Beton (*Indonesia: Badan Standarisasi Nasional*)

[6] ACI 211.1-91 2002 Standard Practice for Selecting Proportions for Normal *Heavyweight, and Mass Concrete.*

[7] Mahasiripan A 2014 A study of Different Sacrificial Anode Materials to Protect Corrosion of Reinforcing Steel in Concrete Thammasat Int J Sci Tech **19**.

[8] ASTM C876-91 R99 2005 Standard Test Method for Half-Cell Potentials of Uncoated Reinforcing-Steel in Concrete (*West Conshohocken, PA: ASTM*).

[9] NACE 2002 RP-0169 2002 (*Texas: NACE International*).

IOP Conf. Series: Materials Science and Engineering **796** (2020) 012014 doi:10.1088/1757-899X/796/1/012014

Layout Lean model of production management based on change management to improve efficiency in the production of packaging in auto parts sector SMEs

J Mancilla-Escobar[1], I Maccasi-Jaurequi[1], C Raymundo-Ibañez[2,4] and M Perez[3]

[1] Ingeniería Industrial, Universidad Peruana de Ciencias Aplicadas (UPC), Lima 15023, Perú.
[2] Dirección de Investigación, Universidad Peruana de Ciencias Aplicadas (UPC), Lima 15023, Perú.
[3]Escuela Superior de Ingeniería Informática, Universidad Rey Juan Carlos, Mostoles, 28933, España.

E-mail: carlos.raymundo@upc.edu.pe

Abstract. This investigation proposes a lean model of production management for a small and micro enterprise (SME) dealing in packaging production in the auto parts sector, with the purpose of increasing efficiency in its production line. To this end, this model integrates study method tools and SMED, seeking to reduce waiting times and unnecessary operations, which have been identified as causing the problem on the basis of an initial diagnosis of the situation. Furthermore, to support and guarantee success in implementing these tools, John Kotter's Change Model is used. After two months of implementation, the results showed a 50% reduction in setup operation time and a 32% decrease in filing operation time, which are bottlenecks in the production process. Thus, with the new indicators, the line efficiency increased by 7% from the initial situation, thus showing improvement.

1. Introduction

In recent years, the auto parts sector has grown commercially around the world. The exports in this sector rose to US $953661 million during 2016, with the United States, Germany, and China as major exporters accounting for exports worth US $413947 million. In Latin America, Mexico, Brazil, and Venezuela are the largest suppliers, totaling US $150320 million. In terms of auto parts export ranking, Peru ranked 52nd in the world and 5th in Latin America, with total sales of US $107 million in 2016 [1]. Later, in 2018, despite the Peruvian automotive trade falling by −3.79% as a result of fewer vehicle sales in the production line of cars, buses, and heavy vehicles, auto parts exports recorded an increase of 90.97% from the sale of tires and spare parts [2]. However, small and micro enterprises (SMEs), who are responsible for most of the exports in this sector, continue to have efficiency and productivity problems in their processes, resulting in low competitiveness and hard business growth. [3]

The SMEs in the auto parts industry together with SMEs from other industries generate roughly 40% of the national GDP, meaning they are the largest drivers of economic growth in Peru [4].

[4] To whom any correspondence should be addressed.

Content from this work may be used under the terms of the Creative Commons Attribution 3.0 licence. Any further distribution of this work must maintain attribution to the author(s) and the title of the work, journal citation and DOI.
Published under licence by IOP Publishing Ltd

Furthermore, they contribute significantly to generating employment for approximately 80% of the labor supply [5] because currently 96.5% of existing companies in Peru belong to the SME sector [6].

This study aims to enable the MSEs of the auto parts sector to solve or reduce problems with respect to production line efficiency. Currently, these problems are caused by high levels of wastage in down times and unnecessary operations that are generating hidden production costs. Moreover, in this sector, employers and workers generally have low qualifications [7]; this situation results in a lower probability of generating profits that provide the possibility of reinvestment in process improvement, thus contributing to an increase in the MSE mortality rate [8]. Therefore, through this article, a production management model based on the use of change management and lean tools is proposed, which through their integration seeks to generate new work methods based on understanding the importance of the need for change for improving economic indicators. As a consequence of this, activities that do not add value can be reduced using the SMED technique and studying the methods will allow reducing the setup time and the time of the filing operation, the latter being bottlenecks in the production process because it is poorly performed.

2. State of the art

2.1. Production management models based on change management.

Many organizations fail to position themselves better in the market because of the lack of a development program within their improvement policies. In addition, when it is implemented, in some cases, expectations are not met because the most ideal method or strategy of change management is not adopted for that organization, resulting in unwanted change [9] [10]. Other results explain that the organization managers differ significantly in their transformational leadership and management parameters with respect to their type of approach to change and organizational structure, the latter being those who will define the degree of contribution of the former in the organization [11] [12]. However, although change management practices have a significant impact on fulfilling objectives in the change program, they do not appear to have an impact on organizational results such as financial results, operational productivity, etc. [13]

2.2. Production management models based on lean manufacturing

Some researchers claim that the implementation of the lean manufacturing philosophy is more successful in larger companies than in SMEs, especially in less developed countries. This is attributed to factors such as the lack of commitment and management leadership, poor communication, and scarce resources. These are considered as barriers to lean adaptation in SMEs [14] [15] [16]. However, other researchers claim that although these factors are present and they represent difficulties in using lean tools, the adaptation of techniques such as TPM, statistical process control, SMED, production leveling and others have a more positive impact on operational performance, obtaining good results even if partially implemented [17] [18]. Furthermore, it has been suggested that in order to propose a successful production model based on lean methodology, it is essential to have as a starting point the elimination of wastage, that is, activities that do not add value to the process. For this purpose, the implementation of the value stream mapping (VSM) technique is recommended; it is a tool that analyzes process inefficiencies and serves as a basis for a lean transformation in the company [19] [20]. However, there are those who consider that the use of overall equipment efficiency (OEE), which is an approach that focuses on operations rather than on the process, serves as an orientation indicator for prioritizing the improvements to be made and the operational control [17].

3. Contribution

This stage will show the contribution according to the papers chosen for each type, where the integration of the contributions for each one of them will be graphed and explained for the construction of the proposed model.

IOP Conf. Series: Materials Science and Engineering **796** (2020) 012014 doi:10.1088/1757-899X/796/1/012014

3.1. Proposed Model

The proposed model shown in Figure 1 is based on the lean philosophy complementing the study methods to reduce the wastage during the production process in auto parts SMEs using John Kotter's change management as a methodology for its implementation with a view to improving efficiency in the production line, through worker leadership.

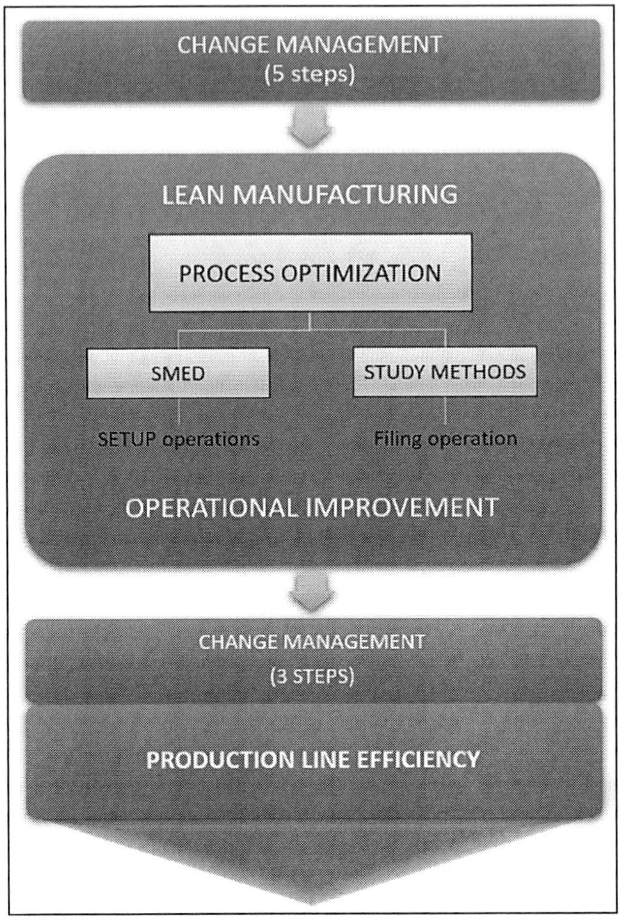

Figure 1. Lean model for production management based on change management by John Kotter

Phase 1 involves generating change in the company through John Kotter's Change Model, whose first five steps are used to create a sense of urgency, form a coalition, develop a vision, communicate the vision to the whole organization, and eliminate obstacles because in order to implement the lean methodology it is essential that workers are convinced about accepting it.

Once the obstacles are eliminated and the workers' acceptance has been achieved, Phase 2 of lean manufacturing will be implemented. The status of the current situation must first be ascertained through VSM and OEE, among other tools, in order to analyze the causes of the inefficiency in the process line. Although this model only applies one lean technique, SMED, the proposed model can be applied to all techniques and its choice will depend on the analysis of causes.

Finally, after Phase 2, in which operational improvement is carried out, the last three steps of change management are implemented, which are to ensure short-term goals and build on the change achieved to ensure a cycle of continuous improvement and anchor the culture of change in the company.

3.2. Proposed Process.

To implement the proposed model, follow the steps shown in the implementation guide, see Figure 2.

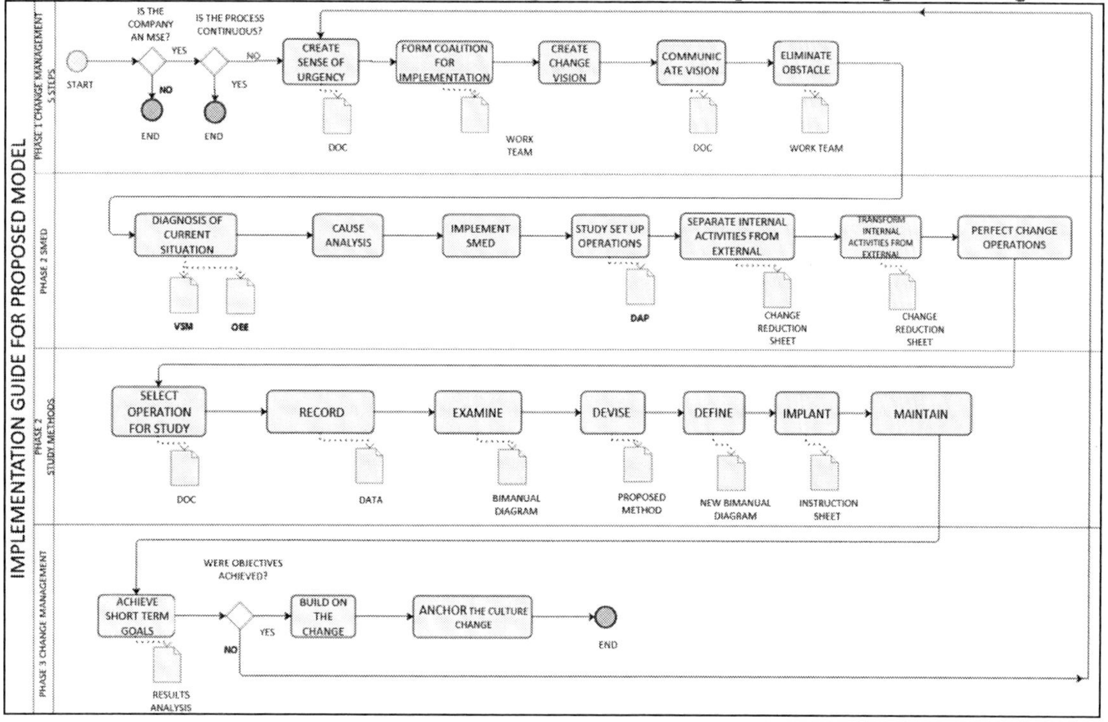

Figure 2. Implementation guide for proposed model

3.3. Indicators

3.3.1. Resistance to change. The purpose of this indicator is to assess the degree of resistance to change existing in the company.

$$Resistance = \frac{change - resistant\ answers}{Number\ of\ workers} \quad (1)$$

The result will indicate the percentage of workers who resist the implementation of the proposed model.

3.3.2. Reduction in filing time (%). The purpose of this indicator is to compare the data before and after the improvement to see the variation in the time it takes to execute the bottleneck operation.

$$Reduction = \frac{Previous\ time\ -\ current\ time}{Previous\ time} \times 100\% \quad (2)$$

The result indicates the percentage by which the time of the filing operation has been reduced with respect to its initial value before the improvement occurred.

3.3.3. Reduction in setup time. As in the previous case, this indicator will compare the initial and final values to see the variation in setup time before and after the improvement.

$$Reduction = \frac{Previous\ time - current\ time}{Previous\ time} \times 100\% \tag{3}$$

The result indicates the percentage of time taken for the filing operations that has been reduced compared to its initial value before the proposed model was produced.

3.3.4. Online eficiency. The efficiency of the production line indicates the percentage of time in which value has been added to the product, that is, the total added value time.

$$L.\,Efficiency = \frac{total\ time\ of\ all\ operations}{No.\,stations \times cycle\ time} \tag{4}$$

The result will indicate the variation of the initial and final values, after having implemented the model.

4. Validation

4.1. Description of the case study.
To validate the proposed model, the company Empaquetaduras Montalvo EIRL, which is an SME dedicated to the manufacture and commercialization of parts, pieces, and accessories for vehicles has been taken as a case study. This company is located in Tablada de Lurín, district of Villa María del Triunfo in Lima, Perú.

Figure 3. Stamping station of the company

4.2. Initial Diagnosis

When making a diagnosis about the situation of the company, it was found that production line efficiency was 52% and that of the sector was 64% as shown in Figure 4; this difference put the company at a disadvantage with its competitors with respect to its ability to deliver to its customers on time, given that the takt time or the rate at which the market demanded products is 3.50 min / packaging time. On analyzing the causes of this problem, there was evidence that the main reasons were wastage in production; for example, the filing operations were at 32.8%, being carried out with a deficient method along with bottlenecking with a very high time of 36.67 min / lot packing, as well as setup time being 30.33%, recorded as 30 min / batch change. The time in which value was not added to the product in these operations was 12 min / lot packaging in the finished product and 20 min / batch change in setup, according to the ideal of 10 min / batch change. Given that the company works in a single 8-hours shift and the amount of batch changes on average is 6 lots per day, the time lost due to all causes, represented approximately 2.8 hours / day. Each unit is sold at S /. 2.00 on average. The scenario described meant that the company kept production costs hidden, entailing an opportunity cost of approximately S /. 44,000 per year, where the economic impact of causes referred to as "others" with a lower percentage was also considered.

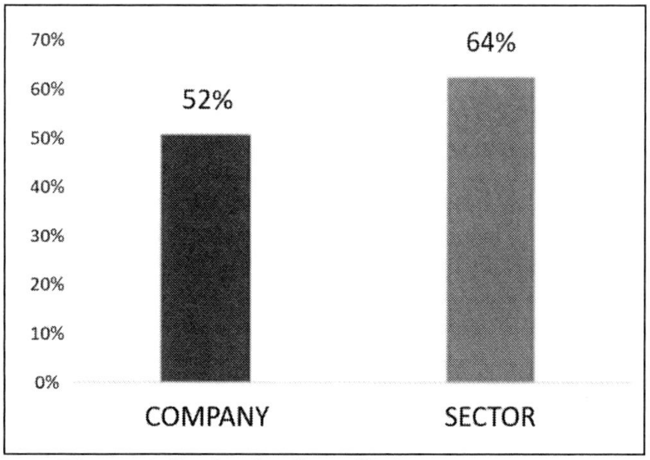

Figure 4. Production line efficiency versus sector

4.3. Application of model to the case study.

4.3.1. Change management

The start of the proposed model, applied in the company under study, was carried out through change management based on John Kotter's plan. Establishing a sense of urgency, workers were trained with the vision of being a benchmark company in the sector through process efficiency and communication with the organization. A coalition was formed by three workers who were in charge of leading the change. Training was carried out to eliminate or reduce resistance to change, which was considered as an obstacle to implementation, based on an understanding of the importance of the issues that the company had and how these affected it, see Figure 5

Figure 5. Kotter's change management skills

We sought to ensure success, implementing SMED improvement tools and study methods for which a milestone diagram was developed in order to control the results achieved.

4.3.2. SMED

To implement the SMED tool, a work duo was formed that received training on the tool and that was responsible for carrying out the exchange operations. After studying the setup operations through a DAP, internal and external operations were identified with the time that each of them took. Following the steps of SMED, both types of activities were separated and then the internal and external activities that were possible were transformed. To reduce setup time, a new operator joined the process, having previously analyzed his workload. This allowed the internal activities to be reduced from approximately 30 minutes to 15 minutes. Specifically, the second operator helped with the activities of finding tools, stamps, cleaning materials, and change, for which he used the help of a transporter cart, see Figure 6.

Before **After**

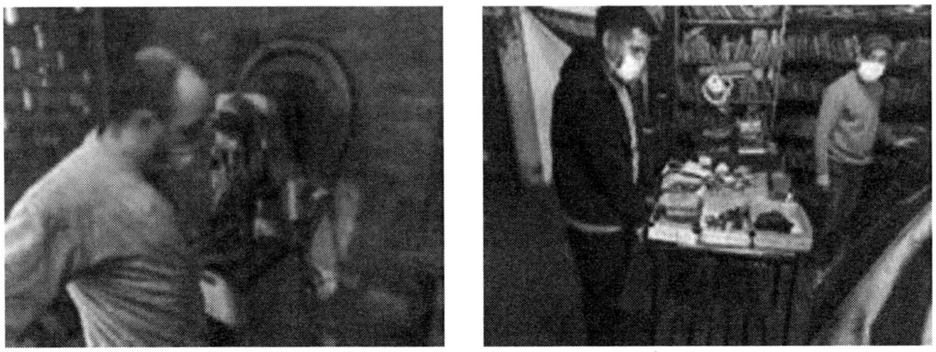

Figure 6. Before and after improvements in setup

4.3.3. Study of methods

The tool for the study of methods in the finished product operation was also implemented. When performing the 7 steps of this technique, the above operation was selected because it was a bottleneck; its activities were examined and recorded and time delayed by a bimanual diagram, as they were activities that did not add value to the product but were done because they believed that higher quality was added. These activities were basically repetitive but unnecessary, such as turning the ring around the filing axis and re-holding it. However, by not having a clear and defined method for this operation, the person in charge of performing this task obtained high and variable times. Therefore, a new method was developed, where unnecessary activities were eliminated and implanted in that workstation, see Figure 7.

Before **After**

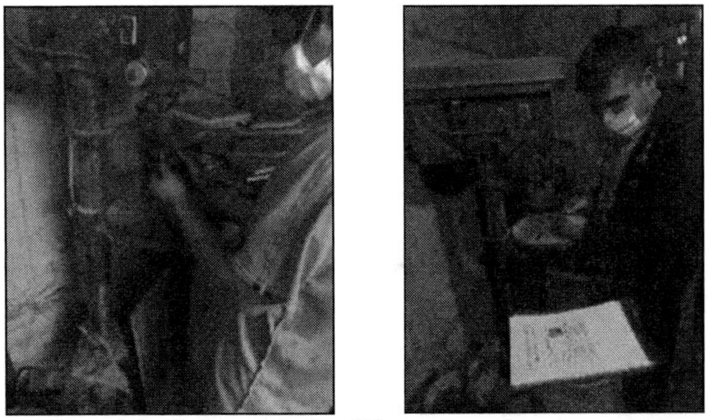

Figure 7. Before and after filing process improvements

4.4. Results

4.4.1. Resistance to change

$$Resistance = \frac{0}{5} \times 100 = 0\% \tag{5}$$

The proposed model obtained positive results, the resistance to change was eliminated because initially, according to the survey conducted, there was a resistance of 15%, but after providing the training, the workers accepted the change completely.

4.4.2. Reduction in filing time (%)

$$Reduction = \frac{(37 - 25)}{37} \times 100 = 32.4\% \tag{6}$$

After implementing the model, by means of the study methods the bottleneck timings and the filing operation were reduced by 32.4%. This allowed the cycle time to be reduced as waiting time.

4.4.3. Reduction in setup time

$$Reduction = \frac{(30 - 15)}{30} \times 100 = 50\% \tag{7}$$

Likewise, with the implementation of SMED, the setup time was reduced by 50%, with respect to the start of the model implementation. This improvement allowed to streamline the process of setting up the press, thereby making the machine more available to produce more work orders.

4.4.4. Production line efficiency

$$New\ production\ line\ efficiency = \frac{74}{5 \times 25} \times 100 = 59\% \tag{8}$$

Finally, as a result of the improvements achieved by the implemented model, the production line increased 7% with respect to its initial condition. However, it still differs from the sector but by a lower percentage, 5%.

5. Conclusions

- The proposed model allowed the efficiency of the production line of the case study to increase by 7% through a 32% reduction in filing time, the bottleneck process, and 50% setup time.
- It is estimated that the improvement achieved will generate savings for a sum of approximately S / 30,000 per year by reducing wastage due to waiting times in the production area.
- The four scenarios that were validated show a general improvement in line efficiency, which allows the feasibility of the proposed model to be applied to other organizations in a similar condition.
- The success of the proposed model is based on the integration of two models, the lean model and change management by John Kotter that focus on operational techniques and on the promotion of human development with growth perspectives, which together make up a strong model.

6. References

[1] Services to the Exporter DdIdM 2017 Specialized Report: Auto parts in the world (*Lima*).
[2] INEI 2019 Technical Report: National Production (*Lima*).
[3] Mares C 2013 Micro and Small Enterprises (MYPEs) in Peru *Juridical Magazine Thomson Reuters* **I**(18) 1-20.
[4] Imagen UdCe. Foncodes. 2018 Retrieved from: http://www.foncodes.gob.pe/portal/.
[5] Barraza BJS 2014 THE MYPES IN PERU. ITS IMPORTANCE AND TAX PROPOSAL *Quipukamayoc* **13**(25) 127-131.
[6] Economy. PQS.pe: The Entrepreneurs' Portal. 2018 Retrieved from: https://www.pqs.pe/.
[7] Arbulú J, Otoya J 2006 SMEs in Perú *PAD. REVISTA DE EGRESADOS* **7** 32-37.
[8] General OGdEdIyEEdlS 2017 STUDY OF THE CURRENT SITUATION OF PERUVIAN COMPANIES (*Lima*).
[9] Al-Haddad S, Kotnour T 2015 Integrating the organizational change literature: A model for successful change *J Organ Change Manag* **28**(2) 234-262.
[10] Skalik J 2016 Strategic orientation in change management and using it when designing a company's development *DE GRUYTER OPEN LTD* **20**(1) 197-210.
[11] Gryazeva-Dobshinskaya VG, Dmitrieva YA 2018 Leadership and management styles: Typological approach to personal resources of change management *International Scientific Conference, FarEastCon (Vladivostok).*
[12] Riquero I, Hilario C, Chavez P, Raymundo C 2019 Improvement proposal for the logistics process of importing SMEs in Peru through lean, inventories, and change management *Smart Innovation, Systems and Technologies* **140** 495-501.
[13] Ranieri AB 2011 Change management practices: Impact on perceived change results *J Bus Res* **64**(3) 266-272.
[14] Yadav V, Rakesh J, Mittal ML, Panwar A, Sharma MK 2019 An appraisal on barriers to implement lean in SMEs. *J Manuf Tech Manag* **30**(1) 195-212.
[15] Bhasin S 2012 Performance of Lean in large organisations *J Manuf Syst* **31**(3) 349-357.
[16] Belhadi A, Touriki FE, El Fezazi S 2018 Lean Implementation in Small and Medium-Sized Enterprises in Less Developed Countries: Some Empirical Evidences From North Africa. *J Small Bus Manag* **56**(S1) 132-153.
[17] Dresch A, Rafael Veit D, Nascimento de Lima P, Pacheco Lacerda D and Cisco Collatto D 2019 Inducing Brazilian manufacturing SMEs productivity with Lean tools. International J Produc Perfor Manag **68**(1) 69-87.
[18] Yadav V, Jain R, Mittal ML, Panwar A, Lyons A 2019 The impact of lean practices on the operational performance of SMEs in India *Ind Manag Data Syst* **119**(2) 317-330.
[19] Ben Fredj-Ben Alaya, L 2016 VSM a powerful diagnostic and planning tool for a successful Lean implementattion: a Tunisian case study of an auto parts manufacturing firm. *Prod Plan Control* **27**(7-8) 563-578.
[20] Goshime Y, Kitaw D, Jilcha K 2018 Lean manufacturing as a vehicle for improving productivity and customer satisfaction: A literature review on metals and engineering industries. Int J Lean Six Sigma **10** (2) 691-714.

Theoretical Study of the Absorption Refrigeration Cycle Using Water-Lithium Bromide as Working Pair for Cold Storage Application

Sarwo Edhy Sofyan [1*], Muhammad Farhan[1], Khairil[1], Jalaluddin[1], Akram[1]
[1]Department of Mechanical and Industrial Engineering, Syiah Kuala University Darussalam – Banda Aceh 23111, Indonesia

E-mail: sarwo.edhy@unsyiah.ac.id

Abstract. This study aims to investigate the performance of a cooling absorption of the water-lithium bromide cycle for a cold storage application which has a capacity of five tonnes of fish. The cooling absorption system utilizes the waste heat of a 6 MW steam power plant as the heat source for the generator. The thermodynamics analysis is carried out to determine both the heat rejection rate of the power plant and the performance of the refrigeration absorption cycle. The performance of the cooling system is investigated by varying the temperature of each component of the refrigeration absorption cycle (i.e generator, absorber, condenser, and evaporator). Also, the influence of temperature on the heat input (to the generator), heat output (from the absorber), and the pump power, is examined. The results show that the Coefficient of Performance (COP) of the refrigeration absorption system increases when the temperature of the absorber, condenser, and evaporator rises. On the contrary, increasing the temperature of the generator leads to the reduction of the COP of the system. It is also found that the pump power increases when the temperature of the absorber rises, while on the contrary decreases when the generator temperature increases. The results also reveal that the amount of heat input into the generator is directly proportional to the temperature of the generator. While the heat output of the absorber decreases when the temperature of the absorber increases.

1. Introduction

The vapor absorption cycle is one of the commonly used cooling systems besides vapor compression. The advantage of the vapor absorption cycle is that it uses heat to increase pressure. The vapor absorption cycle utilizes the waste heat as the input energy of the generator allowing the condition for the weak solution changes to produce the cooling effect. This technology can significantly reduce the electricity used for the cooling system, thereby, reducing the operational cost.

There are several studies on the absorption refrigeration system carried out by the researchers, including the theoretical [1-2] and experimental [3-4]. Manu and Chandrashekar [1] conducted a theoretical study on the single-stage lithium bromide-water vapor absorption heat pump. Maji [2] investigated the performance of solar-assisted absorption refrigeration which includes the ejector into the system with the Lithium bromide-water as the working pair. An experiment was conducted by

* Corresponding author: sarwo.edhy@unsyiah.ac.id

Content from this work may be used under the terms of the Creative Commons Attribution 3.0 licence. Any further distribution of this work must maintain attribution to the author(s) and the title of the work, journal citation and DOI.
Published under licence by IOP Publishing Ltd

Hanriot et al., [3] to study the effect of the technical parameters on the performance of the ammonia-water refrigeration cycle used automotive exhaust gas as the source heat.

This hypothetical study investigates the performance of a water lithium bromide absorption refrigeration, using the waste heat of a 6 MWe steam power plant, for cold storage application. The cold storage has a capacity of five tons of fish. The study is conducted base on the thermodynamic analysis. The performance of the absorption refrigeration cycle is examined considering the various temperatures of each component of the system.

2. Research Methodology

2.1 Cold Storage

The cold storage design is based on Charlie's work [5]. The cold storage has a capacity of five tons of fish. It has dimensions of 6 m long, 2.5 m wide, 2.6 m high and 0.194 m thick. The cold storage cooling load is 115.28 kW. Table 1 shows the design parameters of the cold storage and Tables 2 presents the materials of the cold storage and their thermal conductivities.

Table 1. Design parameter of the cold storage.

Design Parameter	Value
Ambient air temperature (°C)	27.5
Outdoor Air Relative Humidity (%)	75
Wind Speed (mph)	7.5
Indoor air temperature (°C)	-7
Indoor Air Relative Humidity (%)	90

Table 2. Wall materials and their thermal conductivities.

Material	Thickness (m)	Thermal conductivity (Kkal/h °C)	Material	Thickness (m)	Thermal conductivity (Kkal/h °C)
Wall			**Ceiling**		
Carbon steel	0.002	30.95	Carbon steel	0.002	30.95
Glass wool	0.03	0.032	Glass wool	0.045	0.032
Polystyrene	0.05	0.033	Polystyrene	0.1	0.033
Air gap	0.11	0.095	Air gap	0.21	0.095
Aluminium	0.002	176	Aluminum	0.02	176
Floor			**Door**		
Asphalt shingles	0.005	0.053	Carbon steel	0.002	30.95
Pine wood board	0.005	0.126	Glass wool	0.03	0.032
Polystyrene	0.1	0.033	Polystyrene	0.031	0.033
Glass wool	0.04	0.032	Air gap	0.03	0.095
Carbon steel	0.004	30.95	Aluminum	0.002	176

2.2 Heat Source

In this study, the heat source used as the heat input to the generator comes from the waste heat of a 6MWe steam power plant. Figure 1 shows the schematic of the steam power plant and the Rankine cycle.

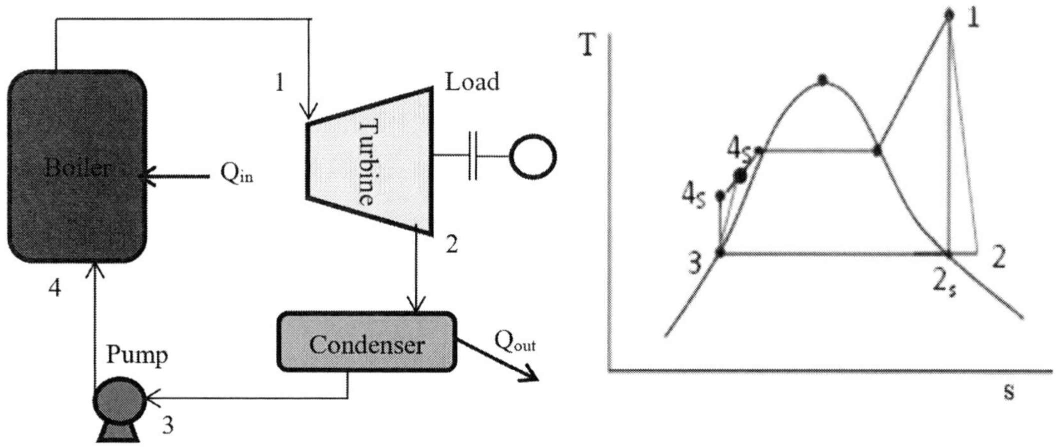

Figure 1. Schematic of the steam power and its thermodynamics cycle

The amount of heat dissipated by the condenser of the steam power plant is estimated by using the thermodynamic analysis given as follows:

$$Q_{out} = \dot{m}(h_2 - h_3) \quad (1)$$

In which, Q_{out} is the heat released by the condenser (kW), $\dot{m}$ is the mass flow rate of the steam (kg/s), h_2 is the actual enthalpy at point 2 (kJ/kg), and h_3 is the saturated liquid enthalpy at point 3 (kJ/kg).

The mass flow rate of the steam is calculated as:

$$\dot{m} = \frac{\dot{W}_{cycle}}{(h_1-h_2)-(h_4-h_3)} \quad (2)$$

In which, $\dot{m}$ is the steam mass flow rate (kg/s), $\dot{W}_{cycle}$ is the net power output of the cycle (W), h_1 is the enthalpy of the superheated vapor at point 1 (kJ/kg), h_2 is the actual enthalpy at point 2 (kJ/kg), and h_3 is the saturated liquid enthalpy at point 3 (kJ/kg), and h_4 is specific enthalpy the pump exit.

The enthalpy at point 2 is determined as follows:

$$h_2 = h_1 - \eta_t(h_1 - h_{2s}) \quad (3)$$

In which, h_2 is the actual enthalpy at point 2 (kJ/kg), h_1 is the enthalpy of the superheated vapor at point 1 (kJ/kg), η_t is the steam turbine efficiency (it is assumed 85 %), h_{2s} is the ideal enthalpy at state 2s (kJ/kg).

where the enthalpy at state 2s is given as:

$$h_{2s} = h_f + X_2 . h_{fg} \quad (4)$$

In which, h_{2s} is the ideal enthalpy at state 2s (kJ/kg), h_f is the saturated liquid enthalpy (kJ/kg), X_2 is the quality of the liquid-vapor mixture, and h_{fg} is the evaporation enthalpy (kJ/kg).

The enthalpy at point 4 is:

$$h_4 = h_3 + \dot{W}_p / \dot{m} \quad (5)$$

In which, h_4 is the enthalpy at point 4 (kg/s), h_3 is the saturated liquid enthalpy at point 3 (kJ/kg), $\dot{W}_p$ is the pump power (kW), and $\dot{m}$ is the mass flow rate of the working fluid (kg/s).

The pump power over the working mass fluid flow rate is calculated as:

$$\frac{\dot{W}_p}{\dot{m}} = \frac{v_3(p_4 - p_3)}{\eta_p} \quad (6)$$

In which, $\dot{W}_p$ is the pump power (kW), and $\dot{m}$ is the mass flow rate of the working fluid (kg/s), v_3 is the specific volume of saturated liquid at point 3 (m^3/kg), p_4 and p_3 are the pressures at points 4 and 3, respectively, η_p is the pump efficiency (it is assumed 85 %).

2.3 Absorption Refrigeration Cycle

The thermodynamic analysis is carried out to examine the performance of the absorption refrigeration cycle. Based on the energy balance on each component of the absorption refrigeration cycle (i.e absorber, condenser, generator, and evaporator), the governing equations are then presented. The thermodynamic analysis of the absorption refrigeration cycle assumes a steady-state and steady flow. Besides, it ignores the pressure drop due to friction. Figure 2 shows a schematic of the absorption refrigeration cycle.

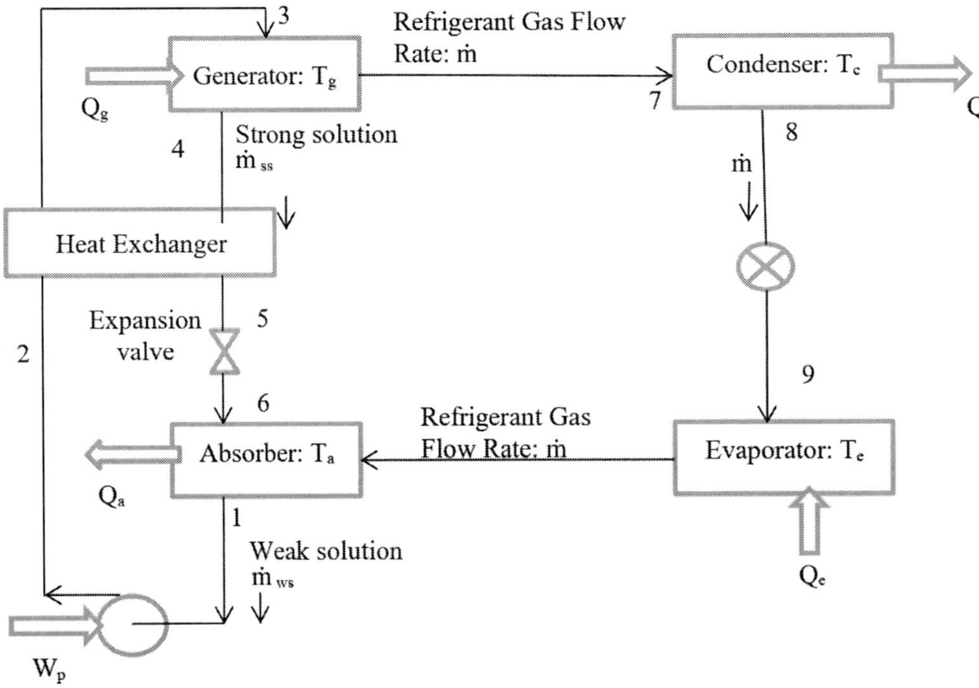

Figure 2. The schematic of the absorption refrigeration cycle.

1) Evaporator

The energy balance at the evaporator is illustrated in figure 3 and represents mathematically as follows:

$$Q_e = \dot{m}(h_{10} - h_9 \quad (7)$$

In which, Q_e is the heat transfer rate of the evaporator (kW), $\dot{m}$ is the mass flow of the refrigerant (kg/s), h_9 and h_{10} are the enthalpy at points 9 and 10 (kJ/kg), respectively.

The mass flow rate of the refrigerant can be calculated by rearranging the Equation (7)

$$\dot{m} = \frac{Q_e}{h_{10} - h_9} \quad (8)$$

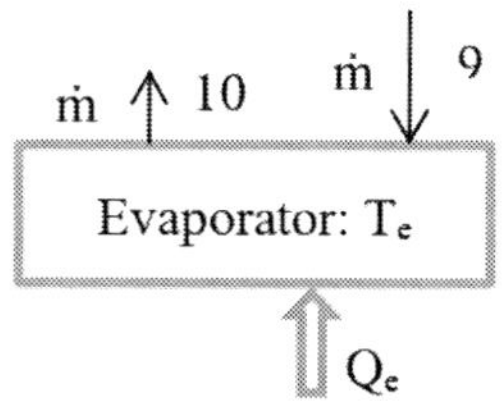

Figure 3. The energy balance at the evaporator.

2) Condenser

The energy balance at the condenser is illustrated schematically in figure 4 and presented mathematically in Equation (9).

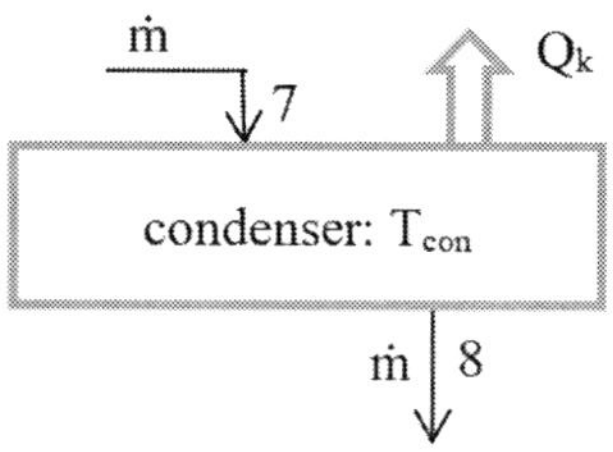

Figure 4. The energy balance at the condenser.

$$Q_c = \dot{m}(h_7 - h_8) \quad (9)$$

In which, Q_c is the heat transfer rate of the condenser (kW), $\dot{m}$ is the refrigerant mass flow rate (kg/s), h_7 and h_8 are the enthalpy at points 7 and 8 (kJ/kg), respectively.

3) Absorber

Figure 5 shows the energy balance at the absorber and mathematically is presented by Equation (10).

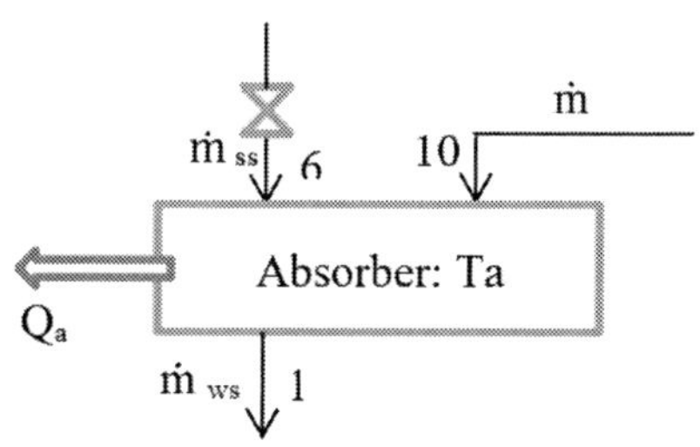

Figure 5. The energy balance at the absorber.

$$Q_a = \dot{m}h_{10} + \dot{m}_{ss}h_6 - \dot{m}_{ws}h_1 \quad (10)$$

In which, Q_a is the heat transfer rate of the absorber (kW), ṁ is the refrigerant mass flow rate (kg/s), $\dot{m}_{ss}$ is the mass flow rate of strong solution (kg/s), $\dot{m}_{ws}$ is the mass flow rate of weak solution (kg/s), h_1, h_6 and h_{10} are the enthalpy at points 1, 6, and 10.

Where

$$\lambda = X_{ws}/(X_{ss} - X_{ws}) \tag{11}$$

The λ can be also calculated as

$$\lambda = \dot{m}_{ss}/\dot{m} \tag{12}$$

The mass flow rate of the strong solution can be calculated by rearranging the Equation (12)

$$\dot{m}_{ss} = \dot{m}\lambda \tag{13}$$

While the mass flow rate of the week solution is given as

$$\dot{m}_{ws} = \dot{m} + \dot{m}_{ss} \tag{14}$$

In which, λ is the circulation ratio, X_{ws} is the percentage of LiBr in the weak solution (%), X_{ss} is the percentage of LiBr in the strong solution (%), $\dot{m}_{ss}$ is the mass flow rate of the strong solution (kg/s), $\dot{m}_{ws}$ is the mass flow rate of weak solution (kg/s), ṁ is the refrigerant mass flow rate (kg/s).

4) Heat Exchanger

A heat exchanger is usually provided to increase the efficiency of the absorption refrigeration cycle. The heat exchanger is used to cool the strong solution (returning from the generator to the absorber) by transferring its heat to the weak solution (pumping from the absorber to the generator). The energy balance of the heat exchanger is shown schematically in Figure 6 and presented as

$$\dot{m}_{ws}(h_3 - h_2) = \dot{m}_{ss}(h_4 - h_5) \tag{15}$$

Rearrange Equation (15), the enthalpy at point 3 is

$$h_3 = \frac{\dot{m}_{ss}(h_4 - h_5)}{\dot{m}_{ws}} + h_2 \tag{16}$$

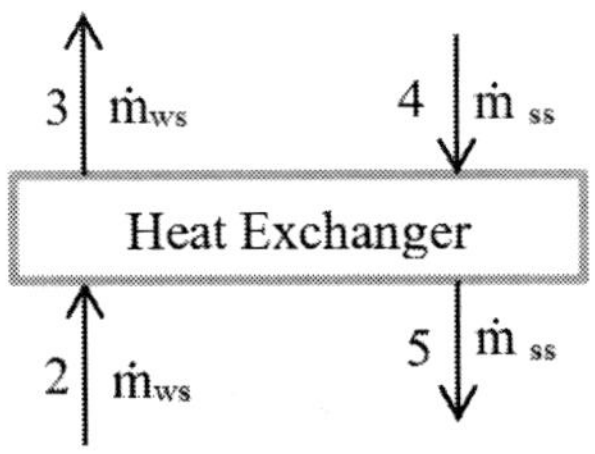

Figure 6. The energy balance at the heat exchanger.

5) Generator

The energy balance at the generator is illustrated in figure 7 and presented by Equation (17).

IOP Conf. Series: Materials Science and Engineering **796** (2020) 012015 doi:10.1088/1757-899X/796/1/012015

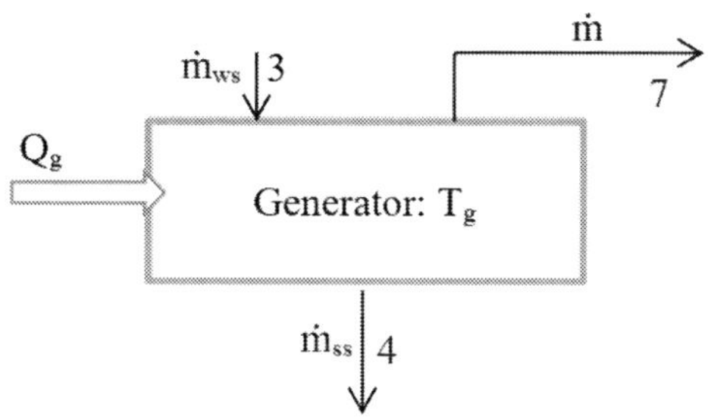

Figure 7. The energy balance at the generator.

$$Q_g = \dot{m}h_7 + \dot{m}_{ss}h_4 - \dot{m}_{ws}h_3 \tag{17}$$

6) Pump

The energy balance for the pump is shown in figure 8 and mathematically presented by Equation (18).

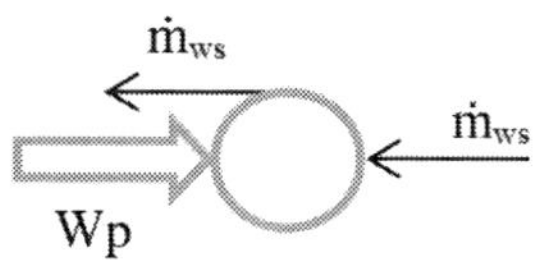

Figure 8. The energy balance at the pump.

$$W_p = \dot{m}_{ws}\,(\Delta H)/\rho \tag{18}$$

In which W_p is the pump power (kW), $\dot{m}_{ws}$ is the mass flow rate of weak solution (kg/s), ΔH is the pressure drop (kPa), ρ is the density (kg/m^3).

The density is calculated as:

$$\rho = \rho(\text{LiBr}) * X_{ss} + \rho(\text{air}) * (100 - X_{ss}) \tag{19}$$

7) Coefficient of Performance (COP)

The Coefficient of Performance of the absorption refrigeration cycle is calculated as follows

$$COP = Q_e \,/\, (Q_g + Wp) \tag{20}$$

In which, COP is the coefficient of performance of the absorption refrigeration cycle, Q_e is the heat transfer rate of the evaporator (kW), Q_g is the heat transfer rate of the generator (kW), W_p is the pump power (kW).

3. Hypothetical Study

The performance of the absorption refrigeration cycle is investigated under specific ranges of temperatures of the refrigeration system's components (i.e generator, absorber, condenser, evaporator). The influence of the absorber and generator temperatures on the power input of the pump is also examined. The simulation of the performance of the absorption refrigeration cycle is conducted initially by utilized parameters as given in table 3. Based on these reference parameters, the performance of the cooling system considering the different component temperature of the cooling system is carried out. Table 4 presents the temperature ranges used in the simulation, as suggested by Sauchin Kausik [6], for each component of the absorption refrigeration cycle.

Tabel 3. The reference temperature of the absorption refrigeration cycle components.

Parameter	Value
Temperature of condenser (°C)	40
Temperature of evaporator (°C)	2,5
Temperature of absorber (°C)	20
Temperature of generator (°C)	85

Table 4. The variable temperature of the absorption refrigeration cycle components.

Parameter	Temperature Ranges
Temperature of Generator (°C)	55-90
Temperature of condenser (°C)	24-46
Temperature of absorber (°C)	16-32
Temperature of evaporator (°C)	2.5-10

The input parameters of the governing equations are determined based on the reference temperature and state condition of the working fluid at each component of the absorption refrigeration cycle. For the cycle with lithium bromide water pair, the water serves as the refrigerant for the system. The saturation pressure for condensation in the condenser at 40 ° C can be obtained from the steam table which is equal to 0.074 bar (7,384 kPa). The generator pressure is assumed to operate at the same pressure as the condenser. The saturation pressure for saturated steam form in the evaporator at 2.5 °C is equal to 0.00737 bar (0.737 kPa), which also corresponds to the pressure at the evaporator. Based on this information, the input parameters (i.e enthalpy) used in the governing equations can be determined. The enthalpy of the saturated water and superheated water vapor at given temperatures can be determined from the steam table. While the enthalpy and concentrations of the mixed solution are obtained from the LiBr-Water graph, pressure-temperature-concentration-enthalpy (P-T-X-h). As an example, the table 5 presents input parameters for the governing equations at reference component temperatures (see table 3).

Table 5. Input parameter at the reference temperature of each component.

Point	Temperature	Pressure		Enthalpy	LiBr Concentration
	(°C)	(kPa)	(mmHg)	(kJ/kg)	%
1	20	0.737	5.53	-178	50
2	20	7.384	55.50	-178	50
3	-	7.384	55.50	-	50
4	85	7.384	55.50	-80	59
5	20	7.384	55.50	-187	59
6	20	0.737	5.53	-187	59

IOP Conf. Series: Materials Science and Engineering **796** (2020) 012015 doi:10.1088/1757-899X/796/1/012015

7	85	7.384	55.50	2651.9	-
8	40	7.384	55.50	167.57	-
9	40	0.737	5.53	167.57	-
10	2,5	0.737	5.53	2505.89	-

4. Result and discussion

4.1. Analysis of potential heat output of steam power plant used for the absorption refrigeration cold storage

In designing cold storage with an absorption refrigeration cycle, it is necessary to calculate the amount of the heat input to the generator to ensure the cold storage properly operates. In this study, the heat supplied to the absorption refrigeration cycle is extracted from the waste heat of a 6MWe steam power plant. The heat output of the power plant is analyzed thermodynamically based on the parameters provided by the Siemen SST-200 steam turbine, as shown in table 6.

Tabel 6. Specifications of Siemen SST-200 steam turbine [7].

Power output	6 MWe
Inlet pressure	120 bar
Inlet temperature	540 °C
Back Pressure	2 bar
Efficiency	85%
Speed	14.600 rpm

From the calculation, it is obtained that the waste heat the steam power plant corresponds to 16.2 MW. According to Osama [8], the potential for waste heat can be used by the absorption refrigeration cycle ranges from 58-62%. These values correspond to 9.2-9.9 MW of the waste heat of the power plant.

4.2. The Effect of generator temperature on the COP of the absorption refrigeration cycle

Figure 9 shows the effect of the generator temperature on the COP of the absorption refrigeration cycle. It is found that the COP of the absorption refrigeration cycle varies with the generator temperature. An increase in generator temperature renders a reduction in COP of the absorption refrigeration cycle. It is observed that the highest COP is produced at a generator temperature of 65 °C. When the generator temperature is increased to 70 °C, It can be seen that the COP of the absorption refrigeration cycle significantly decreases. This tendency is affected by the escalation of the strong solution concentration (Xss) leading to a drastic decline in the circulation ratio (λ) from 50 to 13.16. As a result, it affects the mass flow rate of both the strong ($\dot{m}ss$) and weak ($\dot{m}ws$) solutions and contributes to an increase in generator heat input (Qg). In the temperature range of 70-90 °C, the insignificant escalation of the strong solution concentration (Xss) causes the circulation ratio (λ) to decrease gradually. From the figure, it can be concluded that the higher the generator temperature, the lower the COP of the absorption refrigeration cycle is attained.

IOP Conf. Series: Materials Science and Engineering **796** (2020) 012015 doi:10.1088/1757-899X/796/1/012015

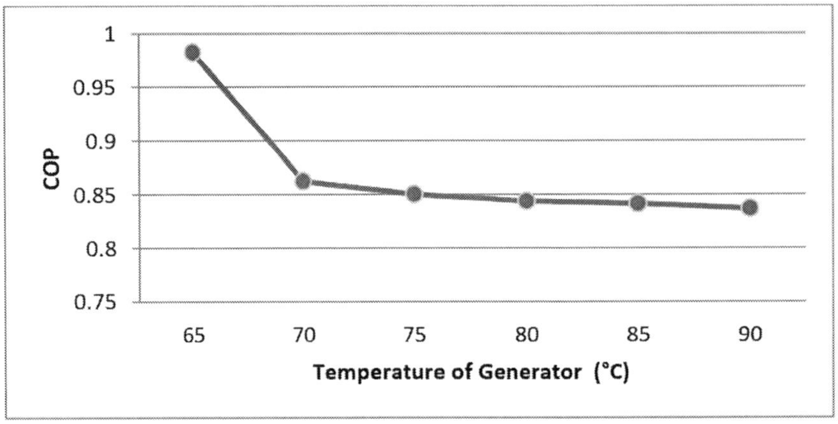

Figure 9. Variation of COP with Generator Temperature.

4.3. The effect of absorber temperature on the COP of the absorption refrigeration cycle

Figure 10 presents the variation in COP of the absorption refrigeration cycle at a specific range of the absorber temperature. From the figure, it can be seen that the lowest COP is generated at an absorber temperature of 16 °C. The COP rises gradually to the highest value at the absorber temperature of 32 °C. The increase in COP value is influenced by the increase in enthalpy of the solution in the absorber (h1 and h2) and also the concentration of the weak solution of the lithium bromide (Xws) due to changes in the absorber temperature. Thus, it can be concluded that the COP of the absorption refrigeration cycle increases directly proportional to the increase of the absorber temperature.

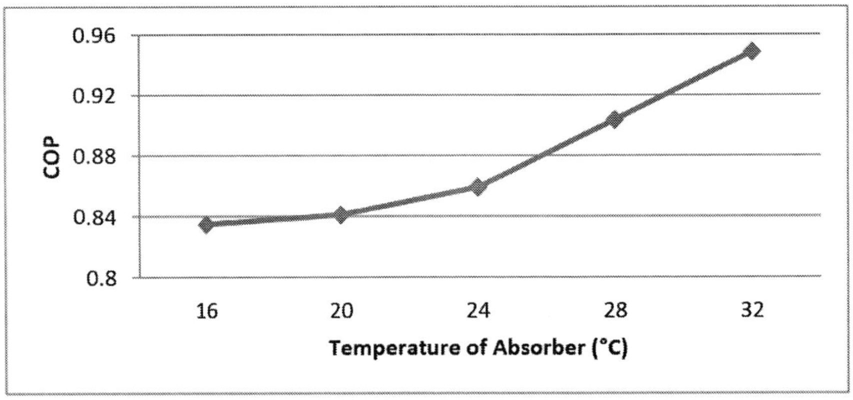

Figure 10. Variation of COP with absorber temperature

4.4. The effect of condenser temperature on the COP of the absorption refrigeration cycle

Figure 11 shows the effect of the condenser temperature on the COP of the refrigeration absorption cycle. From the figure, it can be seen that the COP value of the refrigeration absorptions cycle increases over the condenser temperature. It is observed that the COP of the refrigeration system significantly increases from the temperature of 24°C-36°C. This tendency occurs due to the increase in the saturated temperature of refrigerant during the condensation process in the condenser is in line with the increase in the saturated pressure of the refrigerant in the condenser. The rise of the condenser pressure renders the generator pressure to increase as both the condenser and the generators operate at the same pressure. As a result, it decreases the concentration of the strong solution of the Li Br (Xss), yielding the circulation ratio to increase (λ). The increase in the value of λ is directly proportional to the increase in the mass flow rate of the refrigerant solution ($\dot{m}$), the weak solution ($\dot{m}ws$), and the strong solution ($\dot{m}ss$), consequently decreases the amount of heat input supplied to the generator (Qg). Due to the COP of the absorption refrigeration cycle is inversely proportional to the heat input supplied to the generator (See Equation (20)), thus, the lower the heat input supplied, the higher the COP of the refrigeration cycle attained. It is observed that the COP of the absorption refrigeration cycle insignificantly changes when the temperature of the condenser is increased from 36-40 °C. At the temperature range of 40-46 °C, the amount of heat input supplied to the generator (Qg) is decreased significantly which results in a significant increase of COP of the absorption refrigeration cycle.

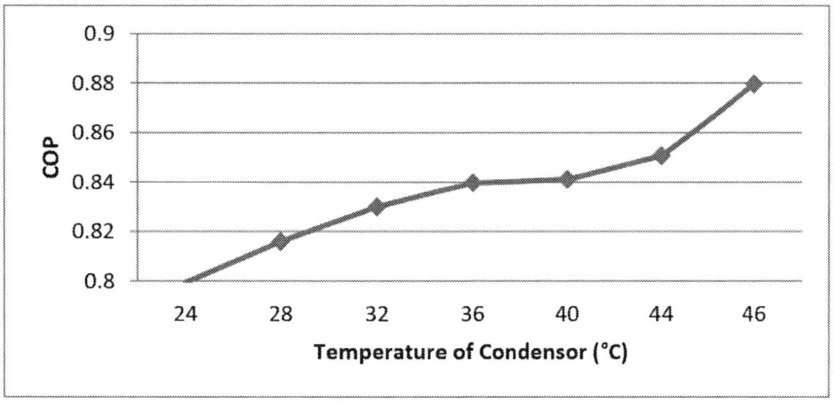

Figure 11. Variation of COP with condenser temperature.

4.5. The effect of evaporator temperature on the COP of the absorption refrigeration cycle

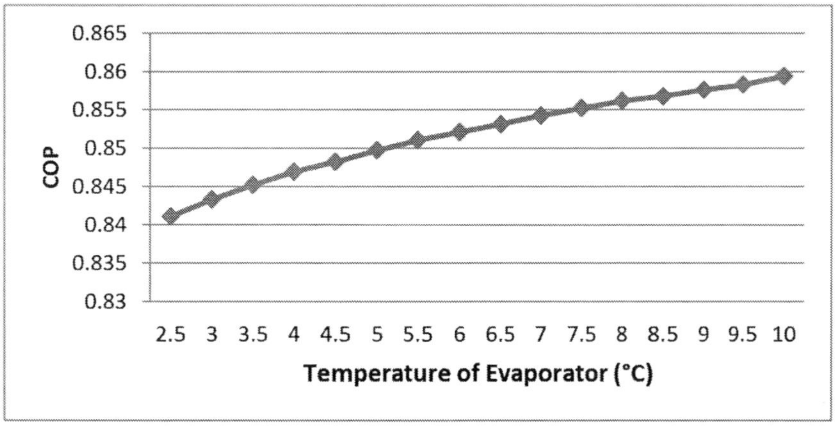

Figure 12. Variation of COP with an evaporator temperature

Figure 12 shows that the variation in the evaporator temperature with the COP of the absorption refrigeration cycle. From the figure, it appears that the increase in the COP value of the absorption refrigeration cycle is directly proportional to the increase in temperature of the evaporator. This is due to the increase in the enthalpy of the refrigerant leaving the evaporator is in line with the evaporator temperature. As a result, it decreases the amount of mass flow rate of the refrigerant needed to be circulated in the refrigeration cycle. A decrease in the mass flow rate of the refrigerant renders a reduction in the mass flow rate of the strong and weak solutions and circulation ratio. Besides, the changes in the evaporator temperature affect the mass flow rate and pressure of the working fluid.

These phenomena affect the pump power and the amount of heat input at the generator that influences the COP of the refrigeration absorption cycle. This result is also confirmed by Senapati [7].

4.6. The effect of generator temperature on the pump power of the absorption refrigeration cycle

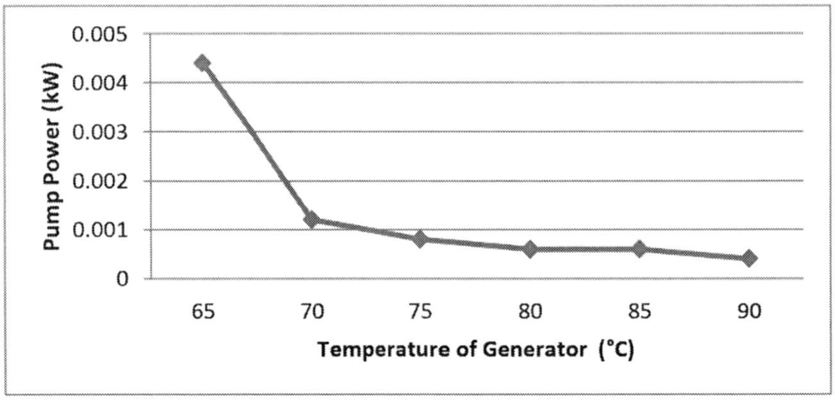

Figure 13. Variation of pump power with generator temperature.

Figure 13 shows the relationship between generator temperature and pump power in the absorption refrigeration cycle. It can be seen that the highest pump power (Wp) is generated at a generator temperature of 65 ° C. The pump power (Wp) drops significantly when the generator temperature rises to 70 °C. This tendency occurs due to the increase in the value of the strong solution (Xss) causing a drastic decrease in the value of the circulation ratio (λ), thereby reducing the mass flow rate of both the strong ($\dot{m}_{ss}$) and the weak ($\dot{m}ws$) solutions. The significant reduction of the mass flow rate of the weak solution ($\dot{m}ws$) implies a remarkable reduction in pump power (Wp) needed to circulate the weak solution ($\dot{m}ws$). While at a temperature of 70 °C-90 °C, it is observed that the pump power (Wp) demand has decreased slowly. This is due to an increase in the concentration of the strong solution (Xss) causes the circulation ratio (λ) to slowly decrease, thereby reducing both the mass flow rate of strong ($\dot{m}ss$) and weak ($\dot{m}ws$) solutions.

4.7. The effect of absorber temperature on the pump power of the absorption refrigeration cycle

From Figure 14 it can be seen that the pump power increases with increasing absorber temperature. The pump requires the lowest power (Wp) when the absorber temperature is 16 °C and the power continues to increase along with the increasing of the absorber temperature. An increase in absorber temperature causes both the flow rate ($\dot{m}ws$) and density (ρ) of the weak solution to increase. Based on the relationship as given by Equation (18), the pump power depends on the value of the multiplication of the mass flow rate of the weak solution ($\dot{m}ws$) and pressure drop (ΔH) divided the weak solution density (ρ). Thus, it can be understood that the higher the flow rate of the weak solution ($\dot{m}ws$), the higher the pump power (Wp) is required. Conversely, the pump power (Wp) is being lower when the density (ρ) of the weak solution is getting higher. Since the numerator value is higher than the denominator, it contributes to the enhancement of the required pump power.

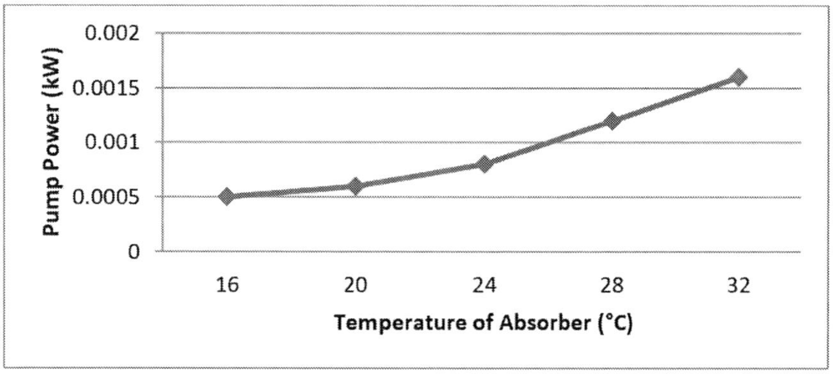

Figure 14. Variation of pump power with absorber temperature.

4.8. The effect of generator temperature on the generator heat input of the absorption refrigeration cycle

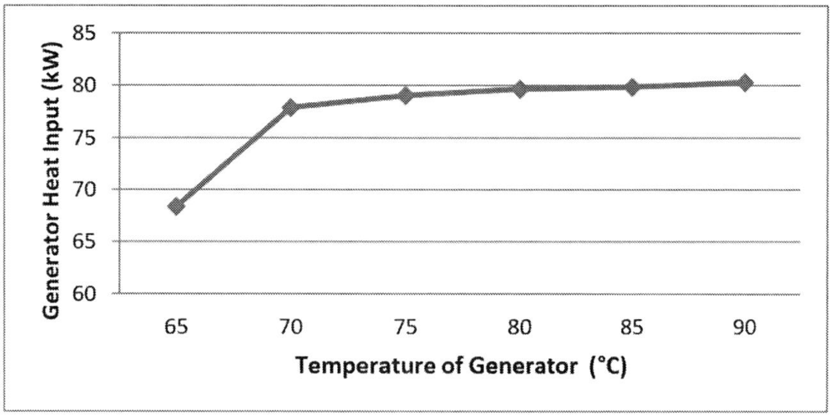

Figure 15. Variation of generator heat input with generator temperature.

In the absorption refrigeration cycle, the largest energy input required for the operation of the system is the heat supplied for the generator. Figure 15 shows that the heat input supplied to the generator (Qg) is getting higher as the temperature of the generator increases. The increase in generator temperature contributes to the increase in the concentration of the strong solution (Xss), enthalpy of the weak solution leaving the heat exchanger (h3), and the enthalpy of the strong solution leaving the generator (h4). Increasing the concentration of the strong solution (Xss) renders the circulation ratio (λ) to decrease. The reduction in circulation ratio (λ) leads to the flow rate of both the weak ($\dot{m}$ws) and the strong ($\dot{m}$ss) solutions to decrease. So that it can be concluded in designing an absorption refrigeration system, the temperature of the generator must be set as low as possible (in the range of generator temperatures given from the literature) so that the system is more efficient.

4.9. The effect of absorber temperature on the absorber heat output of the absorption refrigeration cycle

It can be observed from figure 16 that the amount of heat dissipated by the absorber decreases in line with the increase of the temperature absorber. When the absorber temperature increases, the concentration (Xws) and the enthalpy (h1) of the weak solution would also increase. An increase in the concentration of the weak solution renders the mass flow rates of the weak ($\dot{m}$w) and the strong ($\dot{m}$ss) solution to increase. This phenomenon causes the absorber required to release more heat to operate properly.

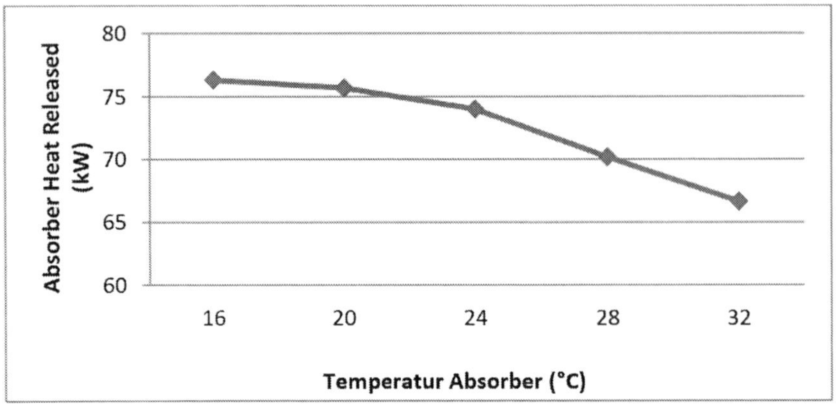

Figure 16. Variation of heat released by the absorber with absorber temperature.

5. Conclusions

The following conclusion can be drawn from the study of the absorption refrigeration cycle for cold storage application:

1. The waste heat released by the condenser of the steam power plant has enormous potential as an energy source to operate an absorption refrigeration cooler.
2. Changes in temperature of the Condenser, Generator, Absorber, and Evaporator greatly affect the Coefficient of Performance of the absorption refrigeration cooler.
3. Changes in generator and absorber temperatures affect pump power. The change in the pump power is relatively small, thus, it can be ignored in calculating the Coefficient of Performance of the absorption refrigeration cycle.
4. The amount of heat required to supply to the generator is greatly influenced by the temperature of the generator. The lower the temperature of the generator, the lower the amount of heat required to operate the system.

Reference

[1]. Manu S and Chandrashekar T K 2016 *A simulation study on performance evaluation of single-stage LiBr–H2O vapor absorption heat pump for chip cooling* (India: Sri Siddhartha Institute of Technology and Mangalore Institute of Technology and Engineering)

[2]. Majdi H S 2016 Performance evaluation of combined ejector LiBr/H2O absorption cooling cycle (Iraq: Al-Mustaqbal University College)

[3]. Hanriot S, Brito P, Maia C and Rego A 2019 Analysis of working parameters for an ammonia-water absorption refrigeration system powered by automotive exhaust gas (Brazil: Pontifical Catholic University of Minas Gerais)

[4]. Iffa R B, Bouaziz N and Kairouani Lakdar 2017 Optimization of Absorption Refrigeration Systems by Design of Experiments method (Tunisia: University of Tunis El Manar)

[5]. Charlie 2010 Cold Storage Planning for Five Tons of Fish Capacity (Banda Aceh: Syiah Kuala University)

[6]. Kaushik S 2014 Thermodynamic Analysis of Vapor Absorption Refrigeration System and Calculation of COP (India: Bipin Tripathi Kumaon Institute of Technology)

[7]. Siemens.com/steamturbines

[8]. Masry O A E 2002 Performance Of Waste Heat Absorpsition Refrigeration System (Riyadh:: Collage of Engineering, King Saud University)

IOP Conf. Series: Materials Science and Engineering **796** (2020) 012016 doi:10.1088/1757-899X/796/1/012016

Lean Production Model Aligned with Organizational Culture to Reduce Order Fulfillment Issues in Micro- and Small-sized Textile Businesses in Peru

B Martinez-Condor[1], F Mamani-Motta[1], I Macassi-Jaurequi[1], C Raymundo-Ibañez[2,4] and M Perez[3]

[1] Ingeniería Industrial, Universidad Peruana de Ciencias Aplicadas (UPC), Lima 15023, Perú.
[2] Dirección de Investigación, Universidad Peruana de Ciencias Aplicadas (UPC), Lima 15023, Perú.
[3]Escuela Superior de Ingeniería Informática, Universidad Rey Juan Carlos, Mostoles, 28933, España.

E-mail: carlos.raymundo@upc.edu.pe

Abstract. This paper proposes an optimization model aimed at increasing production capabilities at a small-sized textile business dedicated to manufacturing polo shirts, while reducing order fulfillment issues, including incomplete orders and late deliveries. Hence, an assessment identified downtimes from unnecessary transportation travel, time spent looking for materials, and excessive losses due to cutting fabric errors. In this light, the study focused on selecting adaptable tools, such as 5S, Plant Layout, and Method Study, which may help improve production capabilities and address these situations. However, to guarantee that the objectives set forth are being met, the organizational culture must also be identified and improved so that it may serve as the foundation for the optimization model. Then, the proposed model will be validated to determine whether the selected operating tools supported by the strengthening of organization culture contribute to increasing production.

1. Introduction

In this globalized age, many countries around the world, led by China, reporting $8,637,540 per month, and followed by Bangladesh and Turkey, have started producing various types of polo shirts. In this industry, Peru only generates $328,638 per month, accounting for only 0.8% of the worldwide production [1]. In recent years, textile production has exhibited a huge decline in contrast to other sectors. The main factor that has affected production for about six years has been the competition with Chinese, Vietnamese, or Indian imports entering the market as their prices are extremely lower than the national production [2]. However, in the last two years, the textile production sector has positively expanded, reporting a 7.53% production increase for the period between January and November 2018 [3]. Despite this positive increase, several micro- and small-sized businesses within this still sector continue to face serious production problems, especially in terms of low production capabilities, which

[4] To whom any correspondence should be addressed.

Content from this work may be used under the terms of the Creative Commons Attribution 3.0 licence. Any further distribution of this work must maintain attribution to the author(s) and the title of the work, journal citation and DOI.
Published under licence by IOP Publishing Ltd

not only constrain and hinder sector growth but also national production levels. Therefore, their production process must be streamlined.

The textile and clothing industry have become prominent in recent decades as it reports one of the highest export rates overall and exerts a demand from other specialized markets. The sector contributes to creating jobs in other industries, such as agriculture and plastics, based on its use of cotton and other inputs. Moreover, this industry accounts for 12% of the national GDP, with the textile sector accounting for 4.7% thereof [4]. These products are renowned in the international market for their use of high-quality fibers, which has also fostered the increasing production trend recently reported along with their entry in the Peruvian market, thus generating revenue for the textile industry, as well as for the entire nation.

As the textile and clothing sector is a major industry in Peru, it is unfortunate that, in comparison to other sectors, its production had decreased since a few years ago, and that, although the sector is recovering, it has not yet established itself as a stable sector. Therefore, this research study proposes improving the polo shirt manufacturing process at a small-sized textile business in the Gamarra fashion center through the implementation of a model comprising the 5S, Plant Layout, and Study Methods techniques, measured and assessed by means of the Denison Organizational Culture Survey. Another important aspect is that this proposal may also be adopted by companies manufacturing different garments, i.e., shirts, pants, polo shirts, and shorts to address their low productivity issues.

2. State of the art

2.1. Organizational Culture-based Production

The production rates are assessed through the organizational culture at a textile company in Portugal with the purpose of evaluating employee performance. Hence, a model comprising an error-detection culture, an error-prevention culture, an organizational culture, and performance rates is created and measured by a survey. In this study, the implemented model successfully improved production management supported by values and attributes [5]. However, in recent years, as organizations evolve within the changing markets, they must start managing learning to remain competitive in the market. Hence, many companies are turning to knowledge management as a viable method for improving their conditions of stability. In this light, the Denison model is used to study organizational culture dimension through the adaptability, mission statement, engagement, and capacity indicators. In fact, managers with greater organizational culture acumen have been able to steer their weakness for awareness and accuracy toward the production area and forecast the process strategies required [6]. However, a study conducted in the finance sector in Kenya determined the influence organizational culture exerted on company performance, thus concluding that organizational culture is also deemed an important source of sustainable competitive advantage within the finance industry [7]. In Malaysia, the PLS-SEM method was used to correlate production to organizational culture at manufacturing companies, as a part of a greater assessment of their organizational culture [8]. Finally, a study applies a theoretical model in Pakistan's manufacturing industry that engages an organizational culture which supports environmental practices for improving production rates. Based on a survey and using Smart-PLS, the study concluded that adaptability, mission statement, and consistency exert a positive impact on environmental performance and, therefore, in its production levels [9].

Within this context, the literature agrees that the Denison model is the best model for assessing organizational culture as it uses four different indicators to take any organization's temperature and determine its current situation.

2.2. Lean Manufacturing-based Production

The following papers discuss different tools and methods of implementing lean manufacturing as a production optimization model. However, all of these papers are consistent when defining the best method used for increasing production rates. For example, in one of these papers, lean manufacturing tools are used to address resource management flexibility at an Indian company, taking employees and machinery as resources [10]. Nevertheless, lean manufacturing is also used to eliminate waste in the packaging area at a small-sized textile company in Perú [11]. Further, value-stream mapping (VSM) is deployed at a car parts manufacturing company in India. In this instance, the 5 Whys method contributes to root causes identification and the implementation of a roadmap for improving productivity within the company in the future. This method successfully increased production in the textile process [12]. However, the different wastes found within a natural fiber garment production line were identified and prioritized using VSM to collect the required data. At the end, seven wastes were identified as per the theoretical model proposed by Taichí Ohno, later using the Super Decision software to prioritize these wastes [13]. As waste must also be removed from the production process, a Lean Six Sigma framework is deployed to improve operational performance in an MRO installation, wherein this methodology is used to dispose wastes and improve production processes by reducing standard production times [14]. Lean Six Sigma benefits are also described, thus evidencing the effective nature of this methodology for identifying and removing waste, and thus improving process flows, strengthening company interactions with suppliers and customers, and fostering faster learning, which were all deemed the main problems to be addressed [15]. Finally, to achieve operational excellence through Lean Six Sigma, a case study was carried out at an automotive company in India using the LSS methodology to improve the performance of the semi-automated transfer line process. The application of the LSS methodology reduced drilling defects by machining injector bodies.

In conclusion, lean manufacturing tools are an effective method for eliminating waste from production processes.

3. Contribution

3.1. Proposed Model

The general model is grounded on lean methodology, which focuses on customer valuation by removing waste to reduce cycle times based on the assurance provided through monitoring and improving organizational culture. Although the model uses lean methodology tools, the supplementary tools are also used to deploy lean production.

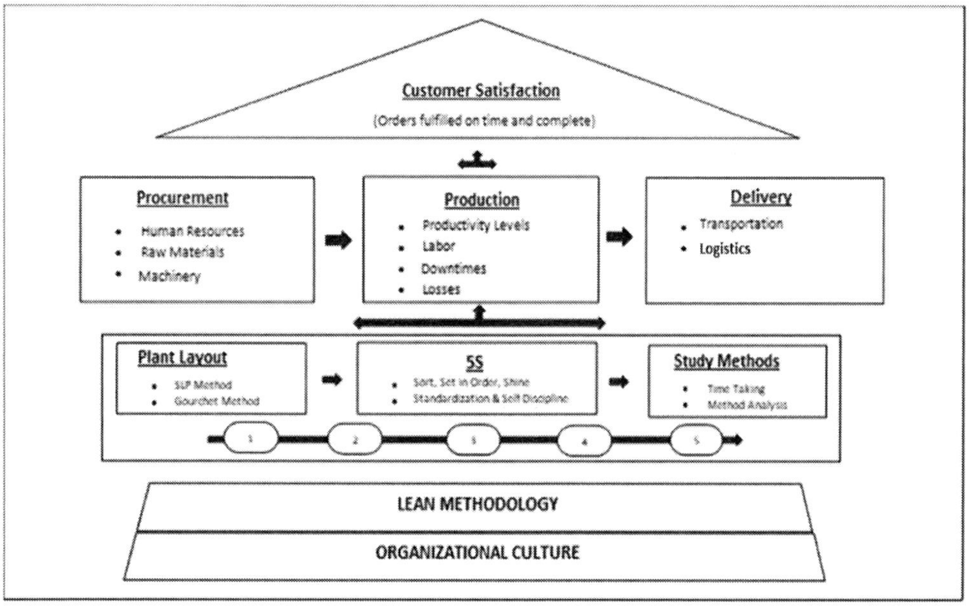

Figure 1. Proposed model

3.2. Proposed Method

The following flowchart shows the activities performed in each implementation stage.

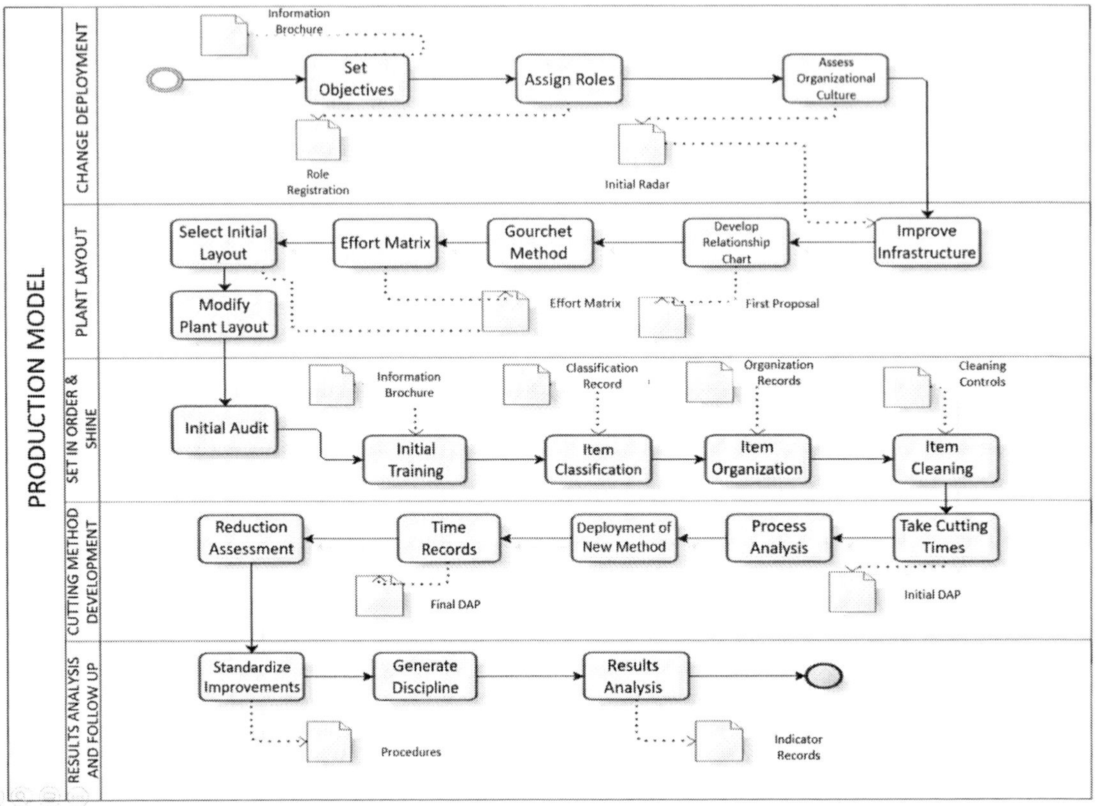

Figure 2. Proposed method

Therein, the process input and output records are the artifacts included in the different model components, which provide information required to determine optimum development levels.

3.3. Indicators

To measure post-implementation results, the following indicators will be used to track lean model implementation and its alignment to organizational culture:

- Organizational Temperature, which provides an initial assessment of the alignment exhibited by staff members with company policies.
- % Decrease in Travel Time, which shows the amount of time reduced in travel time within the production process.

$$\frac{Time_{\,after} - Time_{\,before}}{Time_{after}} \times 100 \tag{1}$$

- % Downtime Reduction, which specified the amount of downtime reduced throughout the polo shirt production process.

$$\frac{Downtime_{after} - Downtime_{before}}{Downtime_{after}} \times 100 \tag{2}$$

- % Losses, which identifies the fabric losses per fabric roll used in polo shirt production.

$$\frac{Losses\ per\ Roll\ in\ Kg}{Roll\ Used\ in\ Kg} \times 100 \tag{3}$$

- % Orders Fulfilled, which measure customer satisfaction levels in terms of successful order fulfillment.

$$\frac{Orders\ fulfilled}{Total\ orders\ accepted} \times 100 \tag{4}$$

- % Plant Capacity, which shows the overall growth in productive efficiency based on the number of products manufactured within a given period of time.

$$\frac{Standard\ Time}{Average\ Time} \times 100 \tag{5}$$

4. Validation

4.1. Description of Test Scenarios

Model implementation was validated though a case study conducted at Monper's, a textile MSE located at the Gamarra fashion center in downtown Lima. However, its manufacturing workshop is located at 201, Ave. Del Aire, in the district of San Luis, Lima. This company manufactures different varieties of polo shirts, with long- and short-sleeved round-neck t-shirts being its flagship product.

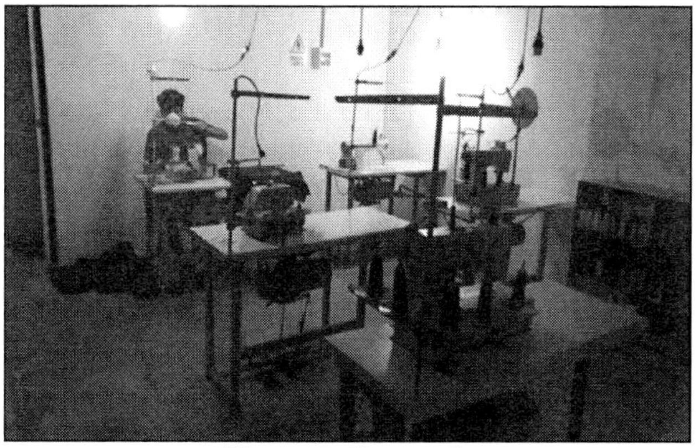

Figure 3. Case study

4.2. Initial Values

In the recent years, the company has reported that 20% of its orders are being fulfilled late. To verify these numbers, a sample of 12 orders with different quantities were taken. For each of these orders, the committed lead times against the actual lead times were recorded, thus determining an approximate gap of one day and a half between the two.

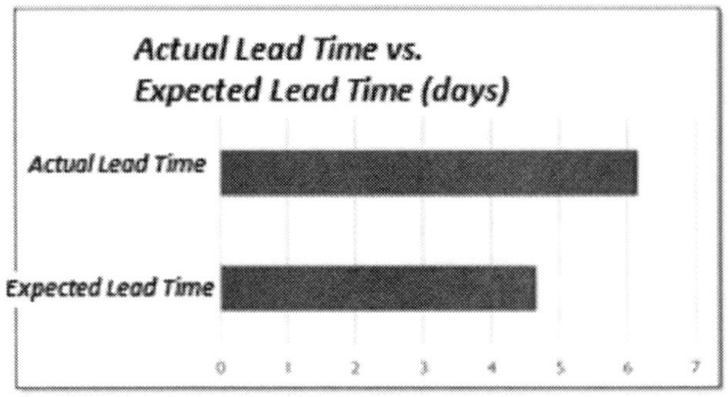

Figure 4. Actual lead time vs. Expected lead time

Actual plant efficiency is at 67.74% as daily production was originally calculated at 31 shirts per day, while the actual production rates only account for 21 shirts per day. Therefore, the company is currently losing 150 soles per day as each shirt is sold at 15 soles.

In terms of downtimes, an average downtime of 2.012 minutes per shirt was recorded. To determine surplus travel times, we recorded standard production times, yielding an average of 0.44 minutes of unnecessary travel time per shirt. Finally, the losses due to inadequate cuts, expressed in percentages, represent between 20% and 30% of the total weight of the roll.

4.3. Contribution Application

4.3.1. Change Deployment. In the first stage, a Denison Organizational Culture Survey is applied to measure current organizational culture levels. Further, staff members receive informative sessions and are assigned roles as part of the new organizational culture deployed.

4.3.2. Plant Layout. Based on the results from the Organizational Culture Survey, work conditions are improved for staff members, as a way to strengthen their engagement and commitment to changes implemented. Moreover, the plant layout is modified based on the SLP model, which identifies optimum Plant Layout based on quantitative and qualitative criteria.

4.3.3. Sort, Set in Order, and Shine. The first three principles of the 5S tool are implemented. First, an initial audit is performed. Then, staff members receive induction and information toolbox talks. Next, the Sort, Set in Order, and Shine principles are implemented.

4.3.4. Cutting Method. To reduce losses and reuse surplus fabric to manufacture more shirts, a new and improved methodology for cutting fabric was introduced. Then, the corresponding reviewed information gathered from shirt manufacturing manuals and guides will be shared with staff members through a Process Analysis Chart. For these purposes, the new material proposed is sulfite paper.

4.4. Results Summary

The study reported improvements in all the items addressed; however, these results must be periodically monitored and recorded so that they may continue to improve over time.

Table 1. Results summary.

Phases	Indicator	Initial	Final
Change deployment	Organizational Culture Index	41%	72.5%
Plant layout	Travel Time Reduction	27%	
Set in order & shine	Reduction of time spent searching for materials	**54.87%**	
Cutting method development	Production losses	15%	11.5%
Results monitoring	Order Fulfilment	70.5%	88.5%

5. Conclusions

As there are three main reasons for late order fulfillment (shortages of raw materials, machine downtimes, and low productivity), this study focused on addressing the latter, which represented 95.86% of the issues reported. In this regard, the shirt manufacturing workshop evidenced an insufficient production rate that did not allow it to fulfil 100% of the orders received. Therefore, this study proposed the implementation of a model to reduce unnecessary travel times and downtimes, as well as to find a suitable fabric cutting method for increasing round-neck t-shirt productivity rates.

6. References

[1] M. Trade 2018 T-shirt and polo shirt production.
[2] Management 2018 Economy of Perú.
[3] INEI 2018 Information Technology, Landscape of Peruvian Economy.
[4] INEI 2018 Information Technology, Landscape of Peruvian Economy.
[5] Araujo R 2019 The Quality Management System as a Driver of Organizational Culture: An Empirical Study in the Portuguese Textile Industry *Qual Innov Prosper - Kvalita Inovacia Prosperita* **23** 1-24.
[6] Gholam A A 2017 Effect of organizational culture on knowledge management based *Soc Behav Sci* **23** 387–395.
[7] Joseph O O 2019 Organizational Culture and Performance: Evidence from Microfinance Institutions in Kenya *Sage* **11** 112-200.
[8] Lee W L 2018 Organizational culture and performance of Malaysian manufacturing firms *Int J Adv Appl Sci* **5** 59-66.
[9] Bakhsh Magsi H, Ong T S, Ho J A and Sheikh Hassan A F 2018 Organizational Culture and Environmental Performance *Sustainability* **10** 2690.
[10] Chauhan G 2016 An analysis of the status of resource flexibility and lean manufacturing in a textile machinery manufacturing company *Int J Organ Anal* **24** 107-122.
[11] Portocarrero V, Quispe G, Nuñez V, Raymundo C, Moguerza J M 2018 Model for the implementation of a waste recycling plant in small and medium-sized companies in the textile industry *CISCI 2018 - Decima Septima Conferencia Iberoamericana en Sistemas, Cibernetica e Informatica, Decimo Quinto Simposium Iberoamericano en Educacion, Cibernetica e Informatica, SIECI 2018 – Memorias* **1** 52-57.
[12] Kumar S, Dhingra A and Singh B 2018 Lean-Kaizen implementation *J Eng Design Tech* **16** 143-160.
[13] Behnam D 2018 Value stream mapping approach and analytical network process to identify and prioritize production system's Mudas (case study: natural fibre clothing manufacturing company) *J Text I* **109** 64-72.
[14] Hill T 2018 The implementation of a Lean Six Sigma framework to enhance operational performance in an MRO facility *Prod Manuf* **23** 22-44.
[15] Randhawa J 2017 Examining the role of 5S practices as a facilitator of business excellence in manufacturing organizations *Meas Bus Excell* **21** 191-206.

IOP Conf. Series: Materials Science and Engineering **796** (2020) 012017 doi:10.1088/1757-899X/796/1/012017

LEAN maintenance model based on change management allowing the reduction of delays in the production line of textile SMEs in Peru

G Arrascue-Hernandez[1], J Cabrera-Brusil[1], P Chavez-Soriano[1], C Raymundo-Ibañez[2,4] and M Perez[3]

[1] Ingeniería Industrial, Universidad Peruana de Ciencias Aplicadas (UPC), Lima 15023, Perú.
[2] Dirección de Investigación, Universidad Peruana de Ciencias Aplicadas (UPC), Lima 15023, Perú.
[3]Escuela Superior de Ingeniería Informática, Universidad Rey Juan Carlos, Mostoles, 28933, España.

E-mail: carlos.raymundo@upc.edu.pe

Abstract. This article examines the problem of production line delays in a textile small- and medium-sized enterprise (SME) that produces polyester fibre from recycled bottles, based on orders. Factors that have resulted in production line delays include prolonged unscheduled maintenance time, and preparations and adjustments prior to operating the equipment. To address the problem, a model was developed applying lean manufacturing tools through change management, with the aim of increasing equipment availability and useful life. To validate the model, a pilot was developed to determine how the increase in equipment availability helps reduce delays in the production line, which eventually improves completion of customer orders.

1. Introduction

The 2009 global economic crisis originated in the United States and had a significant impact worldwide. Peru was no stranger to this situation, and thus, in the following years, the country went through a process of readjustment. Over the last three years, Peru has remained in the 50th place in terms of gross domestic product, and in Latin America, it is in the 6th place. [1] The textile industry in Peru is not properly developed, owing to reasons such as the lack of innovation in its products and processes or the scarce support from the state for small- and medium-sized enterprises (SMEs). [2] The percentage of participation in textile manufacturing has decreased from 3.72% to 3.31% [3], which suggests that the Peruvian textile industry does not have a strong presence in the international market [4]. These types of companies, which use automated machinery, frequently have problems completing their orders on time, resulting in customers seeking other alternatives, mainly abroad.

For this reason, adequate maintenance management based on lean manufacturing tools will benefit production in such a way that it increases the availability of the machines to start the productive process, thus optimizing associated costs.

[4] To whom any correspondence should be addressed.

Content from this work may be used under the terms of the Creative Commons Attribution 3.0 licence. Any further distribution of this work must maintain attribution to the author(s) and the title of the work, journal citation and DOI.
Published under licence by IOP Publishing Ltd

2. State of The Art

2.1. Change Management
People's behaviour is an integral part in any process a company wants to execute or improve. The ADKAR Model serves as an objective-oriented change management model, which can be applied to a personal or organizational context. The model can be applicable to make a successful change for 1 person, 20 persons, 1,000 persons, or for the entire organization, which contributes to standardization. Change management makes a great contribution, close to 80%, to the completion of a project, since it focuses on reducing resistance to change and makes the transition to the new culture smooth and not an imposition. [5] [6]

2.2. Maintenance Management using Lean Manufacturing
Production improvement begins with a good analysis of the production process, and for this purpose, we use lean manufacturing tools, which aim at reducing the delivery time of finished products without significantly affecting the current work systems in the manufacturing industry. [7] [8]

There are several diagnostic tools, which are as follows:

- Value stream mapping (VSM), through which data is collected to know the current state of the industries, identifying the waste that could be eliminated. [9]
- Overall equipment effectiveness (OEE), which depends on process stability and equipment performance, can improve production by identifying unwanted machine downtime and adopting corrective actions at different points in the production cycle. [10]
- However, there are also lean manufacturing tools that help provide a solution to the problem, which are as follows:
- The 5S methodology enables the adoption of new forms of work integrating self-discipline, order, cleanliness, and safety. [11] [12]
- Total productive maintenance (TPM) is a plant improvement methodology that allows for continuous and rapid improvement in the manufacturing process using employee participation, incorporating preventive methods, and maintenance concepts. [13]

The Single-minute Digit exchange of die (SMED) technique aims at reducing waste in the production system. This includes certain improvements to reduce the time required for changing machines, reducing allocated resources, or both. [14]

3. Contribution

3.1. Model Proposed
The model proposed was developed for a textile company for reducing production line delays. This model will be detailed through a design, which was conducted based on the concepts obtained from the literature review and on the designs found in other scientific articles. It encompasses the use of lean manufacturing tools for optimal maintenance and the proposed model is described in Figure 1.

3.1.1. Analysis of the current situation and goals

- Takt Time: Considering the demand, the time necessary to meet such demand will be identified.
- Value Stream Mapping (VSM): All the waste and times that do not add value to the production will be identified. Thus, it will be possible to identify the company's real manufacturing time.
- Ishikawa: It helps find out where the problem is focused and then helps provide a solution.
- Overall Equipment Effectiveness (OEE): It will provide details on the company's production management level through the machines, which will give way to making improvements in the

maintenance area. This will eventually lead to an improvement in the processes along the production line.

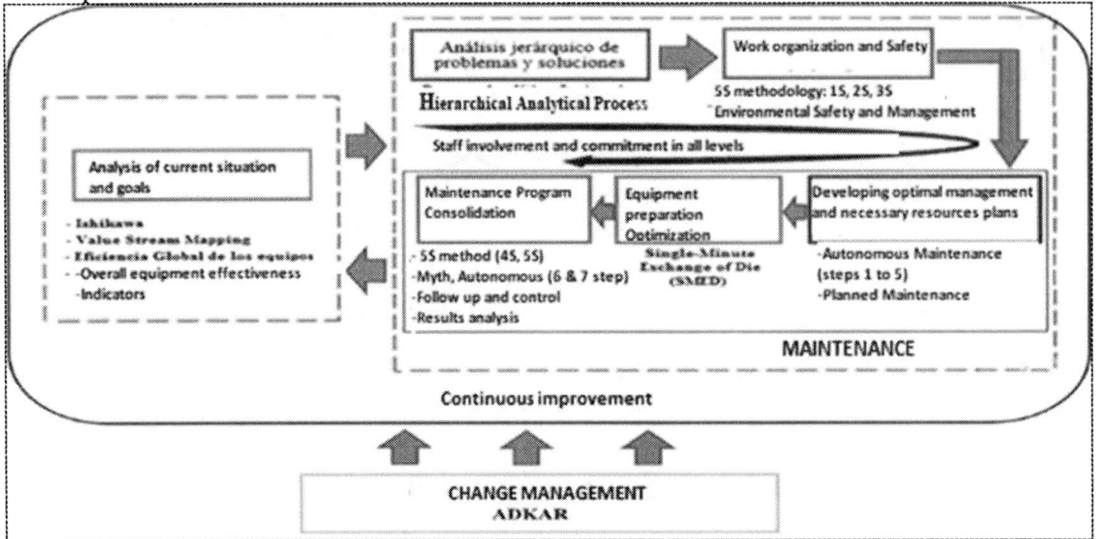

Figure 1. Maintenance management model

3.1.2. Hierarchical analysis of problems and solutions. Analytical Hierarchical Process (AHP Matrix): It will identify the root problems and search for the most effective solutions to these problems associated with maintenance, which will then be presented to management at the start of change management. During this procedure, 6 steps are taken into account, which are decomposition of the problem, weighting of the criteria matrix, normalization of the criteria matrix, comparison of alternatives, decision of the best alternative and, finally, consistency evaluation.

Table 1. Consistency assessment by selected criteria

Alternatives	Sol. 1	Sol. 2	Sol. 3	Sol. n	Average vector	Vector λ max.
Sol. 1				Research data zone		
Sol. 2				Research data zone		
Sol. 3				Research data zone		
Sol. n				Research data zone		
TOTAL				Research data zone		

3.1.3. Work organization and safety

3.1.4. 5S Methodology. Work organization is essentially linked to the development of the 5S, mainly the first three. However, before starting with the implementation of this methodology, it is necessary and extremely important to conduct an audit to see the current company level and the level after the implementation.

- Classify: It consists of categorizing any unnecessary items by means of a list and red cards to be able to remove them more easily.
- Organize: Location of materials by frequency of use, as well as signalling of corridors for the transit of machinery and waste location. This way they will be more easily identified by all company personnel.
- Clean: In this activity, the cleaning will be arranged, the cleaning plan will be prepared along with any elements necessary, and, finally, an evaluation will be conducted.

3.1.5. Safety Pillar. Safety is related to safe work by operators, following written procedures for safe work, receiving periodic training on hazards identification and risk assessment in their jobs, and training in the use and management of personal protective equipment, among others.

3.1.6. Development of optimal management and necessary resources plans

- Autonomous Maintenance (AM): The first five steps of AM will be developed, starting with initial cleaning, removal of pollution sources, inspection and cleaning standards, general inspection, and autonomous inspection. The purpose of this maintenance is to detect and deal promptly with abnormalities in the equipment, which is precisely the objective of good maintenance.
- Planned Maintenance (PM): The purpose of PM is to maximize the lifespan of equipment, project the maintenance frequency to the equipment, and at the same time, introduce all the available data such as technical sheets and fault control sheets.

3.1.7. Optimizing the preparation of equipment. **Single-Minute Exchange of Die (SMED):** First, internal activities should be analyzed and separated from external activities. Second, internal activities should be converted into external activities, and third, internal activities should be reduced.

3.1.8. Consolidating the maintenance program. Finally, consolidation of the maintenance program focuses on performing the 4th and 5th "S" (i.e., standardize and discipline) for the work areas, along with the 6th and 7th steps of the autonomous maintenance plan for equipment in general, supported by follow-ups and controls by management, as well as the operators in charge, to ensure the activities are executed in the most optimal manner.

3.2. Proposed Method

The proposed method to be followed for the successful implementation of the model includes five phases in addition to a preliminary analysis, as shown in Figure 2, which must be applicable to all companies that meet certain considerable criteria, starting from whether they belong to the textile sector to whether they have delays in their production line.

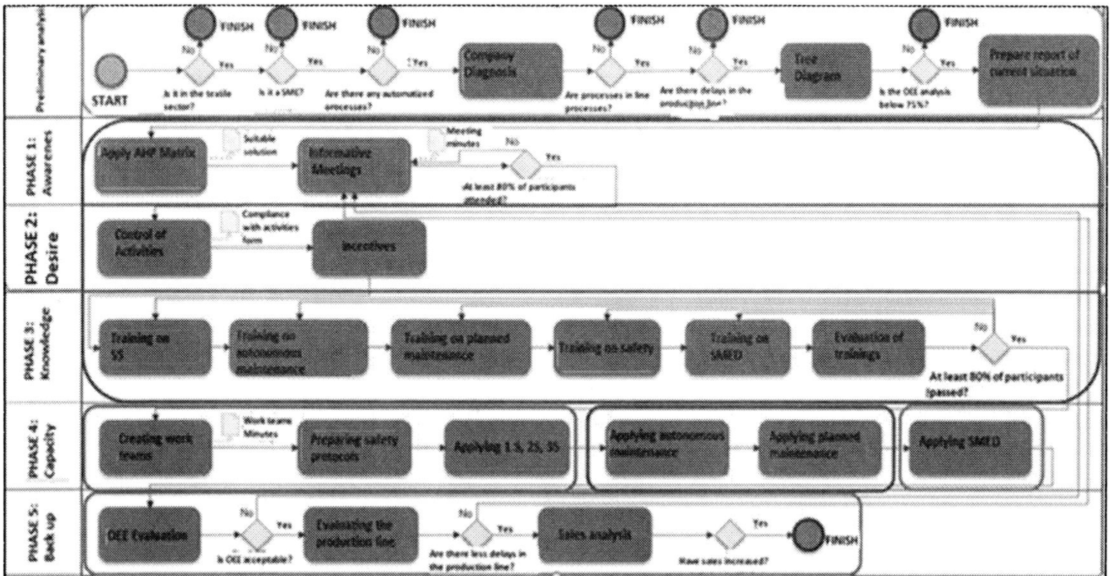

Figure 2. Implementation flowchart

IOP Conf. Series: Materials Science and Engineering **796** (2020) 012017 doi:10.1088/1757-899X/796/1/012017

3.3. Indicators

3.3.1. Availability of equipment. It seeks to measure a level based on the time in which equipment is at the service of the company to be operated, which represents the main problem to be solved.

$$Availability = \frac{Operating\ Time}{Total\ available\ Time} \quad (1)$$

Unacceptable: Less than 70%, **Intermediate:** Between 70% and 90%, **Acceptable:** More than 90%

3.3.2. Overall equipment effectiveness (OEE). It will make it possible to know the level at which the company is with respect to the global indicator and to evaluate the balance sheet or the economic losses incurred.

$$Availability\ (A)\ x\ Effectiveness\ (n)\ x\ Quality\ (q) \quad (2)$$

Unacceptable: Less than 65%, **Intermediate:** Between 65% and 75%, **Acceptable:** More than 75%

3.3.3. Products delivered on time. It will provide data of the level at which the company has improved with respect to the arrival of products to the customer.

$$\frac{\#\ of\ products\ delivered\ on\ time}{Total\ orders} * 100 \quad (3)$$

Unacceptable: Less than 60%, **Intermediate:** Between 60% and 85%, **Acceptable:** More than 85%

4. Validation

4.1. Case Study

The maintenance management model based on change management to reduce production line delays was implemented in a SME in the textile sector, dedicated to the production and marketing of polyester fiber based on 100% recycled PET bottles. The company is located at Av. Tomas Alva Edison No. 215, district of Ate, in Lima, Peru, and it has participated in the national and international markets for more than 15 years. During validation, the artefacts of the proposed model, included in each phase of the implementation guide, were analysed.

4.2. Initial Diagnosis

The company in question operates under the make-to-order model, in other words, at a customer's request. According to an analysis of sales, the company had a decrease in the last five years. This is due to the delay in orders, as only 83% are delivered on time.

Takt Time was calculated according to demand, which turned out to be 68.58 min per bundle. However, the production rate obtained from the VSM performed is 82.65 min per bundle, which is not enough to meet the required demand. The result of the OEE analysis was 61.48%, with availability being the main factor contributing to the problem, which was 65%.

4.3. Implementation

4.3.1. Phase 1: Raising Awareness. Meetings were scheduled with the plant and operations managers to show them the problems and the solution that will be chosen. For this purpose, the AHP matrix was used as support to justify the reason for this decision.

4.3.2. Hierarchical analysis of problems and solutions. The AHP analysis was performed, where the problem was broken down, the criteria matrix was weighted and normalized, the solution alternatives were compared, the best option was taken, and the consistency matrix was evaluated. In this way, the most viable option is proposed. In this case, the 5S and TPM tools are proposed, in addition to SMED. Then, it is presented to management.

4.3.3. Phase 2: Desire. This phase, which belongs to ADKAR change management, was carried out with the aim of motivating operators, who were offered compensation for their good performance. It should be noted that these compensations were previously discussed with management, since they must make the decision to execute it or not.

4.3.4. Phase 3: Knowledge. Once the worker was motivated to participate in the project, training began based on industrial engineering tools. At the end of the training, workers were evaluated to measure their acquired knowledge.

4.3.5. Phase 4: Capacity. Activities for managing proper maintenance began.

4.3.6. Work Organization and Safety. A 5S audit was conducted to obtain the current level of the company. The result was 51%. Then, we started with the implementation of each S.

- Seiri–Classify: Using the red cards to identify dirty pits, an action plan was drafted for these elements.
- Seiton–Sort: The materials were classified by frequency of use, and the work area was organized depending on that classification.
- Seiso–Cleaning: Cleaning plans were implemented, and at the end, the before and after was compared.

Figure 3. Before and after the implementation

Regarding workers' safety activities, safe work procedures were established, and hazards and risk in the company were identified.

4.3.7. Development of optimal plans and necessary resources. Autonomous maintenance (AM): The equipment was cleaned by the operators to recognize any pollution sources and inspect the machine in depth. This is implemented so that the operator has a standard cleaning process and distinguishes between normal and irregular operation of the machine.

Planned Maintenance (PM): The equipment was coded according to the area of work (an abbreviation of name and the position in which they are with respect to the production line), and the technical file of the equipment as well as a maintenance plan were created.

4.3.8. Optimization of equipment preparation. Single-minute Exchange of die (SMED): The activities performed when the machine is stopped were evaluated, and it was proposed that they should be converted to activities with the machine in operation following the safety guidelines. Finally, a comparison was made between the current method and the proposed one, whose improvement is reflected in the optimized time that went from 58.86 to 38.69 min.

4.3.9. Phase 5: Reinforcement

4.3.10. Consolidating the maintenance program. We continued with the steps of the 5S methodology for the work areas as well as the steps of the AM for the equipment, in such a way that it becomes a habit within the company, standardizing the activities and disciplining the workers, with the help of supervisors, as well as with the participation of management.

4.4. Results

The scenario 3 corresponds to the third work shift of the day, which is between 00:00 and 08:00 hours. There are no raw material inputs and inputs, as well as outputs of finished products to the customer. First, from a time taken, an operating time of 25314.24 minutes was obtained, of a total available time of 33600 min. (10 weeks). This operating time represents the availability of the equipment during this turn.

$$Availability = \frac{25314.24\ min}{33600\ min} x100\% = 75.34\%$$

Second, the analysis of the overall efficiency of the equipment for this scenario was as follows:

$$OEE = 75.34\%\ x\ 96.17\%\ x\ 98.01\% = 71.01\%$$

Thirdly, the quantity of products made based on the improvement made, which are placed in the warehouse of finished products and which are dispatched during the first corresponding shift between 08:00 - 16:00 hours, was analysed.

$$\%\ Products\ delivered\ on\ time = \frac{537\ fardos}{624\ fardos} * 100 = 86.06\%$$

Table 2. Summary of indicators

Indicator	Traffic Light			Base Line	Improvement
% Availability	<70%	70% - 90%	>90%	65%	75.34 %
% OEE	<65%	65% - 75%	>75%	61.48%	71.01 %
% Products delivered on time	<60%	60% - 85%	>85%	83%	86.06 %

Table 3. Summary of indicators by scenario

Indicator	Scenario 1	Scenario 2	Scenario 3
% Availability	78.40 %	75.80 %	75.34 %
% OEE	73.88 %	71.59%	71.01 %
% Products delivered on time	89.72 %	86.38 %	86.06 %

The study reported benefits in all the points addressed; however, not all of them achieved the expected result because the time available for implementation was limited.

5. Conclusions

It was possible to reduce the production time per bundle, achieving an increase from 524 to 567 bales. This increase of 43 bales during each month reduced the delays in the production line, which was the company's main problem, increasing the percentage of delivery to customers from 83% to 89.72%.

Once the model is implemented, and assuming demand from previous years persists, an 8.206% increase in sales is projected.

Despite not reaching the expected results with respect to the availability and OEE indicators, a significant improvement was achieved considering the time available to execute it. It should be noted that this is an implementation process and the adaptation time varies from company to company and usually takes more time when it comes to tools such as lean manufacturing. However, the model emphasizes continuous improvement, and thus, if application is ongoing, the results will be much better.

6. References

[1] International Monetary Fund 2018 Retrieved from: http://cort.as/-I0A-.

[2] Barraza B J S 2014 LAS MYPES EN PERÚ. SU IMPORTANCIA Y PROPUESTA TRIBUTARIA, Hernán Yacsahuache Pasapera (SMEs in Peru. Importance and tax proposal).

[3] I. N. D. E. E., INFORMÁTICA 2017 Panorma de la economía peruana (*Lima*).

[4] Textiles A. P. d. T. 2017 Asociación Peruana de Técnicos Textiles (Peruvian Association of Textile Technicians) Retrieved from: http://cort.as/-I0C9.

[5] Da Veiga A 2018 An approach to information security culture change combining ADKAR *Inform Comput Secur* **26** (5) 584-612.

[6] Karambelkar M B S Onboarding is a change: Applying change management model ADKAR to onboarding *Hum Resour Manag Int Dig* **25** (7) 5-8.

[7] Nallusamy S, Saravanan V 2016 Lean Tools Execution in a Small Scale Manufacturing Industry for Productivity Improvement- A Case Study *Indian J Sci Tech* **9** 335-347.

[8] Nallusamy S, Saravanan V 2016 Enhancement of Overall Output in a Small Scale Industry through VSM, Line Balancing and Work Standardization *Int J Eng Res Africa* **26** 176-183.

[9] Rekha P. N 2016 An Optimized Model for Reduction of Cycle Time Using Value Stream Mapping in a Small Scale Industry *Int J Eng Res Africa* **27** 179-189.

[10] Hossen A A 2018 Improvement of overall equipment efficiency of ring frame through total productive maintenance: a textile case *Int J Adv Manuf Tech* **94** 239-256.

[11] Jebadurai D S, Rose A R, Atthisugan I, Baby B 2017 Implementation of 5S in sales warehouse *J Chem Pharm Res* **2** 113-117.

[12] Guillen K, Umasi K, Quispe G, Raymundo C 2018 LEAN model for optimizing plastic bag production in small and medium sized companies in the plastics sector *Int J Eng Res Tech* **11**(11) 1713-1734.

[13] Dhiravidamani P, Ramkumar A S, Ponnambalam S G and Subramanian N 2018 Implementation of lean manufacturing and lean audit system in an auto parts manufacturing industry–an industrial case study *Int J Comput Integ M* **31**(6) 579-594.

[14] Bidarra T and Godina R 2018 SMED methodology implementation in an automotive industry using a case study method *Int J Ind Eng Manag* **9** (1) 1-16.

IOP Conf. Series: Materials Science and Engineering **796** (2020) 012018 doi:10.1088/1757-899X/796/1/012018

Production management model through MPS and line balancing to reduce the non-fulfillment of orders in lingerie clothing MSEs in Peru

K Flores-Andrade[1], R Guardia-Miranda[1], P Castro-Rangel[1], C Raymundo-Ibañez[2,4] and M Perez[3]

[1] Ingeniería Industrial, Universidad Peruana de Ciencias Aplicadas (UPC), Lima 15023, Perú.
[2] Dirección de Investigación, Universidad Peruana de Ciencias Aplicadas (UPC), Lima 15023, Perú.
[3]Escuela Superior de Ingeniería Informática, Universidad Rey Juan Carlos, Mostoles, 28933, España.

E-mail: carlos.raymundo@upc.edu.pe

Abstract. The focus of this research is to establish control and planning management in the sewing production process of lingerie clothing to better prepare companies for demand growth. The lack of improvement tools in this sector, the lack of staff training and a lack of quality culture has led to companies, especially MYPES, not being able to meet the established delivery times and non-fulfillment of orders with the customers, which represents 80% of dissatisfied orders due to the limited production capacity and non-productive time in the plant. This problem is due to limited production capacity, deficient production planning, and lack of materials. In order to solve this problem, industrial engineering tools were used. The application of these tools improved production from 79% to 95%.

1. Introduction

Textile companies have grown exponentially within the informal market, becoming commercially dynamic without the required accompanying strategic development. The textile industry is very important for the country's economy because it accounts for 1.3% of the GDP. Furthermore, this industry is considered the second most important industry within the manufacturing sector, which accounts for 7.4% of the GDP [1]. According to National Society of Industries, 87.5% of textile products are consumed in the domestic market, and demand for 12.5% of textile products is from the external market. Thus, the national market is the biggest consumer of textile products. Most textile industries are made up of microenterprises, which represent 95% of the total number of companies within this sector [1]. Most textile companies engaged in garment production (MSEs) are managed according to the owner's basic knowledge, without proper planning and control.

The companies' production is not aligned with demand, which causes problems related to non-fulfillment of orders [2]. Currently, the market has begun appreciating the time taken and the quality of the product's delivery [3]. This is related not to cost reduction but rather to the elimination of its

[4] To whom any correspondence should be addressed.

Content from this work may be used under the terms of the Creative Commons Attribution 3.0 licence. Any further distribution of this work must maintain attribution to the author(s) and the title of the work, journal citation and DOI.
Published under licence by IOP Publishing Ltd

cause. Non-fulfillment of orders is the most important problem of the sector. Non-fulfillment or delay of orders affect the profitability of the company; as a result of the root causes, they lead to an excess of working hours for the workers and a stagnation of raw material. "Poor management will result in unsatisfied demands and will lead to the failure of organizations" [4].

Lingerie clothing production has increased from 2012 to 2017 by 52% [4]; therefore, an opportunity to improve can be observed, considering that companies must be prepared for the increase in demand for this kind of garments. For this reason, companies, especially MSEs, need to be well organized in order to satisfy growing demand and correct existing problems. Without discipline and culture to shape the company, the results of any search for improvements will be circumstantial and are likely to fail. Therefore, several techniques that contribute to the improvement of fulfilling orders in companies have been used. This model suggests ways for reduction of such non-fulfillment of orders by means of a production planning and control model in Peruvian lingerie clothing MSEs. For such purpose, the value of the customer and the fulfillment of orders should be maximized in order to increase the final value of the products. The Deming cycle (PDCA) and other tools that may contribute to the company should be used in each phase including master production schedule, material requirements planning (MRP), line balancing, and training, focusing on the improvement of quality in organizations. These tools will be used to solve a problem existing in lingerie clothing MSEs of the textile industry.

2. State of the art

2.1. Master planning schedule and material requirements planning

The non-fulfillment of orders is due to poor estimates regarding production capacity, performance of raw material resupply, and production control. Lack of production control also causes thefts, recounting errors, misplacement, and damage of raw materials. This inaccuracy costs 10% of profits in retail companies [5].

In one study, the MRP system is used because it is more flexible and adaptable to the changing conditions of the market. This model allows a faster and more consistent delivery of a mix of products with different volumes [6]. However, other authors have stated that the MRP has some limitations with respect to production capacity and fixed delivery periods. Therefore, they present a model that includes material requirements planning (MPS) for a more competitive production planning [7][8]. Another study applied the MRP tool under a mixed integer linear programming model, which allowed for the estimation of net requisites of products in processes and the amount of batches [9].

2.2. Line balancing

On the one hand, the stage presenting the most deficiencies is thought to be sewing because in this stage, several operations having different cycle times to be performed are involved [10]; on the other hand, some authors claim that the sewing process is the most important and crucial stage among all processes of garment production [11]. As a result, the problematic areas, commonly known as the bottleneck, appear in the production line, work, or operation station. The bottleneck operation has the longest cycle time, and this is where the sewing process slows down because not all the workers are experts in carrying out the whole process. Bottleneck problems are the main reason for efficiency reduction in the production line [10]. The delivery period for this process is fixed according to the time taken to move the material, seeking to eliminate this waste, that is, the waiting time in processes. Reorganizing internal processes in the production process alone is not enough to ensure the fulfillment of orders. Deficiencies within the supply chain should be identified, and strategies must be adopted accordingly [12].

Line balancing is a type of technique used to balance the production line. However, the application of this technique is not ideal without meeting the standards of working faster while achieving better quality [10]. In the study, observation and description of the work method will provide insights regarding how the work is done and identify how to improve the process and how long it should actually take [13].

Line balancing helps to assign tasks in workstations, so as to achieve optimal assignment. With this method, the elimination of bottlenecks and reduction of activities with no added value in each workstation is sought, that is, a method to share tasks. The research methodology includes estimates on the hourly production capacity; the cycle time of each process; and identifying bottlenecks, activities with no added value, and the production line balance by means of the work-sharing method. Line balancing through the work-sharing method provides higher efficiency in the line of production as well as the workforce [14].

3. Contribution

3.1. Proposed model

An outline of the model can be found in Figure 1. The design for the solution model is presented, which is divided into the four phases of PDCA.

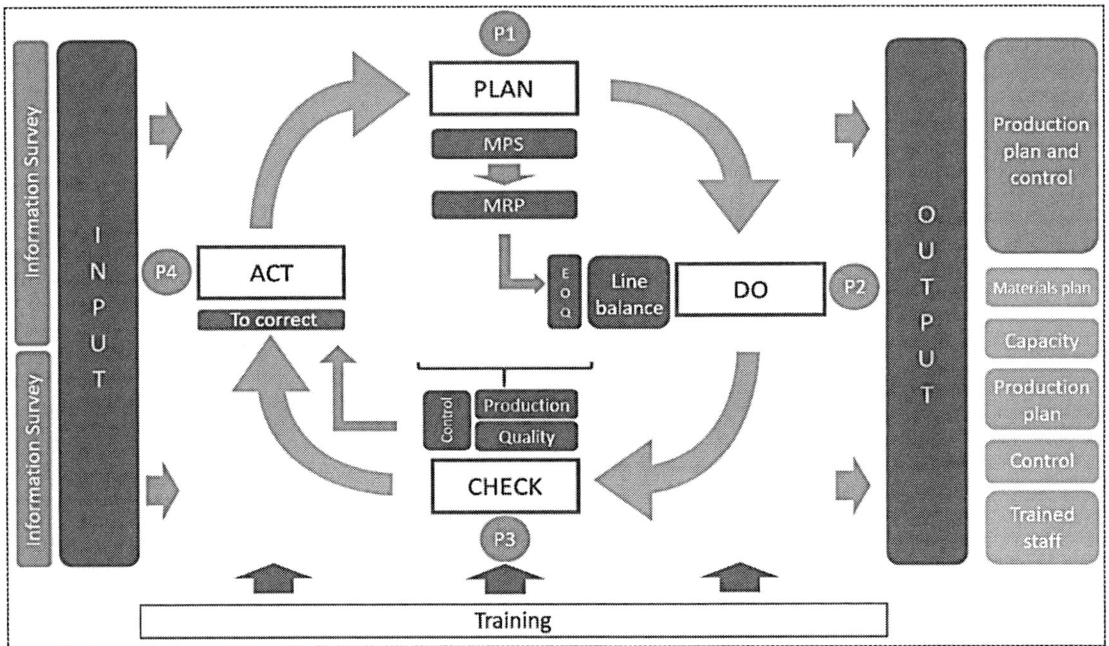

Figure 1. Production plan and control model

3.1.1. Phase 1: PLAN. The members of the senior management and the workers involved in the production process should be committed. Furthermore, an identification of the process answering the five Ws should be made. After its application, the MPS is applied to the preliminary capacity and the MRP, thus achieving the plan for materials, purchases, and production.

3.1.2. Phase 2: DO. The line balancing tool is applied, which will be used from the moment products and orders are requested. With the application of this tool, workers' request and capacity optimization will be achieved. Throughout its application, quality training should be made to raise awareness among workers in order to create knowledge in terms of quality.

3.1.3. Phase 3: CHECK. The third phase involves a verification of production control and quality training. The production control activity is related to the MPS and line balancing, in which planning compliance is controlled. Furthermore, a follow-up of the specifications made by the customers ordering products is carried out, in which surveys are conducted to identify whether workers need to have more training.

3.1.4. Phase 4: ACT. The carried-out process needs to be assessed in order to constantly improve its performance. If any defect, error, or reduction is found, the PDCA stages should be applied in order to solve it.

3.2. Proposed method. The proposed method is presented with a flow chart, which can be seen in Figure 2, of processes describing in detail each activity that needs to be performed.

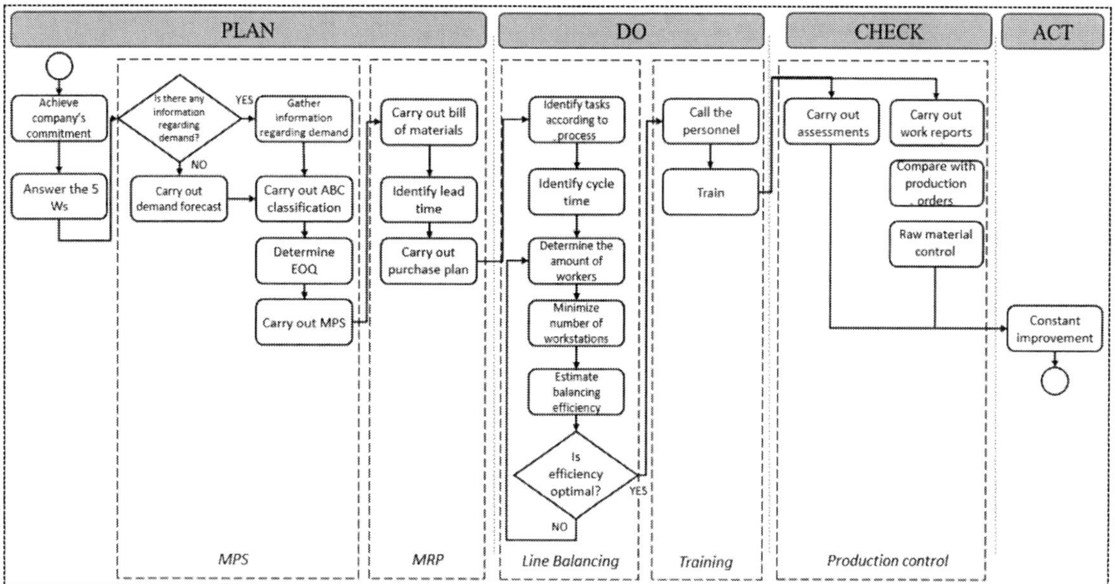

Figure 2. Flowchart of the proposed model

3.3. Indicators

3.3.1. Indicator 1. Fulfillment of Orders (FOO): Fulfillment of orders at a level higher than 90% is expected to be achieved.

$$FOO = \frac{Orders\ fulfilled}{Total\ orders} \times 100 \tag{1}$$

Interpretation: If the result is higher than 85%, it is considered optimal; between 60 and 84%, it is considered regular; and finally, if the result is 59% or lower, it is considered deficient.

3.3.2. Indicator 2. Production Capacity (PC). Production capacity at 95% efficiency is expected to be achieved.

$$\mathrm{PC} = \frac{Actual\ capacity}{Suggested\ capacity} \times 100 \tag{2}$$

Interpretation: If the result is higher than 95%, it is considered optimal; between 70 and 94%, it is considered regular; and finally, if the result is 69% or lower, it is considered deficient.

3.3.3. Indicator 3. Production Efficiency (PE). Production efficiency of 95% is expected to be achieved.

$$PE = \frac{Finished\ products}{Requested\ products} \times 100 \quad (3)$$

Interpretation: If the result is higher than 95%, it is considered optimal; between 70 and 94%, it is considered regular; and finally, if the result is 69% or lower, it is considered deficient.

3.3.4. Indicator 4. Raw Material Handling (RMH). Less than 5% of waste is expected to be achieved.

$$RMH = \frac{Raw\ material\ used}{Raw\ material\ received} \times 100 \quad (4)$$

Interpretation: If the result is lower than or equal to 5%, it is considered optimal; between 6 and 10%, it is considered regular; and finally, if the result is 11% or higher, it is considered deficient.

4. Validation

4.1. Case study

The pilot plan will be applied in a company called Sensualidad y más EIRL, identified with the taxpayer number 20600864387. This company is engaged in the design, production, and sale of female lingerie garments. The key processes are design, cutting, sewing, and packaging. The company's demand is seasonal; it sells more products in January, February, July, November, and December. This company sells products as per request, but it also has three stores located centrally in Gamarra. The production is carried out in a workshop located in the district of La Victoria.

4.2. Initial diagnostic

A diagnostic was carried out in the case study, generating the following initial results:

Table 1. Initial diagnostic

Indicator	Diagnostic
1. FOO	78 %
2. PC	79%
3. PE	89%
4. RMH	11%

4.3. Contribution development within the context

4.3.1. Component 1. The ABC classification is carried out in order to find the most important product:

Table 2. ABC Classification

	Units	Sales (S/.)	% of sales	Accumulated %	ABC Classification
Line 1	25632	S/. 812,021.76	64.00%	64%	A
Line 2	10269	S/. 329,883.84	26.00%	90%	B
Line 3	4110	S/. 126,878.40	10.00%	100%	C
TOTAL	40011	S/. 1,268,784.00	100%		

As can be seen, Line 1 is most important for the company.

After the MPS is carried out, the size of the batch is needed to develop it:

$$Qpoq = \sqrt{\left(\frac{f}{f-d}\right) * \frac{2 * D * S}{H}} = \sqrt{\left(\frac{93}{93-92.8}\right) * \frac{2 * 25632 * 11}{276 * 0.51}} = 1683\ unidades \quad (5)$$

The number of units produced in each period is 1,683. Once the EOQ and initial inventory (421 units) is known, the MPS is carried out. As a result, the monthly MPS should be of 2,000 products.

4.3.2. Component 2

Once the production planning is established, the material requirements planning (MRP) is carried out. The BOM is carried out for materials acquisition.

Afterward, the lead time for each material is carried out:

Table 3. Materials' Lead Time

Material	Requirement	Time
Finished product	1 unit	22.83 minutes
Parts	1 unit	15 minutes
Thread	1.5 meters	1 week
Tulle	2 meters	1 month
Blond lace	2 meters	3 months
Elastic	1 meter	1 week
Underwire	1 pair	1 week
Cup	1 pair	2 months
Straps	20 cm	3 months
Bow	25 cm	1 week
Packaging	1 unit	3 months

Finally, the purchase plan is carried out at this stage, in which the estimated time needed to acquire the raw materials before production starts is three months.

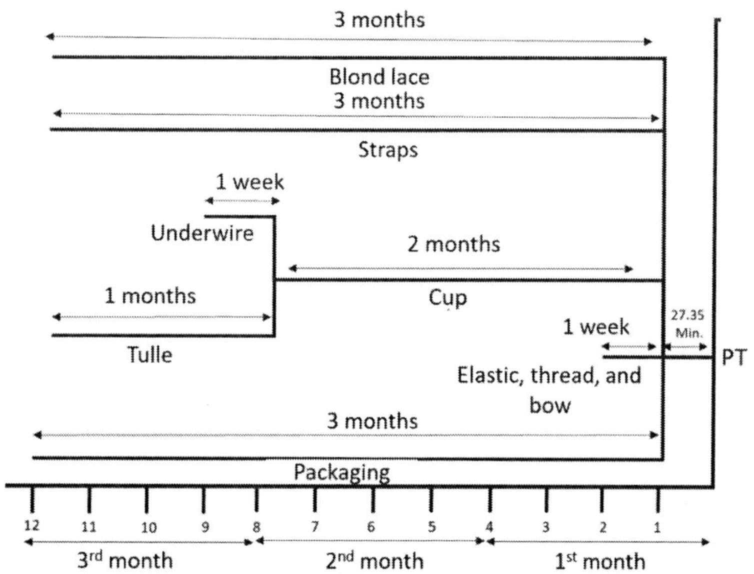

Figure 3. Plan

4.3.3. Component 3

In order to perform line balancing, first the tasks to be carried out in the production process need to be detailed and subsequently grouped.

Table 4. Tasks

Description	Means / Equipment	Distance (m)	Time (min/units)	Precedence
1. Design product	Computer		1	-
2. Inspect and cut fabric	Scissors		0.5	-
3. Transport		2	0.25	-
4. Mend	Mending machine		2.5	2
5. Mend basting stitch	Mending machine		1.5	1
6. Inspect blond lace			1	1
7. Cut blond lace	Scissors		0.5	6
8. Inspect elastic			1	1
9. Cut elastic	Scissors		0.5	8
10. Transport		0.5	0.25	
11. Sew	Regular sewing machine		2.5	9
12. Inspect cup			1	1
13. Line cup			2	12
14. Sew	Regular sewing machine		3	13
15. Inspect straps			1	1
16. Cut straps	Scissors		0.5	15
17. Place straps			0.5	16
18. Inspect bow			1	1
19. Cut bow	Scissors		0.5	18
20. Sew	Regular sewing machine		3	19
21. Transport		0.5	0.25	
22. Sew frayed edge	Flatlock sewing machine		2.5	20
23. Place underwire			1	22
24. Transport		0.5	0.25	
25. Sew straps	Coverstich machine		1.25	23
26. Sew elastic	Coverstich machine		1.25	25
27. Transport		0.5	0.25	
28. Sew bow	Knitting machine		2	26
29. Pack			1.25	28
30. Store			1.25	29
TOTAL:		4	33.25	

The line balancing is carried out to define the optimal number of workers to perform the sequence of tasks.

Table 5. Sequence of Tasks

N°	Task	Name of the task	Minutes	Estimated time (Sec/units)	Units/minutes	Workers	Total Units			EFFICIENCY
1	A	Inspect and cut	7.5	450	0.133	2	0.267	225	310	73%
2	B	Remesh	4	240	0.25	1	0.25	240	310	77%
3	C	Sew	15.5	930	0.0645	3	0.194	310	310	100%
4	D	Sew frayed edge	3.5	210	0.285	1	0.286	210	310	68%
5	E	Packaging	1.25	75	0.8	1	0.8	75	310	24%

It has been recorded that eight workers are needed for a demand of 75 units per shift. With this number of workers, 92 units were produced per shift. Also, an increase in productivity is achieved with 11.5 garments per worker.

Table 6. Productivity

		Workers	Per./hour	Perf. /shift	Productivity
75 8	Units Hours	8	11	92	11.5
63 8	Units Hours	6	8	63	10.5

However, it can be seen that the efficiency of the tasks is not adequate, and therefore, an improvement is needed. Next, the last two activities are grouped in order to be performed by a single worker to improve efficiency:

Table 7. Sequence of tasks

N°	Task	Name of the task	Minutes	Estimated time (Sec/units)	Units/minutes	Workers	Total Units			EFFICIENCY
1	A	Inspect and cut	7.5	450	0.133	2	0.267	225	310	73%
2	B	Remesh	4	240	0.25	1	0.25	240	310	77%
3	C	Sew	15.5	930	0.0645	3	0.194	310	310	100%
4	D+E	Sew frayed edge+ Packaging	4.75	285	0.211	1	0.211	285	310	92%

It has been recorded that seven workers are needed for a demand of 75 units per shift. With this number of workers, 92 units were produced per shift. Also, an increase in productivity is achieved with 13.14 garments per worker.

Table 8. Productivity

		Workers	Perf./hour	Perf./shift	Productivity
75 8	Units Hours	7	11	92	13.14285714
63 8	Units Hours	6	8	63	10.5

By producing 92 units per shift, 2484 garments can be produced per month, with a cycle time of 27.85 minutes/garment, achieving an average efficiency of 85.5%.

$$Cycle\ time = \frac{69{,}195\,\frac{minutes}{month}}{2484\,\frac{garment}{month}} = 27.85\,\frac{minutes}{garment} \tag{6}$$

- **Indicator 1:** According to the measurement, a result of 95.90% was obtained, which indicates that the capacity for efficiency is optimal.
- **Indicator 2:** A result of 89.7% was obtained, which indicates that according to the measurement, the actual production capacity is optimal in respect to the one suggested.
- **Indicator 3:** A result of 95.3% was obtained, which indicates that according to the measurement, the production efficiency is optimal.
- **Indicator 4:** A result of 4.3% was obtained, which indicates that according to the measurement, the raw material handling is optimal.

5. Conclusions

After production planning, it was concluded that according to the PMP, 2,000 units need to be produced monthly for the three months in which demand is higher. The production master plan is used to set the production plan for the company.

Line balancing is carried out, which helps to know the number of workers needed and also to establish the number of workers for each task. According to the results, seven workers are needed for production in the plant.

An efficiency of 71.5% was obtained in the line balancing of the tasks.

Between the current situation and the pilot plan, an improvement with respect to the production capacity and creation of knowledge about the quality of workers in a period of time can be observed. If these levels of efficiency and productivity continue, the expected production levels can be achieved, and customer satisfaction can be met.

6. References

[1] Ministry of Production 2017 Textile Industry and Garment Production: Sector Research study *Retrieved from: http://ogeiee.produce.gob.pe/images/oee/docTrab_Textil.pdf* 157

[2] Ortega G A, Jaramillo K V O, Cabrera J P O and Trejos C A R 2017 Production planning and control model in the mid-term for a textile industry in a make-to-order environment *Revista Ingenierías Universidad de Medellín* **16**(30) 169-193.

[3] Sari D K, Hetharia D, Saraswati D and Marizka R 2019 Design of Flat Shoes Quality Control System using PDCA (Case Study at PT DAT). In IOP Conference Series: Materials Science and Engineering (Vol. 528, No. 1, p. 012073). IOP Publishing.

[4] Nemtajela N and Mbohwa C 2017 Relationship between inventory management and uncertain demand for fast moving consumer goods organisations *Procedia Manufacturing* **8** 699-706.

[5] Miclo R, Lauras M, Fontanili F, Lamothe J and Melnyk, S A 2019 Demand Driven MRP: assessment of a new approach to materials management *Int J Prod Res* **57**(1) 166-181.

[6] Hong P, and Leffakis Z M 2017 Managing demand variability and operational effectiveness: case of lean improvement programmes and MRP planning integration *Prod Plan Control* **28**(13) 1066-1080.

[7] Rossi T, Pozzi R, Pero M and Cigolini R 2017 Improving production planning through finite capacity MRP *Int J Prod Res* **55**(2) 377-391.

[8] Carazas L, Barrios M, Nuñez V, Raymundo C, Dominguez F 2019 Management model logistic for the use of planning and inventory tools in a selling company of the automotive sector in Peru *Advances in Intelligent Systems and Computing* **971** 299-309.

[9] Sadeghi H, Makui A and Heydari M 2014 A simulation method for Material requirement planning supply dependent demand and uncertainty lead-time *Afr J Bus Manag* **8**(4) 127 -135.

[10] Safra I, Jebali A, Jemai Z, Bouchriha H and Ghaffari A 2018 Capacity planning in textile and apparel supply chains *IMA J Manag Math* **30**(2) 209-233.

[11] Haque M T, Hossain M R and Hasan M S 2018 Bottleneck problem reduction of a garment manufacturing industry in Bangladesh by using line balancing technique *Int J Res Adv Eng Tech* **4**(2) 28-32.

[12] Macchion L, Fornasiero R and Vinelli A 2017 Supply chain configurations: a model to evaluate performance in customised productions *Int J Prod Res* **55**(5) 1386-1399.

[13] Džubáková M and Kopták M 2017 Work standardization in logistics processes *Qual Innov Prosper* **21**(2) 109-123.

[14] Kayar M and Akalin M 2015 A Research On The Effect Of Method Study On Production Volume And Assembly Line Efficiency *J Text Apparel/Tekst Konfeksiyon* **24**(2).

IOP Conf. Series: Materials Science and Engineering **796** (2020) 012019 doi:10.1088/1757-899X/796/1/012019

Lean Manufacturing Model for production management to increase SME productivity in the non-primary manufacturing sector

S Flores-Meza[1], J Limaymanta-Perales[1], J Eyzaquirre-Munarriz[1], C Raymundo-Ibañez[2, 4] and M Perez[3]

[1] Ingeniería Industrial, Universidad Peruana de Ciencias Aplicadas (UPC), Lima 15023, Perú.
[2] Dirección de Investigación, Universidad Peruana de Ciencias Aplicadas (UPC), Lima 15023, Perú.

E-mail: carlos.raymundo@upc.edu.pe

Abstract. Currently, there is a large percentage of small and medium-sized enterprises (SMEs) in the Peruvian textile market that show economic loss because of the payment of penalties to customers, which are incurred owing to the delay in the delivery of order batches. This is due to poor production management and a lack of focus. The manufacturing sector is essential because of its high contribution to the country's gross domestic product. Currently, SMEs do not employ methodologies that help improve production and process management as they do not realize how important and necessary the methodologies are, in addition to how complex these may be. Therefore, this paper will propose a production management model designed for SMEs in this sector, based on Lean methodology where the objective is time reductions and production increases as well as exerting changes to the organizational culture. Thus, this model will help organizations to avoid incurring economic losses because of the payment of penalties for orders not delivered on time. To validate the present model, a time simulation was performed in the manufacturing area of a textile company. The result of this project was positive, since there was a 25% increase in productivity and a 20% reduction of takt time with respect to the initial data.

1. Introduction

The textile sector represents a significant share of the world's gross domestic product (GDP). In 2017, the worldwide participation indicator was 3.4%. In Peru, this industry represents 8.9% of the entire manufacturing production and 1.3% of the national GDP (2014). In addition, textile exports had increased by 11.5% over the previous year. SMEs represent 99.5% of Peru's business sector, and of this percentage, 16.5% are companies engaged in manufacturing garments (Ministry of Production, 2015). Further, 47.8% of these companies are located in Lima. [1][2][3] However, a competitive disadvantage for most SMEs in Latin America is their often informal and unprofessional management. This is due to a lack of knowledge of the various business management tools, which reduces effectiveness, profitability, and sustained development. [6]

---

[4] To whom any correspondence should be addressed.

Content from this work may be used under the terms of the Creative Commons Attribution 3.0 licence. Any further distribution of this work must maintain attribution to the author(s) and the title of the work, journal citation and DOI.
Published under licence by IOP Publishing Ltd

This study has an economic justification, since the use of a production model based on Lean tools and process management will allow the company to reduce downtime and fulfill order batches on time, thus avoiding penalties. In the sectorial analysis, it was identified that 77.3% of SMEs deliver 43.5% of all orders late, which is primarily due to deficient production management. [7] Through the literature review, it was found that the Lean manufacturing philosophy has a positive impact on companies. A study found that the development of Lean tools in a textile company managed to reduce the duration of activities that do not add value to the process; this led to economic savings of $25,916,485. In addition, this implementation achieved qualitative benefits, such as the improvement of the work environment, a better image of the work area, the creation of a teamwork culture, and a sense of belonging to the company. [5]

Although there is a considerable amount of information about the application of Lean manufacturing tools in companies, few studies have addressed the impact of Lean on SMEs and even less have focused on process management in textile SMEs. In the Peruvian reality, about 70% of these companies are informal. [6] Therefore, it can be deducted that the implementation of Lean manufacturing or process management is scarce. Based on the aforementioned, this paper proposes the implementation of a production management model based on Lean tools (5S and Kanban), supported by knowledge and change management, with the aim of improving employee skills, increasing profits, and reducing penalties due to late order fulfillment. The model is based on 5S implementation, using Kanban cards, throughout which knowledge and change management will be introduced.

This article has been structured in four main sections. In the first section, an intensive review of the literature on Lean manufacturing tools and process management was conducted. In the second section, the proposed methodology is described, focusing on the contribution of the study. In the third section, the validation of the proposal is developed. Finally, in the fourth section, the conclusions of the study and the bibliographical references are detailed.

2. State of The Art

The literature review has allowed us to learn about techniques and methodologies that have been used by organizations to achieve better production management, which uses a set of tools, methodologies, and resources to maximize process or company productivity levels [11]. Among the techniques used to improve production or processes by reducing cycle times are Lean manufacturing, Six Sigma, and reengineering. [12] [13]

The authors define each of these techniques from different points of view. Lean manufacturing can be defined as a systemic and systematic approach for identifying and removing waste or activities that do not add value through continuous improvement by making the product and the information necessary to produce a quality product flow. Moreover, Six Sigma is an improvement methodology developed to improve process quality with the goal of creating a near-zero defect quality strategy by improving customer satisfaction and financial results. Finally, process reengineering is a means to achieve sweeping changes in performance, taking costs, cycle times, service, and quality as measurements and seeking to improve customer-related processes to add real value. After reviewing all the aforementioned concepts and definitions, we can note that all these methodologies serve the purpose of improving customer-related processes within an organization and that also contribute value to the product offered. However, only the Lean manufacturing methodology can be applied to all companies and their levels regardless of company size since the methodology does not require large investments and results may be observed in the short term. [14] [15] [16] [17]

3. Contribution

3.1. Model Proposed

Based on the research conducted and the comparative analysis of different methodologies, a model was developed wherein the important dimensions of production management models were assessed. As internal MSE processes were considered as essential to product development, we combined the

Lean manufacturing methodology with change and knowledge management to guarantee the successful implementation of the model since change and knowledge management focus on the most important aspect: the human factor.

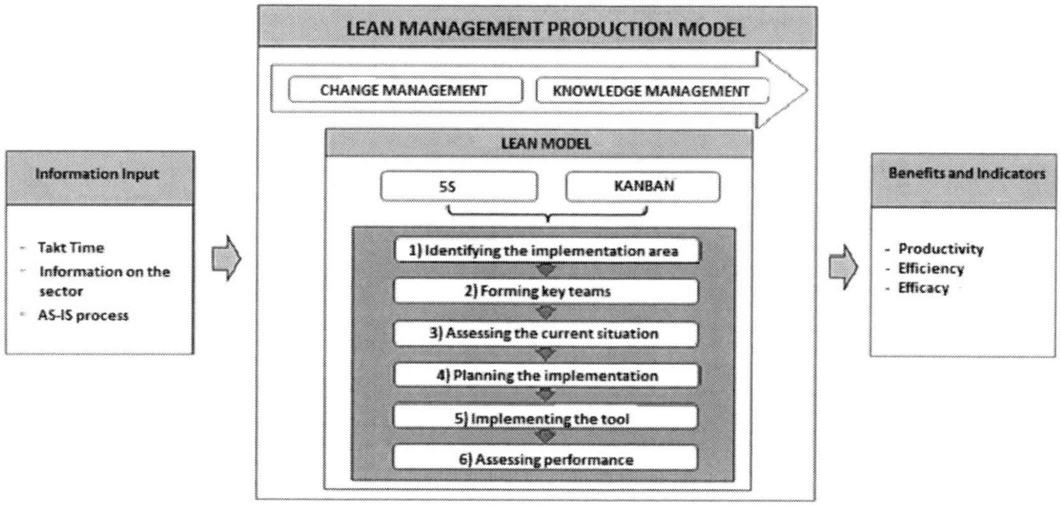

Figure 1. Detailed view of lean management production model

3.2. Components of the Model Proposed

3.2.1. Knowledge Management. Knowledge management is a tool that allows for the possibility of assessing your own knowledge and the importance of new knowledge in an organization. It is necessary to know the area and the personnel that will be involved in the management model.

3.2.2. Change Management. Change management is a structured approach to managing change issues primarily related to people and the organization to achieve initial objectives. The objective of this management is to evaluate and create a contingency plan to deal with changes in work activities, improve communications, and provide training to counteract resistance.

3.2.3. Lean Manufacturing Model

- **First Stage:** Identifying the implementation area. A reduced physical area should be selected to be used as a pilot area. After the pilot test, the implementation may be expanded to other areas of the company.
- **Second Stage:** Forming key teams. A key team will be in charge of the proper development of the implementation project.
- **Third Stage:** Assessing the current situation. An initial audit of the selected area will be conducted to acquire an awareness of the current situation and thus manage current knowledge.
- **Fourth Stage:** Planning 5S and Kanban implementation within the project.
- **Fifth Stage:** Implementing the 5S and Kanban in the work area. In this stage, all phases will be developed for each of the selected tools.
- **Sixth Stage:** In this stage, performance evaluations and final audits will be carried out after 5S implementation. Subsequently, Kanban will be applied in order to verify whether the project reached the desired results.

3.3. Process of the Proposed Model

This section presents a flowchart that will graphically represent the procedures that will be carried out during the development of the project. As mentioned above, tools will be used to help the human factor cope with change. This is extremely important for the success of the implementation of a Lean production model in any type of company since it is oriented to the management of the human factor for the purpose of facilitating the implementation of a Lean management production model.

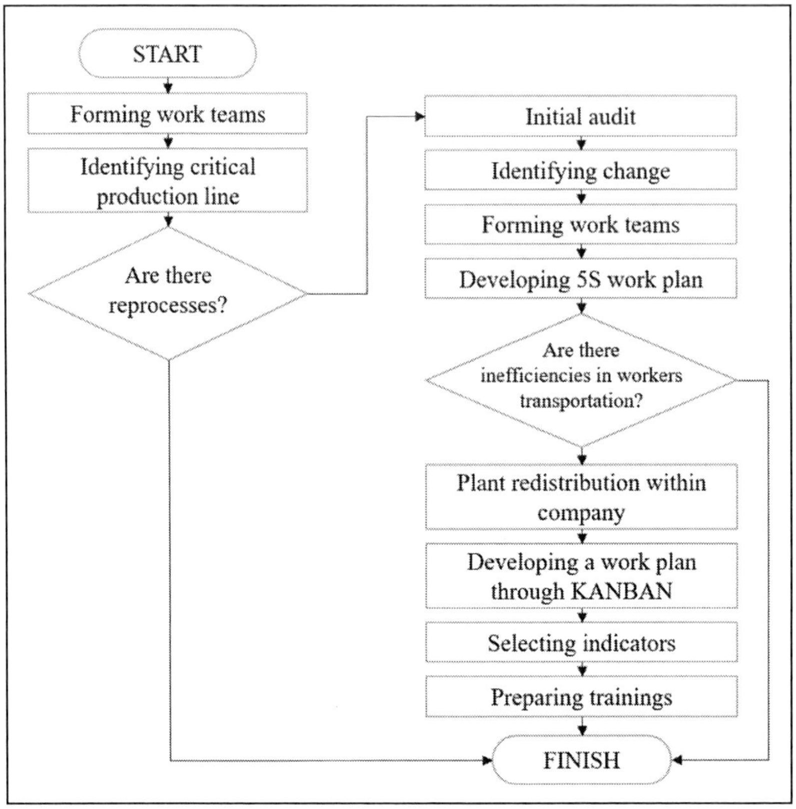

Figure 2. Process flowchart

3.4. Proposed Indicators

- **Productivity:** This indicator will help identify the number of products manufactured per hour and thus identify whether the required demand is being met.

$$Productivity = \frac{Total\ of\ products\ manufactured}{Total\ hours} \quad (1)$$

- **Efficacy:** The purpose of this indicator is to measure performance of the company area under analysis. The formula for calculation is as follows:

$$Efficacy = \left(\frac{Goal\ achieved}{Scheduled\ goal}\right) x \left(\frac{Scheduled\ Time}{Time\ Achieved}\right) x \left(\frac{Client\ Evaluation}{Total\ Evaluation}\right) \quad (2)$$

- **Efficiency:** This indicator will help us achieve the goals with minimum resources. The formula is as follows:

$$Efficiency = \left(\frac{Hours\ worked}{Hours\ available}\right) x \left(\frac{Used\ budget}{Allocated\ budget}\right) \quad (3)$$

4. Validation

4.1. Case Study
The test scenario is the manufacturing area of a textile SME. A study of operating times for each station was carried out. The proposed model will be applied to eliminate downtime. In order to do this, based on the resources of the SME, such as number of operators and number of machines, the Arena software was used to simulate the time it would take to make a batch of 100 pairs of pants. The intention is to comply on time with the scheduled batches and avoid penalties.

4.2. Current Process Simulation
The current flowchart of the path of items was prepared.

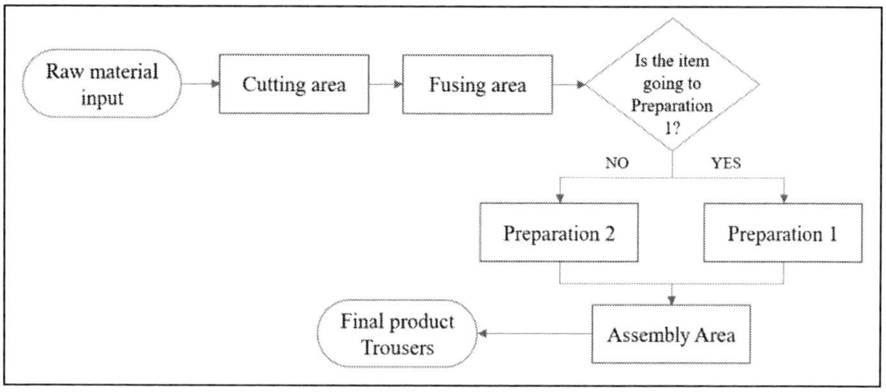

Figure 3. Current flowchart of items

The flowchart of the supervisor's path is presented to give the explanations in each station and for each batch.

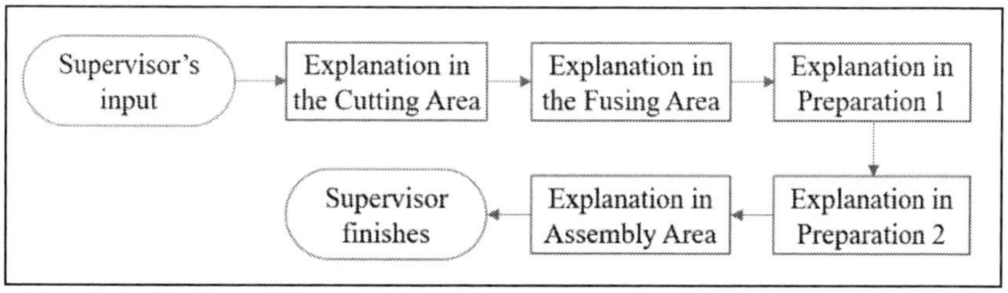

Figure 4. Flowchart of the current path of the supervisor

Using a time recorder (100 data), a PAD is carried out for the process, in which all the activities of each station are described and those that do not generate added value to the product are identified.

Table 1. Time (min) of activities per work station

Work station	Time in minutes		Total time (min/trouser)
	Activities that add value	Acitivities that do not add value	
Cutting	0.21	0.09	0.3
Fusing	0.18	0.08	0.26

IOP Conf. Series: Materials Science and Engineering **796** (2020) 012019 doi:10.1088/1757-899X/796/1/012019

Preparation 1	0.28	0.98	3.26
Preparation 2	2.65	1.14	3.79
Assembly	5.57	2.39	7.96
Total	10.88	4.66	15.5

The simulation is made in the Arena software, in which the average times recorded and the resources for each station are entered. The Arena software simulates production of a batch of 100 pairs of pants.

The following table shows a summary of the results obtained:

Table 2. Summary of results

Description	Data	Unit
Working time	8	Hours/day
Days per month	20	Days/month
Current takt time	0.97	Min/trouser
Av. Production current month	9,897	Trousers/month
Scheduled production of month	12,000	Trousers/month
Batches on time average	18	Batches
Delayed batches average	5	Batches

In one month, there is an average of five late orders. This results in the payment of penalties, which impacts profits. During 2017, 20.3% were late orders and the profit lost was 201,659 Soles. In 2018, the values for late orders and profit lost were 24.5% and 204,264 Soles, respectively.

4.3. Simulation with the Implementation of the Model Proposed

The simulated proposal considers the implementation of the 5S and Kanban Lean manufacturing tools as well as the improvement in the efficiency and productivity of the collaborators owing to change and knowledge management.

The proposed scenario eliminates the following activities:

- Explanation of the characteristics by the supervisor.
- Position of the supervisor.
- The additional time spent searching for tools and materials.
- Unnecessary motion due to disorder in the manufacturing area.

After this, the item path flowchart is as follows:

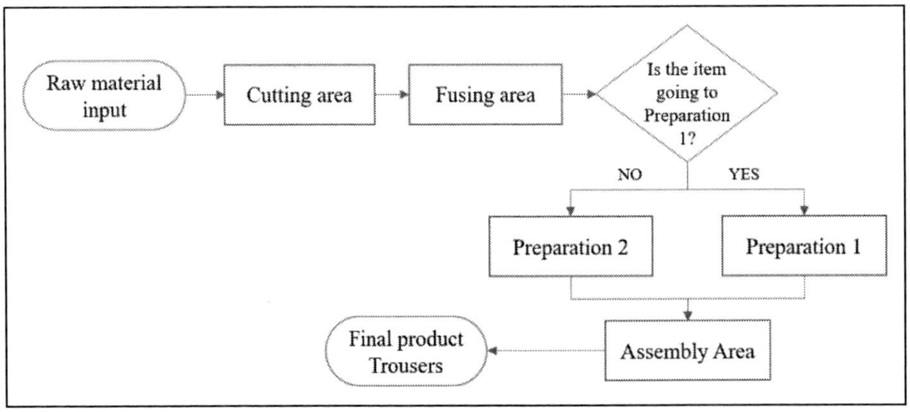

Figure 5. Proposed flowchart of the path of the items.

Although the flowchart for the items is the same, the action flow performed by the supervisor has been eliminated.

Since non-value-added activities are expected to be removed with the implementation of the proposed tools, it is suggested that activities that generate downtimes and delays are eliminated from the Process Analysis Diagram.

Moreover, after eliminating downtimes, the table looks as follows.

Table 3. Time (min) of activities per workstation

Workstation	Time in minutes		Total time (min/trouser)
	Activities that add value	Activities that do not add value	
Cutting	0.21	0.00	0.21
Fusing	0.18	0.00	0.18
Preparation 1	0.28	0.00	0.28
Preparation 2	2.65	0.00	2.65
Assembly	5.57	0.00	5.57
Total	10.88	0.00	10.88

Arena simulates the production process for 100 pairs of pants. After the simulation in Arena, the following results are obtained:

Table 4. Summary of the results of the simulated proposal

Description	Data	Unit
Working time	8	Hours/day
Days per month	20	Days/month
Current takt time	0.78	Min/trouser
Av. Production current month	12,308	Trousers/month
Scheduled production of month	12,000	Trousers/month
BATCHES ON TIME average	23	Batches
DELAYED BATCHES average	0	Batches

With the proposed model, the takt time of the pants is reduced by 20% and the quantity produced per month increases by 25%. This avoids late orders and eliminates the costs of paying penalties.

5. Conclusions

The Lean Production Management Model integrates knowledge and change management together with 5S and Kanban techniques in order to obtain better results in the implementation, taking into account the human factor as the main factor to achieve success. Productivity increased by 24.8% with respect to the productivity initially obtained, from 9,850 to 12,300 pants produced per month; further, takt time was reduced by 20%. With these results, the demand for orders will be covered and the payment of penalties for late orders will be reduced by 84%.

6. References

[1] Aravendan M and Panneerselvam R 2016 Implementation of Hybrid Genetic Algorithm for CLSC Network Design Problem—A Case Study on Fashion Leather Goods Industry *American J Oper Res* **6** 300.
[2] National Institute of Statistics and Informatics 2018
[3] National Institute of Statistics and Informatics 2016

[4] Central Reserve Bank of Peru 2017
[5] National Institute of Statistics and Informatics 2014
[6] National Institute of Statistics and Informatics 2017
[7] PRODUCE 2017 Textile and Clothing Industry: Sectorial Research Study
[8] Benites J M R 2016 Implementación de Lean Manufacturing para mejorar la calidad del producto en la empresa productora de "Calzado Lupita" SA-2016 *Innovacion En Ingeniería* **2**.
[9] Improvement through lean manufacturing tools in a Garment Company, 2014
[10] Diario Gestión, 2019
[11] Santosa W A and Sugarindra M 2018 Implementation of lean manufacturing to reduce waste in production line with value stream mapping approach and Kaizen in division sanding upright piano, case study in: PT. X *MATEC Web of Conferences* **154** 01095.
[12] Santosa W A and Sugarindra M 2018 Implementation of lean manufacturing to reduce waste in production line with value stream mapping approach and Kaizen in division sanding upright piano, case study in: PT. X *MATEC Web of Conferences* **154** 01095.
[13] Stankalla R, Chromjakova F and Koval O 2019 A review of the Six Sigma belt system for manufacturing small and medium-sized enterprises *Qual Manag J* **26** 100-117.
[14] Rafoso-Pomar S and Artiles-Visbal S 2012 Reingeniería de procesos: conceptos, enfoques y nuevas aplicaciones *Ciencias de la Información* **42** 29-37.
[15] Stankalla R, Chromjakova F and Koval O 2019 A review of the Six Sigma belt system for manufacturing small and medium-sized enterprises *Qual Manag J* **26** 100-117.
[16] Becerra A, Villanueva A, Núñez V, Raymundo C and Dominguez F 2019 Lean manufacturing model in a make to order environment in the printing sector in Peru *Adv Int Syst Comput* **971** 100-110.
[17] Patel M and Desai D A 2018 Critical review and analysis of measuring the success of Six Sigma implementation in manufacturing sector *Int J Qual Reliab Manag*.

IOP Conf. Series: Materials Science and Engineering **796** (2020) 012020 doi:10.1088/1757-899X/796/1/012020

Comprehensive Management model for increasing the competitiveness of small and medium artisan jewelry enterprises in Peru

K Mejia-Pajuelo[1], K Solis-Galdos[1], D Mauricio-Sanchez[1], C Raymundo-Ibañez[2, 4] and M Perez[3]

[1] Ingeniería de Gestión Empresarial, Universidad Peruana de Ciencias Aplicadas (UPC), Lima 15023, Perú.
[2] Dirección de Investigación, Universidad Peruana de Ciencias Aplicadas (UPC), Lima 15023, Perú.
[3]Escuela Superior de Ingeniería Informática, Universidad Rey Juan Carlos, Mostoles, 28933, España.

E-mail: carlos.raymundo@upc.edu.pe

Abstract. Today, many small and medium enterprises (SMEs) either go out of business or see their growth being stalled due to poor activity management and organization, which is why they are often unprepared to engage against large companies leading their competitive markets. This study aims to develop a comprehensive management model, exclusively designed for SMEs in the jewellery sector, which may encompass all areas that create value and make them competitive. The development of this comprehensive management model was deployed in 21 steps classified into three tiers based on three major philosophies. This study uses lean manufacturing to identify activities that add value and reduce waste, and lean green was used to provide the environmental care and culture approach required for a friendly and fair workplace environment. The selected scenario is a jewelry workshop, wherein techniques and tools were completely applied and yielded a 20% increase in operating performance and a 15% increase in annual return.

1. Introduction

According to the Jewelry and Goldsmith Analysis [1], the sector's exports experienced a 30% increase against the previous year, with a total revenue of US$ 121.3 million. Although these numbers represent a great contribution to the Peruvian economy, they are considered as low yield as, according to the Institute of Mining Engineers of Peru [2], Peru contains 24% and 5% of the world's silver and gold reserves, respectively, but only less than 1% is used for manufacturing jewellery [3], while the foreign market generates US$ 8,800 million per year. This situation has undoubtedly affected the growth of businesses within the sector.

The business landscape comprises 1,728,777 companies, with 99.5% corresponding to micro and small companies and 0.5% representing large companies [4]. According to the National Business Survey, only 20% of the micro businesses reported having a business plan in place when they began

[4] To whom any correspondence should be addressed.

Content from this work may be used under the terms of the Creative Commons Attribution 3.0 licence. Any further distribution of this work must maintain attribution to the author(s) and the title of the work, journal citation and DOI.
Published under licence by IOP Publishing Ltd

operating, while the remaining 80% usually disappear from the market within a year, as opposed to large companies that remain in business. Many SMEs decide against introducing management within their work guidelines, since these businesses are mostly unaware that applying knowledge in the management, organization, planning, and control of their operations can really create a difference in their results and make them more competitive in the national and international markets. This same scenario is experienced by many local organizations within the country.

All previous works describe comprehensive management models for large serial production line companies, a scenario that does not favour the Peruvian jewellery market. Moreover, no model or research study has even been validated in operating SMEs. Finally, none of the comprehensive management models developed has been applied in the jewellery industry.

Therefore, this study proposes a 3-tier model: business units, comprised by the different areas of the company; technological support, meaning the software used for supporting operations; and management, which is responsible for planning, organizing, directing, and monitoring each business unit based on the three fundamental model philosophies: lean manufacturing, lean green, and culture.

2. State of The Art

For the literature review, we focus on studies that might contribute to the development of different company areas. Starting with sales, and since businesses would not exist without sales, companies must plan, develop, and manage the various channels available to the customer for acquiring the product or service they offer. The multiple-equation model, as proposed by Isaac, Harald, and Scott [5], quantifies the impact and magnitude of the online and offline sales channels of the company, in terms of the expenses generated by both channels. Similarly, Abhishek, Jerath, and Zhang [6] propose a game theory model on reselling against the sales agency. This model also assesses which of the following channels offers the greatest benefits to companies: online sales, offline sales, direct sales, and agency resales.

On the other hand, like with any other jewellery manufacturing company, the production area, where the added value is created for the product, must also be considered. To control the production flow in craft workshops, we used the Cobacabana technique, developed by Matthias Thürer [7], as an alternative proposal to the Kanban technique. Unlike Kanban, Cobacabana adapts to variable workflows, where multiple activities must be performed at the same time, besides assessing delivery times by reducing production delays. Another proposal developed by Mhargareta Gansterer [8] studies aggregate planning impacts on a make-to-order (MTO) environment through a comprehensive hierarchical production planning (HPP) model, which combines the use of a linear mathematical model with a discrete event simulation model capable of generating a positive effect on saving money and resources, planning, and on-time deliveries. In the same manner, Randhawa and Ahuja [9] seek to improve quality, production, cost optimization, moral values, and work culture in general for the manufacturing industry through the successful implementation of the 5S tool.

Moreover, since this is a MTO environment, the activities of the workers must be effectively managed to provide quick response times. For these purposes, Nallusamy developed a comprehensive production management mode to address downtime in companies by applying single-minute exchange of die (SMED) and Kaizen, common lean manufacturing tools, at each workstation. Similarly, Saravanan's lean approach considers that a successful SMED implementation is able to increase productivity and reduce downtime for both machines and operators. On the other hand, author Suganthini Rekha argues that value-stream mapping (VSM) identifies manufacturing wastes to determine and eliminate actions that do not add value, which in turn improves company productivity. Finally, Bocken [10] developed a study that uses VSM to explore for a broader and more sustainable business thinking, which provides a conceptual link to the activities of the company, such as design, production, supply chains, relationships, and distribution channels.

In this light, supply chain management (SCM) has become a key strategic tool for improving performance and guaranteeing the competitiveness of companies in the market [11]. For this reason, Macchion and Fornasiero assessed different configurations based on a discrete event simulation by emphasizing supply chain performance, in terms of delivery times and supply chain inventory volumes, when standard production becomes custom production. On the other hand, Antmann developed a mixed integer linear programming technique that proposes integrating environmental responsibility to the supply chain. Finally, Jukka Hemilä and Jyri Vilko [12] conducted a study based on qualitative interviews and expert group sessions, which provides a holistic perspective of the factors affecting new business development for the service supply chain.

Ultimately, all companies always strive to gear themselves toward continuous improvement in their operations and a constant renewal of goals and objectives to avoid lagging in the market. Continuous improvement is defined through multiple methodologies aimed at removing work module defects, adding value to services or products, and implementing a new work philosophy.

3. Research method

3.1. Operations in Small and Medium Artisan Jewellery Enterprises
The business units of small and medium artisan jewellery enterprises are divided and developed as follows: sales are made from cash points and face to face. Preliminary jewellery designs are outsourced or drawn by hand, based on the item requested and the urgency. The manufacturing process includes the following operations: rolling, cutting, sanding, polishing, washing, and drying. The procurement process, as well as the delivery of the finished products, may be performed by staff members or, in some cases, outsourced. Here technological support is not usually a priority for SMEs because companies usually do not record the data generated by each business unit.

3.2. Comprehensive Management Model for Small and Medium Artisan Jewel Enterprises
The proposed model divides comprehensive management for small and medium artisan jewellery enterprises into three tiers; the first layer is business units, such as sales, design, production, distribution, and supply chain areas, with the latter two usually being outsourced. The second layer is technological support and the third layer is management, which includes three philosophies: lean green, lean manufacturing, and culture. These philosophies serve as comprehensive management

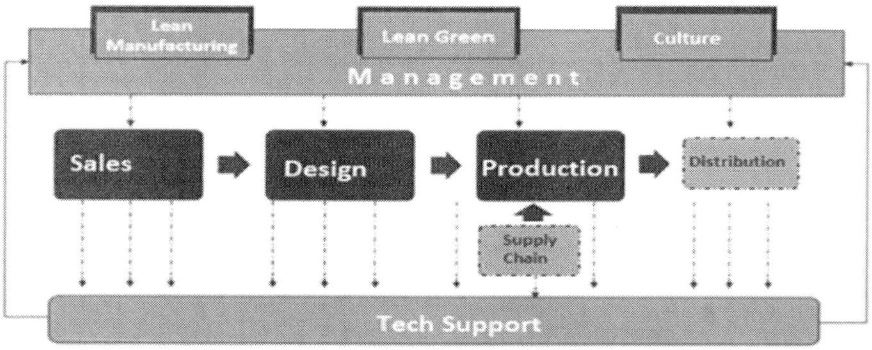

Figure 1. Comprehensive management model

pillars for the business units of the proposed model. The model ultimately aims at streamlining SME management and increasing their competitiveness in the market.

3.3. Operation of the Proposed Model

3.3.1. Comprehensive Sales Area Management. For the management of the sales area, a flowchart reveals the travel path for staff members, customers, and information. On this basis, help tools were designed for the online and physical store salespersons, such as sales speech and the design *sheets*, job profiles, and sales procedure manuals, which facilitate sales closings, increase sales, and proper storage of daily information. Daily information is mainly comprised of daily reports, including daily sales, sales by products, sales by seller type, sales margin, and access to customer information. Based on the results from these reports, the company will be able to set new objectives, follow-up on indicators, and assess its continuous improvement. (Figure 2).

3.3.2. Comprehensive Design Area Management. For the management of the design area, a flowchart reveals the travel path of customer designs from one area to the next, up to proper information storage. Based on this, the following help tools were created for designers: design sheets and job profiles to facilitate design control, staff member results, and proper storage of daily information. Here, daily information is mostly comprised of daily reports, including number of designs, number of modifications per order, appointed designer, costs, and design database. Based on the results from these reports, the company will be able to set new objectives, follow-up on indicators, and assess its continuous improvement (see Figure 3).

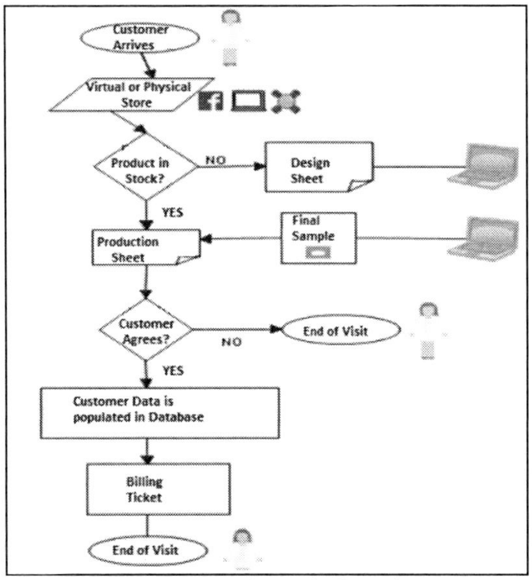

Figure 2. Sales unit flowchart

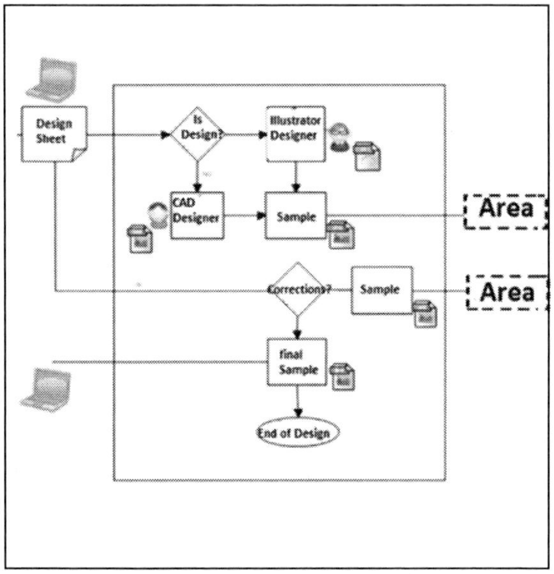

Figure 3. Design unit flowchart

3.3.3. Comprehensive Production Area Management. For the management of the production area, a layout denotes the jewelry manufacturing travel path, from the moment it arrives at the workshop to its final packaging. Based on this, the following help tools were created for artisans: production sheet, Kardex, aggregated plan, and job profiles, which facilitate managing orders, staff member results, and proper storage of daily information. Here, daily information is mostly comprised of daily reports, including daily production, type of jewels produced, weight per jewel, time per jewel, appointed staff member, costs, and manufactured product database. Based on the results from these reports, the company will be able to set new objectives, follow-up on indicators, and assess its continuous improvement (Figure 4).

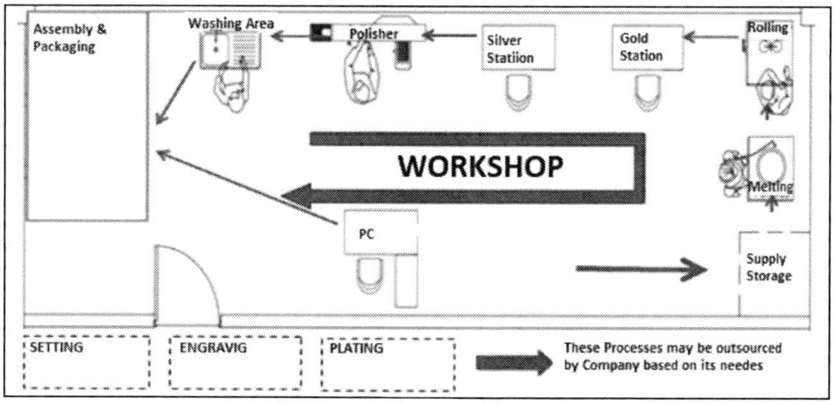

Figure 4. Proposed layout

3.3.4. Model Implementation. An implementation guide was developed to facilitate model execution. Model implementation consists of four stages:

First stage: Plan: Step 1: Company Diagnosis, Step 2: Define objectives and goals, Step 3: Create the work plan, Step 4: Submit work forms to the Board, Step 5: Perform integration and company commitment activities.

Second stage: Organize: Step 6: Work group organization and leader selection by area, Step 7: Perform toolbox talks on new tools, Step 8: Assign tools for working, Step 9: Establish a Company support system, Step 10: Transformation of information.

Third stage: Lead: Step 11: Sales area management, Step 12: Apply designated tools, Step 13: Design area management, Step 14: Apply designated tools, Step 15: Production area management, Step 16: Apply designated tools.

Fourth Stage: Control: Step 17: Reporting by area, Step 18: Continuous improvement.

4. Case Studies

This section uses a case study to discuss the role of the MGI model in SMEs. The case study was conducted with the specialized staff members from an artisan jewelry shop. The objective of this case study is to validate the model presented in the research method section for a successful MGI model implementation in a small company, as previously discussed in Section D. Model.

4.1. Artisan Jewelery Company

To study the application of the model presented in Section 3, we selected the Expresat Artisan Jewelry (JAE) company, established in mid-2013 in downtown Lima, Peru. JAE is a small custom gold and silver jewelry manufacturing company with 12 workers and annual revenue of 80 thousand soles. Besides their MTO shop, the company also runs physical and online stores, where they exhibit their products, such as rings, bracelets, earrings, and necklaces. Further company details are provided in the following section.

4.2. Analysis: The JAE Case

In this section, the company is further characterized through interviews and analysis. The model presented in Section 3 has been used to assess JAE's status in terms of the MGI application. This research study is based on the company data collected through the application of the MGI model supported by interviews with the general manager and assistant manager of the company. These interviews have focused on the company's processes, management support, strategy development, and the integration of areas with supporting technology for the application of the MGI model in a SME.

4.2.1. Management by JAE Business Unit

Comprehensive sales area management
JAE serves more than 100 customers, with ages ranging between 18 and 45. Customer satisfaction means that the company has successfully met the requirements of its customers with quality products, competitive prices, and on-time deliveries. For the sales area, training was performed on virtual reports and tools such as the sales speech and the procedure manual and the design sheets. The company was asked to describe area progress in a virtual report containing information on the number of sales, number of abandoned sales, costs per product, and the types of products sold. At the same time, indicators were improved with support from the comprehensive sales management tools. In general, the company managed to increase its sales and reduce its percentage of abandoned sales and goals. Moreover, the sales force received support through the management tools while the MGI model was being executed.

Table 1. Sales area results table.

RESULTS AND BEHAVIORS CONTROL	Before	After
NO. UNSATISFIED CUSTOMERS	91	2
% ABANDONED SALES	15%	1%
% GOALS ACHIEVED SALESPEOPLE PER STORE	50%	107%
% COMPLAINTS PER SALESPERSON	27%	2%

4.2.2. Comprehensive JAE Design Area Management. For the design area, new designers received training on tools such as the design sheet, the procedure manual, and the design storage. The company was asked to describe area progress in a virtual report containing information on the number of designs, number of reprocesses, costs per design, times, and staff members in charge. In general, the company managed to implement the area and its tools, supported by teamwork. These cultural factors contributed to eliminate obstacles and reduce previously calculated implementation times.

Comprehensive JAE design area management
For the design area, new designers received training on the selected tools, AutoCAD and Adobe Illustrator, in addition to training on tools such as the design sheet, the procedure manual, and the design storage. The company was asked to describe area progress in a virtual report containing information on the number of designs, number of reprocesses, costs per design, times, and staff members in charge. In general, the company managed to implement the area and its tools, supported by teamwork. These cultural factors contributed to eliminate obstacles and reduce previously calculated implementation times. The reasons for introducing technological support to management were to improve efficiency, reduce costs, improve design and sales times, and improve company efficiency to become competitive at local and international level.

Comprehensive JAE production area management
Before starting with the production changes, the aggregate production plan was developed. This tool helped us determine the type of company production, forecast future demands, and define production strategies to meet demands.

It must be considered that during the production area implementation, the company moved its workshop to the downtown area of the city, thus enabling us to apply the SMED tool for reducing workstation changeover times. Our layout follows the same jewelry manufacturing scheme: the first machine placed was the laminator, where the manufacturing process starts. Next, two folders—gold and one silver—were settled and a tool rack was placed between them. Then, the polishing machine

IOP Conf. Series: Materials Science and Engineering **796** (2020) 012020 doi:10.1088/1757-899X/796/1/012020

was placed, followed by a washing and a jewelry assembly table. During the machine layout distribution, there were many changes to test how comfortable it was for operators to move around and procure the required tools. Once the final workshop layout was defined, we proceeded to organize the materials, supplies, products, and tools used in the production processes. For these purposes, we used the 5S System, a lean manufacturing tool. This tool has the following steps: sort and eliminate, straighten and set, shine, standardize for visual control, and sustain discipline and habits. Below, the activities developed for each variable are detailed.

Sort: There was a sorting process for all tools and materials based on their lifespan. Moreover, waste was eliminated from the surrounding areas to the work folder since they only filled the space and obstructed the production process.

Straighten: New racks were acquired for the production instruments. Furthermore, compartments were added to the production table for storing the basic materials used throughout the day. The work folder was also conditioned to keep the melting and acid hoses in order.

Shine: Use protocols were established to maintain the polishing machine in proper operation. This machine should be cleaned before use and after use.

Visual Control: Colored labels were placed to divide workshop components: blue for the tools, red for the three raw materials, and green for materials in general. On the other hand, purple was used for finished products ready for delivery.

Discipline and Habits: Rules were created, and notices were placed in work areas to ask employees to preserve the changes implemented and for creating the habit of maintaining the workshop organized. All employees were also asked to leave their workstations in order, clean all machines, store waste in the designated containers, and dispose the garbage accumulated during the day. For better control purposes, a checklist detailing the spaces that must be cleared at the end of the shift was developed.

The critical production process tool is Cobacabana, which, as per the state of the art, indicates order reception and delivery dates, and their corresponding priority. At the same time, we proposed a special production sheet for the products offered by the company, wherein jewel designs, and details such as weight, chain type, setting, engravings, and business hours, are specified.

5. Results

Next, the main results of the comprehensive management model are reported based on the selected metrics and data extracted from area reports (Table 2).

Table 2. Implementation results for company "X"

MANUFACTURING TIMES & COSTS FOR A NECKLACE WITH NAME							
No.	Description	Time 2017	Time 2018	Optimization In Minutes	Cost in Soles 2017	Cost in Soles 2018	Savings in Soles
1	Supplies	1440	0	1440	20	0	20
2	Transport Supplies to Manufacturing Area	60	0	60	20	0	20
3	Design	1440	5	1395	1	3	-2
4	Rolling	10	0	25	6	0.06	6
5	Cutting	30	15	10	3.9	1.95	1.95
6	Sanding	10	3	7	1.3	0.39	0.91
7	Polishing	15	3	7	1.95	0.39	1.56
8	Washing & Drying	15	3	7	1.95	0.39	1.56
9	Assembly	3	1	2	0.39	0.13	0.26
10	Finishing	5	3	2	0.65	0.39	0.26
11	Supervision	0	3	-3	0	0.39	-0.39
12	Labeling Packaging	10	3	7	1.3	0.39	0.91
TOTAL		3038	39	2959	58.44	7.48	51.02

6. Conclusions

In summary, this study evidences the effectiveness of the comprehensive management model applied in a jewelry manufacturing company, not only increasing sales but also productivity and quality assurance in strategic and operational works. On the other hand, the following conclusions were reached based on the tools and programs implemented.

The development and implementation of the online sales and physical store channels facilitated the work for sellers, offering greater sales opportunities, and strengthening customer relationships. Similarly, the positive use of sales force incentive strategies stimulated area staff members, increasing up to 80% over the sale obtained against the pre-implementation scenario.

The use of lean tools progressively increased productivity and labor efficiency, fully complying with the implementation program within the times allocated in the production area. Moreover, the organization and layout changes applied to the production workshop were critical to strategically connect workstations among themselves and prevent delays from unnecessary movements. Finally, the Cobacabana tool provided greater control of completed orders, eliminated downtimes due to lack of information, and installed quality assurance through a product seal of conformity before shipment.

The internationalization of the design area greatly reduced the time spent in the design in more than 90%, going from a 24-hour waiting time to just 30–60 minutes, thus adding value to the product in terms of customer service and response times.

Further, a relationship was created between the three management philosophies and company staff members. Lean manufacturing was deployed as an inclusive methodology, which encourages all participants to generate value for the company. Similarly, its inclusive nature also creates culture, which leads us to the second philosophy, wherein the organizational climate and behaviors were fostered. This prompted all staff members to work together, provide mutual support, and identify themselves with the values and missions of the company, which was evidenced as profitability levels increased. Finally, the green philosophy generated operator awareness in terms of their work and the possible effects caused by their operations.

7. References

[1] ADEX 2018 Análisis del Sector Joyeria y Orfebreria. Obtenido de Retrieved from: http://contenido.adexperu.org.pe/descargas/estadisticas_sector_joyeria.pdf

[2] IIMP 2018 Instituo de Ingenieros de minas del Peru. Obtenido de Retrieved from: http://www.iimp.org.pe/actualidad/peru-cuenta-con-el-5-delas-reservas-mundiales-de-oro-y-24-de-plata

[3] ADEX, A D 2018 ADEX. Recuperado el 6 de 09 de 2018, de Retrieved from: http://www.adexperu.org.pe/notadeprensa/en-peru-se-destina- menos-del-1-de-oro-y-plata-a-la-elaboracion-de-joyas/

[4] PRODUCE 2015 Anuario Estadístico Industrial, MYPE y Comercio Interno. Recuperado el 30 de abril de 2018, de Retrieved from: https://www.produce.gob.pe/documentos/estadisticas/anuarios/anuarioestadistico-mype-2015.pdf

[5] Dinner I M, Van Heerde H J and Neslin S A 2013 Driving Online and Offline Sales: The Cross-Channel Effects of Traditional, Online Display, and Paid Search Advertising *J Mark Res* **50** 527-545.

[6] Abhishek V, Jerath K and Zhang Z J 2016 Agency Selling or Reselling? Channel Structures in Electronic Retailing *Manag Sci* **62** 2259- 2280

[7] Thürer M, Land M J and Stevenson M 2014 Card-based workload control for job shops: Improving COBACABANA *Int. J. Prod Econ* **147** 180- 188.

[8] Gansterer M 2015 Aggregate planning and forecasting in make-toorder production systems *J Prod Econ* **170** 521-528

[9] Randhawa J A 2017 An investigation into manufacturing performance achievements accrued by Indian manufacturing organization through strategic 5S practices *Int J Prod Perform Manag* 754 - 787.

[10] Bocken N M P, Rana P and Short S W 2015 Value mapping for sustainable business thinking *Int J Prod Res* **32** 67-81.

[11] Zare R, Chavez P, Raymundo C, Rojas J 2018 Collaborative Culture Management Model to Improve the Performence in the Inventory Management of a Supply Chain *Congreso Internacional de Innovacion y Tendencias en Ingenieria, CONIITI 2018 – Proceedings, 8587073.*

[12] Hemilä J and Vilko J 2015 The development of a service supply chain model for a manufacturing SME *Int J Logist Manag* 517-542.

IOP Conf. Series: Materials Science and Engineering **796** (2020) 012021 doi:10.1088/1757-899X/796/1/012021

Lean Manufacturing Production Method using the Change Management Approach to Reduce Backorders at SMEs in the Footwear Industry in Peru

D Dextre-del-Castillo, S Urruchi-Ortega, J Peñafiel-Carrera, C Raymundo-Ibañez[2,4] and F Dominguez[3]

[1] Ingeniería Industrial, Universidad Peruana de Ciencias Aplicadas (UPC), Lima 15023, Perú.
[2] Dirección de Investigación, Universidad Peruana de Ciencias Aplicadas (UPC), Lima 15023, Perú.
[3] Escuela Superior de Ingeniería Informática, Universidad Rey Juan Carlos, Mostoles, 28933, España.

E-mail: carlos.raymundo@upc.edu.pe

Abstract. This article proposes a production method that aims to increase the manufacturing capacity of a footwear small- and medium-sized enterprise (SME) to reduce backorders. Therefore, an assessment is carried out and delays in production processes, excess product transport time, defective products, and inefficient work methods are identified. This article proposes designing a Lean manufacturing method using the change management approach, whose methodology is composed of six phases. In phase 0, change management is carried out; in phase 1, the company's current situation is reviewed using the Value Stream Mapping (VSM); in phase 2, the work area is reorganized (implementing SLP and 5S); in phase 3, production is balanced (implementing Line Balancing); in phase 4, continual improvement is established using the Kaizen tool; and finally, in phase 5, the results are evaluated. Through validation, it was possible to confirm that Lean manufacturing tools along with change management increased order deliveries by 82%.

1. Introduction

Worldwide, footwear production reached an estimated 45,000 million pairs in 2018, a 3% increase over the previous year. The majority of shoe manufacturing is carried out in Asia, where 86% of the world's shoes are produced. The top shoe production countries are in Asia, and include China, Vietnam, and Indonesia. In 2018, the footwear industry in Peru had an impact equivalent to 2.9% of the Gross Domestic Product (GDP) and represents 4.1% of all employment [1]. Today, there are several problems in the production of footwear small- and medium-sized enterprises (SMEs), such as lack of quality in their processes and products, low productivity, high rate of defective products, poorly trained personnel, and more. These problems have led to a high rate of backorders, which has consequently generated less economic income, high production costs, and dissatisfied customers.

[4] To whom any correspondence should be addressed.

Content from this work may be used under the terms of the Creative Commons Attribution 3.0 licence. Any further distribution of this work must maintain attribution to the author(s) and the title of the work, journal citation and DOI.
Published under licence by IOP Publishing Ltd

In January 2019, footwear manufacturing production had a percent variance of -29.33%. Companies that produce footwear are primarily micro enterprises with 96.7% of the market, while small enterprises account for 3.2%, and medium- and large-sized enterprises account for 0.1%. These data are calculated based on the Ministry of Production's 2018 Industrial Census [2].

Using these improvement tools is recommended for SMEs, as they have major problems such as a high percentage of backorders, low production capacity, a high percentage of defective products, and more, which jeopardizes their ability to stay in business. Currently, there is not much information about integrating Lean manufacturing and change management in small- and medium-sized enterprises. Therefore, the reason for writing this article is to provide a model design that implements Lean manufacturing focused on change management, considering success factors, to address the problems that SMEs in the footwear industry face [3] [4].

2. State of The Art

2.1. Lean Manufacturing

Authors Lande, Chauhan & Randhawa agree that Lean manufacturing is recommended for SMEs, as they have major problems such as a high percentage of backorders, low production capacity, and a high percentage of defective products. Therefore, Lean tools can increase a company's production capacity and improve its sustainability [5] [6] [7]. Lean Manufacturing tools include flexibility in managing an Indian company's resources, meaning workers and machinery [8]. In addition, lean manufacturing is used to eliminate waste in the cutting area of a small footwear company in Colombia. Using the "5 Whys," Value Stream Mapping (VSM) in a car parts manufacturing company in India identified root causes and implemented a future roadmap to improve productivity within the company [9].

2.2. Change Management

According to Budhiraja, there is a vast amount of literature on change management practices at large companies to process all the changes necessary for growth. However, within SMEs, it is necessary to develop an ecosystem to improve adaptability to changing circumstances, to increase their competitiveness [10][11]. According to the literature, in the change process, senior management is enthusiastic and willing to provide infrastructure. However, the process is limited by employee motivation levels, as there is a negative attitude toward accepting change [12][13]. Knol's research makes use of the elements inherent to organizational change to establish how organizations can address this phenomenon. Using conceptual reflections, developing and interpreting this concept is described over time. In addition, it shows that organizations understand that the only constant over time is change and that it has become a factor that can cause them to be unstable [14].

3. Input

3.1. Proposed Method

The method has the following five phases: change management, review of the current situation, reorganization of the work area, balancing production, and continual improvement. (See Figure 1). The purpose of the proposed method is to reduce backorders for footwear SMEs, thereby reducing cycle times in bottlenecks and therefore increasing production capacity. In comparison with other models, this includes an initial awareness using change management to avoid problems with implementing Lean tools.

3.1.1. Phase 0: Change Management. Change management is a structured orientation and aims to help management, employees, and sponsors accept change in their business environment.

3.1.2. Phase 1: Review of the Current Situation. In this phase, the processes are studied, times are recorded, and the plant's capacity is determined. In addition, the production process's waste or changes are found. Thus, using the VSM tool is proposed (Figure 2).

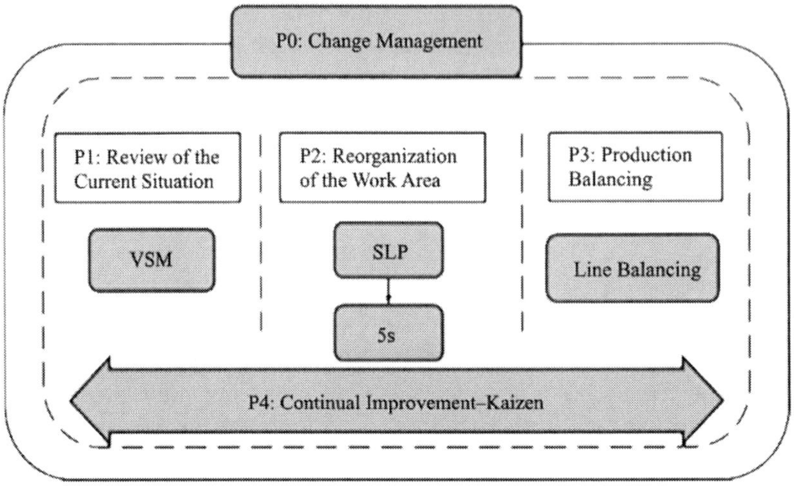

Figure 1. Proposed method

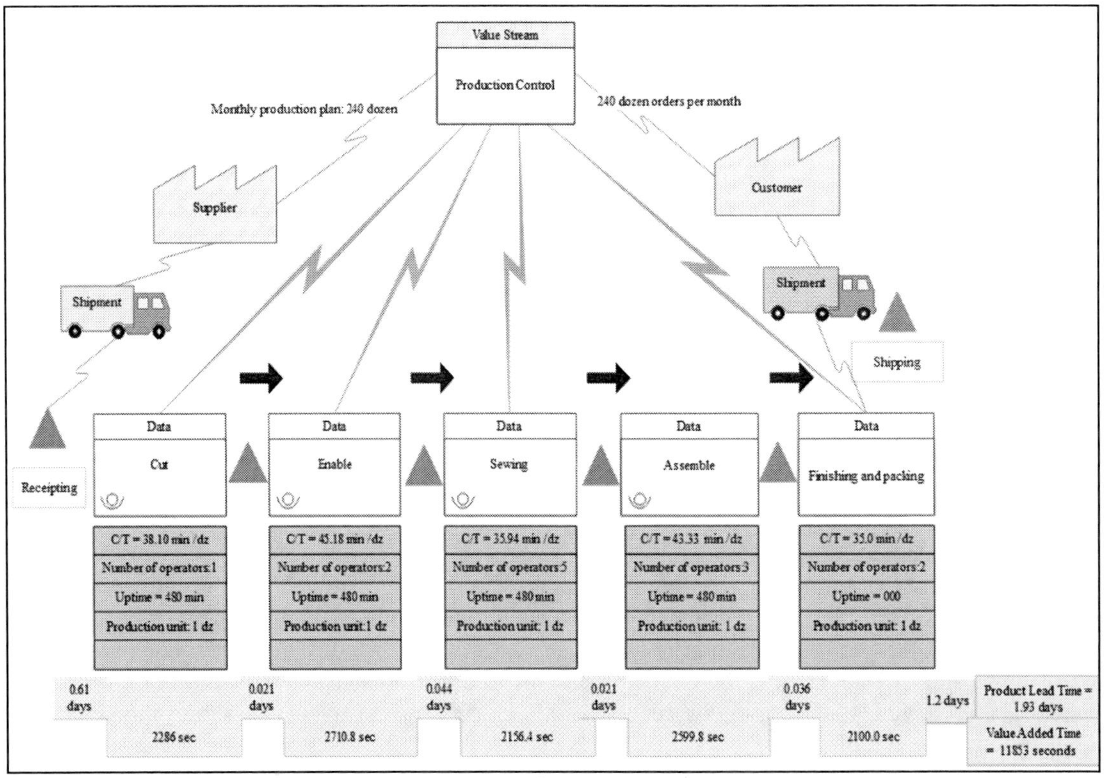

Figure 2. Value Stream Map

3.1.3. Phase 2: Reorganization of the Work Area. This phase aims to improve the work area to reduce unnecessary time searching for tools, unnecessary movements, and routes between areas. We propose two tools: SLP and 5S.

The objective of the Systematic Layout Planning (SLP) tool is to reduce the distance that materials travel, to logically structure processes and to make the plant's distribution more flexible for future modifications. The objective of the 5S tool is to prevent the work environment from becoming disorganized, lack of instructions, unnecessary movements, and more.

3.1.4. Phase 3: Production Balancing. This phase aims to balance the workload of each station, to minimize the times in the bottlenecks that exist in the production process of the shoes, since the idle times between processes lead to high production costs. The main objective of the Line Balancing tool corresponds to the equalization of working times at all stations in the process and the optimal distribution of workers.

3.1.5. Phase 4: Continual Improvement. This phase promotes the culture of continual improvement and constantly improving production processes. To do so, the Kaizen tool was proposed. The objective of implementing Kaizen is to eliminate exceptional changes identified using Kaizen events.

The Kaizen philosophy implemented consists in the incentive of teamwork through weekly meetings of all operators with the heads of each area of the SME. In these meetings, the operators give ideas to improve the processes and the changes to be made are decided. In addition, the weekly production objectives are established so that each operator strives to achieve its objective.

3.2. Proposed Method (Flowchart)

An implementation guide was described to optimally implement this method. The activities in pink are the change management steps and those in green are from 5S, the tools that comprise the basis for this study.

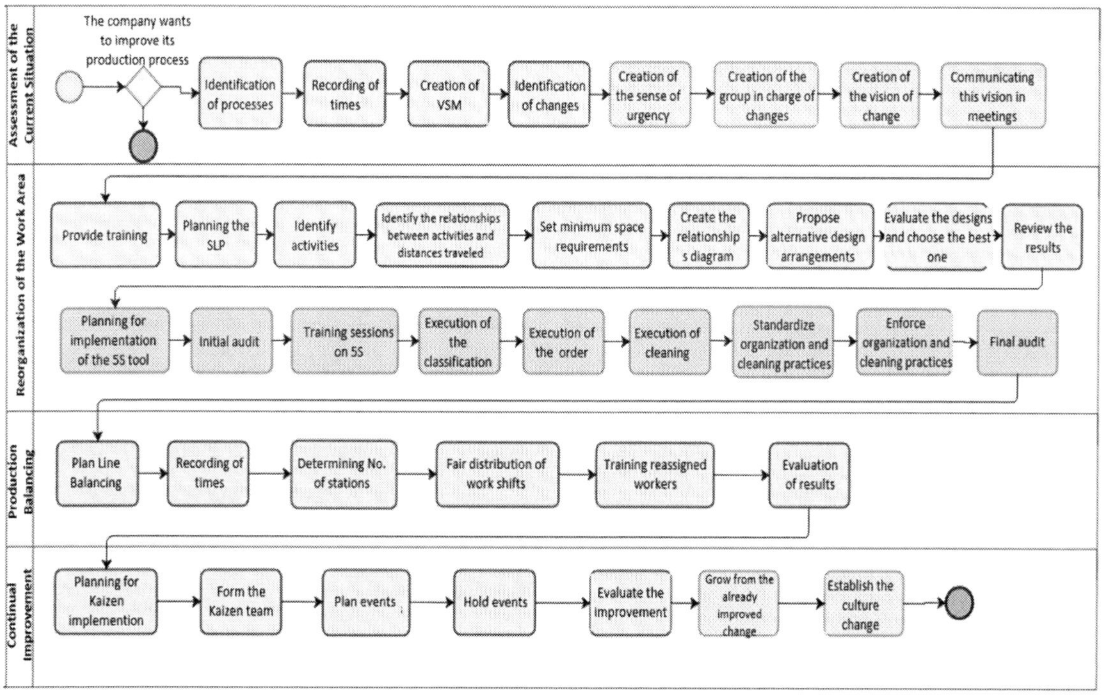

Figure 3. Proposed implementation model

3.3. Indicators

3.3.1. Production capacity percentage

$$\%\ de\ PC = \frac{PC\ Obtained - Initial\ PC}{PC\ Obtained} \tag{1}$$

3.3.2. Excess product transport time (movement)

$$Excess\ time = \frac{Initial\ Movement - Movement\ Obtained}{Initial\ Movement} \tag{2}$$

3.3.3. Percentage of defective products

$$\%\ of\ DP = \frac{Initial\ \%\ of\ DP - \%\ of\ DP\ Obtained}{Initial\ \%\ of\ DP} \tag{3}$$

3.3.4. Wait time between processes

$$Wait\ time\ between\ processes = \frac{Initial\ WT - Final\ WT}{Initial\ WT} \tag{4}$$

3.3.5. Percentage of on-time deliveries

$$\%\ of\ OTD = \frac{Final\ \%\ LD - Previous\ \%\ LD}{Current\ \%\ of\ DP} \tag{5}$$

4. Validation

4.1. Case Study Description

To validate the previously described proposal, the method was implemented by studying M&F, a small shoe distribution company that has been manufacturing and marketing wholesale and retail footwear since 2006. It is known for its high quality and innovative products. The production line has equipment, machinery, and manual operations. The company is located at San Pablo Street N° 588, Belaunde, Comas. The shoe company only works with three lot sizes: 40 dozen, 60 dozen, and 80 dozen. The company only accepts these sizes because of its knowledge of purchases in these sizes.

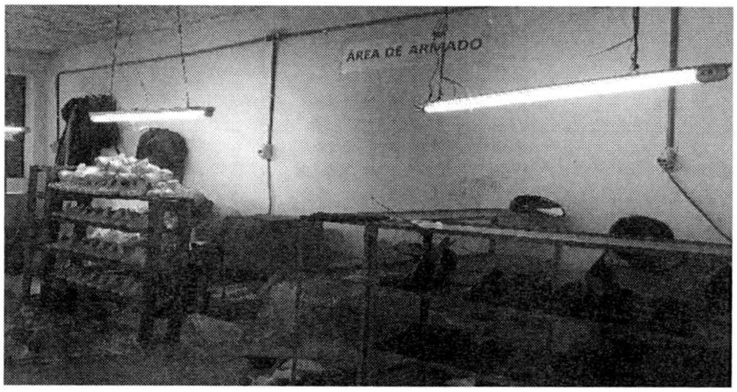

Figure 4. M&F company

4.2. Assessment

The company has a cycle time of 45.18 minutes per dozen and its sales revenue was 770,880 Peruvian Soles last year. The assessment was carried out in five phases.

Table 1. Initial result

Indicator	Initial result
% order delivery	50.88%
Production capacity	212 doz/month
Excess transport time	41.40 min/doz
% of defective products	10%
Wait time between processes	13.32 min/doz

4.3. Applying the Method to the Case Study

4.3.1. Phase 0: Change Management. The first results when validating change management were presenting the economic benefits, instructing the work area to reduce the risk of accidents, reducing material transport time, and reducing cycle time. Finally, the team must be maintained so that they can continue to implement more changes.

4.3.2. Phase 1: Review of the Company's Current Situation. The company was asked for the 2018 information required to be able to complete the VSM, to observe the flow of information and materials from the production process, obtaining a cycle time of 45.18 minutes per dozen.

4.3.3. Phase 2: Reorganization of the Work Area. To validate this phase, two Lean tools were implemented: SLP and 5S. The first tool was able to reduce the product's total transport time by 70% and reduced the cycle time from 45.18 to 44.9 minutes per dozen. A final audit was carried out using the second tool and the score obtained was 107 out of 156 points, which represents a 38% improvement. This tool reduced cycle time from 44.9 to 44.56 minutes per dozen.

4.3.4. Phase 3: Production Balancing. The tasks that must be fulfilled to make the company's products and their sequential order were determined to create the station priority diagram. With the new stations, the production area's cycle time was reduced from 44.56 to 42.8 minutes per dozen. As a result, production capacity increased from 53 to 56 dozen per week.

4.3.5. Phase 4: Continual Improvement. When the improvements were implemented, teams were formed to promote the Kaizen philosophy among all workers. This tool reduced cycle time from 42.8 to 40.5 minutes per dozen, or a 5% reduction.

4.4. Indicators

Table 2. Results Obtained

Indicators	Initial Results	Results Obtained	Expected Results
Percentage of deliveries	50.88%	82%	80%
Production capacity	212 doz/month	237 doz/month	230 doz/month
Excess product transport time	41.4 min/doz	12.42 min/doz	4.06 min/doz
% of defective products	10%	4%	3.5%
Wait time between processes	13.32 min/doz	4.06 min/doz	4.70 min/doz

Table 2 shows that, in most cases, the expected results were exceeded, with the exception of the "wait time between processes" indicator. Its expected result of 4.70 minutes per dozen was not reached, obtaining 4.06 min/doz.

5. Conclusions

The general objective was achieved, as backorders were reduced, increasing order deliveries by 82%. In addition, within the specific objectives, production capacity increased from 213 to 237 dozen per month. This article's contribution was to reduce wait time between processes from 13.32 minutes per dozen to 4.04 minutes per dozen. By applying SLP, a more organized and fluid work environment was obtained through personnel shifts. With the 5S tool, it was possible to improve organization and cleaning practices by 38%.

6. References

[1] National Institute of Statistics and Information Technology (INEI) 2018 Panorama of Peru's Economy (*Lima*).

[2] National Institute of Statistics and Information Technology (INEI) 2018 Panorama of Peru's Economy (*Lima*).

[3] Central Reserve Bank of Peru (BCRP) 2018 (*Lima*).

[4] Pearce A, Pons D and Neitzert T 2018 Implementación de Lean-Outcomes de estudios de casos de SME [Implementation of Lean Outcomes in SME Case Studies] *Perspectivas de Investigación de Operaciones* **5** 94–104.

[5] Lande M, Shrivastava R L and Seth D 2016 Critical success factors for Lean Six Sigma in SMEs (small and medium enterprises) *The TQM Journal* **28**(4) 613–635.

[6] Chauhan G 2016 An analysis of the status of resource flexibility and lean manufacturing in a textile machinery manufacturing company *Int J Organ Anal* **24** 107–122.

[7] Randhawa J 2017 Examining the role of 5S practices as a facilitator of business excellence in manufacturing organizations *Meas Bus Excell* **21** 191–206.

[8] S. Kumar 2018 Lean-Kaizen implementation *J Eng* **16** 143–160.

[9] Redeker G A, Kessler G Z and Kipper L M 2019 Lean information for lean communication: Analysis of concepts, tools, references, and terms *Int J Inform Manag* **47** 31-43.

[10] Budhiraja S 2019 Organizational readiness for change: an inherent concern for Indian small and medium enterprises (SMEs) *Dev Learn Org Int J* **12** 15–26.

[11] Martínez Bustos E, Carrasco Sagredo C and Bull M T 2018 Propuesta metodológica para implementar la primera fase del modelo de gestión del cambio organizacional de Lewin [Methodological proposal to implement the first phase of Lewin's organizational change management model] *Estudios Gerenciales* **34** 88–98.

[12] Alvarez Valley A T, Proaño Córdova T D and Cruz Lascano ME 2017 Estructura, Cultura Y Cambio Organizacional Cultura - Cambio - Forma - Fuerza [Structure, Culture and

Organizational Change Culture - Change - Form - Force] *Revista Científica Hermes* **18** 304 -324.

[13] Riquero I, Hilario C, Chavez P, Raymundo C 2019 Improvement proposal for the logistics process of importing SMEs in Peru through lean, inventories, and change management *Smart Innov Syst Tech* **140** 495-501.

[14] Knol W, Slomp J, Schouteten R and Lauche K 2018 Implementing lean practices in manufacturing SMEs: testing ‘critical success factors’ using Necessary Condition Analysis *Int J Prod Res* **56** 3955–3973.

IOP Conf. Series: Materials Science and Engineering **796** (2020) 012022 doi:10.1088/1757-899X/796/1/012022

Production Management Model Based on Lean Manufacturing and Change Management Aimed at Reducing Order Fulfillment Times in Micro and Small Wooden Furniture Companies in Peru

R Castro-Chara[1], R Valenzuela-Leandro[1], P Chavez-Soriano[1], C Raymundo-Ibañez[2, 4] and F Dominguez[3]

[1] Ingeniería Industrial, Universidad Peruana de Ciencias Aplicadas (UPC), Lima 15023, Perú.
[2] Dirección de Investigación, Universidad Peruana de Ciencias Aplicadas (UPC), Lima 15023, Perú.
[3]Escuela Superior de Ingeniería Informática, Universidad Rey Juan Carlos, Mostoles, 28933, España.

E-mail: carlos.raymundo@upc.edu.pe

Abstract. This research study seeks to identify and prioritize the causes of order fulfillment delays in a small wooden furniture manufacturing company. The authors propose a 5-phase Lean Optimization model to address and reduce this problem. Post-implementation results yielded a 54.87% reduction in material search and transportation times, a 32.86% reduction in travel times between stations, and a 19.81% increase in line efficiency. In addition, order fulfillment percentages increased from 12.5% to 60%.

1. Introduction

Around the world, the wooden furniture industry is the second most important business activity in wood product manufacturing, of which China is the leading producer with a 41% global market share, and Latin America only reports a 3% market share. In addition, furniture accounts for only 0.7% of the total foreign trade in the world, amounting to US $134 billion [1]. Regarding Peru, although sector production has been increasing in recent years, this activity only contributes 1%–2% on an average to the non-primary goods GDP and between 0.2% and 0.3% to the total national GDP [2]. Likewise, neither the wood industry nor the furniture industry is sufficiently developed in Peru. In fact, about 97% of the furniture industry is traditionally family-owned, using artisan methods and characterized by limited technological progress and low dissemination of production techniques [3]. This situation directly affects competitive levels against other countries for this sector. For example, according to a study that assesses the competitive profile of the wooden furniture industry, Peru exhibits a low competitive level in terms of improving production processes and lines, thus evidencing a disadvantage against top reference countries such as China, Brazil, and Chile [4]. Therefore, micro and

[4] To whom any correspondence should be addressed.

Content from this work may be used under the terms of the Creative Commons Attribution 3.0 licence. Any further distribution of this work must maintain attribution to the author(s) and the title of the work, journal citation and DOI.
Published under licence by IOP Publishing Ltd

small wooden furniture companies usually lack the productive development required to adequately meet the order fulfillment deadlines requested by their customers.

The wooden furniture industry, with a 78% market share in the wood manufacturing sector, represents one of the industries with higher added value within the wooden products currently moved in the Peruvian market. In addition, 44% of the entire wooden furniture production industry in Peru is in the city of Lima [4]. Further, this industry accounts for 47% of the employment generated in the wood manufacturing sector, being able to generate approximately 700,000 jobs [5]. However, this industry, despite having a vast amount of forest resources available (78 million hectares), imports more wood products than it exports, thus generating a trade balance deficit and making this sector's inability for meeting internal demands evident [6]. Likewise, in exports, Peru is surpassed by other countries with fewer timber resources, such as Chile, which exports 31 times more wood products than Peru [4]. Consequently, the combination of a small-scale production, the limited response capabilities against large orders, and the need for change within the organizational core of the different micro and small companies that comprise the sector renders Peru unable to position itself as a prominent power in South America and the rest of the world in terms of wooden furniture production. Therefore, it becomes a competitive market. For these reasons, the improvement of processes and production lines in micro and small wooden furniture companies to reduce order fulfillment times through efficient production techniques is the first step toward fostering sector development and growth.

The underlying motivation of this project is to generate a viable and functional model by integrating change management with different lean manufacturing tools adapted to the productive environment of the micro and small wooden furniture companies in Lima, Peru, which is aimed at efficiently solving order fulfillment delays as per the needs of the sector. Furthermore, this research study seeks to prove the effectiveness of combining two approaches, such as lean manufacturing and change management, for reducing order fulfillment times in micro and small wooden furniture companies.

This paper is divided into the following sections: introduction, state of the art, contribution, and validation and discussion. Finally, the paper denotes the conclusions from the research study and lists all the references consulted.

2. State of The Art

2.1. Production Management Models Based on Change Management

The change management model proposed by John Kotter is often used as a catalyst for steering efforts throughout change adaptation in an organization. However, some research studies underscore that positive results depend on the scope defined for the tool, the interactive communication among staff members, and the strategic approach proposed [7][8]. In turn, these studies also argue that this tool exhibits high levels of effectiveness at multiple management levels within a company, as long as it is supported by leadership and organization techniques [9]. Additionally, this tool is mostly used for maximizing impacts from the results and steering the project toward establishing permanent changes over time, whether small or large, thus guaranteeing successful transformation of the services, productive resources, and staff members of the organization [8][9]. Nevertheless, in some cases, changes can be successful even when model steps and sequential order are bypassed [10].

2.2. Production Management Models Using Lean Manufacturing

Some studies that focus on the application of lean manufacturing tools propose an improvement plan to reduce clutter and disorganization in micro and small companies with the purpose of increasing production, organizing workstations, reducing time spent searching for tools, and reducing preparation times [11][12]. Likewise, other authors argue that lean manufacturing can also be applied to optimize SME plant layouts and reduce travel distances between workstations, eliminate bottlenecks, and generate a continuous production flow. For this, manufacturing cells must first be designed on the basis of production process flows before their application according to the result obtained [13].

Other studies report that lean manufacturing tools have yielded successful and effective results in terms of an equitable workload distribution and the optimization of human and machine resource allocation [14][15]. These results are obtained because lean tools focus on increasing production capacity, productivity, and line efficiency [16][17][18]. These different studies consistently agree that lean manufacturing tools are used in companies to reduce unnecessary activities, cycle times, production process times, and order fulfillment times.

3. Contribution

3.1. Proposed Model

The general model depicted in Figure 1 uses lean methodology to promote customer satisfaction by eliminating activities that do not add value, which, in turn, reduces cycle times, improves line efficiency, and increases production capacities. Although this model uses lean techniques, it uses change management as a catalyzing tool to educate and prepare staff members during the transition toward a lean manufacturing model.

- Establish and Measure Change: Its objective is to establish and generate the need for change in the organization through the deployment of the first six steps of John Kotter's change management model.
- Workstation Organization: The 5S methodology is implemented for establishing order in workstations.
- Production Line Optimization: Through the work study, the most efficient method is identified and applied. Then, line balancing is used to determine the optimum number of workstations.
- Layout Optimization: In this phase, considering the results obtained in the previous phases, an AFP analysis is conducted for grouping stations into manufacturing cells and determining the ideal workflow.
- Results Analysis: Finally, the model applies the last two steps from the Kotter model, which consolidates improvements and changes in order to polish and present the results to the company. Likewise, the new culture is anchored in the company.

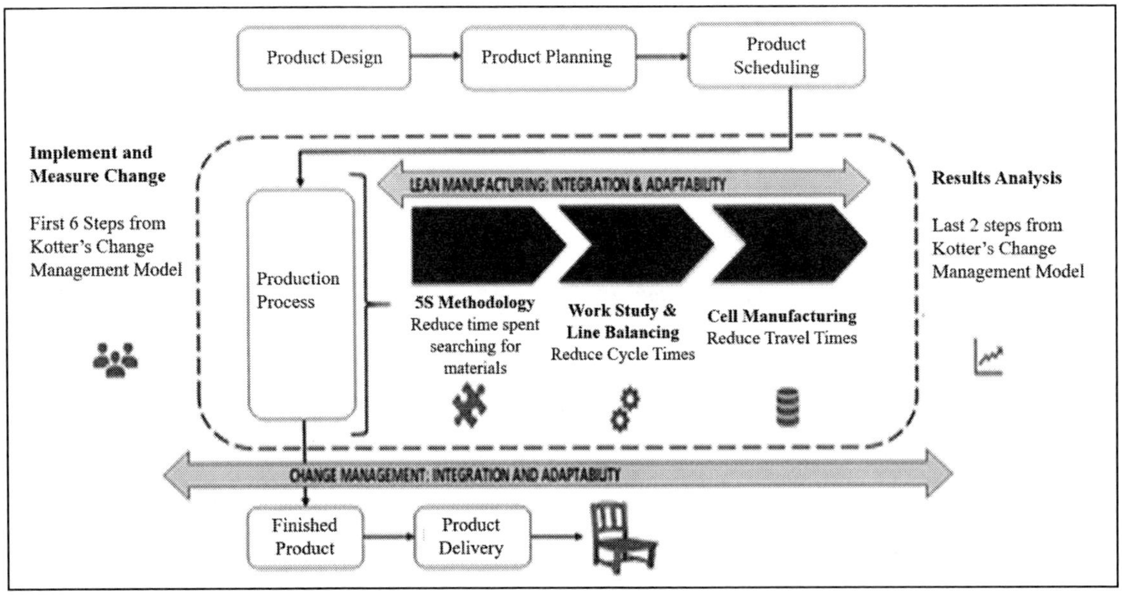

Figure 1. Proposed production model

3.2. Proposed Method

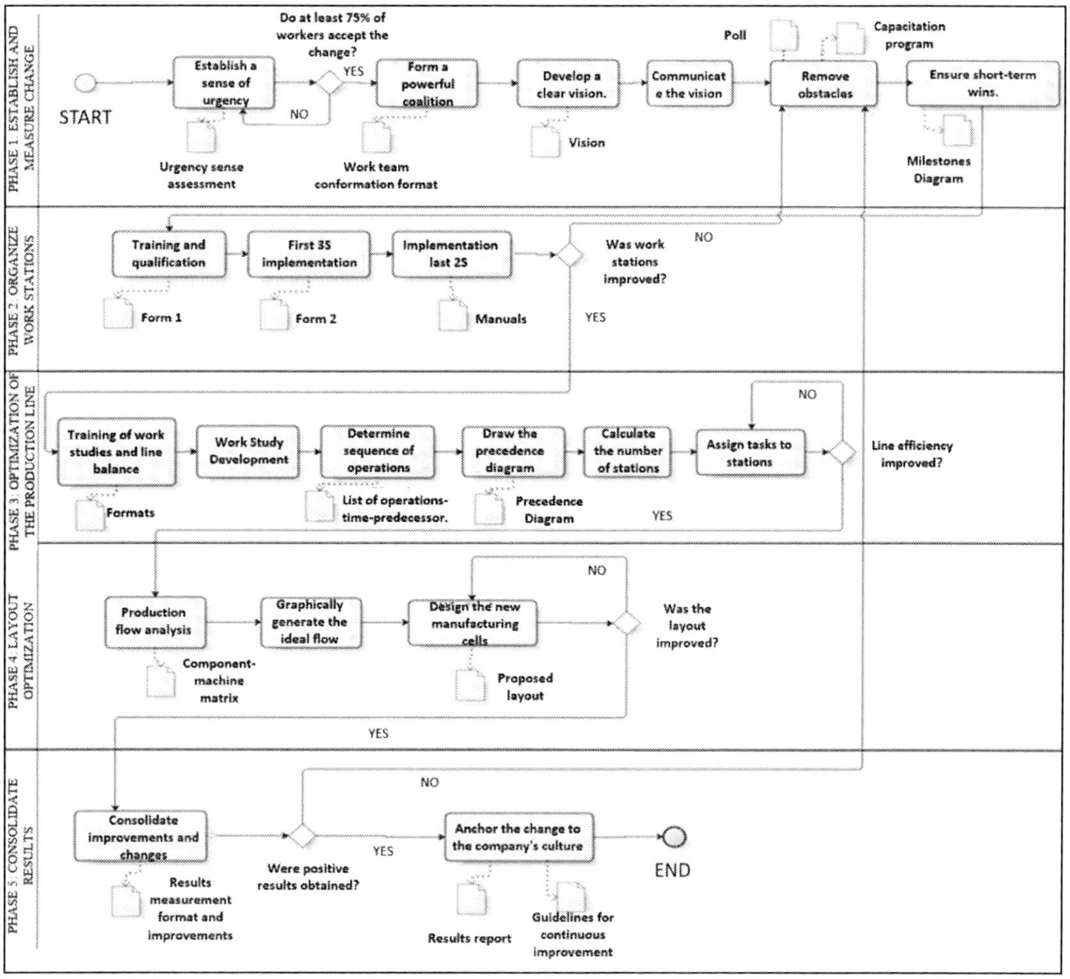

Figure 2. Model application flowchart

The method proposed for the implementation of the optimization model comprises five phases, which detail the corresponding sequence of activities.

3.3. Indicators

3.3.1. Production Capacity. Target: To measure how much the company's production capacity has increased after applying the improvements proposed.

$$\frac{Total\ Production\ Time}{Cycle\ Time} \tag{1}$$

3.3.2. Line Efficiency. Target: To measure how much line efficiency has increased after reducing cycle times and optimizing the number of workstations.

$$\frac{Total\ Production\ Time}{Cycle\ Time\ \times\ No.Stations} \times 100\% \tag{2}$$

3.3.3. Orders Fulfilled on Time. Target: To assess how much compliance with order fulfillment deadlines has increased after the proposed model was applied.

$$\frac{Orders\ Fulfilled\ on\ Time}{Total\ Number\ of\ Orders} \times 100\% \quad (3)$$

4. Validation

4.1. Case Study
The research study was performed in La Casa del Buen Maestro E.I.R.L., a small wooden furniture production company located in the Villa El Salvador industrial park in Lima, Peru.

4.2. Initial Assessment
After assessing the sales reported for each product, chairs were selected as the flagship product of the company. Therefore, a chair production line was selected for this study.

- Order Fulfillment Delays: 2–4 days
- Delay Percentages: 87.5%
- Cycle Time: 53.6 minutes
- Takt time: 37.3 minutes
- Processing Time: 229.3 minutes
- Line Efficiency: 61.11%

This assessment evidences high production times as the main cause of the problem identified. In turn, this cause is generated by the following sub-causes:

- % High Cycle Times: 36.66%
- % Inadequate Layout: 32.21%
- % Delays due to Time Spent Searching for Materials: 18.47%

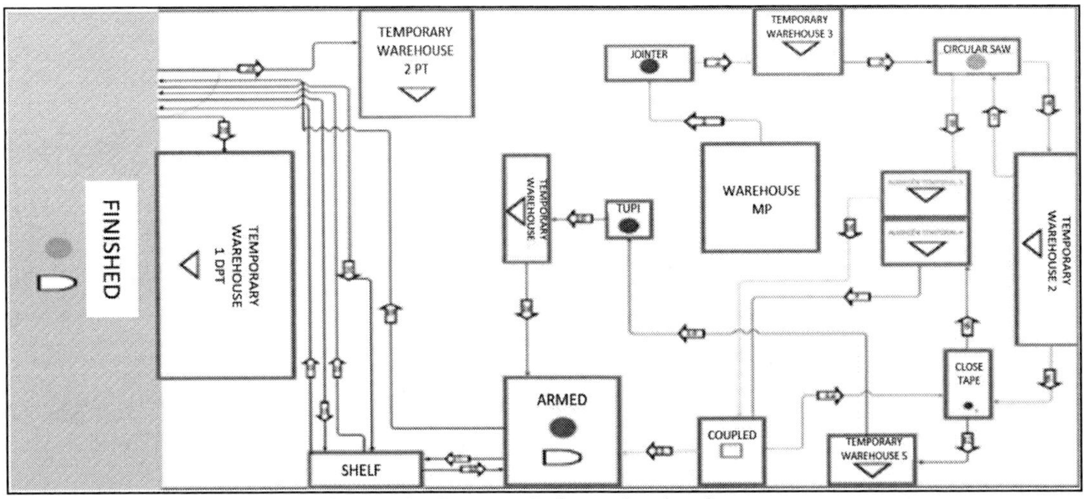

Figure 3. Current chair production layout

Figure 3. denotes the current plant layout, which evidences crossings between workstations and long travel distances between stations, which translates into high transportation times.

4.3. Model Implementation in the Case Study

4.3.1. Phase 1: Establishing and Measuring Change. According to production staff survey respondents, 100% of the workers support the changes implemented. Likewise, training sessions were conducted to reduce the index of resistance to change, which decreased from 32.14% to 3.57%.

4.3.2. Phase 2: Workstation Organization. During this phase, the 5S methodology was successfully deployed in order to sort, set in order, and shine the work environment for optimum performance, with overall positive results. However, only regular results were reported for the third S, as the company did not have a cement floor. In other words, the floor was completely covered by sand, wood shavings, and sawdust, which prevented the floor from being completely cleaned.

Before	**After**
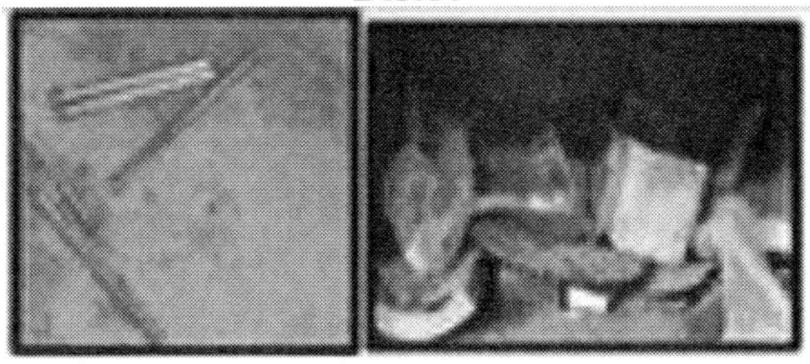	
Materials and tools are laying on the floor in Production Area covered by sawdust, wood shavings and places of wood	Materials and tools are stored in 2 shelves, labelling the corresponding name of each shelf space

Figure 4. Before and after 5S implementation

4.3.3. Phase 3: Production Line Optimization

Work Study
Times were recorded for straight cut activities because this operation was identified as a production bottleneck. Then, the proposed improvements, including the corresponding procedure sheet, were implemented.

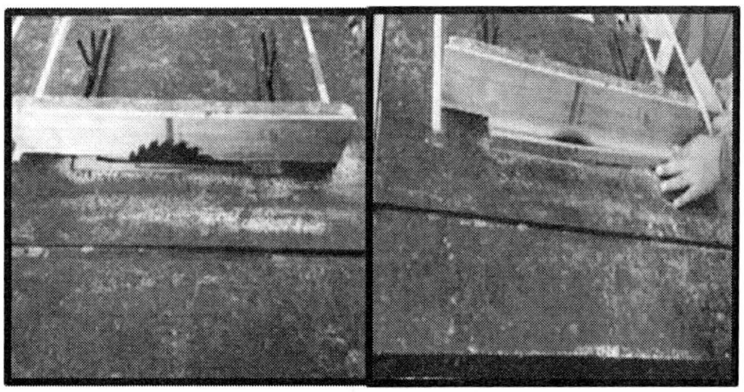

Figure 5. Improvements made to the Cutting Station

After implementing these improvements, cycle times were reduced by 19.96% when stopping tracing wood with a tape measure.

Line Balancing
Staff members are reorganized to distribute the workload of the operator in the station where cycle times originate.

4.3.4. Phase 4: Layout Optimization. Cell Manufacturing eliminates crosses and shortens travel times between stations.

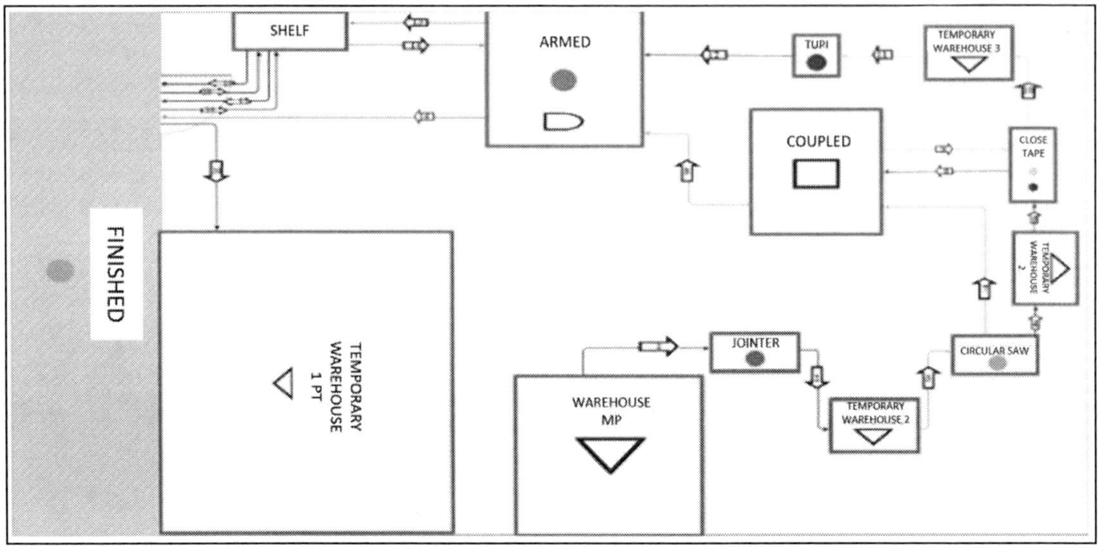

Figure 6. Proposed layout with cell manufacturing

4.4. Results Analysis

Table 1. Before and after indicator comparison.

Indicators	Before	After	% Variation
Resistance to Change Index	32.14 %	3.57 %	
Cycle Times	53.6 minutes	42.9 minutes	-19.96 %
Search for Materials and Tools	8.2 minutes	3.7 minutes	54.87 %
Time between Stations	14.3 minutes	9.6 minutes	32.86 %
Processing Times	229.3 minutes	208.2 minutes	9.20 %
% Efficiency	61.11 %	80.92 %	
% Compliance in Order Fulfilment	12.5 %	60 %	47.5 %

In this phase, the company appoints a team responsible for future growth and continuous improvement.

5. Conclusions

By implementing the proposed model, the time spent searching for materials was reduced by 54.87%. In addition, cycle times reported a 20% decrease in chair production, from 53.6 min/chair to 42.9 min/chair. Further, a 32.14% reduction was reported in travel times between workstations. As per these results, on-time order fulfillment experienced a 47.5% increase (from 12.5% to 60%).

Model implementation also increased line efficiency by 19.81%, and production capacity grew from 189 chairs/month to 236 chairs/month.

John Kotter's change management model was used as a catalyst, proving to be a proper support tool for managing the transition from a traditional model to a lean model.

6. References

[1] de la Torre R 2018 Latin America and Spain are a small appendage in the Furniture Industry Retrieved from: http://masmadera.net/industria-del-mueble-en-el-mundo/.

[2] Ministry of Production 2016 Industrial Statistical Yearbook, SMEs and Local Trade.

[3] J. Cc. Tello 2012 Male and female entrepreneurs of the wooden furniture industry in south lima Retrieved from: http://www.citemadera.gob.pe/.

[4] Narro W M A, Cárdenas M C G, Reto R I R and Chup J H Z 2017 "Strategic Planning for the Wood Furniture Industry in Peru (*Pontificia Universidad Catolica del Perú*).

[5] CITEMADERA 2018 The Wood Industry in Peru.

[6] Ca Vega, Kg Echevarría, Rm Quispe, And Pt Sánchez 2018 Strategic Plan of the Wood Furniture Industry in Peru (*Pontifical University Católica Del Perú*).

[7] Baloh J and Ward M M 2017 Implementing team huddles in small rural hospitals: How does the Kotter model of change apply? *J. Nurs. Manag.* **26** 571-578.

[8] Wheeler T R and Holmes K L 2017 Rapid transformation of two libraries using Kotter's Eight Steps of Change *J. Med. Libr. Assoc.* **105** 276–281.

[9] Maclean D F W and Vannet N 2016 Improving trauma imaging in Wales through Kotter's theory of change *Clin. Radiol.* **71** 427–431.

[10] Riquero I, Hilario C, Chavez P and Raymundo C 2019 Improvement proposal for the logistics process of importing SMEs in Peru through lean, inventories, and change management *Smart Innov Syst Tech* **140** 495-501.

[11] Luz E M S 2016 Implantação de ferramentas Lean em pequenas e médias empresas: estudo de caso em uma indústria moveleira (Implementation of Lean tools in small and medium sized enter-prises: a case study in a furniture industry) *Espacios* **37**.

[12] Johnson A, Prasad S and Sharma A K 2017 Manufacturing Lead Time reduction in a Scaffold Making Industry using Lean Manufacturing Techniques - A case study *Int. J. Mech. Eng. Technol.* **8** 137–148.

[13] Nyemba W R and Mbohwa C 2017 Process Mapping and Optimization of the Process Flows of a Furniture Manufacturing Company in Zimbabwe Using Machine Distance Matrices *Procedia Manuf.* **8** 447–454.

[14] Nallusamy S and Adil Ahamed M A 2017 Implementation of Lean Tools in an Automotive Industry for Productivity Enhancement - A Case Study *Int. J. Eng. Res. Africa* **29** 175–185.

[15] Nallusamy S 2016 Efficiency Enhancement in CNC Industry using Value Stream Mapping, Work Standardization and Line Balancing *Int. J. Performability Eng.* **12** 413–422.

[16] Suhardi B, Sari A P and Laksono P W 2016 Implementation of Standardization Work to Improve Productivity in Indonesian Furniture Industry *J. Eng. Appl. Sci.* **11** 2602–2606.

[17] Luchese J 2016 Implantação Da Lógica Tambor-Pulmão-Corda Em Uma Empresa Da Indústria Moveleira *Holos* **8** 262.

[18] Saravanan V, Nallusamy S and George A 2018 Efficiency Enhancement in a Medium Scale Gearbox Manufacturing Company through Different Lean Tools - A Case Study *Int. J. Eng. Res. Africa* **34** 128–138.

IOP Conf. Series: Materials Science and Engineering **796** (2020) 012023 doi:10.1088/1757-899X/796/1/012023

Lean production management model under the change management approach to reduce order fulfillment times for Peruvian textile SMEs

L Durand-Sotelo[1], M Monzon-Moreno[1], P Chavez-Soriano[1], C Raymundo-Ibañez[2,4] and F Dominguez[3]

[1] Ingeniería Industrial, Universidad Peruana de Ciencias Aplicadas (UPC), Lima 15023, Perú.
[2] Dirección de Investigación, Universidad Peruana de Ciencias Aplicadas (UPC), Lima 15023, Perú.
[3]Escuela Superior de Ingeniería Informática, Universidad Rey Juan Carlos, Mostoles, 28933, España.

E-mail: carlos.raymundo@upc.edu.pe

Abstract. Currently, small- and medium-sized enterprises face order fulfillment issues, thus generating reduced service levels. In addition, these companies are usually not aware of the importance of continuous improvement tools or of training staff as a mitigation strategy for this situation. Within this framework, the authors performed a literature review to compile production models through which downtimes could be reduced. The production model designed therefrom comprises Lean Manufacturing and work study tools within a Change Management approach. This design focuses on model implementation by small companies without requiring large investment, cutting-edge technology, or qualified personnel. Finally, an application case study was conducted in a small textile manufacturing company located at the Gamarra Fashion Center in Lima, Peru. The results that were reported revealed that late order fulfillment instances reduced by up to 18%, which had an impact on downtimes, unnecessary movements, and in-process inventory levels, thus increasing productivity by 85%.

1. Introduction

In recent years, Asian companies have ventured into different international markets by offering lower prices than those offered by the existing suppliers in those geographies, thereby exerting a negative impact on the exports from different local markets, which, in turn, has left the sector in crisis [1]. Consequently, companies are seeking to improve their production processes and increase their service levels by identifying the different issues affecting their operating costs. Hence, to compete with the Chinese approach to sales, more value had to be added to the process in order for it to be passed on to customers [2]. In light of this, companies strived to remain competitive by implementing different techniques of manufacturing quality products, while delivering them within the deadlines set forth by customers. However, a survey conducted on companies operating within the sector revealed that three

[4] To whom any correspondence should be addressed.

Content from this work may be used under the terms of the Creative Commons Attribution 3.0 licence. Any further distribution of this work must maintain attribution to the author(s) and the title of the work, journal citation and DOI.
Published under licence by IOP Publishing Ltd

out of five companies were struggling to fulfill their orders in time [3]. This affected mostly small- and medium-sized companies, where the lack of production planning, training, and organizational culture compromised service levels. In addition, several studies note that this problem is mainly caused by downtimes and defects, with the latter usually causing serious production delays [4]. Further, other studies claim that these issues are rooted in delayed deliveries of raw materials on site, as well as frequent replacements of defective garments. Therefore, companies must focus on developing new operating procedures to improve product yield and quality through engineering techniques.

Therefore, this paper proposes a production model that is based on Lean Manufacturing and work study techniques focused on Change Management aimed at reducing late order fulfillment times at a textile company. This model seeks adaptability at all production levels within small companies because of their large numbers within the field.

2. State of The Art

Using this framework as a point of departure, a literature review compiled production models that could be applied to small companies attempting to reduce downtimes and improve quality to offset late order fulfillment times.

Small companies usually tend to carry out production activities empirically and in a disorderly manner. Therefore, different authors argue that the implementation of production models within a company is a critical effort in the mitigation of these issues. A study conducted in a textile manufacturing company noted that an eclectic production model was able to increase production capacities by up to 50% [5]. In addition, a 35% reduction in operating costs was reported after eliminating activities that did not add value to the process through production planning and control activities [6] [2]. Conversely, other authors propose that Lean Manufacturing-based production models be used to identify and eliminate waste [7]. A study conducted in a spinning company where Lean Manufacturing was implemented reported 20% reductions in operating costs and a 15% reduction in late order fulfillment [8].

However, some authors suggest that even better results are achieved by prioritizing the company's workforce first before making any changes to the production processes. For example, a study carried out in a textile manufacturing company reported that Change Management contributed toward reducing production times by 15% [9]. Finally, several European and Asian authors assert that using a proper combination of Lean Manufacturing and Change Management in production models would beget great benefits for companies wishing to significantly reduce production times. In light of this, a combination of these two methodologies in a small automotive company increased productivity by 34% while reducing downtimes by 20% [10]. An additional study conducted in another automotive company also proved that efficiencies improved from 40% to 65% when these two methodologies were combined [1].

3. Contribution

The proposed production model is based on Lean Manufacturing and work study tools. Likewise, a Change Management approach is proposed throughout the model to foster efficient adaptation of the new techniques by company employees. Through Change Management, the production model may guarantee adequate production cycle times, thus reducing late order fulfillment instances without compromising on quality and production costs.

Consequently, the model starts with preparation for change, where staff members are trained in adapting a continuous improvement initiative. Change is then nurtured throughout the company on the basis of the 5S philosophy with the clear purpose of reducing product defects by process standardization. (Fig. 1).

In addition, the Theory of Constraints, Line Balancing, and Method Time Measurement (MTM) are set forth as pillars of change, reducing production cycle time by identifying and balancing critical process tasks. Likewise, these pillars also reduce unnecessary movements and improve ergonomics of the staff members engaged.

IOP Conf. Series: Materials Science and Engineering **796** (2020) 012023 doi:10.1088/1757-899X/796/1/012023

Finally, the model is standardized by performing the last Change Management and 5S steps to commit staff members to the new production model.

Therefore, the proposed method consists of thirteen steps (Fig. 2).

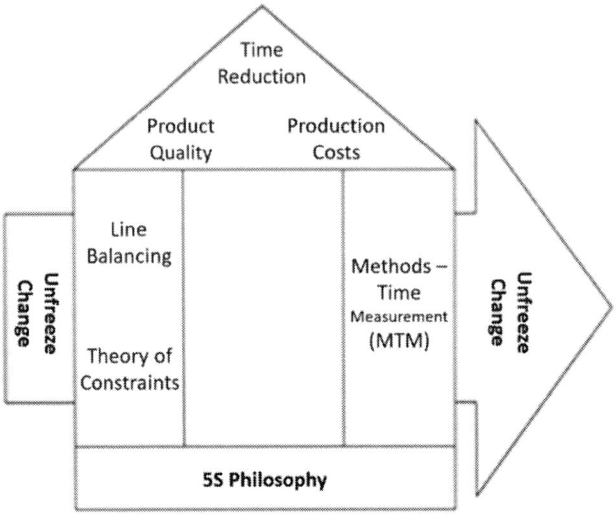

Figure 1. Proposed production model

3.1. Indicators

The proposed methodology dictates that the production process must be controlled. Therefore, the following indicators are defined:

3.1.1. Late Order Fulfillment Instances. Indicator that controls the number of times a product order is fulfilled after the established deadline.

$$IT = \frac{Number\ of\ Orders\ Fulfilled\ Late}{Number\ of\ Orders\ Fulfilled} x100\% \quad (1)$$

3.1.2. Productivity. Determines the number of products made on a daily basis.

$$PDT = \frac{Number\ of\ Garments}{Day} \quad (2)$$

3.1.3. Cycle Time. This indicator controls the number of garments made with each minute that passes.

$$TC = \frac{Number\ of\ Garments}{Minute} \quad (3)$$

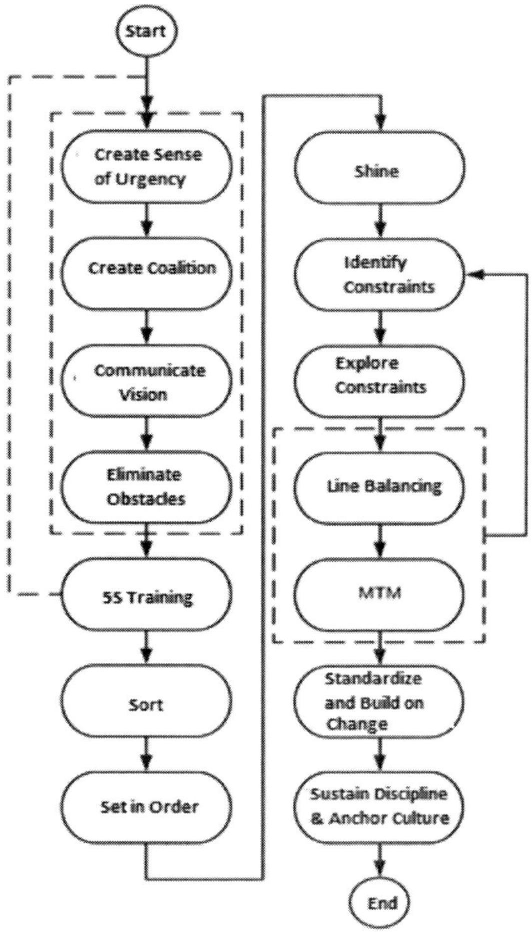

Figure 2. Proposed methodology

4. Validation

This section discusses the implementation of the production model in a textile manufacturing company located in the Gamarra Fashion Center, which manufactures sportswear, school uniforms, etc. This company reports order fulfillment delays of up to 5 days after the proposed deadline, as well as a 13% increase in operating costs. To validate the effectiveness of the production model, pre-implementation indicators will be compared against post-implementation indicators.

Figure 3. Textile manufacturing company

4.1. Implementation of a Production Model

4.1.1. Change Management. A production model strategy that is focused on engagement and collaboration by all company staff members, from the owner to the operators. Therefore, awareness training sessions were held to encourage staff members to own the problem and adopt a new production model. To this end, we relied on a team member who exerted influence on other staff members, thus fostering greater model adaptability. In addition, change adoption was monitored through surveys.

4.1.2. 5S Audit. Based on the level of awareness and commitment displayed by team members regarding the new production model, the initial 5S audit was performed to identify compliance with the pillars of this philosophy. The results were not favorable as at least 54% of the staff members neither acknowledged nor complied with the criteria assessed.

4.1.3. Organization. This first stage of the philosophy required participation by all team members to use red cards in order to eliminate agents, considered foreign to the production process.

4.1.4. Order. At this second stage, workplaces were defined by yellow lines to avoid cluttering in-process products within the area, thereby reducing unnecessary movements and accidents.

4.1.5. Cleaning. At this stage, all machinery, wiring, etc. were inspected. Similarly, cleaning standardization policies were established for the area.

4.1.6. Execution of Theory of Constants. Upon organizational culture standardization, bottlenecks were identified. With regard to this, closures and pockets were singled out as critical manufacturing activities. Therefore, all other activities were subordinated to these activities, scheduling operator lunch and break times so that the process never stops.

4.1.7. Line Balancing. To offset production cycle times in stations, two workstations that used the longest time or had the lengthiest candidate criteria were eliminated.

4.1.8. MTM Application. Having changed process flows, the model focused on operator motion to assess the time lost because of unnecessary motion from poor posture. The study determined that the tables were not adequate because they did not allow the operator's arms to rest at the proper height of 96 cm above the floor. Therefore, the worktables were replaced. In addition, the new tables requested had drawers (as requested) wherein work tools could be stored.

4.1.9. Standardize and Build on Change. At this stage, collaborators became more engaged in the face of the new production model; hence, the purpose here was to standardize activities though signposting and toolbox talks.

4.1.10. Sustain Discipline and Anchor Culture. This last stage monitored the current status of the production model, undertaking staff member surveys and audits to measure model compliance.

The post-production model implementation results obtained are as follows.

Table 1. Comparison of case study indicators

Concept	Pre-Implementation	Post-Implementation	Variation
% Late Fulfillment	39%	18%	+19%
Productivity	13 min / garment	24 min / garment	+85%
Cycle Times	26.8 min	20.1 min	-33%

As can be observed, after the new production model was implemented, a 19% reduction was explained by the main problem reported by the company. Similarly, productivity increased by 85% for the sweat suit process while cycle times decreased by 33%. On the other hand, operating costs decreased by 20% because of the reduction in overtime caused by unnecessary motion.

5. Conclusions

By implementing a production model based on Lean Manufacturing and work study tools with a Change Management approach, late order fulfillment incidences were reduced by 18% even as product quality improved.

The transversal Change Management approach deployed in this model was crucial for the implementation and adaptability of engineering techniques in a small company at the Gamarra Fashion Center. In addition, the staff members had always performed their jobs on the basis of customary practices instead of relying on technical preparation. Therefore, nurturing a new mindset was critical for fostering model adaptation.

In conclusion, the study proved that workers required continuous training in Lean Manufacturing. Likewise, tools must be continuously monitored to guarantee long-term implementation, which could allow new hires to easily adopt the production model.

6. References

[1] Durán Lima, J E and Pellandra, A 2017 La irrupción de China y su impacto sobre la estructura productiva y comercial en América Latina y el Caribe *Serie Comercio Internacional CEPAL Naciones Unidas* **131**

[2] Kayar M and Akalin M 2016 Comparing Heuristic and simulation methods applied to the apparel assembly line balancing problem *Fibres Text East Eur* **2** (116) 131-137.

[3] Baeza R 2016 REDUTEX: a hybrid push–pull production system approach for reliable delivery time in knitting SMEs *Prod Plan Control* **27** (4) 263 -279.

[4] Behnam, D, Ayough, A and Mirghaderi, S H 2018 Value stream mapping approach and analytical network process to identify and prioritize production system's Mudas (case study: natural fibre clothing manufacturing company *J Text Int* **109**(1) 64-72

[5] Cui H Yan 2017 The optimization design of uniform's hanging production line *Int J Cloth Sci Tech* **27** (3) 370-389.

[6] Ministerio de Comercio Exterior y Turismo 2017 Perú: Medidas Antidumping aplicadas a los textiles de procedencia China (*Informe Coyuntural N° 004*).

[7] Chávez J, Osorio F, Altamirano E, Raymundo C, Dominguez F 2019 Lean production management model for SME waste reduction in the processed food sector in Peru *Adv Int Syst Comput* **971** 53-62.

IOP Conf. Series: Materials Science and Engineering **796** (2020) 012023 doi:10.1088/1757-899X/796/1/012023

[8] Naciones Unidas, Departamento de Asuntos Económicos y Sociales – División de Estadística 2017 Clasificación Industrial Internacional Uniforme de todas las actividades económicas (CIIU) (*Revisión 4 ST/ESA/STAT/SER.M/4/Rev.4. Nueva York, Estados Unidos de AméricaI*).

[9] Slović D, Tomašević I, Radović M 2016 Improving Productivity in the Apparel Industry Through Gain Sharing and Continuous Process Improvement: The Case of a Serbian Manufacturer *Fibres Text East Eur* **2** (116) 15-22.

[10] Rahman S, Laosirihongthong T and Sohal A S 2018 Impact of lean strategy on operational performance: a study of Thai manufacturing companies J Manuf Technol Manag **21** 839 -852.

IOP Conf. Series: Materials Science and Engineering **796** (2020) 012024 doi:10.1088/1757-899X/796/1/012024

Lean Manufacturing Production Management Model focused on Worker Empowerment aimed at increasing Production Efficiency in the textile sector

V Sosa-Perez[1], J Palomino-Moya[1], C Leon-Chavarri[1], C Raymundo-Ibañez[2,4] and F Dominguez[3]

[1] Ingeniería Industrial, Universidad Peruana de Ciencias Aplicadas (UPC), Lima 15023, Perú.
[2] Dirección de Investigación, Universidad Peruana de Ciencias Aplicadas (UPC), Lima 15023, Perú.
[3]Escuela Superior de Ingeniería Informática, Universidad Rey Juan Carlos, Mostoles, 28933, España.

E-mail: carlos.raymundo@upc.edu.pe

Abstract. For companies operating within the garment manufacturing industry, having frequent downtimes in their production flows is an extremely common issue. In this context, a balanced production line is required to prevent high waiting times due to limited productive capacity. A well-balanced assembly line allows products to be produced in an optimum time while using less resources, such as machines, materials, or labour, since the right number of products is produced with the exact amount of resources, thus generating savings in production costs. This paper seeks to foster optimum resource allocation through the line balancing tool. Finally, to define a work methodology, best practices were selected, and a procedures manual was developed focusing on Standardization. Both tools were implemented after implementing changes to the company culture by means of the Employee Empowerment tool. As a result of this implementation, workers acquired greater accountability and control over the resources, methods, and equipment of their work areas. After the proposed improvements had been deployed, the company reported an increase of over 20% in production line quality, performance, and efficiency.

1. Introduction

In the last decade, the Peruvian economy has reported an average annual growth of 4.3% [1]. One of the reasons for this feeble growth is that local companies, whether micro-, small-, or medium-sized enterprises, face different problems that affect their productivity levels, which in turn exerts a negative impact on their competitiveness at international levels. In this context, textile and clothing MSMEs play a leading role in the Peruvian business fabric since they represent one of the main non-extractive activities at the national level, accounting for 1.3% of the national GDP and 8.9% of the manufacturing production in 2014 [2]. Therefore, the textile and clothing sector has become the second most important sector within the manufacturing GDP, only surpassed by the precious and non-

[4] To whom any correspondence should be addressed.

Content from this work may be used under the terms of the Creative Commons Attribution 3.0 licence. Any further distribution of this work must maintain attribution to the author(s) and the title of the work, journal citation and DOI.
Published under licence by IOP Publishing Ltd

ferrous metals industry. However, despite its production value growing approximately 14.8% between 2009 and 2014, the textile industry exhibited, from January to August 2017, a cumulative negative growth rate of 6.5%, while for the last 12 months, the variation was negative, approximately 6.0% [1]. Therefore, the root causes for this issue must be determined.

This study seeks to describe current sector performance and propose an improvement model to avert downtimes and raise the productive efficiency of the company. In addition, this study may also serve as an example for other garment manufacturing companies toward better production controls and development. To this end, Value Stream Mapping and the 7 Wastes were used as assessment tools in the identification of the main problem the company under study is facing. Finally, a line balancing model was developed and implemented to appropriately allocate resources according to each workstation.

2. State of The Art

2.1. Lean Manufacturing in the Textile and Clothing Sector

According to past research, the application of Lean Manufacturing techniques has had positive results in various production processes, improving efficiency, productivity, and quality. Since the Lean philosophy supports the adaptation of different techniques, actions or activities that do not add value to production processes may be easily identified for their subsequent reduction or elimination, which leads to lower production costs and increased product quality [3]. In this light, Psomas, E., and Antony, J. [4] argue that Lean Manufacturing can address different problems arising during the garment production process, such as overproduction, reprocessing, excess inventory, movements, unnecessary transportation, defective products, and process delays. In addition, this study engages two additional tools: Line Balancing and Standardization

Line Balancing focuses on grouping activities that follow a work sequence at a production plant, which, in this case, is a textile production plant; this is meant for achieving maximum use of resources. According to Oksuz, M.K. [5] and Türkmen, A. [6], one of the main objectives of line balancing is eliminating bottlenecks and production delays. Among the most relevant line Bblancing models are the Moddie Young and the Hoffman methods. Furthermore, K. Syahputri [7] and M. Kayar [8] compare the efficiency and assembly line balancing losses at a textile manufacturing company in Asia, reporting a 15% gap between the Hoffman model and the Moddie Young model when measuring line balancing efficiencies.

On the contrary, work standardization is the main premise for achieving accurate quality results. Therein, a sequence of steps through which identical results can be achieved in terms of time, procedure, work technique, product, and quality are designated. Standardization is an essential part of any viable production. Processes (work methods) at different complexity levels are known as standard work methods.

2.2. Worker Empowerment Applied to the Textile and Clothing Sector

KH Blanchard and JP Carlos [9] argue that Worker Empowerment is a management strategy that allows employees to develop their own skills for becoming competent in making their own decisions. In fact, one of the studies reviewed developed a visual combination model based on resources, Signalling theory, and Experiential Learning theory. This model is then validated through its implementation in 653 employees working in textile production companies. These findings suggest that a culture of organizational learning significantly strengthens team building relationships and worker empowerment in employee competencies. [10] The results conclude that companies must provide sufficient training time for employees to develop an appropriate culture and effective communications within the company.

3. Contributions

3.1. Proposed Model

- Phase 1: Value Stream Mapping diagram, the production flow of the company. This tool identifies lead time along with the times that add and do not add value.
- Phase 2: Line Balancing to optimally allocate resources and balance production lines.
- Phase 3: Standardization to establish the appropriate work methods.
- The above is based on Worker Empowerment, which seeks to generate total work area ownership and accountability in operators.

Each component added to this production management model not only fosters production efficiency but also implementation sustainability. That is, it guarantees that the implementation will remain operational over time.

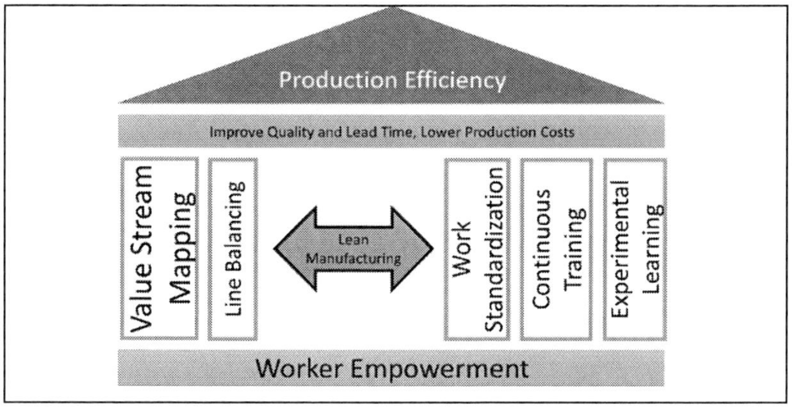

Figure 1. Proposed model

3.2. Proposed Method
This item explains the steps required for deploying the production management model as explained below (Figure 2).

3.3. Indicators

3.3.1. Production Line Losses and Efficiency. These indicators identify the current status of production line flows to determine whether bottlenecks have been successfully removed.

$$LB = [(nC - \Sigma\, Co)/nC)]\ 100 \qquad (1)$$
$$LE = (1 - LB)\ 100$$

3.3.2. Overall Efficiency. Overall Equipment Efficiency, or OEE, is an indicator which measures the efficiency with which a piece of equipment or process works.

$$OEE = Availability \times Performance \times Quality\ Rate \qquad (2)$$

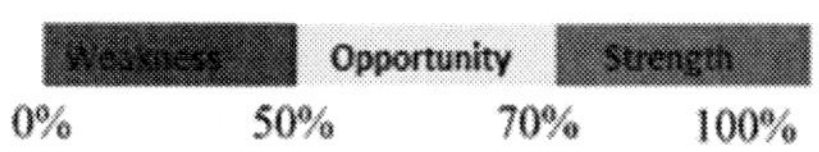

3.3.3. Performance. This aspect assesses the speed percentage with which machines actually operate against their design speed.

$$Performance = \frac{\frac{Cycle\ Time}{TO}}{Total\ Units\ Produced} \qquad (3)$$

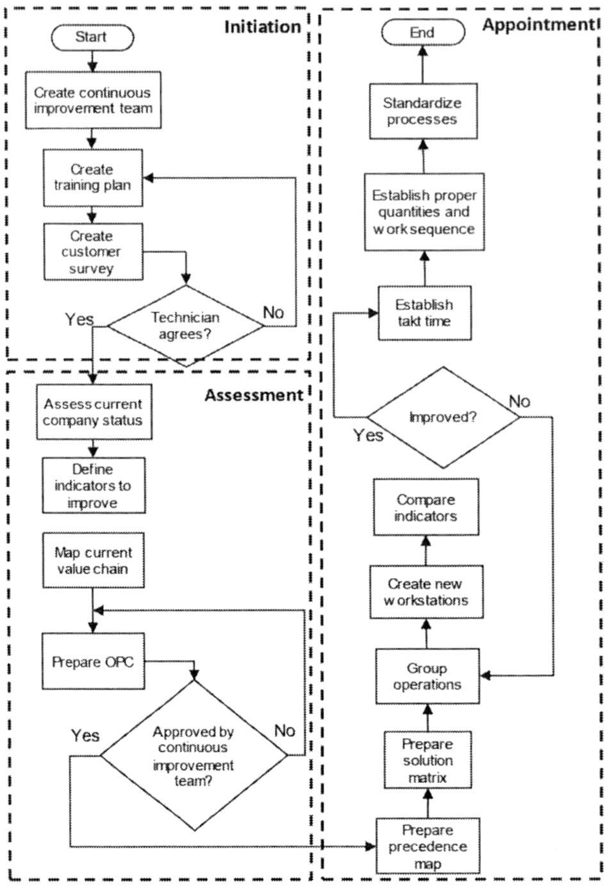

Figure 2. Proposed method

3.3.4. Quality. This aspect determines the number of products manufactured within the required quality levels and which are not rejected.

$$Quality = \frac{Comforming\ Units}{Total\ Units\ Produced} \tag{4}$$

3.3.5. Availability. This aspect assesses the percentage of time that is actually worked against the time scheduled for working.

$$Availability = \frac{TPO - Unscheduled\ Stops}{Total\ Work\ Time - Scheduled\ Stops} \tag{5}$$

4. Validation

4.1. Description

The company under study manufactures and markets garments for babies, boys, and girls. Its facilities include a work area of 500 m2, wherein 23 workers operate about 30 machines between sewing machines, overlock sewing machines, embroidery machines, and others. Located in the district of San Luis in Lima, Peru, the company was established in 2013, which evidences that it is still a developing company (Figure 3).

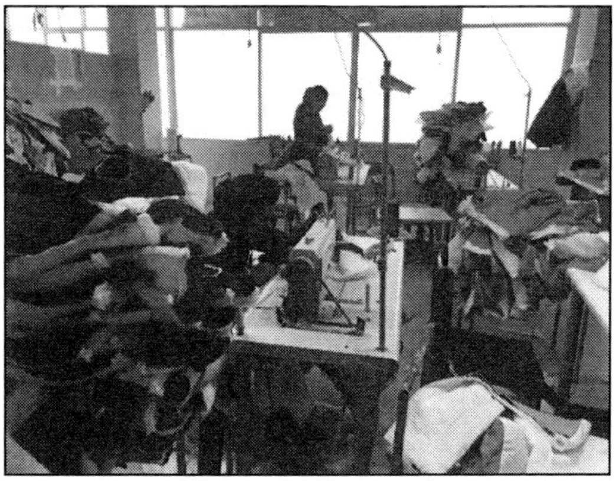

Figure 3. Case study

Recently, this company has been in dire economic straits in terms of achieving its objectives. Although, in recent years, the company sales have increased from s./5,041,388 to s./6,119,531, that is a 21% increase. Costs increased in over 33%, which is a matter of concern for the company.

4.2. Assessment

Since the company produces a large number of products, the ABC methodology was used to determine which products within the same production line represented higher revenue in sales for the company. Based on our findings, polo shirts and blouses were selected as the flagship products. Next, the Value Stream Mapping diagram was plotted, establishing that from a lead time of 13.5 days, a total of 6.5 days was found not to add value to the production process. These days are associated to the sewing process, which generates frequent bottlenecks at this station.

Finally, to identify the causes of the evidenced discontinuous production flow, the 7 Wastes tool is used, which identifies overstock inventories and reprocesses as the most likely causes.

4.3. Method Implementation

Step 1: Preparation of the Operations Process Chart

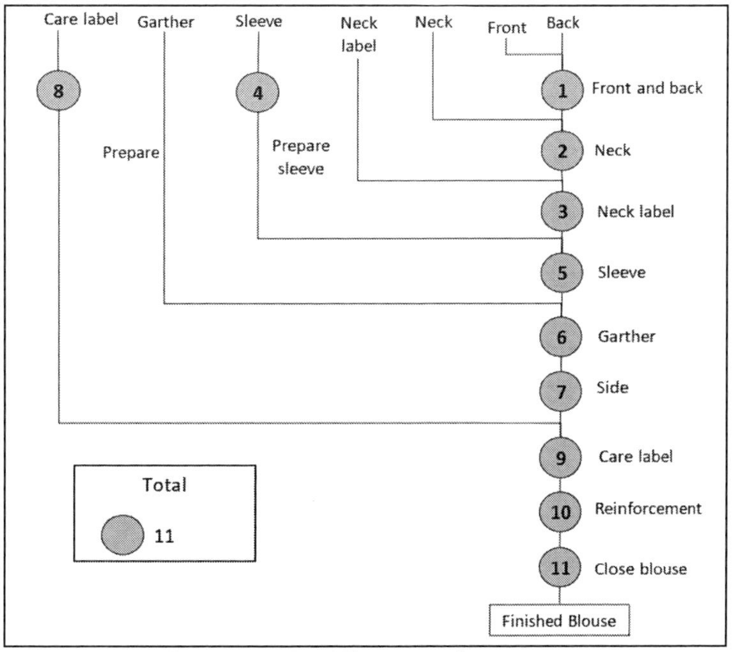

Figure 4. OPC-Blouse sewing process

IOP Conf. Series: Materials Science and Engineering **796** (2020) 012024 doi:10.1088/1757-899X/796/1/012024

Step 2: Calculate balance losses with the standard times.

$$LB = (11x0.98\text{-}19.84) / (11x0.98)\ X\ 100 = 44.16\% \tag{6}$$

$$LE = (1\text{-}0.4416)\ x100 = 55.84\% \tag{7}$$

Step 3: Leopold Matrix, grouping unprecedented activities.

OP	1	2	3	4	5	6	7	8	9	10	11
1		1									
2			1								
3					1						
4					1						
5						1					
6							1				
7									1		
8									1		
9										1	
10											1
11											
	0	1	1	0	2	1	1	0	2	1	1

Figure 5. Leopold matrix

Step 4: Reduce activities through a priority matrix.

	Activity	Time	Total	Difference
1	1	0.74	0.74	0.24
2	2	0.52	0.72	0.26
	10	0.2		
3	3	0.25	0.75	0.23
	6	0.5		
4	4	0.48	0.86	0.12
	8	0.38		
5	5	0.62	0.62	0.36
6	7	0.65	0.65	0.33
7	9	0.7	0.7	0.28
8	11	0.98	0.98	0

Figure 6. Activity matrix

IOP Conf. Series: Materials Science and Engineering **796** (2020) 012024 doi:10.1088/1757-899X/796/1/012024

Step 5: Standardize the new process line through a procedure manual.

Manual de procedimiento de la confección de una blusa

La base para la confección de una blusa, es el trazo de 5 líneas incluyendo la línea imaginaria, llamadas líneas de construcción que son:

1. Línea imaginaria; esta se traza midiendo 2 cms. de la orilla del papel, hacia abajo.

2. Línea de caída de hombro; se traza de la línea imaginaria 4 ó 5cm. hacia abajo.

3. Línea de sisa; se traza de la línea de cintura, hacia arriba, se anota la medida de costado.

4. Línea de bocamanga; se traza a la mitad de lo que mide la línea de caída de hombro y la línea de costado (busto).

5. Línea de cintura; esta se traza de la línea imaginaria, hasta donde mide el largo de espalda o de frente.

Ejemplo de las líneas para la confección de la blusa

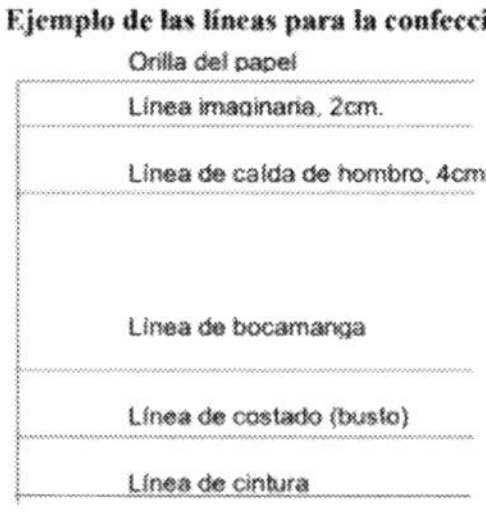

2. Medidas básicas para confeccionar una blusa:

- Trasero:

Son los centímetros que mide la parte de atrás, iniciando siempre en la parte alta del cuello hasta llegar a la cintura.

Se toman dos medidas:

a) Ancho de espalda; referido a la distancia en cms. que hay entre los hombros. Para ello se inicia a tomar la medida del hombro izquierdo hasta finalizar en el hombro derecho.

Figure 7. Sewing procedure manual

- With the line balancing, three activities were deemed unprecedented, which led to line balancing of almost 77%.
- With Standardization, a procedure manual was designed for the sewing area. Then, the manual was disseminated among all operators in the area as printed newsletters.
- Worker Empowerment was carried out through Autonomous Maintenance, one of the pillars of Total Production Maintenance. Here, by implementing actions to clean, adjust, and lubricate, workers were able to take ownership and become accountable over their work areas.

The following table shows the improvements achieved:

Table 1. Before and after results

Tool	Indicators	Before	After
Standarization	Quality	54.79%	83.67%
TPM	Availability	97.78%	95.00%
Line Balancing	Performance	85.03%	95.25%
OEE		43.22%	75.62%

OEE: A 75.62% OEE rate was achieved, which ranks within the acceptable range for productive efficiency.

5. Conclusion

- The proposed model identified high in-process inventories and the quantity of reprocesses within the manufacturing area as the main causes for the problem addressed. This problem was evident when the discontinuous flow of garment production reported a lead time of 13.5 days, whereas the market average does not even exceed 10 days.
- Therefore, the application of the line balancing and Standardization Lean tools to the garment product process removed the two wastes which exerted the greatest economic impact on the company. In addition, production line efficiency also increased from 55.88% to 76.79%.
- Further, the application of the Standardization tool minimized response times for a specific problem, which allowed us to increase our individual efficiency and effectiveness ratios at the company level.

6. References

[1] Ministry of Production 2015 Economic Studies Offices Retrieved from http://demi.produce.g.,ob.pe/estadistica/sectorial.
[2] I National Statistics Institute (INEI) 2015 Retrieved from https://www.inei.gob.pe/media/MenuRecursivo/indices_tematicos/pbi_act_econ_n54_kte_2007-2015_1.xlsx
[3] Bellido Y, Rosa A L, Torres C, Quispe G, Raymundo C 2018 Waste optimization model based on Lean Manufacturing to increase productivity in micro- and small-medium enterprises of the textile sector", CICIC 2018 - Octava Conferencia Iberoamericana de Complejidad, Informatica y Cibernetica, Memorias. 1, pp. 148-153.
[4] Psomas E and Antony J 2019 Investigation of gaps in lean manufacturing: a systematic review of the literature Int J Qual Manag Reliab.
[5] Oksuz M K, Buyukozkan K and Satoglu S I 2017 U-shaped assembly line worker assignment and balancing problem: A mathematical model and two meta-heuristics *Comput Ind Eng* **112** 246-263.
[6] Türkmen A, Yesil Y and Rahman M 2016 Heuristic production line balancing problem solution with MATLAB software programming *Int J Cloth Sci Tech* **28** (6) 750-779.
[7] Zhang L, Narkhede B E and Chaple A P 2017 Evaluating lean manufacturing barriers: an interpretive process *J Manuf Tech Manag* **28** (8) 1086-1114.
[8] Syahputri K, Sari R M, Rizkya I, Leviza J and Siregar I 2018 Improving Assembly Line Balancing Using Moodie Young Methods on Dump Truck Production *IOP Conf Ser: Mater Sci Eng* **288** 012090
[9] KH Blanchard & JP Carlos 2000 Leadership by the book *Executive excellence* **17** (3) 4-5.
[10] Rodrigues, S A P 2016 Impactos da liderança transformacional e do empowerment na orientação empreendedora: análise das unidades hoteleiras do norte de Portugal (*Doctoral dissertation*)

Application of QR codes as a new communication technology and interactive tourist guide in Jaboi, Sabang

Marwan[1], M Yanis[1], Muzakir[1] and G S Nugraha[2]
[1] Department of Geophysical Engineering, Faculty of Engineering, Universitas Syiah Kuala, Darussalam, Banda Aceh, 23111, Indonesia.
[2] Department of Geological Engineering, Faculty of Engineering, Universitas Syiah Kuala, Darussalam, Banda Aceh, 23111, Indonesia

E-mail: yanis@unsyiah.ac.id

Abstract. The Indonesian volcano is only limited for the electrical energy capacity, but in developed countries, the volcano is not only used for power plan but also developed for the tourism industry. Jaboi is one of the volcanoes located in Sabang with a capacity of 80 MW. The five potential of geothermal spots can be used as geo-tourism tours. Besides to improve the economic sector, the main purpose of the geo-tourism is the management of the tours that will be provided for educative lessons about the friendly life in a volcanic area. In this paper, we produced the educational videos that placed on each tourist spot; the video contains including the history of the site, geological background, and also information about the risks of living in a volcanic area. The QR code program is used as a medium for data transfer that can be scanned through a smartphone. This application is widely used in tourism as an economical alternative in technology. In the several spots that difficult to access, we also produced a safe route with a QR code program. The tracks will be overlaid on the smartphone with the 'gpx viewer' application, which possible to download from Google Play or App Store.

1. Introduction

Aceh is one of the regions with the most extensive geothermal potential in Indonesia (1310 MW) located in 17 places [1], one of them is in Jaboi, Sabang. Generally, the geothermal has been closely related to electrical energy with an 80 MW capacity power plant built in Jaboi. Furthermore, the geothermal also possible to develop for the tourism industry; it can have a direct impact on economic people around the volcano area [2]. Referring to [3], ten categories of attractions can be used in a geothermal area for tourism development. Among them, five classes are available at the Jaboi volcano, namely fumaroles, hot springs, volcanic sedimentation, and hot tubs. Figure 1 shows the potential distribution of tourist spots in Jaboi. In another side, the village is confronted with the Malacca Sea, which presents the beautiful panorama and becomes a very complex geo-tourism potential. The problem in geo-tourism management is how the trip can be structurally managed by the local community to produce a financial income, and what the effort to become an educational trip for the tourists [4]. One of the traditional solutions is the training knowledge about the geo-tourism to the local guides. However, this method requires a relatively long time, and belongs to an expensive [5]; the guide also must be trained in English to target world tourists. The QR (quick response) application codes is an economical solution from the side of technology; the method is being able to quickly connect the data from the offline world to online content, and receive all designed files [6].

Content from this work may be used under the terms of the Creative Commons Attribution 3.0 licence. Any further distribution of this work must maintain attribution to the author(s) and the title of the work, journal citation and DOI.
Published under licence by IOP Publishing Ltd

The QR codes serve as a bridge between the physical and digital worlds. Thus, this method is often applied to museums or other underfunded cultural institutions [7]. In this paper, we make the lessons learned video from the tourism spot that presented in QR code as an interactive tourist guide in Jaboi Sabang.

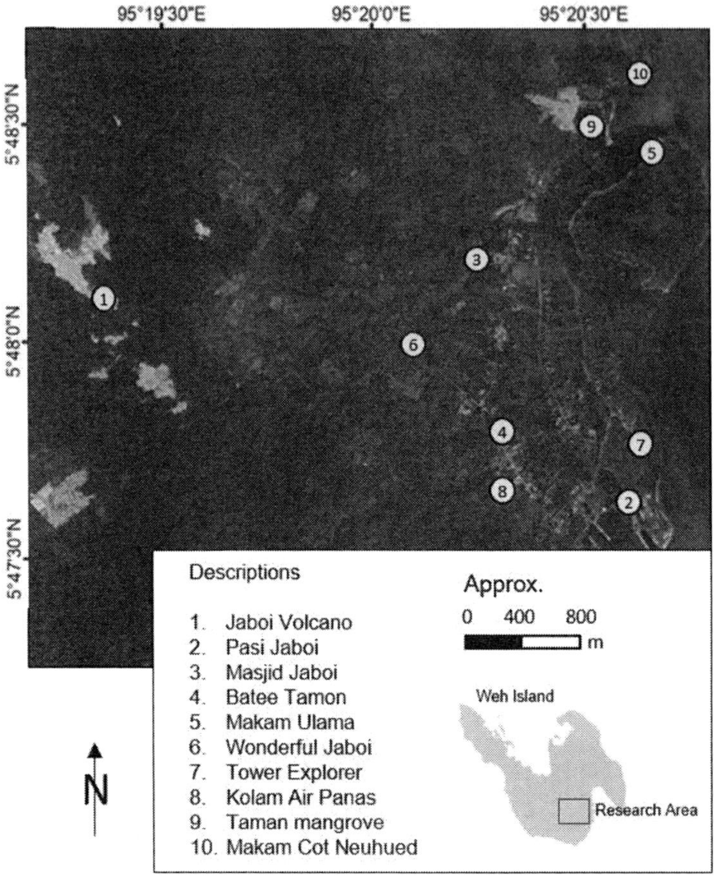

Figure 1. The distribution map of Jaboi destination that will be potentially developed for geo-tourism.

2. History and QR Code Structure

Quick Response Code is a type of 2-dimensional barcode that can be read with smartphones and web applications. Although the Japanese company Denso Wave developed it in 1994 for industrial use, QR codes have become widely used in the advertising and packaging of products after being commercialized in 2011 [9], [6]. The QR Code is a cell-matrix composed of dark module patterns (logical '1') and bright pattern (logical '0'). Each QR code pattern consists of function patterns and encoding regions which are divided into several module patterns, where each design has a specific purpose for processing data readings such as position detection, timing, alignment pattern, format and information of the version that can be referred to ISO / IEC18004: 2006.

Based on the ISO, the minimum module resolution that can be read by most optical devices is 4 px (pixels) with a printout of 300 dpi (dots per inch). Referring to [10] the data on the 3x3 module can only be read through a high resolution camera, which has an accuracy rate of 95% for 5 MP (megapixel), and 100% in 8 MP. However, the smartphones in the market today have been used the high camera resolution over 10 MP. In terms of security [11] stated that QR code is a safe system, it caused the tourist only get the links on the front-end of the server through the internet. Thus it will not interference with the back-end database of an information system.

In general, tourism is considered an innovative industry. Therefore the QR codes have become a popular system; for example, the Cleveland of Art Museum and Bologna's of Archaeology uses a QR code for audio information in its museum collection. The same thing was done at the Natural History Museum in Washington, DC, that uses QR codes as part of the exhibition [12]. Moreover, in Manchester city, all of the streets, castles, vineyards, rivers, parks all open-air venues can inform the tourist with QR code.

3. Material and Methodology

Sabang Island is one of the city which more visited by local and foreign tourists every year. In this preliminary study, the research area is only focused on Jaboi as the location of the volcano. Figure 1 shows the tourism spot where the QR codes will be installed. The QR code pattern is printed in the form of stickers and mounted on board with dimensions of 40 × 60 cm. Furthermore, the board information also contains a short description, another destination of Jaboi and also includes the logos of several collaborative research that funded the program, e.g., Syiah Kuala University, Sabang Government, and the Ministry of Education and Technology. When a smartphone scans the QR code, it will be directed to geosabang.com as the website management and linked to the video that has been made before. This process is called the dynamic barcode, in which the contents in the barcode are possible to change by using the Hypertext Mark-up Language (HTML) without the need to reprint out the barcode. This QR code is generated from https://qrcode.studio that required the video link as a parameter input in the barcode system, the barcode dimension used in this system is 3×3 in order to obtain the accuracy of reading patterns from the smartphone camera.

The video that will be shown on each spot is including the history of the site, geological background, and the lessons learned about the risks of life in the volcanic zone. Figure 2 describes the development system and flow process in the QR code.

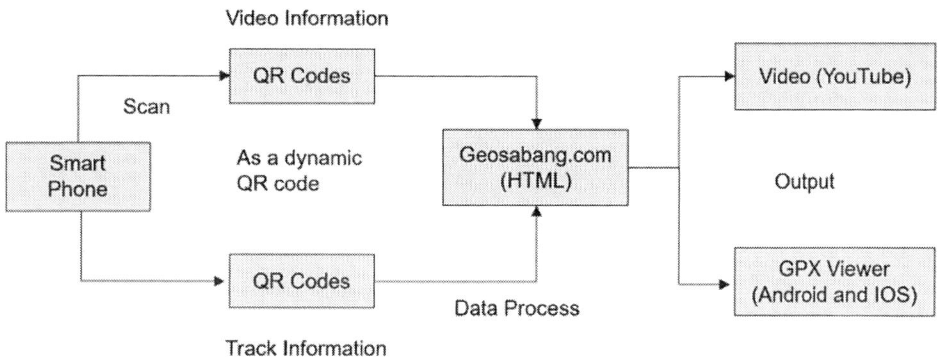

Figure 2. The flow process of dynamic QR code in Jaboi geo-tourism. The video destinations were firstly uploaded on YouTube account for an un-weighted server in the data process.

In another side, the tourism spot located in the forest area that makes it inconvenient to access. For example, Jaboi volcano and three craters, we also developed a safe tracking based on the QR code program. The process of the road is using the tracking feature on the Garmin GPS, where each of the craters will be pointed, and the safe tracking will be recorded. Briefly, the flow of making this QR code is shown in Figure 3. Furthermore, before scanning the barcode, the tourist needs to install the application like GPX Viewer on Google Play or App Store to open the '*gpx' type of data that gets from the QR code scanner.

4. Result and Discussion

The video information is one of the fastest and most economical methods used to achieve the lessons learned in the tourism industry. In this study, we used the barcode technology as a quick transfer of information from the offline board to the online source in the internet server. Based on the observations, it was shown that the local community has been aware and familiar with the tourism industry, this also evidenced by the existence of a small group that manages tourism in the village. Generally, the tourism spots in Jaboi can be categorized into two; the spots related to geological and cultural tourism. Moreover, although the culture is not correlated to geo-tourism; the collaborated of geological landscape and cultural aspect can help and increase the local people to prefer the geo-tourism. The board information was installed in 10 spots location, which 6 of them are related to geothermal tourism. For example (1) the hot spring, is mostly used by Jaboi people for swimming, (2) Batee Tamon is an underwater tourism spot that also has a hot spring, (3) tower explorer as a top view of Jaboi landscape and (4) the Jaboi volcano as the first spot in geo-tourism. Figure 3 shows the hot spring tourism in Jaboi Village.

Figure 3. One of the geological spots that installed the information board, (a) barcode information board, (b) photo documentation at site.

The tracking tour to the Jaboi Volcano and the craters have been a favorite spot for tourists. But for several years, the visitors have been reduced; it caused by unstructured tourism management. In another side, the local community may not always have time to guide the tourists. So that it causes minimal tourist visits. Furthermore, one of the efforts to increase the tourism visit in Jaboi Volcano and craters through provided for the safe road to the location. In this research, we also develop a tracking guide system that used a QR code program. The system is installed in the tourism spot that difficult to access by road. Figure 4 shows a description of the 'get track system ' in the volcano area. The get the track system was first made by tracking to the Jaboi volcano through choice a relatively for the safe route, the route was made by using the Garmin GPS. However, in each of the crater, the coordinates were saved for marker information in the map. The GPX viewer application can be downloaded free through the Google Play Store for Android and the App Store for Apple. Figure 6.c is an overview of the overall track results, starting from the scanning of the barcode (shown as 'You Are Here') to the 4th crater.

After the QR code was installed in each of the spots, we analyze the response of local people related to the implementation of the barcode in Jaboi tourism, then the questionnaires were given to 30 people which distribution to village officials, tourism awareness groups and another local people.

Figure 5 shows the local people responses about the barcode and the video content in Jaboi tourism. Based on the result, it can be analyzed that the 23 or 76.6% of the local people are choosing the implementation of the QR code was a very useful technique for geo-tourism development in Jaboi, it caused by most of the community are interest to technological implementation in tourism. In addition, there were also 5 or 16.67% of the people chose no opinion and 2 people chose it as a bad idea for tourism. The next question is how the quality of the video content that placed in the QR code can provide an information to tourists and local people, the result showed that many of the people (33%) chose the video was very informative, 50% said it was good and there also were 5 people or 17% neutral about the quality of the video. Generally, the local community stated that the video could be as a media transfer for the lesson learned in the geotourism spot, it caused the video is related to historical and geological activities of the spot.

Figure 4. Get the track system on the volcano spot. Figure (a) the get the track system made by a QR code, and (b) is the gpx result on the GPX viewer application.

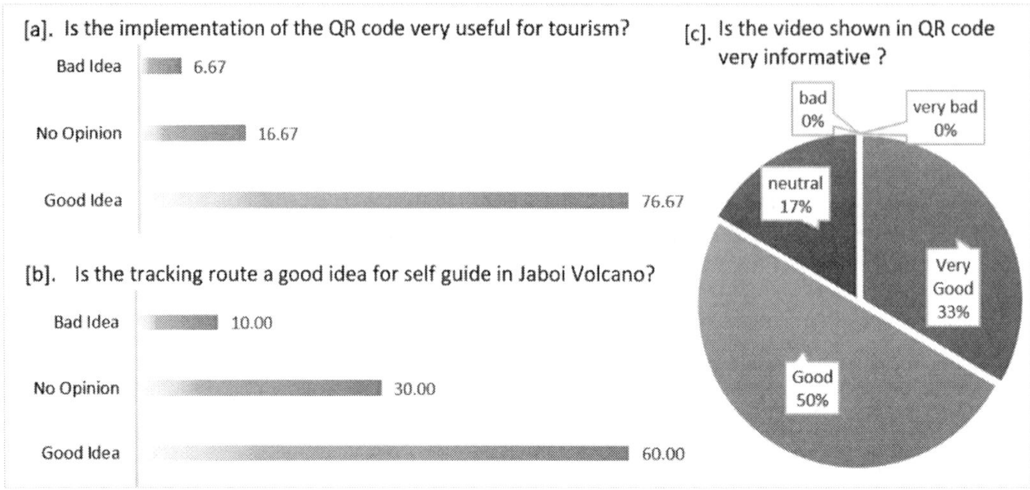

Figure 5. The responses of local people to use the barcode and video content for tourism development.

In another side, the barcode is used for an automatic route to explore the Jaboi volcano. Based on the questioners, it can be showed that 60% of the local people chose the barcode as a good idea for the self-guide to explore the volcano and their craters, but as many as 30% of the people chose no opinion on the system, and 10% chose for a bad idea, this was caused by some people who are not familiar with the modern technology in Jaboi tourism.

5. Conclusion

Technology and tourism are the two most entangled areas in modern daily life. This study has identified the usage of QR codes in various geo-tourism areas. Based on the results, it can be concluded that the implementation of the video application at the tourist spots is the most effective way to encourage the community towards geothermal tourism. The video can also be used as a media lesson learned for public people about the risks of life in the volcanic zone. Furthermore, the used of QR code as a video transfer medium is an efficient, economical, and modern technology.

References

[1] Hochstein M P and Sudarman S 1993 Geothermal resources of Sumatra *Geothermics* **22** 181-200
[2] Wang L, Tian M and Wang L 2015 Geodiversity, geoconservation and geotourism in Hong Kong Global Geopark of China *Proceedings of the Geologists' Association* **126** 426-37
[3] Erfurt-Cooper, Patricia, and Cooper 2015 *Volcano and Geothermal Tourism: Sustainable Geo-Resources for Leisure and Recreation* (Routledge)
[4] Bird D K, Gisladottir G and Dominey-Howes D 2010 Volcanic risk and tourism in southern Iceland: Implications for hazard, risk and emergency response education and training Journal of Volcanology and Geothermal Research 189 33-48
[5] Alshattnawi S 2012 Effective Use of QR Codes in Religious Tourism. In: 2012 International Conference on Advanced Computer Science Applications and Technologies (ACSAT), pp 497-501
[6] Brabazon, T., Winter, M., and Gandy, B. 2014 Digital Wine "How QR Codes Facilitate New Markets for Small Wine Industries" (Springer, London)
[7] Waters, J, 2012, QR Codes For Dummies (John Wiley &Sons, Inc.: New York)
[8] Solima L and Izzo F 2018 QR Codes in cultural heritage tourism: new communications technologies and future prospects in Naples and Warsaw Journal of Heritage Tourism 13 115-27

[9] Fino E R, Martín-Gutiérrez J, Fernández M D M and Davara E A 2013 Interactive Tourist Guide: Connecting Web 2.0, Augmented Reality and QR Codes Procedia Computer Science 25 338-44

[10] Tarjan L, Šenk I, Tegeltija S, Stankovski S and Ostojic G 2014 A readability analysis for QR code application in a traceability system Computers and Electronics in Agriculture 109 1-11

[11] Jun-Chou C, Yu-Chen H, and Hsien-Ju K 2010 A Novel Secret Sharing Technique Using QR Code, International Journal of Image Processing, **4**, Issue 5

[12] Ali P, Stephen C B 2012 The expectations of quick response (QR) codes in print media: An empirical data research anthology UW-L Journal of Undergraduate Research, **15**, 1-13

Analysis of electricity energy savings with Grid-Tie system connected to utility grid.

Y Away, A Adria, Syahrizal, M S Rizal and H Walidainy
Department of Electrical and Computer Engineering, Syiah Kuala University, Banda Aceh, Indonesia

E-mail: yuwaldi@unsyiah.ac.id

Abstract. The microgrid consists of at least one energy source that is connected to a load in a relatively small area. In practice, microgrid are divided into two types, off-grid systems or often said to be standalone systems and on-grid systems that are connected to utility networks. The use of renewable energy sources, such as photovoltaic, has become an alternative energy for the problem of electricity energy deficits. This research examines the possibility of saving power consumption from the network using a grid-tie inverter (GTI) power converter in a microgrid configuration so synchronization can be made between renewable energy systems and utility networks. The integration of photovoltaic energy resources with the grid system will increase the value of electricity capacity while reducing the use of fossil fuels, as well as a positive impact where the use of renewable energy can reduce energy use and bills to be paid. Integration of grid tie testing carried out without a load can convert the power generated by almost 90% distributed in the utility network, while testing with a load using a grid tie system, can save power up to 70% of the energy supplied by the utility network.

1. Introduction

Renewable energy sources have increased in many countries with renewable energy policies and due to emission factors. Among the renewable energy sources is solar energy which is an alternative energy that is being developed and has a reliable system. The supply of electricity used by power plants is still dominated by fossil fuels, especially from coal and oil, making renewable energy the best solution at the moment. Solar energy is one of the cleanest forms of energy available, converting energy from the sun into electricity without hazardous waste or by-products. This system can be installed on grid and off grid. In the off-grid system the energy produced by the solar panels is stored first to the battery using a charge controller and using an inverter to be used in AC loads, the system is not dependent on grid utility networks and is free from power problems and electricity costs [1]. While the on-grid system requires a main electricity network (cannot stand alone) that requires a reference voltage as a reference for its working system, resulting in the solar panel system not being used when the main source network is dead so the building cannot utilize energy produced by solar panels [2]. The use of grid tie inverter systems will increase the value of electricity capacity, while at the same time reducing the use of fossil fuels, as well as having a positive impact where the use of renewable energy can reduce the use of energy and the bill that must be paid. The use of two-way electricity also makes consumers a power producer if electricity produced by renewable plants owned by consumers has excess power then it can be sold to the state electricity company.

Content from this work may be used under the terms of the Creative Commons Attribution 3.0 licence. Any further distribution of this work must maintain attribution to the author(s) and the title of the work, journal citation and DOI.
Published under licence by IOP Publishing Ltd

2. Photovoltaic System

The efficiency of solar cells is a very important factor. Currently, the efficiency that can be produced by solar panels is still around 18.3%, but the efficiency of the solar panel is relatively low, because of the space between cells in the panel [3]. System efficiency considers all components of the system performance. Several factors such as solar cell electrical performance, degradation factor, ability to generate power, environmental factors and cell operating temperature determine the performance of an array. The method being developed is a solar tracker. This method follows the sun's movements throughout the day, which allows the PV panel to capture more exposed sunlight, so the power generated is more optimal.

Figure 1. Photovoltaic Dual Axis.

Tracker is a general term used to describe devices that orient various charges to the sun. By maintaining the consistency of direct exposure from the sun to the panel, the tracker can increase the PV system output by up to 40 percent. Various tracking systems have been developed with many methods, algorithms and various types of sensors. Figure 2 is the result of a comparison between diesel with and without a tracker system [3].

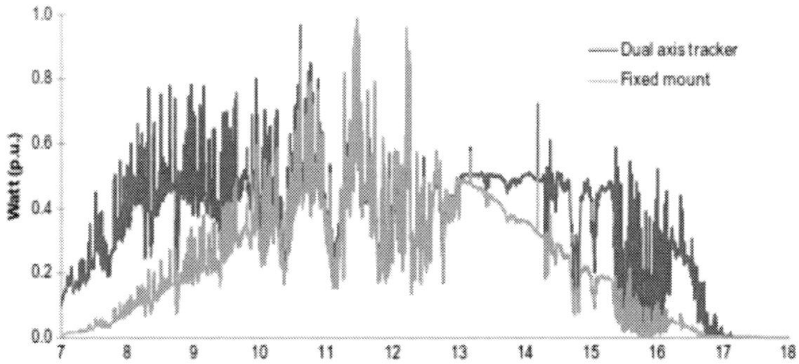

Figure 2. Power output of Photovoltaic comparison.

IOP Conf. Series: Materials Science and Engineering **796** (2020) 012026 doi:10.1088/1757-899X/796/1/012026

3. Trend of Microgrid

Microgrid system refers to the modernization of the electrical energy processing system so that it can be monitored and optimize the operation between the interconnection elements, starting from the generation center to the consumer. Microgrids are characterized by the bidirectional flow of electricity networks. The energy produced by renewable energy is usually different from the grid utility. It requires an inverter to synchronize the voltage and frequency produced so that the system can be connected to the grid [4]. Microgrid network system is divided into on-grid and off-grid system. The off-grid system is a system that is not connected to a network utility or often called a standalone system. While the on-grid system is a system that is connected to a utility network that allows two-way electricity, which is imports and exports between consumers and producers.

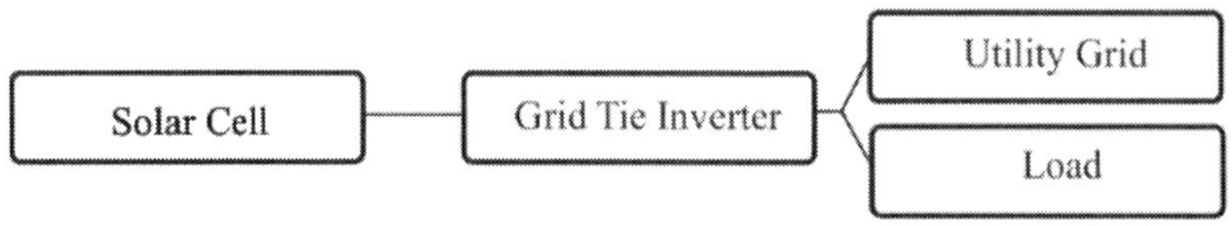

Figure 3. Micro grid flow diagram.

The energy produced by the solar panel array is still in the form of DC before being converted and filtered into the form of an AC grid injection. The inverter used needs to be considered, minus the loss across the inverter, transferred through the system and into the electricity network. The flow of power to the grid is strongly influenced by the phase shift to the inverter and the voltage amplitude of the inverter with the grid [7]. The energy produced by a PV system can be calculated from the average power times the time [8]. The total energy for a certain period will be calculated using equation (3).

$$E = \int_{0}^{t} p(t) \tag{1}$$

The electrical energy generated per year can be calculated using this equation:

$$E_{Year} = E * 365\ Day \tag{2}$$

The stored energy can be calculated using this equation:

$$E_{saving} = E_{Year} * Electrical\ Tariff \tag{3}$$

With the increasing development of renewable energy, energy storage components are becoming very important. For micro grid using solar energy the use of batteries as storage is generally considered suitable. With the many technological advances that occur throughout the world capital from a battery technology, in general, has decreased prices. Battery life depends on the type of service the provider provides, which ranges from 5 to 15 years, depending on the number of cycles, the charge, and discharge rate and how the temperature of the battery.

4. Research Method

In this study, the micro grid system uses a 100 Wp polycrystalline solar cells with a dual-axis tracker system connected with a 200W grid tie inverter, this on-grid system is connected to the Indonesian State Electricity Enterprise network. The scope of research is shown in Figure 4 below.

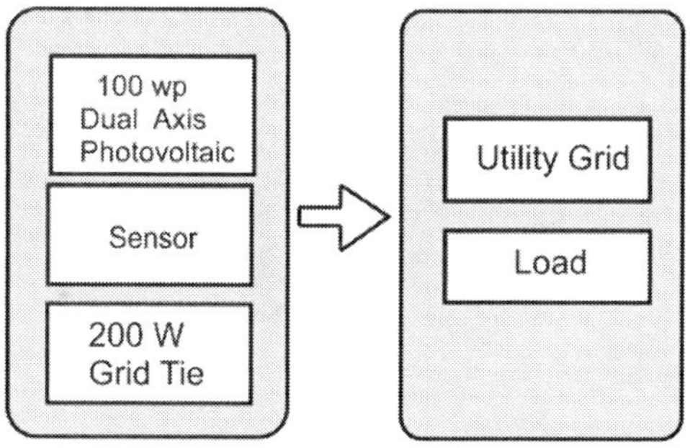

Figure 4. Research block method.

It is known that solar cells with tracking methods have advantages in terms of capturing sunlight so that the power that can be supplied to the grid also increases. While the focus of this research is to look at how the effect of saving electricity by integrating the grid tie system on power usage and utility grid (PLN) costs. Measurements are done in two steps, without load, and with the load. Without the load testing, it means that GTI is directly connected to the utility network, and then measurements from the side of DC input power and AC power supplied to the network [6]. Figure 5 is a series of tests with local loads that must be carried out to see the power distribution mechanism generated by grid-tie on the power supplied to the grid network. Load measurement is parallel with being supplied by two sources so that the current supplied to the utility network can be counted.

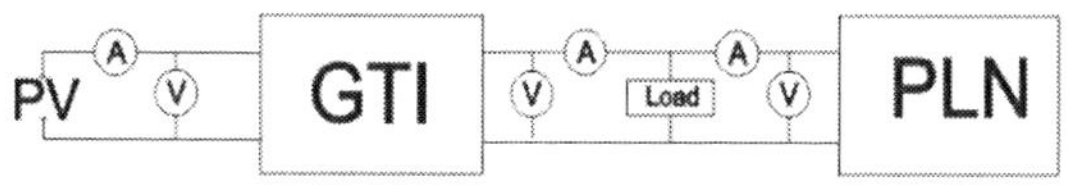

Figure 5. GTI connected to utility grid with load circuit.

5. Result and Discussion

5.1 Solar Energy Potential

Monitoring data on the photovoltaic system includes voltage, current, and power to see how the potential of solar energy generated in the area of Banda Aceh. The average solar panel starts generating power from 6:00 to 18:00. The peak power time is from 10:00 to 16:00 with some fluctuations depending on weather conditions, Figure 6 shows fluctuations in solar energy potential with sunny weather conditions, with an average maximum power reaching 80 W.

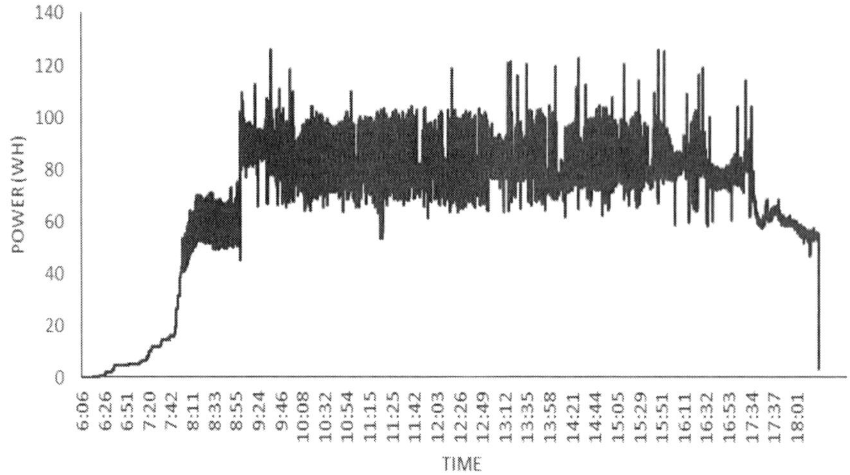

Figure 6. Solar energy potential.

5.2 Energy Saving Analysis

The use of a grid-tie system has a function that can make electricity flow in two directions, exports and imports. By using a grid-tie inverter (GTI), electricity generated by solar panels can be directly channelled into the home load installation system that is connected to a utility grid, this on-grid system can reduce the use of electricity supplied by PLN (Indonesia State Electric Enterprise). Data collected during working hours around (8 hours), starting at 8:00 to 6:00 pm with very sunny weather conditions. The higher power generated by the solar cell, the more power can be channelled to the grid. Figure 9 shows how the power at a load that uses PLN power was saved by adding a grid-tie system. PLN power used to supply loads has decreased as solar cell output power has increased. As a sample the grid savings test is carried out with one of the 35 W household load.

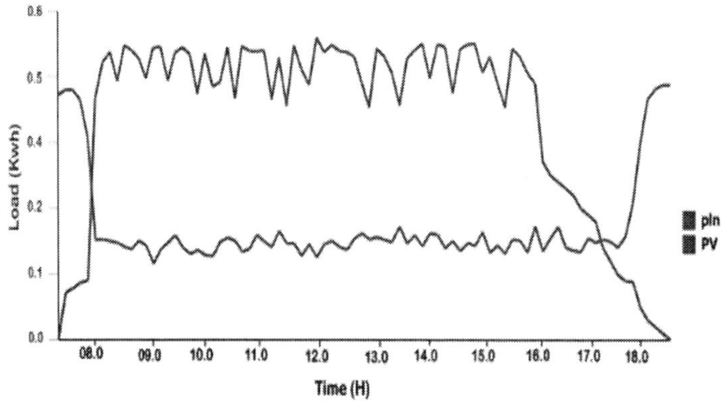

Figure 7. PLN power comparison.

Table 1 shows the comparison of load power conditions when connected with and without a grid tie system (only uses PLN). The power stored by adding grid tie can be calculated using the following equation.

$$P_{saved} = P_{load} - P_{out}GTI \tag{4}$$

Table 1. Power comparison with and without grid tie.

Power (W)			
P_{PLN} without grid tie	Pout GTI	P_{PLN} with grid tie	Power Saved by PLN
35.48	8.7386	28.7414	8.7386
35.48	9.066	26.414	9.066
35.48	36.6876	0	-1.2076
35.48	42.2382	0	-6.7582
35.48	42.294	0	-6.814
35.48	43.736	0	-8.256
35.48	39.3162	0	-3.8362
35.48	44.77	0	-9.29
35.48	44.064	0	-8.584

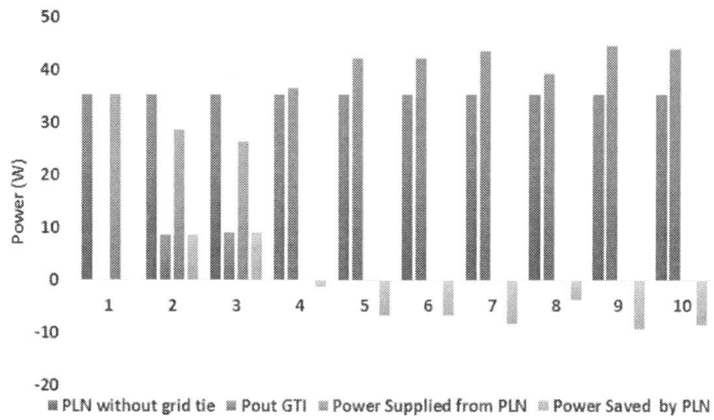

Figure 8. Power comparison with and without grid tie.

As shown in Table 1 when the system is not connected to the grid-tie, the 35 W load is supplied by PLN completely, after installing the grid-tie, the power supplied by PLN can be saved according to the output produced by the grid-tie. Of the 35 W of power needed by the load, only about 8 to 9 watts of power load is supplied from the utility grid, even excess power can be exported into the utility grid, the minus data in table 1 shows the changing power flow leading to export into the utility grid. The power used from the utility grid decreases as photovoltaic produce greater power. It shows by using a grid tie configuration, the electric power required for the charge that should be supplied from the grid can be reduced by up to 70% or even fully supplied by the grid-tie, depending on the power that can be generated by the grid-tie. It is known that PV can only work for 12 hours and produce electrical energy per day with the average power supplied to the grid is 80 W, using equation (4), the energy produced by PV 100 Wp reaches 960 Wh per 12 hours. While the energy produced during a year can be calculated using equation (5), it will produce 175 kWh, but this power is not constant because the fluctuation of PV output depends on weather conditions.

5.3 Cost Saving Analysis

The electrical capacity used in this study is 900VA with a sample load testing of 35 W. Therefore, the energy cost is followed by the tariff adjustment mechanism. For March 2019 the 900 VA tariff was

IDR 1,352 (Indonesian Rupiah), but this rate may change at any time depending on PLN's policy. With a sample load of 35 W, this load takes around 0.42 kW for 12 hours, if calculated for a month the cost paid using the utility grid is an average of IDR 17,232,538. Figure 10 shows how the effect of using grid tie on the cost utility grid for a month, with the addition of the PV grid-tie system paid to PLN, reduced to IDR 8,249.52, which means that the PV grid-tie system could save as much as IDR 8,983,018.

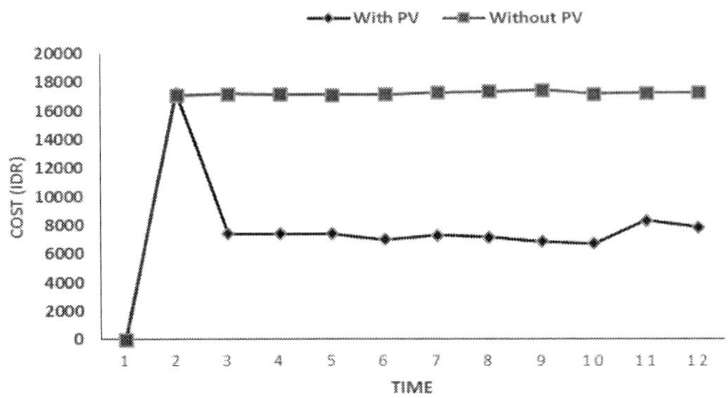

Figure 9. Cost comparison with and without PV system.

6. Conclusion

Grid-tie will operate if it is connected to the network. When there is no supply from the network, it will stop sending power to the utility network. Grid-tie will supply power to the local load and excess power will be distributed to the network. Conversely, if the grid-tie output power is minimum or unable to supply, then the load will be supplied by PLN, this depends on the grid-tie power output. The measurement results show, when the zero load test, shows that almost 90% of the power generated by the grid-tie is distributed to the PLN grid. This shows that the electric power supplied to the network can be used to replace the use of grid electricity when there is no supply of energy sources generated from the main generator.

Costs for PLN electricity consumption can be reduced, test results using local loads and microgrid configurations show that by using a grid-tie system. The electrical power requirements for loads that should be supplied from the grid can be reduced by 70%, or even fully supplied by grid-tie, depending on the power that can be generated by the grid-tie, this also results in lower electricity costs.

Acknowledgment

The author would like to express his gratitude for the support of Syiah Kuala University through the professor's research assistance for 2019. We also expressed our gratitude to the research assistant of the PUSKMATIK (Center for Automation and Robotic Research), Muhammad Ikhsan and Ikhramuddin.

References

[1] Karthikeyan V, Rajasekar S, Das V, Karuppanan P and Singh A K 2017 *Smart Energy Grid Design for Island Countries: Challenges and Opportunities,* ed F M R Islam, *et al.* (Cham: Springer International Publishing) pp 125-57

[2] "Sistem Off-Grid, On Grid PLTS," Solar Surya Indonesia, [Online]. Available: http://solarsuryaindonesia.com/info/sistem-off-grid-on-grid-tie. [Accessed 29 April 2018]

[3] Away Y and Ikhsan M 2017 Dual-axis sun tracker sensor based on tetrahedron geometry *Automation in Construction* **73** 175-83

[4] S. A. Indrima Upadhyay 2018 Integration of photovoltaic into microgrids: future energy system *GE-International Journal of Engineering Research* **6**
[5] Santoso H and Budiyanto B 2013 Microgrid development using a grid tie inverter *Makara Seri Teknologi* **7** 121-127
[6] Shahjehan M, Shahjehan W and Arbab M N 2017 Grid-Tied PV system with energy optimization *International Journal of Engineering Works* **4** 184-189
[7] Partha C G I, Wijaya I W A, Janardana I G N, Budiastra N and Weking A I Sistem on grid pembangkit listrik tenaga matahari menggunakan maximum power point tracking
[8] Rhapsody M R, Arfianto A Z dan Utari D A 2017 Penggunaan IoT untuk telemetri efisiensi daya, Seminar Master
[9] Ghazali S and Nazir R 2016 Performance and energy saving analysis of grid-connected photovoltaic in West Sumatera *International Journal of Power Electronics and Drive System (IJPEDS)* **7** 1348-1354
[10] Varia D 2014 DC to AC converter as grid tie inverter & smart home application IASSVI

IOP Conf. Series: Materials Science and Engineering **796** (2020) 012027 doi:10.1088/1757-899X/796/1/012027

Design and testing of portable laser-based viscometer

E Yusibani[1,4], I R Sasmita[2], L Ardiah[1], E Yufita[1], M S Surbakti[1] and Rahmi[3]
[1] Department of Physics, Faculty of Mathematics and Natural Sciences, Universitas Syiah Kuala, Kopelma Darussalam, Banda Aceh 23111, Indonesia
[2] IRSLAB, Cigadung Raya Tengah rd., Cigadung Bandung, 40191, Indonesia
[3] Department of Chemistry, Faculty of Mathematics and Natural Sciences, Universitas Syiah Kuala, Kopelma Darussalam, Banda Aceh 23111, Indonesia

E-mail: e_yusibani@unsyiah.ac.id

Abstract. Viscosity is one of a physical parameter of a fluid. Viscosity shows a coefficient of resistance in the fluid to reduce the speed of the fluid in a system. In the present study, design and testing has been carried out on a portable laser-based viscometer with the falling sphere method. The main mechanical system consists of two components: an acrylic tube with an inner diameter of 2.335 cm with a length of 28 cm, and a glass sphere diameter of 11.58 mm with the density of 2521 kg·m^{-3}. Two lasers with a wavelength of 650 nm are used to determine the traveling time of a sphere in a range of 11 cm in the sample. Two photodiodes are used to capture the light signal and then forwarded it to the Arduino Uno R3 microcontroller to calculate the falling velocity of the sphere and the viscosity of the sample then displayed into a LCD (2×16). Reproducibility of the viscosity measurement data was confirmed with the standard deviation varied from 0.08 to 1.9 mPas. The sensitivity and the precision of the present design increased with the viscosity value.

1. Introduction

Viscosity is one of an important mechanical property of fluids. Viscosity is defined as the internal fluid's resistance to flow. Measuring viscosity can be done by several methods, such as the capillary tube method [1][2], rotating cylindrical method [3], oscillating disk method [4], straight vibrating wire method [5], curved vibration wire method [6], torsionally vibrating crystal [7] and the falling body method. In the latest method, the objects that is fall into the sample can be a sphere [8][9] or a solid/hollow cylinders [10]. The falling sphere method is one of the earliest, simplest and easiest methods. The basic principle of this method is to measure the falling velocity of the sphere into a sample. This method was first described by George Stokes in 1845 and known as The Stokes Law as described in equation (1) [11].

$$F_f = 6\pi\eta r_s v \tag{1}$$

where η is a dynamic viscosity (Pa·s), r_s is the radius of the sphere (m), and v is the falling velocity of the sphere (m·s^{-1}). Based on the equation (1), Stokes's law states that the greater the friction force, the greater the viscosity value. The Stoke law is valid with very small Reynolds numbers (<200), in steady

[4] To whom any correspondence should be addressed.

Content from this work may be used under the terms of the Creative Commons Attribution 3.0 licence. Any further distribution of this work must maintain attribution to the author(s) and the title of the work, journal citation and DOI.
Published under licence by IOP Publishing Ltd

state and infinite medium. In a falling sphere method, a solid sphere which has a density ρ_s and a radius of r_s is dropped without an initial velocity into the fluid which has a density of ρ_f, where $\rho_s > \rho_f$. It is known that the sphere first gets gravitational acceleration, but for a moment after moving far enough the sphere will travel at a constant speed (v). This constant speed results in the presence of another force in the sphere, which is buoyant force so that Newton Law II applies. Thus, there are three forces that influence the sphere in the fluid, namely Archimedes force, Stokes force, and gravity, so the equation (2) is applies.

$$6\pi\eta r_s v + m_f g = m_s g \tag{2}$$

where m_f is the mass of the fluid (kg), m_b is the mass of the sphere (kg) and g is the local acceleration of the sphere (m·s^{-2}). If we defined mass as density multiply by the volume (V) then equation (2) becomes equations (3) and (4).

$$6\pi\eta r_s v = (\rho_s - \rho_f) g V_s \tag{3}$$

$$\eta = \frac{2 g r_s^2 (\rho_s - \rho_f)}{9v} \tag{4}$$

Based on the above equations, the physical parameters that must be measured are the densities of the sphere and sample fluid, the radius and the velocity of the sphere. In a simple terms, we can measure the velocity of the sphere by measuring its traveling time and distance in the fluid. Many ways have been developed to calculate traveling time by sensors.

A measurement of viscosity using lasers and photodiodes has been done by the previous researcher. The sphere was made from steel with a mass, radius and the traveling distance of 14 g, 7.5 mm, and 12 cm, respectively. The sample measured was edible oil and lubricant. Data is taken analogy in the form of a voltage change through a parallel port by a computer. The results showed that measurements using lasers and photodiodes showed more accurate than manual measurements, the percentage of average error of the viscosity measurement was 2.3% better than manual measurements [12]. Aryanto et al. uses seven photodiodes to capture light signals from a beam blocked by a falling sphere, the specified distance is 5 cm for each photodiode, and the sphere has a diameter of 16.5 mm with a mass weighing of 5.42 g. They use ATmega16 microcontroller to analyse the signal of glycerine sample fluid [13]. Two infrared diode lasers as lighters and a phototransistor have also been used by Warsito et al. to measure the viscosity of glycerine. The main system consists of two components: a glass tubes with a diameter of 4.7 mm, a length of 17 cm, and an iron sphere with a diameter of 1.2 cm, a density of 7643 kg·m^{-3}, the output signal from the comparator is forwarded to the AT89C51 microcontroller [14]. The measurement using a sensor to detect travel time in the falling sphere method will increase in measurement accuracy.

Viscosity is widely used as a standard for engineering fluid material, food, health and pharmaceutical products. The application of viscosity in product standardization can be found in the milk industry, the paint industry and the pharmaceutical industry. While the application of viscosity in the health sector, among others, is to determine the viscosity of blood in the body and the viscosity of urine. Therefore, to develop a low cost portable viscometer with high precision is desirable. In this present study, the authors will describe a design and testing an inexpensive portable laser-based viscometer by falling sphere method.

2. Method

To measure velocity is the main difficulty when using falling sphere method. The largest contribution to measurement uncertainty comes from the velocity measurement. The measurement of the sphere's diameter is also one of the largest contributors to the uncertainty. In the present design, the mass and diameter of the glass sphere is 2.059 g and 11.58 mm, respectively. The traveling distance was 10 to 16 cm. Samples will be put in an acrylic cylinder with inner diameter of 2.335 cm and the height of about 28 cm, so that the volume of the sample will be as much as 127 cc. The present viscometer consists of two photodiodes and lasers. The laser transmitted a red colour with the wavelength of 650 nm then the light signal will be captured by photodiode. When the sphere was traveling in between, then light from the laser will discontinue and the timer will start to ON, when the ball traveling to the other sensor, then the timer will start to OFF. The results of this time measurement will be calculated as a velocity and finally viscosity by Arduino Uno R3 microcontroller. In the present design, the traveling time and the viscosity values will be displayed in the LCD keypad (2x16). The temperature of the fluid is at 303 K measured by RTD PT100 at atmospheric pressure. A schematic diagram and pictures of the apparatus set up in the present design can be found in Figure 1.

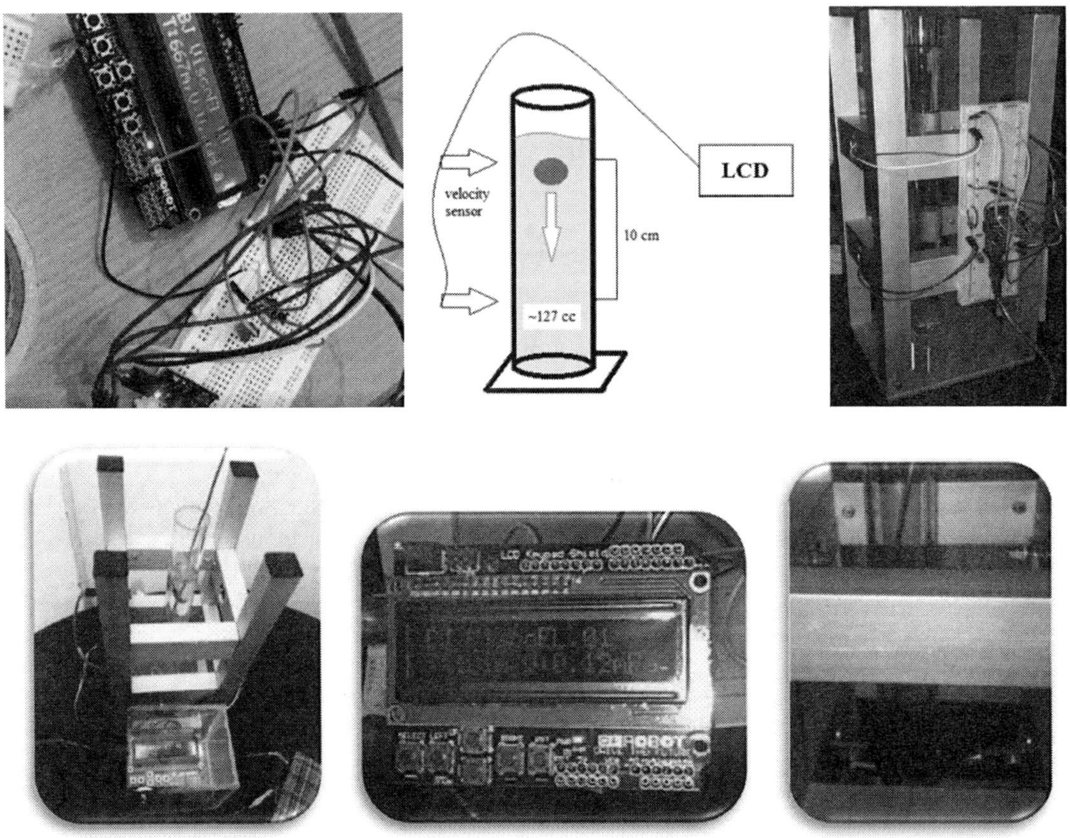

Figure 1. The pictures of portable laser-based viscometer (BJ ViscoEL.01) and the experimental set up with a display of LCD 2×16 pixels.

To calculate viscosity, equation (4) is applied. This equation is only valid if the terminal velocity is reached, and if the sphere falls in an infinity medium without inertial effects. For fall within a cylindrical tube with a radius r, the correction of the wall effect (W) is given by Faxen [15] as show in equation (5). Considering a traveling distance in the cylindrical container of height H (equation (6)), thus the modification of equation (4) to correct for effects on the velocity of the sphere due to its interaction with container walls (W) and ends effect (E) is given in equation (7).

IOP Conf. Series: Materials Science and Engineering **796** (2020) 012027 doi:10.1088/1757-899X/796/1/012027

$$W = 1 - 2.10444\left(\frac{r_s}{r}\right) + 2.08877\left(\frac{r_s}{r}\right)^3 - 0.94813\left(\frac{r_s}{r}\right)^5 + \ldots \quad (5)$$

$$E = 1 + 3.3\left(\frac{r_s}{H}\right) + \ldots \quad (6)$$

$$\eta = \frac{2gr_s^2(\rho_s - \rho_f)}{9v}\frac{W}{E} \quad (7)$$

The wall correction was empirically derived and is valid for $0.16 \leq r_s/r \leq 0.32$. Beyond this range, the effects of container walls significantly impair the terminal velocity of the sphere, thus giving rise to a false high viscosity value [16]. Measurement of density of a sample fluid was carried out using a 50 ml piknometer.

In the present measurement, the sphere diameter will be calibrated by measuring pure water viscosity in ambient condition based on references obtained from REFPROF [17]. From the measurements, it was found that the effective diameter of the sphere was decided as much as 6.1 mm, which is about 47% smaller than the actual size. Thus, the effective diameter then will be used to calculate the viscosity. Several liquids, pure and mixture will be measured to evaluate the performance of the apparatus. All of the samples were obtained from Aceh province, Indonesia.

3. Result and discussion

Figure 2 shows the viscosity measurement of pure water by varying the traveling distance (L) of the sphere, i.e. 10, 12, 14 and 16 cm. To estimate traveling time of the sphere, two ways have been done i.e. manually and by the sensor (laser based). Manual means that the traveling time value was obtained by stopwatch (L = 12, 14 and 16 cm) while at L = 10 cm has been done by the sensor. The use of two measurement methods (manually and by sensor) is intended to see how the sensor influences data retrieval. Increasing traveling distance will increase the accuracy; however reproducibility of the measurement data done by sensor is better. The average standard deviation of five measurements is as much as 2 to 45% manually (● ▲ ▼) and 0 to 13% by the sensor (■) compared to REFPROP (0.1 MPa, 303 K). It can be seen that the measurement using a sensor to detect traveling time with the present design is increasing the measurement accuracy. Run number for the *X*-axis means number of experimental data retrieval.

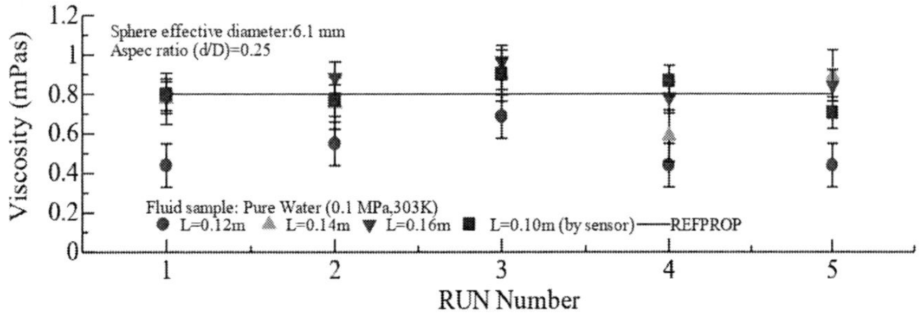

Figure 2. A viscosity measurement of water by the present design compared to REFPROP.

Several viscosity measurements (>15) are taken along the acrylic cylinder for various pure and mixture samples. Traveling distance (*L*) of the measurement is deciding as much as 11 cm. Figure 3 shows traveling time and velocity of pure liquid (lubricant oil SAE40, palm oil, patchouli oil (A from Aceh Selatan, B from Gayo Lues and C from Aceh Besar), kerosene) measures by the present design apparatus. Lubricant oil give the highest traveling time value of about 50 ms while kerosene for the short time of around 20 ms. The falling velocities calculated respectively to the traveling time is up to 6 m/s. Kerosene has bigger value relatively to the others and followed by patchouli oil, palm oil and finally lubricant oil. Reproducibility of the measurement is confirmed with the standard deviation of viscosity value varied from 0.3 to 1.8 mPas.

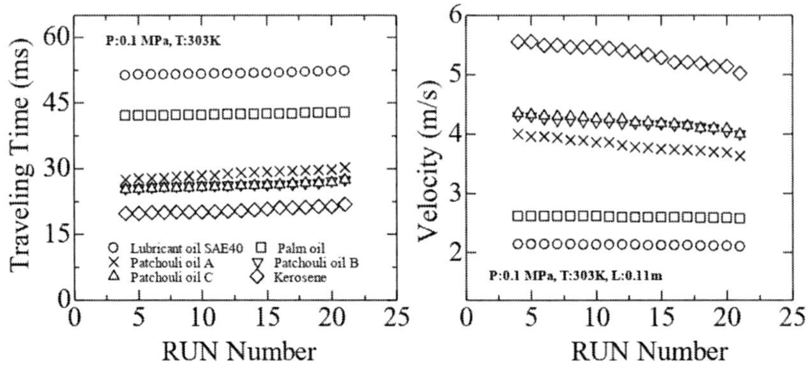

Figure 3. Traveling time and velocity of the various pure samples in the present measurement by sensor.

Figure 4-6 show a traveling time and velocity measurement of lubricant oil SAE40, palm oil and kerosene, respectively, which is mix by patchouli oil C with the composition of 100:0; 75:25; 50:50; 25:75 and 0:100 (v/v). The reproducibility and the sensitivity of the present design apparatus are good especially for a relatively high viscosity. The precision of the present design is decreased at traveling time around 20 ms. This short time measurement give a scatter value of the velocity, however the tendency still recordable as can be seen in Figure 6.

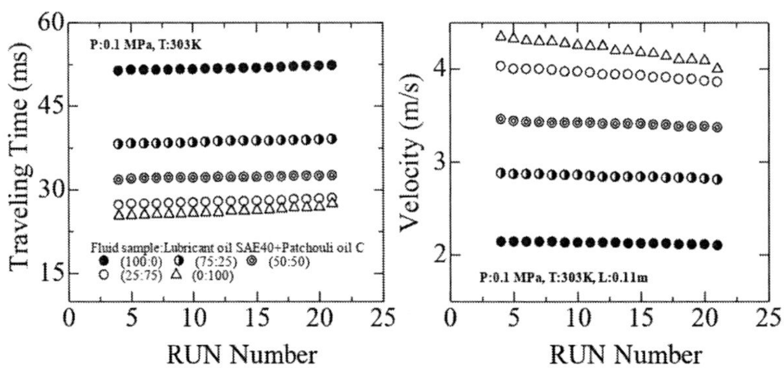

Figure 4. Traveling time and velocity for mixture of lubricant oil SAE40+Patchouli C in the present measurement by sensor.

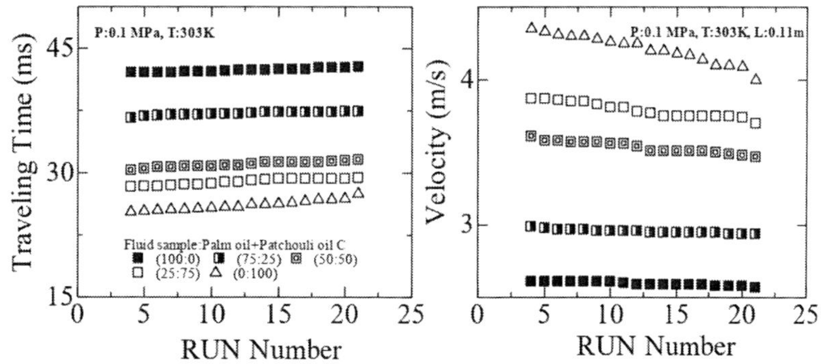

Figure 5. Traveling time and velocity for mixture of Palm oil+Patchouli oil C in the present measurement by sensor.

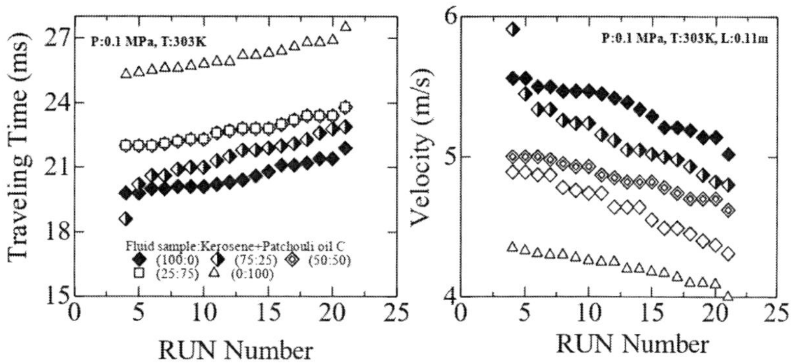

Figure 6. Traveling time and velocity for mixture of Kerosene+Patchouli oil C in the present measurement by sensor.

To calculate viscosity, the correction derived by Faxen was applied. The *W/E* value in the present design is as much as 0.49. The result of the all viscosity measurement can be seen in Figure 7 and Table 1. The table contains the results of density and viscosity measurements by the present viscometer from several samples compared to the available references. The sample consists of pure and mixture fluid to show the sensitivity, precision and accuracy of the present design. It is clearly observed that the sensitivity and the precision of the present design are good. The discrepancy is noticeable when patchouli oil C and kerosene is mixture by increasing volume of 25% of the sample. However, the accuracy of the present design still needs to be improved.

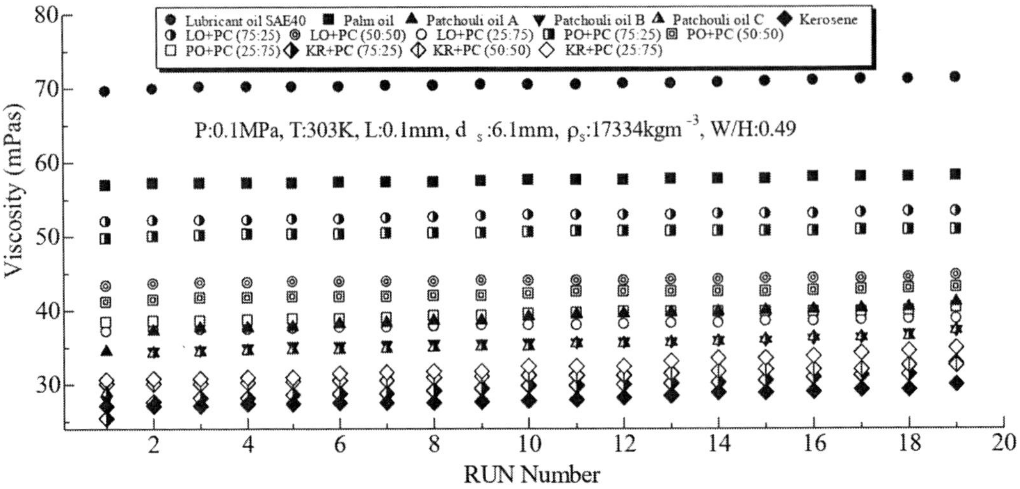

Figure 7. Overall measurement of viscosity with the present design (Note: LO:Lubricant Oil SAE40; PO:Palm Oil; PC:Patchouli Oil C; KR:Kerosene).

Table 1. Density and viscosity measurements of the present design for several samples.

Sample	Density, ρ (kgm^{-3}) P:0.1MPa,T:303K	Viscosity, η (mPas)	References ρ (kgm^{-3}); η (mPas)
Patchouli oil A	942	38.9 ± 1.4	
Patchouli oil B	934	35.3 ± 1.9	946.6; 9.2933 (313K) [18]
Patchouli oil C	925	35.1 ± 1.7	
Palm oil	895	57.5 ± 0.3	899.6;40.78 (CPO, 313K) [18] 877.13;4.71 (edible oil, room T) [19] 973;0.78 (edible oil, 302K) [20]
Palm oil (75%) + Patchouli oil C (25%)	903	50.3 ± 0.9	
Palm oil (50%) + Patchouli oil C (50%)	913	42.1 ± 0.6	
Palm oil (25%) + Patchouli oil C (50%)	918	39.3 ± 0.6	
Lubricant oil SAE40	848	70.5 ± 0.4	898 (288 K);13.14 (373K) [21] 900;270 (298K) [22]
Lubricant oil (75%) + Patchouli oil C (25%)	868	52.7 ± 0.5	
Lubricant oil (50%) + Patchouli oil C (50%)	886	44.0 ± 0.5	
Lubricant oil (25%) + Patchouli oil C (50%)	908	32.3 ± 1.4	
Kerosene	785	28.1 ± 0.9	ns;1.64 (300 K) [23]
Kerosene (25%) + Patchouli oil C (25%)	884	35.1 ± 1.7	
Kerosene (50%) + Patchouli oil C (50%)	851	31.0 ± 1.0	
Kerosene (75%) + Patchouli oil C (50%)	827	29.0 ± 2.9	

ns: not specified

4. Conclusions

A simple and inexpensive portable laser-based viscometer by a falling sphere method has been designed and tested to measure the dynamic viscosity of liquid at atmospheric pressure and temperature. Lubricant oil SAE40, patchouli oil, palm oil, and kerosene have been measured and analysed. Not only pure material, but also mixed fluids have been measured with the composition of 100:0; 75:25; 50:50 and 25:75 (v/v) for lubricant oil, palm oil and kerosene with patchouli oil C. Reproducibility of the viscosity measurement is good with the standard deviation varied from 0.3 to 1.9 mPas. The sensitivity of the present apparatus increased with viscosity value. The present deviations with respect to the existing references suggest that with more refinements it may be possible to take viscosity measurements by the present device with an accuracy of less than 5%.

Acknowledgments
This research is funded by Universitas Syiah Kuala, Indonesia (H-Index Project number 1445/UN11/SP/2017) under Kementrian Riset, Teknologi dan Pendidikan tinggi (KEMENRISTEKDIKTI) Indonesia.

References
[1] Yusibani E, Nagahama Y, Shinzato K, Fujii M, Kohno M, Takata Y and Woodfield P L 2011 *Int. J Thermophys.* **32**(6) 1111-1124
[2] May E F, Berg R F and Moldover M R 2007 Int. J Thermophys. 28 (4) 1085
[3] Diller D E 1965 J. Chem. Phys. 42 6
[4] Kestin J and Leidenfrost W 1959 Physica 25 1033
[5] Tough J T, McCormick W D and Dash J G 1964 Rev. Sci. Instrum. 35 13451348
[6] Yusibani E, Woodfield P L, Shinzato K, Takata Y and Kohno M 2013 Exp. Therm. and Fluid Sci. 47 1–5
[7] Mason W P 1947 Trans. ASME 69 359
[8] Brizard M, Megharfi M, Verdier C and Mahé E 2005 Rev. Sci. Instrum. 76 (2) 025109
[9] Tang J X 2016 Rev. Sci. Instrum. 87 054301 doi:10.1063/1.4948314
[10] Rudenko N S and Slyusar V P 1968 Ukrainian Phy. J. 13 656-659
[11] Resnic H, Walker J and Halliday D 1960 Fundamentals of Physic (oxford) Ed. Halliday D
[12] Surtono A and Susanto E 2007 J. Sains MIPA 13(3) 251-256
[13] Aryanto D, Saptaningrum E and Wijayanto 2012 Jurnal Fisika dan Aplikasinya 8(8) 1-5
[14] Warsito, Suciyati S W and Isworo D 2012 Jurnal Natur Indonesia 14(3) 230-235
[15] Faxen H 1923 Astronomi och Fysik 27(17) 1-28
[16] Webster J G 1999 The Measurement, Instrumentation and sensors handbook Leblanc G E et. al. Viscosity Measurement Copyright 2000 (CRC Press, IEEE Press)
[17] REFPROP-Reference Properties- NIST-USA Reference Fluid Thermodynamic and Transport Properties Database (REFPROP v.8)
[18] Atabani A E, Mahlia T M I, Masjuki H H, Badruddin I A, Yussof H W, Chong W T and Lee K T 2013 Energy 58 296-304
[19] Sutiah K, Firdausi K S and Budi W S 2008 Berkala Fisika 11(2) 53-58
[20] Yusibani E, Hazmi N A and Yufita E 2017 Jurnal Teknologi dan Industri Pertanian Indonesia 9(1) 28-32
[21] Anonymous 2009 total motor oil sae (brochure)
[22] Souza R J M and König A 2012 Ano 28 147-161 ISSN 1807-16006
[23] Engineering ToolBox 2008 Dynamic Viscosity of common Liquids. [Online] Available at: *https://www.engineeringtoolbox.com/absolute-viscosity-liquids-d_1259.html* [Accessed 7 July. 2019]

User requirements elicitation in web-based Participatory Geographic Information System interface design

Z Sofyan[1,2], Zulkifli[2] and Bustami[3]

[1] Department of Information Technology Education, Faculty of Teacher Training and Education UIN Ar-Raniry, Banda Aceh, Indonesia
[2] Development and Planning Agency, Banda Aceh Municipality
[3] Department of Information Technology, Faculty of Science and Technology UIN Ar-Raniry, Banda Aceh, Indonesia

E-mail: zuhrasofyan@gmail.com

Abstract. This ongoing work is aimed to collect a set of specific user requirements to develop a participatory web-based Geographic Information System using the User Centered Design approach. By using several techniques of requirement specification as the part of the User Centered Design such as observation of user behavior, interview session and paper prototyping design for testing user interaction, this initial stage in the development of the application could identify the important features to increase the usability of the application. These key features needed by the user are divided into three categories; the type of available maps, the data input features, and the data management features. The categories founded will help the system developer prioritize the important features to create an application to support spatial-based planning and to support the coordination between government agencies in Banda Aceh Municipality.

1. Introduction

Banda Aceh municipality is the capital of Aceh Province in Indonesia. Being the most developed city in Aceh province, Banda Aceh should improve the quality of service towards its citizen.

To achieve the goal of efficiency and effectiveness of the service in the public sector such as the government agency, the utilization of the Information Technologies, and all internet related mean of communication is preferred [1]. Thus, the e-government is then occurred as the term to explain the system to facilitate the communication between citizen and the government [2] and to deliver information and service to the citizens [3].

The City Development Agency of Banda Aceh Municipality (Bappeda Kota Banda Aceh), as the agency that is directly involved in the development planning in Banda Aceh is responsible for creating an e-government system innovation that promotes the implementation of more efficient community services, and to achieve this goal, the system needs an active participation from all stakeholders [4], in this case related to the provision of the information relating to the city development and coordination between other agencies, as such the active participation system is still rarely adopted by the government, although it has very positive benefits for improving community services [5].

Content from this work may be used under the terms of the Creative Commons Attribution 3.0 licence. Any further distribution of this work must maintain attribution to the author(s) and the title of the work, journal citation and DOI.
Published under licence by IOP Publishing Ltd

IOP Conf. Series: Materials Science and Engineering **796** (2020) 012028 doi:10.1088/1757-899X/796/1/012028

To improve the service for Banda Aceh citizens, an e-government application was developed with the aim of building a web-based focal point Geographic Information System (GIS) application that can be used as a reference for agencies in Banda Aceh city and answering the need of participative cooperation from government agency of Banda Aceh to coordinate and share city planning and development spatial object and location, wherein later the data can be accessed openly to the public.

In an application development focused on the user needs, to be able to produce a good system requires the process of gathering the user requirements, so that the developers can focus on developing the features that are truly used and easy to use by users [6]. Therefore, as an initial stage in creating the application, the approach in user requirements elicitation at the beginning of the design is very helpful for shaping the final product of the system.

In our work, we address research questions to find answers of the user need for a participatory web-based GIS application, using various methods in the user requirement elicitation, and then to test it in a simple prototype to improve and refine the user's needs, before entering the system development stage.

The entire paper has the following structure. The first section provides the background of this research. In the second part we present previous research related to this research, then in the third part explain the use of several User Centered Design methods in determining user requirement elicitation, and in section four we show the implementation of the methods that we use. Section five explains the evaluation process of the results obtained by testing it in the prototyping stage, and the sixth section concludes the paper.

2. Related works

Implementation of spatial data in the development of e-government systems has been carried out worldwide. In Indonesia alone, several researchers have tried to develop spatial based systems aimed at government needs.

RA Maryam in her research developed a web-based GIS application aimed at accommodating information about port facilities and infrastructure in Tanjung Perak Surabaya [7]. ME Kusuma developed a web-based web application for tourism purposes [8], while Danang Kusnadi and Jamal Ma'ruf also researched a GIS system that was used to map the potential of regions in Java, Indonesia [9].

From the overall research, it was found that the spatial-based web applications that were developed were only used for one specific service function, and the content could not be used for the purposes of coordination between government agencies. In addition, the resulting applications also have many shortcomings where the previous authors feel that even though they achieve the desired results but lack other features that should also exist in the application they developed.

To fill the gap that still existed in the previous research, the focus of this system development is to look for the needs of users using the user centered design method, which will make the features that applied to the system are only the desired and useful features used by the users [10].

3. Method

In the software development process, user centered design is one of the method that emphasizes the design process iteratively by involving users at each stage of the research process and design techniques [11]. The advantage of involving the user in the development of the system is that the results of the process will produce a system design that is very relevant to the target user and the product created will be very useful and in accordance with user needs without having to create features that are not or rarely used [12].

The form of the system development process using the User Centered Design method can be seen in the following figure:

IOP Conf. Series: Materials Science and Engineering **796** (2020) 012028 doi:10.1088/1757-899X/796/1/012028

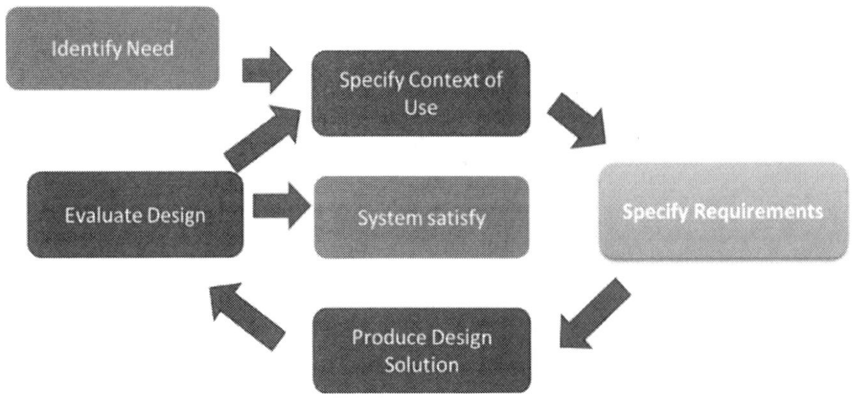

Figure 1. User centered design process [13].

To increase the effectiveness of the application development, the research in this paper is focused at the beginning of the process as depicted from figure above, namely to specify user requirements. Before the user requirements elicitation conducted, the first process according to the user centered design process is to understand the goal of the targeted user in using the application. This is called the context of use specification [14]. After the usage context has been determined, then the user requirements data collection can be carried out to support the process of producing the design solution based on the list of the features gathered from users.

In gathering user requirements, several techniques can be implemented. The first technique is to conduct background initialization interviews, which aim to collect data related to the needs and expectations of users with the system. After that, the second technique that can be used is field observation. This technique is used to observe the tools used and how users behave in collecting and creating spatial data, as well as how each of these government agencies coordinates in the data exchange activity. Then, from the two techniques, the data collected can be revised again using the Forum Group Discussion technique, sequential interview or questionnaire [15].

As the last stage in the determination of user requirements, data sets derived from the user needs in the previous activity can then be translated in the form of paper prototypes to test how the user interacted with the interface and the fulfillment of the user needs in the application.

4. Implementation

From the initialization stage, it was determined that the GIS system to be built must be able to assist the activity of spatial data coordination and sharing between agencies in the Banda Aceh City. In addition, to produce spatial data that is always up-to-date, active participation of users in filling-up the content to the application will be the main goal that must be met by this system. From these primary goals, it can be derived that the users of this system are primarily the agencies in the Banda Aceh city environment, as well as the community and those who want to know the development map and development planning in the city of Banda Aceh. These goals are then translated into the context of use of the system.

After the context of use is determined, the collection of user requirements is done through several stages, namely initialization interviews with the user who initialized the idea to build the system, then by observing patterns and tools used in the user coordination activities before the system is built, then improving user requirements with further detailed interview, and ended with the prototype paper process to test the eligibility of user requirements before entering the system implementation stage.

4.1. Initial interview

The first stage of the user requirement elicitation is to conduct an initial interview with the main user of this application. This interview was held in Bappeda Kota Banda Aceh, where this agency was the

IOP Conf. Series: Materials Science and Engineering **796** (2020) 012028 doi:10.1088/1757-899X/796/1/012028

initiator of this system innovation. From the results of the initial interview, we obtained the information regarding the system requirements, where users can use the map to enter the coordinates along with supporting information at each point. This data should be stored online and used in each coordination meeting between government agencies to determine the development area within the city of Banda Aceh.

4.2. Direct observation

To find out how coordination between agencies related to spatial data was conducted, we conducted direct observations by participating in inter-agency coordination activities at the Banda Aceh City government office. From the results of these observations, we get the user behavior data in inputting coordinates and spatial data information using the Google Earth application. It is known that using this tool only stored the spatial data offline, in the related agencies, and each agency then collects other agencies data if necessary. This certainly hinder the coordination process since it reduces the completeness of the available data and overlapping the development planning data between agencies.

From this stage, we also observe the Google Earth tool which will later be considered in the process of designing the system interface to make it easier for users to translate their knowledge from the previous tool into the newly built system. Also, some features used by users on Google Earth are also accommodated as requirements in the system. Those features are:

- The form of the map display
- Basic layer that they can add freely to the map
- Detailed area boundaries between sub-districts and villages
- Availability of area and distance measurement tools

4.3. Follow-up interview

To refine user requirements, the authors then collected further data by conducting a short interview with 15 respondents who came from several agencies in Banda Aceh City, such as officers from the Bappeda Kota Banda Aceh, the Public Works and Public Housing (PUPR) office and the Office of Investment and One-Stop Integrated Services (DPM-PTSP).

Through this interview stage, we explained to the respondents the system interface to be built and list the features they use from the old model they are using, as well as the features they will use later.

4.4. Paper prototype

From the requirements obtained in the three previous stages, we transformed the result in the form of prototype paper. The same respondents can interact with the system and propose the layout of each interface component. In addition, respondents could also propose their thoughts about requirements that were not previously listed but they felt important for the system.

Figure 2 shows an example of how users can manipulate interface components and freely add the features that they feel will be used in the system by using a paper prototype.

IOP Conf. Series: Materials Science and Engineering **796** (2020) 012028 doi:10.1088/1757-899X/796/1/012028

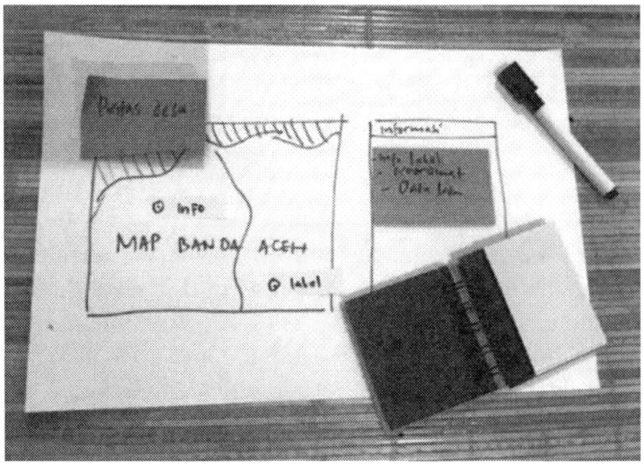

Figure 2. Paper prototype.

5. Result

From the implementation of the process of user requirements elicitation, three main user interface category sections should be available to increase the participation of user for using and sharing data in the application. The three categories are the help layers on the map, which is classified in the map category, the input category features on the map, and the data management category.

5.1. Map category

The availability of the various basic map to facilitate the user in finding the location of the requested spot is deemed important for the supporting features by the user. The respondents' agreement on the type of map and map-related features can be seen in the following figure:

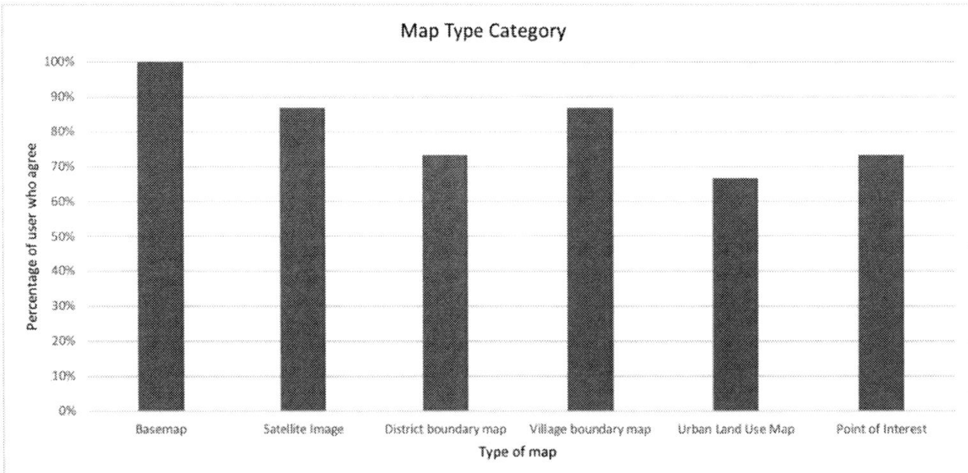

Figure 3. User requirements, map type category.

5.2. Input features category

The next feature that should be available in the spatial data input is explained in Figure 4, where all respondents agree that when inputting location, the ability to point-and-store coordinates and information related to these coordinates are two main features that absolutely must be present, while other features such as layer grouping, and the ability to take measurements on the map is considered relatively important because it is in the range above 50%.

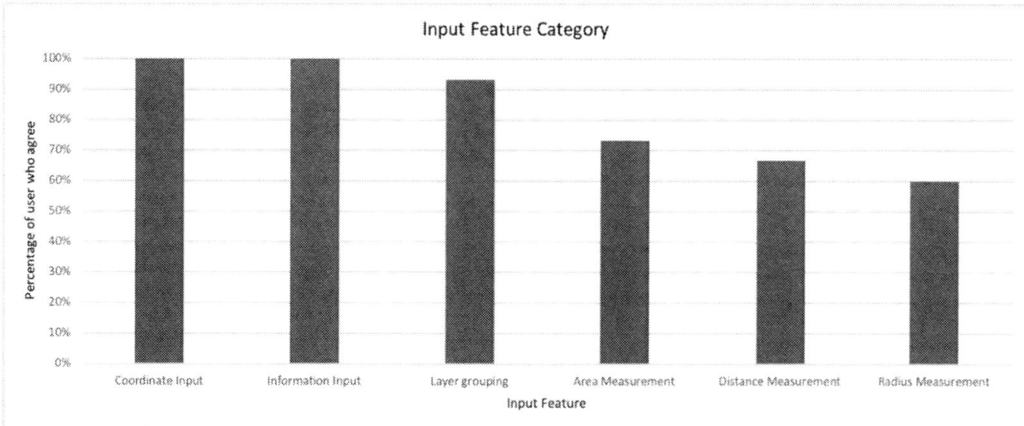

Figure 4. User requirements, input feature category.

5.3. Data management category

In addition to input features and maps that are directly related to the use of the system, the data management features also become an important part that must be present in this participatory application. These features are needed to support the coordination between agencies through the generated data. Some features related to data management can be seen in the following figure:

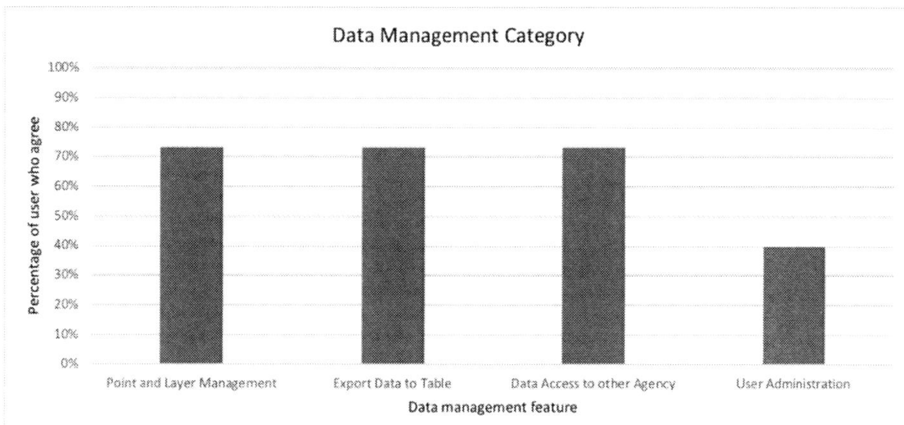

Figure 5. User requirements, data management category.

From the three categories of the interface components, the developer can determine which features are the priorities in developing the interface design. Some components with a smaller percentage compared to other features can be considered less important to be the main features and developed later.

6. Conclusion

From the results of investigations using requirement specification techniques in user centered design, a list of user requirements can be collected and used as a reference in developing the Web-Based Participatory Geographic Information System (GIS) interface. For the participating users in the user requirement elicitation process, there are three categories identified as the key features to help increase their participation in using the Web GIS application. The first category is the user requirement in map type category, that should include the basemap, satellite map and several boundary and Point of Interest map layers. The second category is the input category where the user needs some tools of

IOP Conf. Series: Materials Science and Engineering **796** (2020) 012028 doi:10.1088/1757-899X/796/1/012028

measurement to precisely input the new coordinate point into the map, along with additional information and ability to group the point into custom layers. The last category is the data management where the user able to manage the existing coordinate data, able to access other agencies data and export data to table format to easily collaborate with other agencies.

Finally, the categories derived from user requirements elicitation should be able to clearly identify features that will become the main functionality of the system in the development phase of the User Centered Design process. By focusing only on the required features, it is expected that the development phase of the system can be conducted efficiently.

References

[1] Jeong C H I 2007 *Fundamental of development administration* (Puchong: Scholar)
[2] Palvia S C J and Sharma S S 2007 E-government and e-governance: definitions/domain framework and status around the world In *Int. Conf. on E-governance* vol 5, pp 1-12
[3] Nations U 2014 United Nations E-Government Survey 2014: E-Government for the future we want *United Nations Department of economic and social affairs*
[4] Brabham D C and Guth K L 2017 The deliberative politics of the consultative layer: Participation hopes and communication as design values of civic tech founders *Journal of Communication* **67(4)** 445-475
[5] Ganapati S 2011 Uses of public participation geographic information systems applications in e-government *Public Administration Review* **71(3)** 425-434
[6] International Organization for Standardization 2019 *Ergonomics of human-system interaction: Part 210: Human-centred design for interactive systems* ISO
[7] Maryam R A and Handayani H H 2015 Studi Pengembangan Webgis Sarana Dan Prasarana Pelabuhan (Studi Kasus: Tanjung Perak Surabaya) *Geoid* **10(2)** 120-128
[8] Kusuma M E 2015 *Aplikasi Google Maps API Dalam Pengembangan Sistem Informasi Geografis (SIG) Pariwisata Berbasis Web (Studi Kasus: Kabupaten Sidoarjo)* (Doctoral dissertation, Institut Teknologi Sepuluh Nopember)
[9] Kusnadi D and Ma'ruf J 2017 ELECTRONIC GOVERNMENT PEMBERDAYAAN PEMERINTAHAN DAN POTENSI KELURAHAN (Studi Kasus: Kelurahan Pringsewu Selatan, Kecamatan Pringsewu, Pringsewu) *Jurnal Technology Acceptance Model 5* 37-44
[10] Garrett J J 2010 *The elements of user experience: user-centered design for the web and beyond* (Pearson Education)
[11] User centered design [Internet] The Interaction Design Foundation 2019 [cited 27 July 2019] Available from: https://www.interaction-design.org/literature/topics/user-centered-design
[12] Norman D 1988 *The design of everyday things* (New York: Doubleday) p 188
[13] User-centered design basics [internet] Usability.gov. 2019 [cited 27 July 2019] Available from: https://www.usability.gov/what-and-why/user-centered-design.html
[14] Maguire M 2001 Methods to support human-centred design *International journal of human-computer studies* **55(4)** pp 587-634
[15] Abras C, Maloney-Krichmar D and Preece J 2004 User-centered design *Bainbridge, W. Encyclopedia of Human-Computer Interaction* (Thousand Oaks: Sage Publications) **37(4)** 445-456

IOP Conf. Series: Materials Science and Engineering **796** (2020) 012029 doi:10.1088/1757-899X/796/1/012029

Implementation of SMS Gateway in the Flood Notification System using Raspberry Pi

D Satria[1], S Yana[2], R Munadi[3], S Syahreza[4], E Yusibani[4] and T Hidayat[1]
[1] Department of Computer Engineering, Universitas Serambi Mekkah, Aceh, Indonesia
[2] Department of Industrial Engineering, Universitas Serambi Mekkah, Aceh, Indonesia
[3] Department of Electrical Engineering, Universitas Syiah Kuala, Aceh, Indonesia
[4] Department of Physics, Universitas Syiah Kuala, Aceh, Indonesia

E-mail: dedisatria@serambimekkah.ac.id

Abstract. Flood disaster has become a national disaster that often occurs in the world today. Indonesia is one country that has a lot of flooding. The cause of the flooding is human error in environmental care. To overcome floods in the short term is to build a flood disaster mitigation system to anticipate more victims. This study aims to create an SMS gateway-based flood notification system. This mitigation system was built using ultrasonic sensors and rain sensors as input components, Raspberry Pi mini-computers as processors and GSM modems as outputs. This research resulted in a prototype of a flood notification system that can send SMS when floods occur. Tests have been carried out by varying the height of the water, which is between 0cm to 20cm. SMS will be sent to users when the water level is at 5cm, 10cm, 15cm, and 20cm. In prototype testing used a scale between 0 cm to 20 cm. From several tests conducted, the system has been able to send SMS to users. This research has provided innovations in the field of flood disaster mitigation using GSM communication systems and Raspberry Pi-based data processing.

1. Introduction

Flood disaster is one of the natural disasters that often occur in Indonesia today. Global Warming is one of the causes of flooding, such as rain that is not well distributed. But in addition to Global Warming, human error is another factor causing floods. The case looks like a forest when water reserves begin to be destroyed by some irresponsible people. Situations like this may be due to economic factors that make people choose shortcuts in making decisions to exploit forests illegally. The effect of illegal forest exploitation causes damage around the forest and can create a flood effect that afflicts nearby residents.

As for the causes of floods that are affected by forest destruction, for city residents also feel the same thing as residents around the forest that is flooding. Based on the problem of flooding around the city, garbage accumulation is one of the impacts of the flooding that occurred. The amount of garbage in the city sewers can cause sewers and floods.

The problem of flooding that often occurs in the community today is the lack of public awareness of the environment. The role of government for the people is needed to be aware of the importance of awareness of the importance of forests as a sustainable water storage area. As well as public awareness of the importance of throwing garbage in its place is a long-term program. The role of the government

Content from this work may be used under the terms of the Creative Commons Attribution 3.0 licence. Any further distribution of this work must maintain attribution to the author(s) and the title of the work, journal citation and DOI.
Published under licence by IOP Publishing Ltd

in anticipating long-term public perception is to carry out routine socialization activities on environmental awareness. And the role of government in the short term is to mitigate disasters to anticipate the coming flood.

Based on the importance of development disaster mitigation, the government nationally provides positive responses for private and government agencies to be able to contribute to building disaster mitigation tools or models [1]. The disaster mitigation model or tool is expected to be able to contribute to reducing flood victims in the future.

Based on the importance of developing tools and models to anticipate flood disasters, this study explains the design of flood disaster notification tools. The device built will send an SMS Gateway to the community around the area that has the potential to flood. The prototype was developed using a Raspberry Pi mini-computer as a processor in sending SMS to the public.

2. Related Works

Various research centers are currently developing tools or prototypes in disaster mitigation. Based on research conducted by previous researchers, some research results related to flood mitigation as a flood information system are the ability to send data from a flood detector using GSM data communication. This study provides information to users in the form of flood disaster information in data on date, time, location, and water level [2]. In another study, the use of the SMS gateway is to build a data communication service system based on disaster mitigation systems made by displaying flood and fire data information in the form of web and SMS [3][4][5][6]. In other studies, environmental management information systems have used SMS in sending data such as waste information management. The system provides information to the garbage officer to handle the waste [7], field construction information system [8], the student information system [9], and security monitoring [10]. In addition to using SMS Gateway based data communication in natural disasters, there is an Internet of Things (IoT) based communication system with the concept of exchanging data on a client and internet-based server such as a home fire information system [11][12] and flood information systems [13][14]. From several information systems developed, the average interface is a built-in web form. In another study, a disaster victim information system was built using a mobile application that aims to provide flood victims information to disaster management officers [15]. In previous studies, the use of internet networks in sending data is the main communication system in the process of distribution of disaster information.

3. Method

The research method in making tools begins with analyzing how the system works that the public receives aims to provide information on how the tool works until the flood information. Then the system analysis is developed into a design tool in the form of a block diagram.

In the analysis of how the system works begins with the admin inputting the user's phone number data, then the flood detection system monitors the floods and rain continuously. If there is a flood, then the detected flood is sent to the user according to the telephone number found on the flood detector via the GSM operator, as shown in Figure 1.

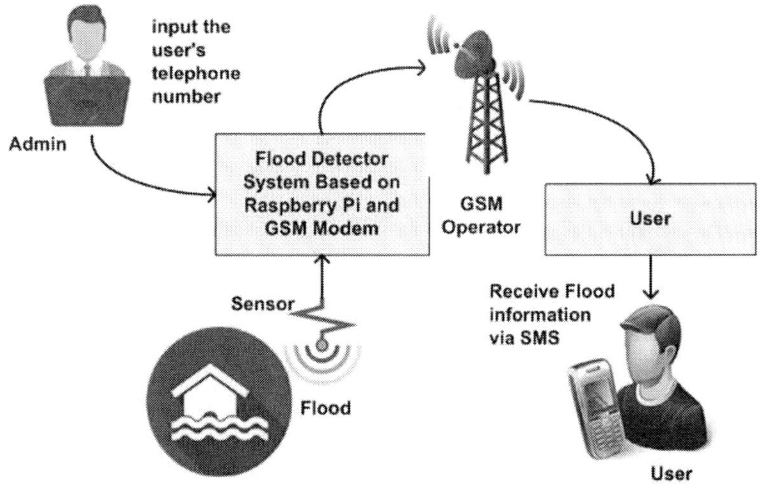

Figure 1. Analysis of how the system works.

The design of the tool system was built using a block diagram, as shown in Figure 2. The system consists of three blocks. Namely the input block which contained an ultrasonic sensor as a water level detection and rain sensor to detect rain. Whereas in the block was blocking data on water level and rain using a Raspberry Pi mini-computer. And the third block is the GSM modem Wavecom as the sender of SMS to users. Raspberry Pi was built using the Raspbian operating system. And the system program was developed using the Python and Gammu applications as a bridge application between the modem and the MySQL database.

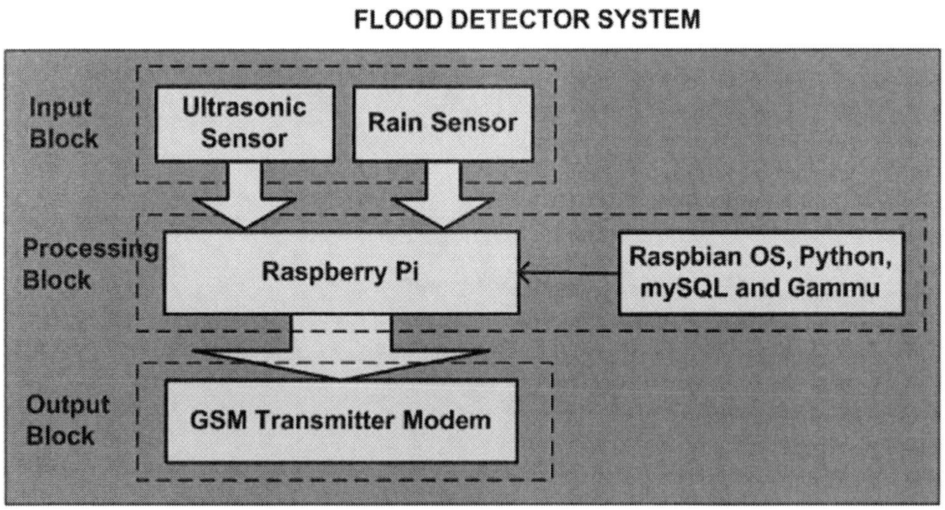

Figure 2. System block diagram.

4. Results and Discussion

Based on the analysis and design of a flood notification system using SMS Gateway, the prototype of the flood notification system follows in Figure 3. In the prototype SMS Gateway based flood notification system uses several main components, namely rain sensor and ultrasonic sensor as a component input, Raspberry pi as data processing or processing component. While the GSM modem as an output component that sends SMS to system users. In the flood detection pipe, there are rain sensors and ultrasonic sensors on the top side. And on the Raspberry Pi, there is a GSM Modem. In the SMS Gateway-based flood notification system construction was built in a condition as a prototype. The prototype in question is a condition in the form of laboratory experiments.

In Figure 4 is the result of SMS received by users of the flood notification system using the SMS gateway. The editorial format sent from the SMS gateway based flood notification system is "Flooding occurred with a height of 5cm in rain conditions". And on the other SMS, you can see the editor "There was a flood with a height of 10cm in conditions not raining". Of the two SMS sent, when the SMS was sent, information consisted of conditions when floods and rain occurred.

The test is carried out using water placed in a container with the water level changed, as shown in Table 1. The test is carried out to find out the status of the SMS sent, and the time the user's Mobile received the SMS. In this test, the rain sensor only sends information that there is rain or no rain on the SMS. Also, SMS sent at flood height 5cm, 10cm, 15cm and 20cm.In this case, the program flow runs every multiple of 5 cm. As seen in numbers 5 cm and 6 cm that at the height of 5cm, the status of the SMS sent with the time of receiving an SMS is 6 seconds. While at the height of 6 cm not sent because the height is not a multiple of 5 cm. This test applies up to a height of 30cm. While testing is carried out at the height of 20cm. While the SMS time received, there are several time differences, and this is due to the processing time on the system of the cellular operator. In this test, when an SMS is received, it is not part of the error of the flood notification system that was built but is influenced by the service provider itself.

From this research, it is expected to provide the latest information on prototype disaster mitigation engineering by implementing ultrasonic sensors, rain sensors, Raspberry Pi, and GSM modems in sending flood information to users. And this research is useful for reducing disaster victims in general

Table 1. Testing time for receiving SMS based on flood height.

No.	Flood Height in cm	Status of Sent SMS	Time the SMS was Received
1	0	Not sent	-
2	1	Not sent	-
3	3	Not sent	-
4	5	Was sent	6 second
5	6	Not sent	-
6	11	Was sent	4 second
7	15	Was sent	8 second
8	20	Was sent	4 second

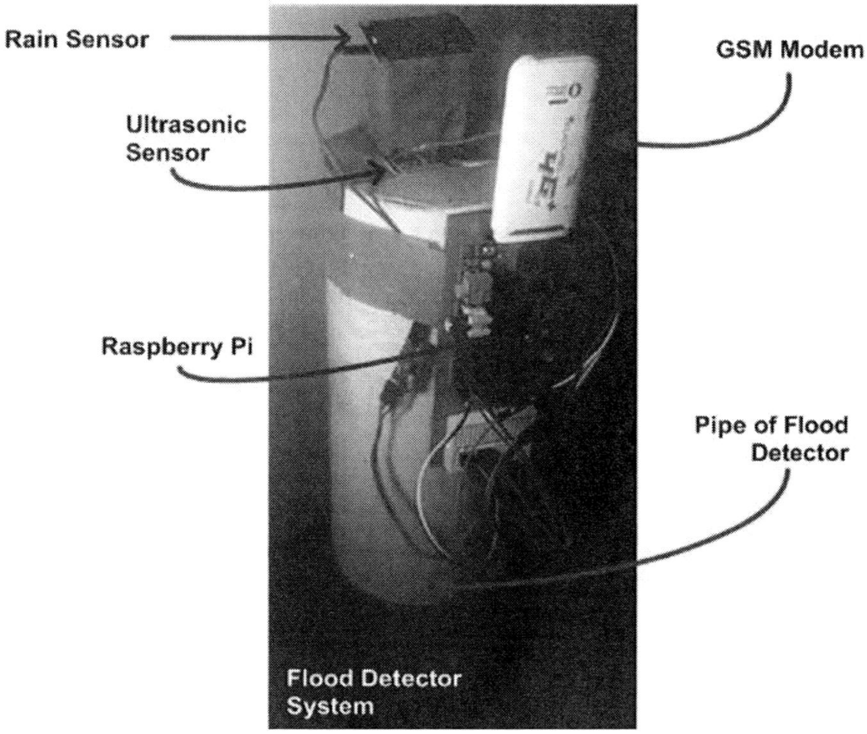

Figure 3. Results of prototype SMS gateway based flood notification system.

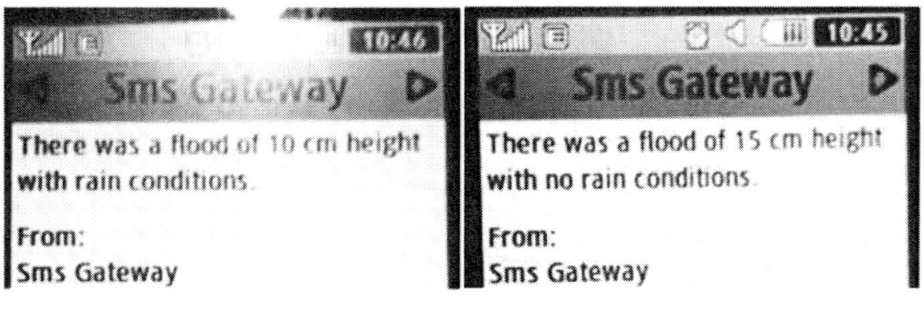

Figure 4. Results of SMS received by users.

5. Conclusions

Based on the importance of developing an SMS Gateway-based flood notification system, this research has produced a prototype of a flood notification system and has been able to send flood notifications via SMS. The system was built using ultrasonic sensors and rain sensor as input, Raspberry Pi as processor and GSM modem as output. From the test results, sending SMS is sent every height of 5cm, 10cm, 15cm, and 20cm. While the greater distance of 5cm and smaller 10cm does not send SMS. The prototype only sends an SMS for every multiple of 5 cm. Besides that, there are also some differences when receiving an SMS. The difference is not due to an error from the prototype system. The difference in reception time is due to the cellular operator system used. From the test method carried out it can be concluded that the prototype system has been able to provide meaningful information for the development of flood disaster mitigation systems.

Acknowledgments
This research was funded by the Ministry of Research and Higher Education (Ristekdikti) through a college collaboration research grant (PKPT) number 119 / SP2H / LT / DRPM / 2019. And we thank the Research Team from Serambi Mecca University and the partner team from Syiah Kuala University for providing technical research consultancy and laboratory facilities.

References
[1] Mercado R J M 2016 Design of wireless sensor networks using embedded Programmable System-on-Chip (PSoC) as applied to community-based flood early warning systems (CBFEWS). In: *2016 International Conference on Advances in Electrical, Electronic and Systems Engineering (ICAEES),* pp 214-23
[2] Satria D, Yana S, Munadi R, Syahreza S and Munawir 2019 Implementation of GSM based Flood Data Communication in the Flood Disaster Location Information System IOP Conference Series: Materials Science and Engineering **506** 012046
[3] Satria D, YanaS, Yusibani E, Syahreza S and Zulfan 2019 Implementation of the SMS gateway in the flood early warning information system for village warning and community information *Int. J. Eng. Adv. Technol.* **8(6)** pp. 4005–4009
[4] Azid S, Sharma B, Raghuwaiya K, Chand A, Prasad S and Jacquier A 2015 SMS based flood monitoring and early warning systems," *ARPN J. Eng. Appl. Sci.* **10(10)** pp. 6387–6391
[5] Kuantama E, Mardjoko P and Saraswati M A 2013 Design and Construction of Early flood warning system through SMS based on SIM300C GSM modem. In: *2013 3rd International Conference on Instrumentation, Communications, Information Technology and Biomedical Engineering (ICICI-BME),* pp 115-9
[6] Piller O, Deuerlein J, Gilbert D and Weber J-M 2015 Installing Fixed Sensors for Double Calibration and Early-warning Detection Purposes *Procedia Engineering* **119** 564-72
[7] Satria D, Zulfan, Munawir and Hidayat T 2019 Implementation of wireless sensor network (WSN) on garbage transport warning information system using GSM module *Journal of Physics: Conference Series* **1175** 012054
[8] Ahmadian H, Satria D and Kurniawan S 2000 Progress of construction project information system based on SMS gateway *J. Nas. Komputasi dan Teknol. Inf.* **1(2)** pp. 114–122
[9] Satria D, Hidayat T, Hidayat M A and Zakaria 2018 Application of SMS gateway on attendance detection systems using RFID *J. Nas. Komputasi dan Teknol. Inf.* **1(2)** pp. 106–113
[10] Satria D and Ahmadian H 2018 Designing home security monitoring system based Internet of Things (IoTs) model *J. Serambi Eng.* **III (1)** pp. 255–261
[11] Dewi S S, Satria D, Yusibani E and Sugiyanto 2018 Design of web based fire warning system using ethernet Wiznet W5500 *Malikussaleh International Conference on Multidisciplinary Studies (MICoMS 2017)* pp. 437–442
[12] Dewi S S, Satria D, Yusibani E and Sugiyanto D 2017 Prototipe sistem informasi monitoring kebakaran bangunan berbasis Google Maps dan modul GSM *Jurnal Teknol. Inf. dan Komunikasi* **1(1)** pp. 33–38
[13] Satria D, Yana S, Munadi R, and Syahreza S 2017 Sistem peringatan dini banjir secara real time berbasis web menggunakan Arduino dan Ethernet *Jurnal Teknol. Inf. dan Komunikasi* **1(1)** pp. 1–6
[14] Satria D, Yana S, Munadi R and Syahreza S 2018 Design of information monitoring system flood based Internet of Things (IoT) in *Malikussaleh International Conference on Multidisciplinary Studies (MICoMS 2017)* pp. 629–639
[15] Bahagia, Satria D and Ahmadian H 2017 Perancangan sistem informasi manajemen data korban bencana berbasis mobile Android *J. Manaj. dan Akunt.* **3(2)** pp. 22–30

IOP Conf. Series: Materials Science and Engineering **796** (2020) 012030 doi:10.1088/1757-899X/796/1/012030

Investigation into time evolution of chlorine emission in pulsed CO_2 laser induced breakdown spectroscopy (LIBS)

A M Sari[1], K Lahna[2], M Ramli[3], K Kurihara[4] and N Idris[2]
[1] Graduate School of Mathematics and Applied Sciences, Universitas Syiah Kuala, Banda Aceh 23111, Indonesia
[2] Department of Physics, Faculty of Mathematics and Natural Sciences, Syiah Kuala University, Jl. Syech Abdurrauf No. 3 Darussalam, 23111 Banda Aceh, Indonesia
[3] Department of Chemistry, Faculty of Mathematics and Natural Sciences, Syiah Kuala University, Jl. Syech Abdurrauf No. 3 Darussalam, 23111 Banda Aceh, Indonesia
[4] Department of Physics, Faculty of Education, University of Fukui 3-9-1, Bunkyo, Fukui 910-8507, Japan

E-mail: nasrullah.idris@unsyiah.ac.id

Abstract. Detection of chlorine (Cl) using LIBS technique is greatly influenced by experimental conditions including the delay time of the detecting device. Time evolution of Cl emission in plasma produced on polyvinyl chloride (PVC) sample using CO_2 laser with variation of delay time has been carried out. The plasma was produced at atmospheric pressure under different environment of helium and air, respectively. It was found that when using the pulse energy of 750 mJ under helium surrounding, strong emission line of Cl at 479.45 nm can be observed until 4 μs of delay time, while when using laser energy 2500 mJ strong Cl emission line can be observed until 7 μs. The higher signal to background ratio was observed at delay time of 1 μs. However, Cl emission line cannot be detected under air environment.

1. Introduction

LIBS is a technique for analyzing various samples, both solid, gas and even liquid samples. In general, LIBS is an atomic emission spectroscopic technique, which qualitative and quantitative analysis are made by monitoring the intensity and wavelength of emission lines from the consequent breakdown. LIBS is powerful analytical tool for analysis of environmental samples. This is because LIBS has many advantages including being able to detect many elements in one measurement, minimal sample preparation and only requires a small sample size [1-2], thus capable to fast monitoring of harmful elements in the environment [3-5].

Cl is a dangerous element at high concentration and it can promote corrosion in steel bones of buildings [6]. Cl content in the environment can increase through industrial waste and natural disasters. Cl detection in environment especially soil samples by LIBS is still a big challenge. Environmental samples have very complex matrices, while Cl generally present as a trace element. Therefore, before carrying out Cl detection in the actual environmental samples, Cl emission dynamic in LIBS was studied using PVC sample, containing Cl as host element.

It was reported detection of Cl with LIBS using a fundamental Nd-YAG laser of 1064 nm, the Cl atomic emission line at 837.5 nm and 134.72 nm is commonly used [7-11]. However, detector system

Content from this work may be used under the terms of the Creative Commons Attribution 3.0 licence. Any further distribution of this work must maintain attribution to the author(s) and the title of the work, journal citation and DOI.
Published under licence by IOP Publishing Ltd

with ICCD camera is less sensitive for infrared (IR) region and expensive, while a special optical fibre is required for detection at ultraviolet (UV) wavelengths [12-13]. In other study, Cl detection using a transversely excited atmosphere (TEA) carbon dioxide (CO_2) laser has successfully detected ionic Cl at visible wavelength from plastic sample [14]. However, Cl cannot be detected in soil sample even though the soil sample have intentionally been added by Cl as much as 1000 ppm [15]. This is because the soil sample has a very complex matrix and Cl is one of the elements difficult to detect because its high excitation energy. Experimental conditions such as laser pulse energy, delay time of the detector and ambient gas greatly affect LIBS measurement. This study was conducted to investigate Cl emission in the CO_2 laser induced breakdown from a PVC sample since Cl is the main constituent in the sample. To search for the optimum conditions, the experimental parameters namely delay time of detection and pulse energy were varied.

2. Experimental procedure

LIBS experiment setup used in this study is similar to the previous research [15]. Lasers used to generate plasma is the TEA CO_2 laser with different pulse energy of 750 mJ and 2500 mJ, respectively. PVC samples have been cut to slide with a dimension of 15 mm x 15 mm and a thickness of 0.73 mm. The sample was attached to the metal plate sample holder and placed in the chamber. Emission spectra from the plasma was recorded by varying the gate delay of the optical multichannel analyser (OMA) system ranging from 0.1 to 10 µs. Measurements was carried out in helium and air environments at a pressure of 1 atm. For Cl detection, the wavelength of each emission line will be fitted to database published by national institute of standard and technology (NIST), United State of America (USA) [16].

3. Result and discussion

The problem in selecting ionic emission lines for carrying out a qualitative and quantitative analysis is that ions only form at the beginning of plasma formation and then decay due to the effects of recombination and cooling of the plasma. Thus, variations in delay time are needed to determine the optimum spectrum recording time.

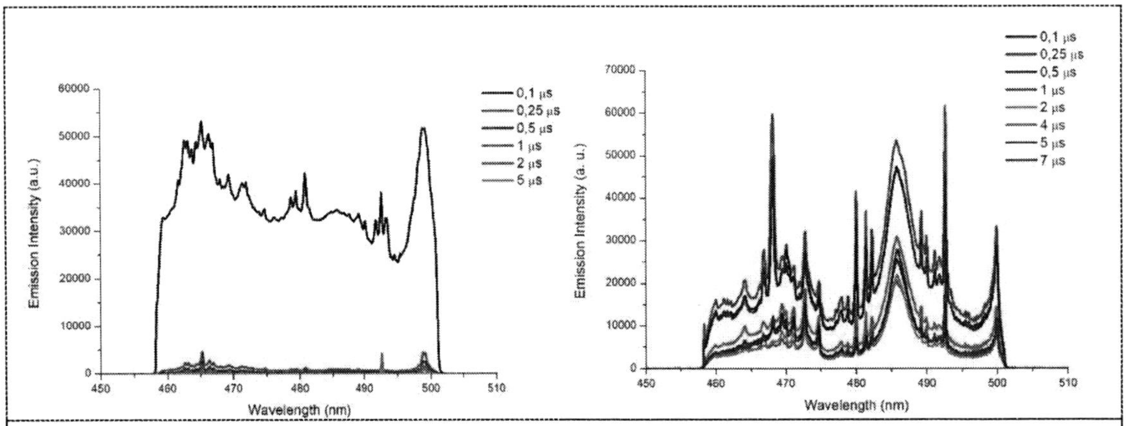

(a) (b)

Figure 1. Spectrum emission of PVC samples in (a) air and (b) helium environments with variation of delay time.

Figure 1 shows emission spectrum detected from plasma produced on PVC sample using the TEA CO_2 laser at different delay time under (a) air and (b) helium surrounding gas, respectively. In case air surrounding gas, Cl ionic emission line cannot be detected. While in the previous work, it was confirmed that Cl emission lines, namely ionic emission lines can clearly detected in this wavelength

IOP Conf. Series: Materials Science and Engineering **796** (2020) 012030 doi:10.1088/1757-899X/796/1/012030

region [14-15]. When air was replaced by helium, Cl ionic emission lines in the visible region appear clearly started since the delay time of 0,1 μs. However, at early regime the emission spectra are suppressed by high background. Cl emission line still can be observed after delay time of 7 μs. At the early stages of plasma formation, signal emissions from governing species are still covered by continuum emission with very high background. Continuous emission is caused by two possible processes, namely recombination radiation and bremstrahlung [17]. Over time, both continuum and signal emissions will decay, however continuum emissions decay faster than signal emissions [18]. It takes the right delay to produce optimum spectrum with high signal intensity and low background. At the beginning of the nanosecond regime, plasma has not yet been formed and the emission lines of all species cannot be observed.

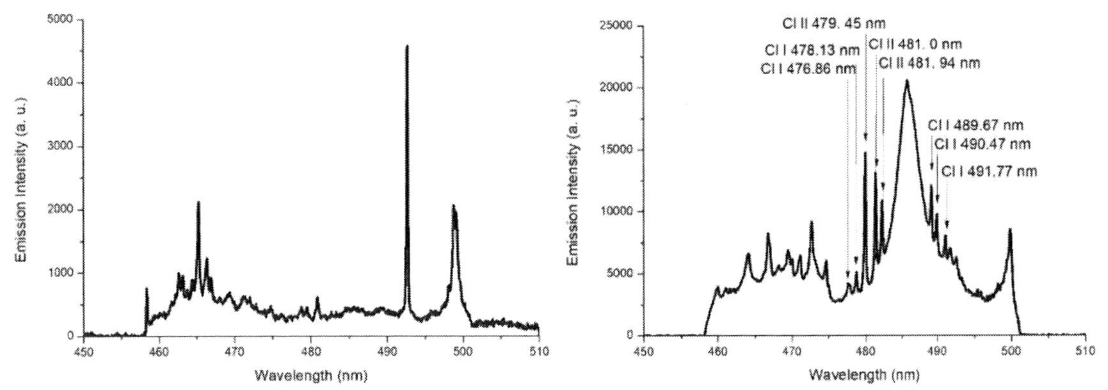

(a) (b)

Figure 2. Spectrum Emission of PVC samples in (a) air and (b) helium environments.

Figure 2 shows the spectrum obtained in the air and helium gas environments using laser pulse energy of 2500 mJ and a delay time of 2 μs. It can be seen that there is no any Cl emission line can be detected when measurement was made under air environment. Nevertheless, the laser pulse energy was increased largely. Meanwhile, when measurements is made under helium environment, Cl emissions line can be detected with strong and clear intensity. The emission lines are ionic Cl lines, namely Cl II 476.86 nm, Cl II 478.13 nm, Cl II 479.45 nm, Cl II 481.0 nm, Cl II 481.94 nm, Cl II 489. 67 nm, Cl II 490.47 nm and Cl II 491.77 nm. The Cl II 479.45 nm emission line is the strongest line that can be observed.

Figure 3 is the signal and background emission intensity of emission line Cl II 479.45 nm with variations of delay time at energies of 750 mJ and 2500 mJ, respectively. It can be seen that the signal intensity once increases and decay at later time. When measurement was made using laser pulse energy of 750 mJ laser energy, the signal and background emission are very high at 1 μs delay time. After 1 μs, the emission intensity continues to decrease and it cannot be detected after 5 μs. While when using a 2500 mJ laser, the highest intensity of the emission and background lines is achieved after a 0.25 μs delay time. However, with 2500 mJ of energy, the emission intensity can be observed longer, up to 7 μs and after approximately 10 μs, the Cl signal becomes undetectable.

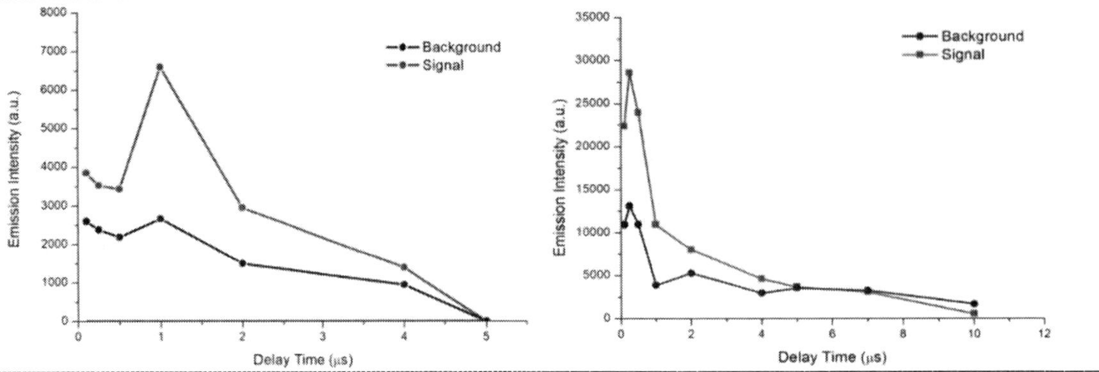

(a) (b)

Figure 3. Experimentally observed temporal variation of the chlorine and background signals in helium surrounding gas at (a) 750 mJ, (b) 2500 mJ laser energy.

To obtain the optimum measurement results, it is necessary to calibrate the intensity of the emission signal against the background. From Figure 4 it can be seen from the two laser energy conditions, having the same graph pattern. The signal intensity ratio continues to increase and reaches the highest intensity at a delay of 1 μs, then continues to decrease to 5 μs and 10μs with laser energy 750 mJ and 250 mJ, respectively. In measurements with a delay time of 0.1 μs, even though the signal emission is high, the background signal emission is still very high so that in this situation the measurement results are not optimum. In the experimental conditions with a delay of 1 μs with 750 mJ laser energy, the signal emission intensity ratio reaches 2.5. while for the use of laser energy 2500 mJ the ratio of signal emission intensity to the background is higher, which is almost reaching 3.

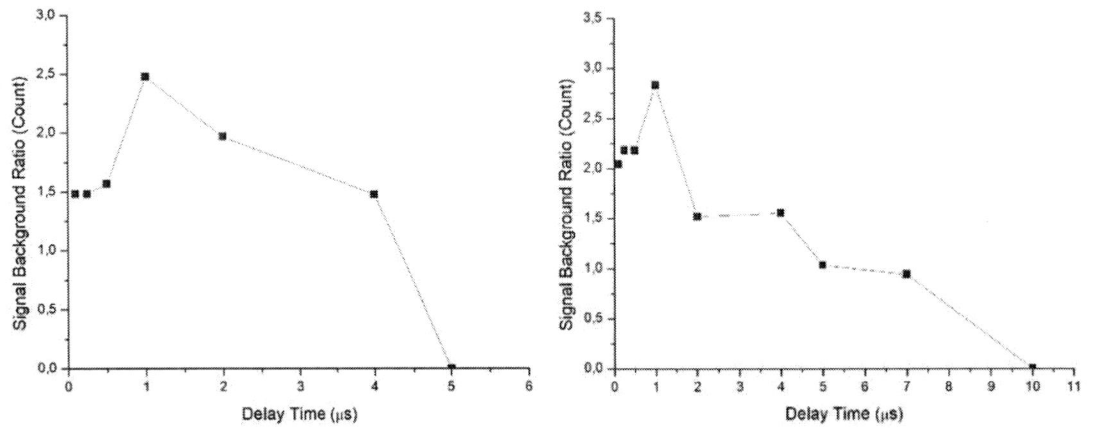

(a) (b)

Figure 4. Signal-to-background ratio (S/B) of Cl II 479,45 nm at (a) 750 mJ and (b) 2500 mJ laser energy.

4. Conclusion

Cl measurements using the TEA CO_2 LIBS technique has produced optimum measurements at 1 μs delay in helium environment. As for measurements in air environment, Cl emission line can not detected. Cl signal emission can be observed longer with 2500 mJ laser energy compared to using a laser with a lower energy of 750 mJ.

References

[1] Winefordner J D, Gornushkin I B, Correl T, Gibb E, Smith B W and Omenetto N 2004 *J. Anal. Atom. Spectrometry* **19** 1061-1083
[2] Miziolek A W and Palleschi V 2006, Schechter I, *Laser-Induced Breakdown Spectroscopy (LIBS), Fundamentals and Applications*, Cambridge University Press.
[3] Idris N, Kagawa K, Sakan F, Tsuyuki K and Miura S 2007 *Appl. Spectrosc.*61 (**12**) 1344-1351
[4] Xiaodong Y, Yang L, Xiaofeng G, Jiming B, Huizhong Y, li S 2014 *Environmental monitoring and assessment* 186 (**12**)
[5] Mingyin Y, Hui Y, Lin H, Tianbing C, Gangfu R, Muhua L 2017 *Applied Optics* 56(**14**):4070
[6] Maslehudin M, Al-Zahrani M M, Ibrahim M, Al-Mehthel M H, and Al Idi S H 2007 *Construction and Building Materials* 21 (**8**)
[7] Gehlen C D, Wiens E, Noll R, Wilsch G and Reichling K 2009 *Spectrochimica Acta Part B* 64 : 1135–1140
[8] Weritz F, Schaurich D, Wilsch G 2007 *Spectrochim. Acta Part B, 62*: 1504–1511
[9] Wilsch G, Weritz F, Schaurich D, and Wiggenhauser H 2005 *Construction and Building Materials, 19* : 724–730
[10] Tran M, Benjamin W S, David W H and James D 2001 *Appl. Spectrosc.* 55 :1455-1461
[11] Kaski S, Hakkanen H, and Korppi-Tommola J 2004 *J. Anal. At. Spectrom.* 19: 474-478
[12] Asimellis G, Hamilton S, Giannoudakos A, and Kompitsas M 2005 *Spectrochimica Acta Part B* 60 : 1132 – 1139
[13] Weritz F, Schaurich D, Wilsch G 2007 *Spectrochim. Acta Part B* 62: 1504–1511
[14] Idris N, Suliyanti M M, Kobayashi T, Kurniawan H, Lie T J, Lee Y I, and Kagawa K, 2005 *Journal - Korean Physical Society* 47 (**2**) : 256-262
[15] Sari A M, Ramli M, Kurihara K, Idris N 2018 *Journal of Physics: Conference Series* 1116 (**3**) 32028
[16] NIST Atomic Spectra Database https://physics.nist.gov/PhysRefData/ ASD/ lines_form.html
[17] Anabitarte F, and Lopez-Higuera J M 2012 *Laser Induced breakdown Spectroscopy: Fundamentals, Applications, and Challenges*. ISRN Spectroscopy
[18] Cremers D A and Radziemski L J 2013 *Handbook of Laser Induced Breakdown Spectroscopy Second Edition* (United Kingdom:John willwy and sons.)

Spatial analysis for calculating closest distance of operators' location to Base Transceiver Station (BTS) in Banda Aceh city

Nizamuddin[1], R P F Afidh[1], Ardiansyah[2], H Sofyan[3] and N Anisah[1]
[1] Informatics Department, Faculty of Mathematics and Life Sciences, Syiah Kuala University, Darussalam, Banda Aceh, 23111, Indonesia
[2] Electrical and Computer Engineering Department, Faculty of Engineering, Syiah Kuala University, Darussalam, Banda Aceh, 23111, Indonesia
[3] Statistics Department, Faculty of Mathematics and Life Sciences, Syiah Kuala University, Darussalam, Banda Aceh, 23111, Indonesia

E-mail: niz4muddin@unsyiah.ac.id

Abstract. Base Transceiver Station (BTS) is used to generate auxiliary signal for cellular uses. BTS will support all the activities regarding data exchange such as messaging and web browsing. The spatial data plays important role to support BTS's signal coverage area. The spatial data also can be used for designing of new cellular networks and construction new BTS. This research purpose is to help operator manager find the distance of BTS distribution that will be used to make group of BTS position for the operators. In this research Geographic Information Systems (GIS) has been utilized to analyse the nearest distance and to create group of BTS. Spatial network analyst uses network data path to find out the nearest distance from operators' location to all of BTS in Banda Aceh by using tools closest facility, and to analyse the operators' location grouping based on the mean value BTS distance by using buffer method. The result of this research has calculated the average distance from the operators' location to all BTS is 6.2 Kilometres. Based on the average distance the first team will manage 7 BTS and the second team will manage 14 BTS.

1. Introduction

Base Transceiver Station (BTS) as an auxiliary signal generator is used for cellular purposes. Every activity for data exchange such as web browsing and others cannot be separated from the role of BTS. To support the smooth communication of data through BTS to be more optimal, the role of spatial data is needed. With spatial data, cellular network development such as new BTS development can be considered and plotted. The position of a BTS has been calculated based on the range of the transmitted signal. The number of base stations in urban areas is greater than in rural areas because each operator wants full signal coverage so that it can attract consumers to use the operator [1].

Base Transceiver Station (BTS) as an additional signal generator is commonly used for cellular purposes. Every activity for data exchange for example web browsing and others cannot be separated from the role of BTS.

Geographic Information System (GIS) as a computer-based system is used to collect, store, organize, transform, manipulate and analyse geographic data [2]. The GIS capability is expected to help provide complete and easy information for BTS operators in Banda Aceh. A geographic

Content from this work may be used under the terms of the Creative Commons Attribution 3.0 licence. Any further distribution of this work must maintain attribution to the author(s) and the title of the work, journal citation and DOI.
Published under licence by IOP Publishing Ltd

information system is utilized for displaying data on the distribution of BTS locations and for providing the complete information of BTS. It also can be used by operator manager to easily find the distance and distribution of BTS. The results of the distance search are also used to make groups of BTS positions. These groups will be used for managing the BTS's operator.

Network analyst is a method of spatial analysis used to analyse the movement of a resource from one location to another and to find the closest distance to that location. Network analysts utilize segment or line features in conducting analysis. Network analyst application in determining the shortest path from one object to another is processed through the arithmetic process of connecting lines that have attributes [3]. In network analysis, one of the line features that can be used is road feature [4]. This study will find the closest distance from the operators' location to the BTS using the road feature.

2. Research Method

2.1. Geographic Information System (GIS)

Geographic Information System (GIS) is a computer-based system to collect, store, organize, transform, manipulate, and analyse geographical data [2]. GIS can be defined as a combination of computer hardware and software that allows to manage, analyse, map spatial information and attribute data with cartographic accuracy [5]. GIS is all types of manual and computer-based procedures for storing and manipulating geo-referenced data [6]. GIS uses spatial data as input in doing its work. Spatial data is data that refers to the position, objects, and relationships between them in the earth's space. Spatial data is one item of information, in which there is information about the earth including the surface of the earth, below the surface of the earth, and under the atmosphere [7].

2.2. Building a Topology

The initial stage before conducting network analysis is to check the road data used by the building topology. Topology is used to find out the relationship between one object with another object. In GIS topology there are data characteristics such as lines, polygons and points. Each data characteristic has certain rules, these rules have been provided by ArcGIS software [8]. The topology has been created to easily find out the lines that are not connected to each other, the lines that exceeded the limit and others. This topology will be included in the Banda Aceh City road map to find out which roads have been cut off. The rules used in this topology are "Must not have dangles", which has several functions including: "a). Extend: to connect the dangle at the end of a line segment to the feature in front of it, as long as the spanning distance is met.", "b). Trim: to remove line features if the dangle point at the end of the intersection line falls within the specified spanning distance tolerance.", "c). Snap: to unite the dangle line to the nearest line that falls within the spanning distance tolerance,". So that there will be no more topological error for all road features that will be used in network analysis.

2.3. Creating a Dataset Network

A road network has rules about how an object passes through it. For example, a road network is a combination of one-way and two-way roads. The rules used in the network dataset include: a). At mileage use kilometres. b). One-way streets are regulated with separate columns added to the attribute table named "ONEWAY" with the following rules as shown in Table 1.

According to the table, the rules of a unidirectional or bi-directional road are very much determined by digitizing road line elements. Especially roads that can only be traversed in one direction can be seen from the digitization of certain road line elements. If in the same direction, the FT (From-To) value is the opposite if it is given the TF (To-From) value [9]. Based on the network dataset rules, the Banda Aceh City road network determines the direction and location of the destination by identifying and verifying the location of the BTS and the operators' location house on the map. Each road segment is given rules, for example for two-way roads are coded TF or FT, so the Banda Aceh City road network can be used at a later stage to conduct network analysts [8].

Table 1. Types of road rules, modified from [8].

Types of road rules	Value
The trip can only be made from the starting line to the finish line, which is the same as the direction at which the line was digitized	FT
The trip can only be made from the finish line to the starting line, which is the opposite direction at the time the line is digitized	TF
Travel can be done Other values in both directions	Value does not need to be filled

2.4. Network Analyst

Spatial analysis is utilized to perform analytical functions using GIS. Spatial analysis leads to many kinds of operations and concepts including simple calculation, classification, structuring, and cartographic modelling [4]. According to Cromley et al [10] one of spatial analysis is network analysis that refers to the analysis of spatial data points or lines as an inseparable network. The network analysis tools used in this study is an extension "closest facility" to find the closest facility. Banda Aceh city road feature has been utilized as input data. This analysis finds the closest distance between the location of the operator's house with all BTS in Banda Aceh city. There are four parameters, which are facilities, location, barriers and routes. Facilities location consists of initial facilities to start the trip and the destination location. The barriers are something that blocks route between two directions. The route is the result of the final route. The results of the closest facility produced the mileage attribute from the operator house to all existing BTS in Banda Aceh city.

2.5. Buffer

The results of the distance using the network analyst was obtained from the process of finding the average value of the BTS distance. After the result is put into a buffer including the average values, then the a BTS grouping is formed.

3. Result and Discussion

The network analyst applying the closest facility tool using the Banda Aceh city road network has found the closest distance from the location of the operator house to all BTS. In this study there are two operator houses that manage all BTS in Banda Aceh City. The following analysis has produced the closest distance from each home operators' location to each BTS.

3.1. Analysis of the Closest Distance of the First Team

There are 21 BTS that spread throughout the Banda Aceh City. The search results for the closest distance from the location of the first operator's house in Darussalam area to all BTS in the city of Banda Aceh has produced the shortest distance to the BTS at Lhok Bangka Darussalam area by 0.9 km and the farthest distance to the BTS at Surien Empeerom area with a distance of 12.2 km. After getting the mileage results to all BTS in Banda Aceh city, the average value of the mileage was searched. The first team's average score is 7.9 Km.

3.2. Analysis of the Closest Distance of the Second Team

The search results for the closest distance from the location of the second operator's house in Geuceu Complex area to all BTS in the city of Banda Aceh has produced the shortest distance to the BTS at Banda Raya 2 area by 0.7 km and the farthest distance to BTS at the Lhok Bangka Darussalam with the distance of 10.7 km. After getting the mileage results to all BTS in Banda Aceh city, the average value of the mileage was searched. The second team average score is 4.9 Km. The following network analysis map for the first and the second team can be seen in Figure 1.

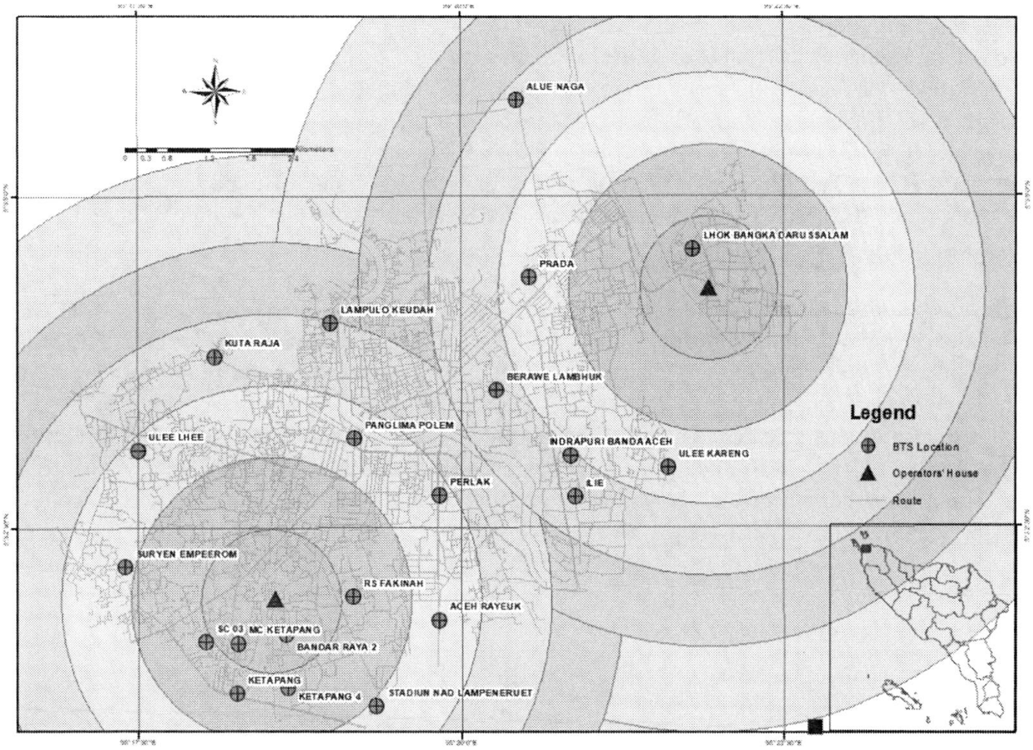

Figure 1. Spatial network analysis and buffer for the first and the second team.

3.3. Average Distance of the Two Teams

Based on the optimal path results, the first team had an average value of 7.9 km, while the second team had an average value of 4.5 km. So, we can conclude that the results of the two teams' travel paths have an average value of 6.2 km, then the average results will be entered into the buffer method. Table 2 shows the average scores of the two teams.

Table 2. The average scores of the two teams.

Operator	Minimum (km)	Maximum (km)	Mean (km)
First Team	0,9	12,2	7,9
Second Team	1	10,7	4,5
Average (km)		6,2	

3.4. First Team Buffer

The results of the network analysis process obtained an average value from both teams, then the buffering process is based on the average value of the two teams. Based on the results of the first team buffer with a range of 6.2 kilometres. As the result the first team will manage 7 BTS. At 1 kilometre range there is 1 BTS, namely Lhok Bangka Darussalam BTS, at 2 kilometres there is no BTS, at 3 kilometres range there are 2 BTS namely Ulee Kareng BTS and Prada BTS, while at 4 kilometres range there are 4 BTS namely Meuraxa BTS, Berawe BTS , Ilie BTS and Alue Naga BTS and range 5, 6 and 6.2 kilometres have no BTS.

3.5. Second Team Buffer

Based on the results of the first team buffer with a range of 6.2 kilometres. As the result the first team will manage 14 BTS. At 1 kilometre range there is 1 BTS, namely Ketapang MC BTS, at 2 kilometres there are 5 BTS, namely SC 03 BTS, Ketapang BTS, Ketapang 4 BTS, Banda Raya BTS and RS Fakinah BTS, at 3 kilometres range there are 6 BTS namely Surien Emperom BTS, Stadiun Lampenerut BTS, Aceh Rayeuk BTS, Perlak BTS, Panglima Polem BTS, and Ulee Lheue, while at 4 kilometres range there are 2 BTS namely Kuta Raja BTS and Lampulo Keudah BTS and range 5, 6 and 6.2 kilometres have no BTS.

3.6. Results of Both Team Buffer

The results of grouping the two teams using the buffer method can be seen in Figure 2. The buffer can only be inputted with the range of distance, it cannot be inputted with the number of points that is wanted to group. The grouping of operators' locations is based on the mean value of the distance which is 6.2 km in this research. The result of the blue colour is 1 km range, green is 2 km range, pink is 3 km range, red is 4 km range, purple is 5 km range, black is 6 km range and green are 6.2 km range. From the grouping of the two teams, the results obtained are the first team will manage 7 BTS while the second team will manage 14 BTS.

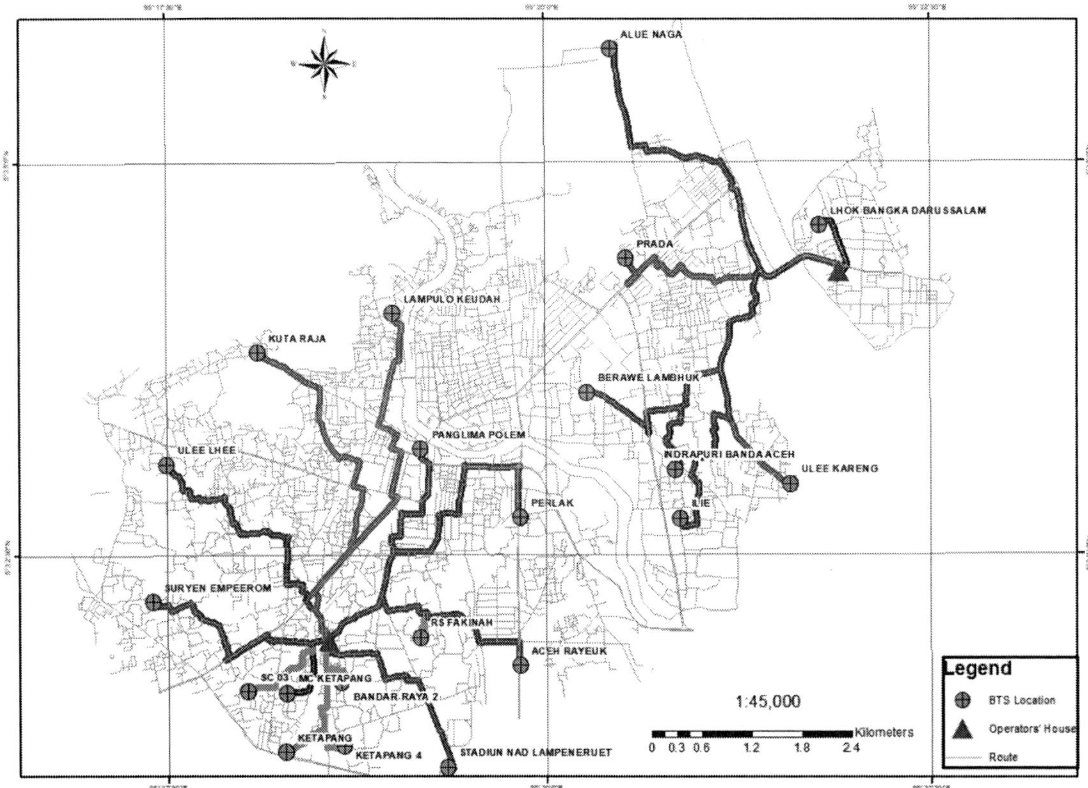

Figure 2. The results of grouping the two teams using the buffer method.

The results of network analysis for group 1 will take route around 19,964 meter or 20 km from tower no. 1 to tower no. 9 and group 2 will take route around 28,869 m or 28 km from tower no. 1 to tower no 15 as shown in Figure 3.

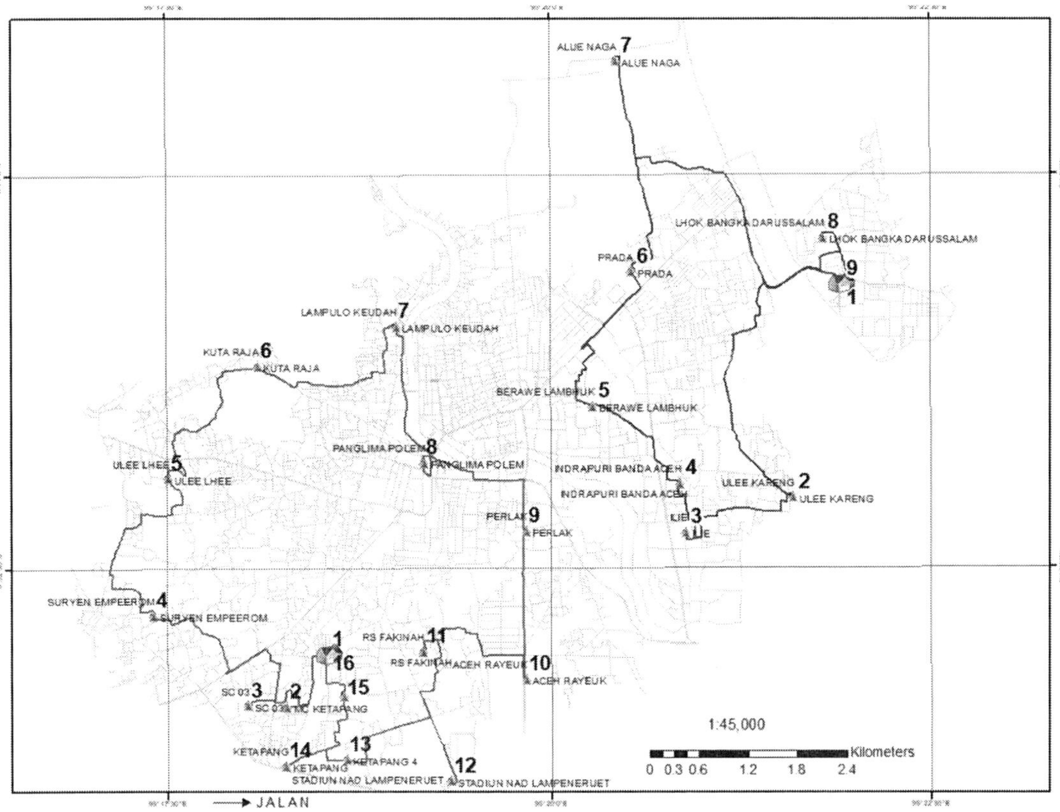

Figure 3. The results of network analysis for the route of group 1 and group 2.

4. Conclusion

The results of the first team's travel path have an average value of 7.9 km while the second team has an average value of 4.5 km, and the results of the average value of the two teams is 6.2 km. Based on the grouping using the buffer method, the results is obtained, the first team will manage 7 BTS located in Syiah Kuala sub-District, Darul Imarah sub-District and Kuta Alam sub-District. While the second team will manage 14 BTS located in Kuta Raja sub-District, Banda Raya sub-District, Meuraxa sub-District, Lueng Bata sub-District, Jaya Baru sub-District and Baiturrahman sub-District.

Acknowledgments

Authors would like to acknowledge ESRI Indonesia that provided ESRI Grand License for Syiah Kuala University including (ArcGIS Server, ArcGIS Desktop, ArcGIS Pro and Geodatabase Spatial Engine) and PT. MANN Tower that provided data for this research. This research is also partially funded by Hibah Laboratorium Terpadu Unsyiah.

References

[1] Hakim F, Anggraeni W dan Pribadi A 2012 Optimasi perencanaan jumlah Base Transceiver Station (BTS) dan kapasitas trafil BTS menggunakan pendekatan goal programming pada sistem 44 telekomunikasi seluler berbasis GSM, Fakultas Teknologi Informasi (ITS),

Surabaya http://download.portalgaruda.org/article.php?article= 54057&val=4186 accessed 01 Juni 2019.
[2] Yousman, Y. 2004. Sistem Informasi Geografis dengan MapInfo Professional, Dani, Yogyakarta.
[3] Nizamuddin, Aldika W A, Ardiansyah, Hizir and Muzailin 2019 WEBGIS Application for Planning the Tsunami Evacuation Route *IOP Conference Series: Earth and Environmental Science* **273** 012051
[4] Demers M N 1997 *Fundamentals of Geographic Information Systems* (New York: Jhon Wiley & Sons.)
[5] Prahasta E 2005 *Sistem Informasi Geografis* (Bandung: Informatika)
[6] Aronoff S 1989 *Geographic Information System: A Management Perspective* (Ottawa: WDL Publication)
[7] Rajabidfard A dan Williamson I P 2000 *Spatial Data Infrastructures: Concept, SDI Hierarchy and Future Directions* Melbourne, Victoria: 45 Spatial Data Research Group, Department of Geomatics, The University of Melbourne.
[8] Sudomo O 2011 *Membangun Geodatabase* (Jakarta: PT. Duta Informatika)
[9] Puntodewo A, Dewi S, dan Tarigan J 2003 *Sistem Informasi Geografis untuk pengelolaan sumber daya alam* Bogor Barat: Center for International Forestry Research.
[10] Cromley, Ellen, K, McLafferty, Sara, L 2002 *GIS dan public health* (New York:The Guilford Press)

IOP Conf. Series: Materials Science and Engineering **796** (2020) 012032 doi:10.1088/1757-899X/796/1/012032

The Relation between Rice Productivity in Ingin Jaya Subdistrict and NDVI (Normalized Difference Vegetation Index) from Surface Reflectance data of LANDSAT 8

Ardiansyah[1], Nizamuddin[2] and S Amalia[2]
[1] Electrical and Computer Engineering Department, Faculty of Engineering, Syiah Kuala University, Darussalam, Banda Aceh, 23111, Indonesia
[2] Informatics Department, Faculty of Mathematics and Life Sciences, Syiah Kuala University, Darussalam, Banda Aceh, 23111, Indonesia

E-mail: ardiansyah@unsyiah.ac.id

Abstract. Rice has an important role as the staple food for most of Indonesian people. Therefore, to ensure the sufficiency of rice stocks, it is necessary to estimate crop production from existing rice fields before they are harvested. This research was conducted to find the relationship between Normalized Difference Vegetation Index (NDVI) and rice productivity of rice fields in Ingin Jaya Subdistrict, Aceh Besar District, Aceh Province. The NDVI was calculated from the surface reflectance data of Landsat 8. The results showed that the NDVI in the study area ranged from 0.4 to 0.84. These results were higher than the NDVI calculated from the Landsat 8 L1 data, which ranged from 0.26 to 0.49. The productivity of rice fields were calculated for 21 locations of sample plot. Based on the calculations, the lowest productivity was 2.047 kg/m2, and the highest was 10.462 kg/m2. The relation between NDVI and the productivity was tested using the Pearson correlation, but the results showed that there was no significant relation between the NDVI and the rice field productivity in Ingin Jaya Subdistrict.

1. Introduction

Rice is a crop that plays a very important role in meeting the food needs of the majority of Indonesian people. Therefore, the estimated amount of rice production is very important for predicting the adequacy of rice supplies at the regional and national levels. The estimation of rice production is conducted by several agencies, including the Logistics Agency (BULOG), the Central Statistics Agency (BPS) and the Ministry of Agriculture. The method used by each agency is different, so the results obtained are different [1]. Although the methods used are different, most of these estimates include the productivity of rice plants in the calculation. One of the common methods in calculating rice productivity is to use the Crop Cutting Experiment (CCE) [2]. CCE is a statistical survey method that estimates yields of crop production based on yields in several small plots of land [3]. This estimation method cannot be carried out before harvest [1].

Current technological developments have made it possible to monitor the condition and development of plants using remote sensing satellite data. One way is to use imagery from satellites that have been processed into a vegetation index [4]. One vegetation index that is widely used is the Normalized Difference Vegetation Index (NDVI). The relationship between NDVI values and productivity has been shown in several studies [1] [4] [5].

Content from this work may be used under the terms of the Creative Commons Attribution 3.0 licence. Any further distribution of this work must maintain attribution to the author(s) and the title of the work, journal citation and DOI.
Published under licence by IOP Publishing Ltd

Landsat images data that are normally used for producing NDVI is available for free download at different product levels according to the processing. The lowest level is Level 1 (L1), which is Landsat image data that has undergone the geometric and radiometric correction process [6]. To use this product, the Digital Number of the L1 data is usually converted into the Top of Atmosphere Reflectance to minimize the effects of the atmosphere. Nevertheless, the effect of aerosols on Visible and Near Infra-Red spectral is still difficult to repair. Surface Reflectance, which is a Level 2 data, has been proven to provide better NDVI results [7].

This paper presents the results of research which aimed at investigating the relationship between NDVI values and rice productivity in the subdistrict of Ingin Jaya, Aceh Besar. This subdistrict is one of the top 10 rice producers in Aceh Besar District with a planting area of 2814 hectares and a production of 15632 tons in 2016. NDVI values were calculated from Surface Reflectance Landsat 8 data. The results of these calculations were also compared with NDVI calculated using L1 data.

2. Research Method

2.1 Data

Features and phenomena on the earth surface can be represented using two geospatial data models, namely vector and raster. Vector data model is based on points (point) and coordinate values (x, y) to construct spatial objects. Raster model represents geographic objects as grid cell structures called pixels (picture elements) [8]. This study used both types of data. The vector data included administrative boundary and land use of Aceh Besar District, while the raster data are Landsat 8 surface reflectance imagery acquired on July 31, 2016. This study also used non-spatial data, which is the rice productivity data collected from the crop cutting experiment.

2.2 NDVI calculation

The level of greenness of a plant can be differentiated by utilizing the unique properties of plants which reflect and absorb electromagnetic waves differently than other objects. This method is the basis for distinguishing vegetation objects from objects other than vegetation. Normalized Difference Vegetation Index (NDVI) is an index that describes the level of greenness of a plant. Vegetation index is a mathematical combination between the red band and the NIR band which has long been used as an indicator of the presence and condition of vegetation [9]. NDVI is calculated from the ratio of Near Infra-Red band (NIR) and Red (R) band as described by Eq. (1).

$$NDVI = \frac{NIR - R}{NIR + R} \tag{1}$$

The range of NDVI values is between -1.0 and +1.0. Values greater than 0.1 usually indicate an increase in the degree of greenness and intensity of the vegetation. Values between 0 and 0.1 are generally characteristic of rocks and vacant land, while values less than 0 may indicate ice clouds, water vapor clouds and snow. Many factors affect NDVI values such as photosynthesis activity in plants, the amount of plant cover, biomass, moisture of plants and soil, and stressed plants (unhealthy).

The NDVI of subdistrict Ingin Jaya was calculated from Landsat 8 Surface Reflectance data according to Equation (1). The data were cropped to the subdistrict boundary before being used in the calculation. The calculation was carried out by using Raster Calculator tool in ArcGIS.

2.3 Crop Cutting Experiment

Crop cutting is an objective method to estimate crop yield from plots of rice plant which is selected using appropriate sampling methodologies. This method involves the actual harvesting of crops from a small area within the farmer's plot [2]. The size of a plot is 2.5 m x 2.5 m. The total weight of harvested rice from each plot was measured and recorded [10]. This study recorded the yield from 21 plots. The coordinates of each plot were recorded and then used for obtaining the NDVI value of the plot. The NDVI and productivity of each plot were then used for calculating the Pearson correlation value.

2.4 Pearson correlation calculation
Simple correlation is a statistical technique for determining the type and measuring the strength of the relationship between 2 variables with quantitative results. The type relationship between two variables can be positive linear or negative linear correlation, while the strength of the variable relationship is categorized into strong, weak, or no relation. The simple correlation coefficient (r) between two variables, x and y, can be calculated by using the Pearson Product Moment formula, which is shown as equation (2).

$$r = \frac{n\sum xy - \left(\sum x\right)\left(\sum y\right)}{\sqrt{\left(n\sum x^2 - \left(\sum x\right)^2\right)\left(n\sum y^2 - \left(\sum y\right)^2\right)}} \quad (2)$$

Two variables are said to be associated if the behaviour of one variable influences the other variable. If it has no effect, the two variables are called independent [11].

This study used the Pearson Product Moment formula to measure strength of relationship between the NDVI values and the productivity of sample plots from several rice fields. A statistical test was conducted to determine whether or not the relationship is statistically significant. The null hypothesis (H_0) of the test was that there was no relationship between the NDVI values and the productivity.

3. Result and Discussion
The NDVI of Ingin Jaya subdistrict is shown in Figure 1. The figure shows NDVI obtained from the calculation which used Landsat 8 surface reflectance data. The NDVI was overlaid with village boundaries and location of plots for crop cutting experiment. The NDVI values range from -0.74 to 0.97. The maximum value is higher than the maximum value of NDVI resulting from a calculations which use the digital number directly from a Level 1 Landsat product.

The white points in the map shown in Figure 1 represent the location of plots for crop cutting experiment. The NDVI values at the location of each points were recorded on a table such as shown as Table 1. In the table, NDVI (L1) column contains NDVI values at the calculated from the Landsat Level 1 Data, while NDVI (SR) column contains NDVI values based on Landsat surface reflectance data. It can be seen that NDVI calculated from surface reflectance has higher values.

The productivity (in Kg/m^2) at each plot location was obtained from crop cutting experiment on the rice field at the time of harvest. The correlations between productivity of a the plot (*ubinan*) and NDVI values were calculated using Phearson correlation. The results are shown in Figure 2. The calculation shows that there is no significant correlation between the NDVI values and rice productivity. The results show that no matter which NDVI used, the result is still no significant correlation. This is contrary to the result obtained by [1] and [4]. This may be due to inaccuracy of the results from the crop cutting experiments. Other researches on the same topic are necessary to confirm this result.

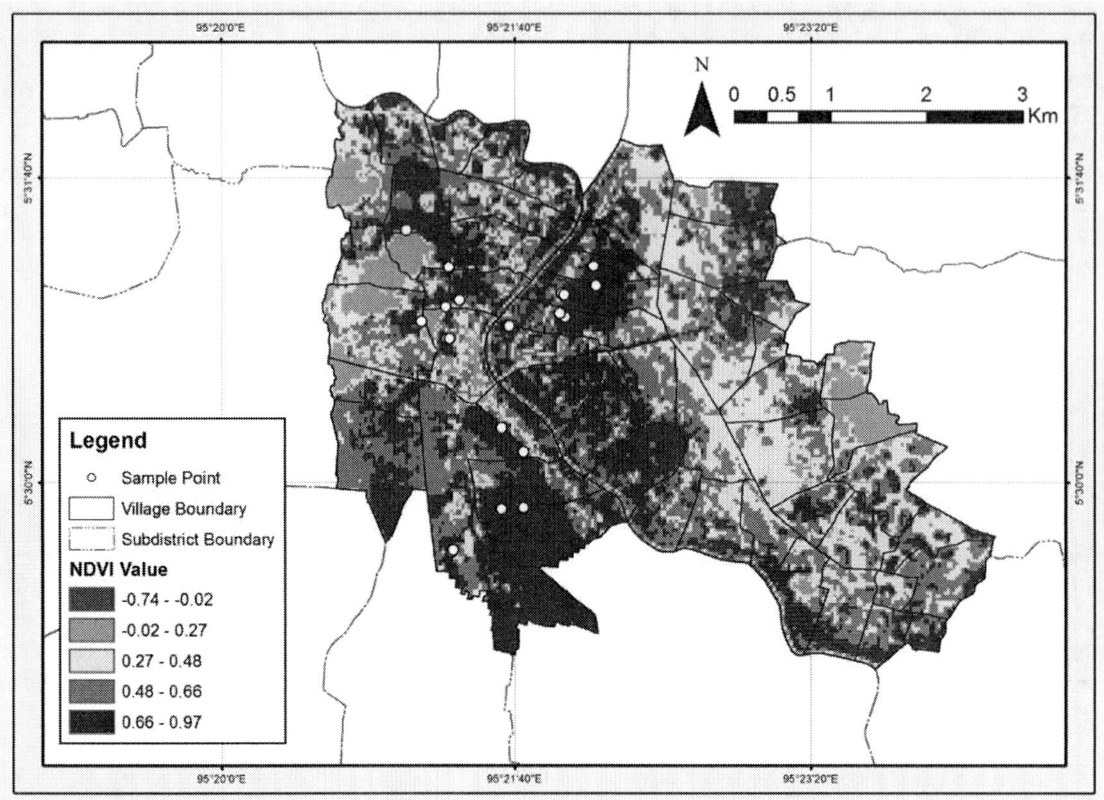

Figure 1. Map of NDVI values (using Surface Reflectance) with points representing location of plots for Crop Cutting Experiment in Ingin Jaya subdistrict.

Table 1. Coordinates, NDVI values, and productivity at sample plot location.

No	Latitude	Longitude	NDVI (L1)	NDVI (SR)	Productivity (Kg/m2)
1	5.515099	95.365905	0.42	0.71	4.972
2	5.515465	95.365383	0.29	0.53	6.75
3	5.517279	95.365541	0.32	0.58	4.05
4	5.517140	95.365829	0.49	0.84	2.587
5	5.517995	95.368738	0.45	0.81	6.75
6	5.519766	95.368502	0.44	0.81	8.167
7	5.514358	95.360442	0.46	0.78	2.047
8	5.514303	95.360588	0.46	0.77	3.442
9	5.493678	95.355218	0.44	0.80	4.725
10	5.493817	95.355350	0.40	0.72	2.284
11	5.513151	95.354761	0.49	0.82	2.925
12	5.513144	95.355004	0.48	0.79	4.331
13	5.505001	95.359867	0.48	0.82	2.925
14	5.502831	95.361904	0.46	0.81	3.375
15	5.497580	95.359867	0.40	0.76	5.4
16	5.497693	95.361904	0.43	0.80	4.59
17	5.514467	95.352459	0.16	0.20	10.327
18	5.514726	95.352280	0.26	0.40	4.86
19	5.522906	95.350635	0.11	0.12	4.05
20	5.523044	95.350910	0.28	0.41	3.654
21	5.520108	95.352693	0.24	0.32	6.142
22	5.519662	95.354937	0.34	0.59	4.62
23	5.516631	95.355910	0.31	0.50	10.462
24	5.516042	95.354633	0.45	0.75	7.087

Correlations

		Ubinan	NDVI_SR
Ubinan	Pearson Correlation	1	-.278
	Sig. (2-tailed)		.223
	N	21	21
NDVI_SR	Pearson Correlation	-.278	1
	Sig. (2-tailed)	.223	
	N	21	21

(a)

Correlations

		Ubinan	NDVI_L1
Ubinan	Pearson Correlation	1	-.338
	Sig. (2-tailed)		.134
	N	21	21
NDVI_L1	Pearson Correlation	-.338	1
	Sig. (2-tailed)	.134	
	N	21	21

(b)

Figure 2. Results of Pearson Correlations calculations between the rice field productivity of harvest plots and their corresponding NDVI values calculated using (a) SR data and (b) L1 data of Landsat 8.

4. Conclusion

The results showed that the NDVI in the study area ranged from 0.4 to 0.84. These results were higher than the NDVI calculated from the Landsat 8 L1 data, which ranged from 0.26 to 0.49. The productivity of rice fields were calculated for 21 locations of sample plot. Based on the calculations, the lowest productivity was 2.047 kg/m2, and the highest was 10.462 kg/m2. The relation between NDVI and the productivity was tested using the Pearson correlation, but the results showed that there was no significant relation between the NDVI and the rice field productivity in Ingin Jaya Subdistrict.

Acknowledgments

Authors would like to acknowledge ESRI Indonesia that provided ESRI Grand License for Syiah Kuala University including (ArcGIS Server, ArcGIS Desktop, ArcGIS Pro and Geodatabase Spatial Engine). This research is also partially funded by Hibah Laboratorium Terpadu Unsyiah.

References

[1] Wahyunto, Widagdo and Heryanto B 2006 Pendugaan Produktivitas Tanaman Padi Sawah melalui Analisis Citra Satelit *Informatika Pertanian* **15** 853-869

[2] Rotairo L, Durante A C, Lapitan P and Rao L N 2019 *Use of Remote Sensing to Estimate Paddy Area and Production* (Manila: ADB)

[3] Kabir M S, Paul D N R, Hossain M I and Rahman N M F 2016 *Bangladesh Rice J.* **20** **1** 11-16

[4] Parwati E, Prasasti I, Arief H, Harsanugraha W K, Harini S, Effendy I, Suhartini T and Nugroho G 1999 Model Estimasi Produksi Tanaman Padi Berdasarkan Indeks Vegetasi *Majalah LAPAN* **1** **1**

[5] Thiruvengadachari S and Skathivadivel R 1997 *Satellite Remote Sensing for Assessment of Irrigation System Performance : A Case Study in India* (Sri Lanka: International Irrigation Management Institute)

[6] Anonymous 2019 *Landsat 8 Data Users Handbook* (United States: USGS)

[7] Vermote E, Justice C, Claverie M and Franch B 2016 Preliminary Analysis Of The Performance Of The Landsat 8/Oli Land Surface Reflectance Product *Remote Sensing of Environment* **185** 46-56

[8] Puntodewo, A, Dewi S, dan Tarigan J. 2003. Sistem Informasi Geografis Untuk Pengolahan Sumber Daya Alam. Bogor : Center for International Forestry Reseacrh

[9] Lillesand, T.R. 1997. Penginderaan Jauh Intepretasi Citra. Terjemahan.Yogyakarta: Universitas Gajah Mada Press

[10] Yuniarto A E, Prasetyo Y, and Haniah 2015. Analisis Tingkat Produktivitas Padi Berdasarkan Metode Ndvi (Normalized Difference Vegetation Index) Dan Lswi (Land Surface Water Index) Menggunakan Citra Landsat Tahun 2007 dan 2009 (Studi Kasus: Kabupaten Karanganyar,Jawa Tengah) *Jurnal Geodesi UNDIP* **4** **3** 26-34

[11] Mattjik, A. A. 2000. *Perancangan Percobaan dengan Aplikasi SAS dan Minitab. Jilid I.* (Bogor: IPB Press)

IOP Conf. Series: Materials Science and Engineering **796** (2020) 012033 doi:10.1088/1757-899X/796/1/012033

A Genetic Algorithm to Determine Research Consultation Schedules in Campus Environment

R Idroes[1,2], A Maulana[3], T R Noviandy[3], R Suhendra[4], N R Sasmita[5], A Lala[1] and Irvanizam[3*]

[1] Department of Chemistry, Faculty of Mathematics and Natural Sciences, Universitas Syiah Kuala, Kopelma Darussalam, Banda Aceh 23111, Indonesia

[2] Department of Pharmacy, Faculty of Mathematics and Natural Sciences, Universitas Syiah Kuala, Kopelma Darussalam, Banda Aceh 23111, Indonesia

[3] Department of Informatics, Faculty of Mathematics and Natural Sciences, Universitas Syiah Kuala, Kopelma Darussalam, Banda Aceh 23111, Indonesia

[4] Department of Electrical and Computer Engineering, Faculty of Engineering, Universitas Syiah Kuala, Kopelma Darussalam, Banda Aceh 23111, Indonesia

[5] Computational and Applied Statistics Research Group, Department of Statistics, Faculty of Mathematics and Natural Sciences, Universitas Syiah Kuala, Kopelma Darussalam, Banda Aceh 23111, Indonesia

E-mail: irvanizam.zamanhuri@unsyiah.ac.id

Abstract. Scheduling algorithm is one of computation problem with high complexity. One of the problems is a difficulty to determine an optimal scheduling alternative that is adjusted with the constraints needed. This study aims to test whether the Genetic Algorithm (GA) program that was built was successful for maximizing the value of fitness. The data used in this study is data simulation from the GA Program. Based on the result of this study, we can concluded that the GA program can be used for scheduling because it has good accuracy. This can be seen from the fitness value of 1 and there is no clash between one schedule and another schedule and there are no restrictions that are violated.

1. Introduction

Scheduling algorithm is one of computation problem with high complexity [1]. One of the problems is a difficulty to determine an optimal scheduling alternative that is adjusted with the constraints needed, resulting in almost impossible task to solve with normal mathematics calculation.

Scheduling process needs a wide searching space and some parameter with various condition [1]. Multiple kinds of condition during the process result in calculation and determination difficulty. This is

[3] *Coressponding author : irvanizam.zamanhuri@unsyiah.ac.id

Content from this work may be used under the terms of the Creative Commons Attribution 3.0 licence. Any further distribution of this work must maintain attribution to the author(s) and the title of the work, journal citation and DOI.
Published under licence by IOP Publishing Ltd

due to the conditional constraints with high variety and complexity [2]. For example, in class scheduling process, there are instructors who are only available to teach at certain days and times, followed by minimal number and capacity of the classroom. These conditions make scheduling process becoming ineffective in providing optimal scheduling alternatives.

One of the solutions to determine the suitable scheduling is by exploiting the advantages of Genetic Algorithm (GA). As one of its the advantages, GA is widely used for the best random searching by trying some random solutions while recording the best solution [3]. GA can also be used to search non-linear big solution space that is hardly to be encoded [4].

This algorithm is widely used by the researcher to solve the scheduling process. Sapru et. al. used GA to determine the time table scheduling for employing guided mutation. They exploited selection scheme based on ranks to ensure that the produced timetable schedule is optimal and proper compared to stagnant solution configuration related to the roulette selection scheme [4]. Reddy et. al. used GA to determine fast, suitable, and easy to use train time scheduling [3]. Kristiadi & Hartanto [5] used GA to determine lecturing scheduling optimization. They built a library with GA concept, arranged with several operators such as Tournament Selection, Uniform Crossover, Weak Parent Replacement, and two mutation method which are Interchanging Mutation dan Violated Directed Mutation. The library is designed to be able to define custom constraint or scheduling conditions which have not been accommodated in the library's core modification program [5].

In university, schedule in research guidance is generally about adjusting lecture and the students, as well as placing them at the specific time in the class. But there are several problems such as different schedules of the adviser (the lecturer) and the student, as well as the number of class and overlapping schedules.

In this study, research guidance schedule problems in the university will be solved. As mentioned in the second paragraph, the most effective approach to solve schedule problems is by using GA. GA is the most popular algorithm and mostly used to solve schedule problems. The programming language used to solve this problem is Perl.

2. Literature Review

2.1 Genetic Algorithm

GA is a powerful optimizing and heuristic algorithm with the principles of evolution as its basis [6]. Importantly, GA is based on Darwin's theory on Survival of The Fittes [7,8]. Processes in GA include selection, crossover, mutation, and elitism repeated until the conditions are met [5]. GA can be done with complete operation process or by combining it with several computation methods [9,10]. Using GA in scheduling can optimize the results effectively [2,6,9,11]. GA has successfully used in several cases of research guidance schedule. Corne et al. used conventional method where each gene represents time where particular guidance happened with crossover and mutation operator. This system has been used in the University of Edinburgh with enhanced time-saving. While Paechter took different methods where the genes for each guidance do not only determine when the time it should be used but also how to search a new period if, after crossover, the schedule conflict occurs. If the research guidance could not be included in one period, it would be left as unscheduled, this is different from previous system which enables unfit schedule [6]. GA flowchart in this study can be seen in Figure 1.

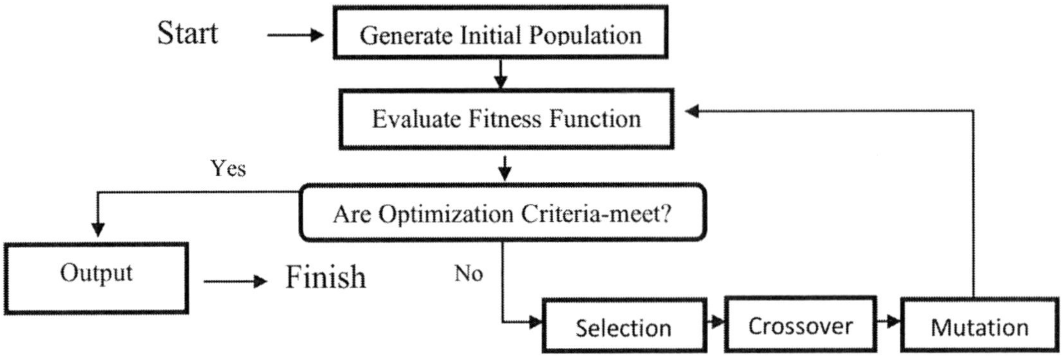

Figure 1. Genetic algorithm stages.

2.2 Problem Definition
The definitions in this study include :

D = $d_1, d_2, d_3, d_4, ..., d_n$

H = $h_1, h_2, h_3, h_4, ..., h_n$

R = $r_1, r_2, r_3, ..., r_n$

J = $j_1, j_2, j_3, j_4, ..., j_n$

D represents *Lecturer* while $d_1, d_2, ..., d_n$ represents the name of **D**. **T** represents *Day* while $t_1, t_2, ..., t_n$ represents the name of *Days*. **R** represents *Room* while $r_1, r_2, ..., r_n$ represents the name of the *Room*. And J represents *Time* while $j_1, j_2, ..., j_n$ represents the name of *Time*. In this study, there are 20 *Lecturers*, 5 *Days*, 1 *Room*, dan 4 *Times* used, and can be seen in Table 1-3

Table 1. Schedule Table.

Schedule	Day	Time	Room
J1	Monday	T1	R1
J2	Monday	T2	R1
J3	Monday	T3	R1
J4	Monday	T4	R1
J5	Monday	T1	R1
J6	Tuesday	T2	R1
J7	Monday	T3	R1
J8	Monday	T4	R1
J9	Wednesday	T1	R1
J10	Wednesday	T2	R1
J11	Wednesday	T3	R1
J12	Wednesday	T4	R1
J13	Thursday	T1	R1

J14	Thursday	T2	R1
J15	Thursday	T3	R1
J16	Thursday	T4	R1
J17	Friday	T1	R1
J18	Friday	T2	R1
J19	Friday	T3	R1
J20	Friday	T4	R1

Table 2. Time Table.

Code	Time
T1	08.00-10.00
T2	10.00-12.00
T3	14.00-16.00
T4	16.00-18.00

Table 3. Lecturers Table.

No	Lecturer
1	D1
2	D2
3	D3
4	D4
5	D5
6	D6
7	D7
8	D8
9	D9
10	D10
11	D11
12	D12
13	D13
14	D14
15	D15
16	D16
17	D17
18	D18
19	D19
20	D20

2.3 Constraints

In this study, There are several rules required for the schedule determination so that the fitness function in GA can run properly, they are:

1. Only one room used for guidance
2. There is not any overlapping schedule among lectures
3. One lecturer for Each student

3. Methods

3.1 Data

To see if GA successfully runs, this study uses the generated data from the authors' generated results which include 1 Room, 5 Days, 4 Times, dan 20 Lecturers.

3.2 Genetic Algorithm

For scheduling problems in this study, the supervisor/promotor/lecturer is determined by the head of the department and written in the guidance list slot, in order to make schedule table filled. If there is not an optimal solution, then GA algorithm will be used to provide an optimal schedule. Discussion about scheme encoding and GA operator used in this section will be discussed in this subsection

3.2.1 Chromosome Structure

Chromosome representation used is bit string. All information that might present should be inserted into the chromosome. The forming chromosomes genes are consisted of four components, which are lecturers **D**, Days **H**, Room **R**, dan Time **J**. Table 1 shows 20 possible combinations, therefore in this study a chromosome consists of 20 genes that describe occurred scheduling condition. The chromosomes structures used in this study is [9]:

$$f : D \times H \times R \times J \ \{0,1\} \tag{1}$$

Chromosome representation used is bit string. In Equation (1) the value of $f(d, h, r, j)$ are 1 if the lecturer, day, room, and time are available, and will be 0 if unavailable

3.2.2 Fitness Value

Fitness function used in this study is optimized fitness function (highest fitness value) so that the will be no overlap among schedule, which is [9] :

$$f = \frac{1}{1+\alpha} \tag{2}$$

Based on Equation (2) where $\alpha = D + H + R + J$. D is lecturer, where it will give 0 if and only if d lecturer presents on the meeting date t in room r and at meeting time j. In contrary, it will give 1 if the overlapping happens for meeting date t, room r, meeting time j, or the overlapping between the lecturers. These conditions are applied for other criteria.

4. Result and Discussion

The results of this study were to test whether the GA program that was built was successful for maximizing the value of fitness. The first step was to change the representation data from the chromosome into a bit string.

The representation of the D value in binary encoding for Lecture was:

$$D_1 = 00000, D_2 = 00001, D_3 = 00010, D_4 = 00011, ... \quad (3)$$

The representation of the H value in binary encoding for Days was:

$$H_1(Monday) = 000, H_2(Tuesday) = 001, H_3(Wednesday) = 010, H_4(Thursday) = 100, H_5(Friday) = 111 \quad (4)$$

The representation of the J value in binary encoding for Times was:

$$T_1 = 0000, T_2 = 0010, T_3 = 0100, T_4 = 1000 \quad (5)$$

And the representasi of the J value in binary encoding for Room was:

$$R_1 = 00 \quad (6)$$

Based on equation 3-6, the chromosome sample that will be formed were:
$\{D_1, H_2, T_3, R_1\}$

D_1 = 00000
H_2 = 001
T_3 = 0100
R_1 =00

Chromosome = 00000001010000 → The total of gens in the *chromosome* was 14 gens

Table 4. The result of scheduling consultation.

Time	Consultation {Lecturers, Days, Times, Room}	Fitness
1	([D1,H1,T2,R1][D2,H1,T1,R1][D3,H1,T3,R1][D4,H2,T3,R1][D5,H1,T4,R1][D6,H2,T2,R1][D7,H2,T1,R1][D8,H3,T1,R1][D9,H3,T2,R1][D10,H4,T1,R1][D11,H2,T4,R1][D12,H4,T2,R1][D13,H3,T4,R1][D14,H3,T3,R1][D15,H5,T3,R1][D16,H5,T1,R1][D17,H5,T2,R1][D18,H5,T4,R1])	1.000
2	([D1,H1,T2,R1][D2,H1,T2,R1][D3,H1,T3,R1][D4,H2,T3,R1][D5,H1,T4,R1][D6,H2,T2,R1][D7,H2,T1,R1][D8,H3,T1,R1][D9,H3,T2,R1][D10,H4,T1,R1][D11,H2,T4,R1][D12,H4,T2,R1][D13,H3,T4,R1][D14,H3,T3,R1][D15,H5,T1,R1][D16,H5,T1,R1][D17,H5,T2,R1][D18,H5,T4,R1])	0.5000
3	([D1,H1,T1,R1][D2,H1,T1,R1][D3,H1,T3,R1][D4,H	0.3333

	1,T3,R1][D5,H1,T3,R1][D6,H2,T4,R1][D7,H2,T1,R1][D8,H3,T1,R1][D9,H3,T2,R1][D10,H4,T1,R1][D11,H4,T1,R1][D12,H4,T2,R1][D13,H3,T4,R1][D14,H3,T3,R1][D15,H5,T1,R1][D16,H5,T1,R1][D17,H5,T2,R1][D18,H5,T4,R1])	
4	([D1,H1,T1,R1][D2,H1,T1,R1][D3,H1,T3,R1][D4,H1,T3,R1][D5,H2,T3,R1][D6,H2,T4,R1][D7,H2,T1,R1][D8,H3,T1,R1][D9,H3,T2,R1][D10,H4,T1,R1][D11,H4,T1,R1][D12,H4,T2,R1][D13,H3,T4,R1][D14,H3,T3,R1][D15,H5,T1,R1][D16,H5,T1,R1][D17,H5,T2,R1][D18,H5,T2,R1])	0.2500
5	([D1,H1,T1,R1][D2,H1,T1,R1][D3,H1,T3,R1][D4,H1,T3,R1][D5,H2,T3,R1][D6,H2,T4,R1][D7,H2,T1,R1][D8,H3,T1,R1][D9,H3,T2,R1][D10,H4,T1,R1][D11,H4,T1,R1][D12,H4,T2,R1][D13,H3,T4,R1][D14,H3,T3,R1][D15,H5,T1,R1][D16,H5,T1,R1][D17,H5,T2,R1][D18,H5,T2,R1])	0.2500
6	([D1,H1,T1,R1][D2,H1,T1,R1][D3,H1,T3,R1][D4,H1,T3,R1][D5,H2,T3,R1][D6,H2,T4,R1][D7,H2,T1,R1][D8,H3,T1,R1][D9,H3,T2,R1][D10,H4,T1,R1][D11,H4,T1,R1][D12,H4,T2,R1][D13,H3,T4,R1][D14,H3,T3,R1][D15,H5,T1,R1][D16,H5,T1,R1][D17,H5,T2,R1][D18,H5,T2,R1])	0.2500
7	([D1,H1,T1,R1][D2,H1,T1,R1][D3,H1,T3,R1][D4,H1,T3,R1][D5,H2,T3,R1][D6,H2,T4,R1][D7,H2,T1,R1][D8,H3,T1,R1][D9,H3,T2,R1][D10,H4,T1,R1][D11,H4,T1,R1][D12,H4,T2,R1][D13,H3,T4,R1][D14,H3,T3,R1][D15,H5,T1,R1][D16,H5,T1,R1][D17,H5,T2,R1][D18,H5,T2,R1])	0.2500
8	([D1,H1,T1,R1][D2,H1,T1,R1][D3,H1,T3,R1][D4,H1,T3,R1][D5,H2,T3,R1][D6,H2,T4,R1][D7,H2,T1,R1][D8,H3,T1,R1][D9,H3,T2,R1][D10,H4,T1,R1][D11,H4,T1,R1][D12,H4,T2,R1][D13,H3,T4,R1][D14,H3,T3,R1][D15,H5,T1,R1][D16,H5,T1,R1][D17,H5,T2,R1][D18,H5,T2,R1])	0.2500
9	([D1,H1,T1,R1][D2,H1,T1,R1][D3,H1,T3,R1][D4,H1,T3,R1][D5,H2,T3,R1][D6,H2,T4,R1][D7,H2,T1,R1][D8,H3,T1,R1][D9,H3,T2,R1][D10,H4,T1,R1][D11,H4,T1,R1][D12,H4,T2,R1][D13,H3,T4,R1][D14,H3,T3,R1][D15,H5,T1,R1][D16,H5,T1,R1][D17,H5,T2,R1][D18,H5,T2,R1])	0.2500
10	([D1,H1,T1,R1][D2,H1,T1,R1][D3,H1,T3,R1][D4,H1,T3,R1][D5,H2,T3,R1][D6,H2,T4,R1][D7,H2,T1,R	0.2500

1][D8,H3,T1,R1][D9,H3,T2,R1][D10,H4,T1,R1][D1 1,H4,T1,R1][D12,H4,T2,R1][D13,H3,T4,R1][D14,H 3,T3,R1][D15,H5,T1,R1][D16,H5,T1,R1][D17,H5,T 2,R1][D18,H5,T2,R1])

Table 4 was the result of the GA program which is based on the best fitness value. The result of GA programs showed that the best fitness value is 1,000. The number of iterations given was 10,000, but the iteration process in this study stopped at the 51st iteration with the total of chromosomes was ten chromosomes. The results of this study indicated that the combination of D×H×J×R would produce a schedule that does not conflict with other lecture schedules.

5. Conclusion

Based on the result of this study, we can concluded that the GA program can be used for scheduling because it has good accuracy. This can be seen from the fitness value of 1 and there is no clash between one schedule and another schedule and there are no restrictions that are violated.

Reference

[1] Abbaszadeh M and Saeedvand S 2014 A Fast Genetic Algorithm for Solving University Scheduling Problem *IAES Int. J. Artif. Intell.* **3** 7
[2] Hou M and Chen Q 2010 Course scheduling system design and implementation based on genetic algorithm. In: *2010 International Conference On Computer Design and Applications,* pp V3-611-V3-4
[3] Sai Satyanarayana Reddy S, Prasada Reddy G S, Hemanth P V. and Chatterjee P 2017 Train time scheduling using genetic algorithm *Int. J. Civ. Eng. Technol.* **8** 410–3
[4] Sapru V, Reddy K and Sivaselvan B 2010 Time table scheduling using Genetic Algorithms employing guided mutation. In: *2010 IEEE International Conference on Computational Intelligence and Computing Research,* pp 1-4
[5] Kristiadi D and Hartanto R 2019 Genetic Algorithm for lecturing schedule optimization *IJCCS (Indonesian J. Comput. Cybern. Syst.* **13** 83
[6] Burke E, Elliman D and Weare R 1994 A Genetic Algorithm Based University Timetabling System *Interface* 35–40
[7] Abdullah S and Turabieh H 2008 Generating university course timetable using Genetic Algorithms and local search *Proc. - 3rd Int. Conf. Converg. Hybrid Inf. Technol. ICCIT 2008* **1** 254–60
[8] Wang Y Z 2003 Using genetic algorithm methods to solve course scheduling problems *Expert Syst. Appl.* **25** 39–50
[9] Ahmad I R, Sufahani S, Ali M and Razali S N A M 2018 A Heuristics Approach for Classroom Scheduling Using Genetic Algorithm Technique *J. Phys. Conf. Ser.* **995** 10–5
[10] Kumar S V S, Karthik J and Srinivas Rao D 2018 A computational heuristics approach for classroom scheduling using genetic algorithm technique *Int. J. Pure Appl. Math.* **119** 167–71
[11] Limanto S, Benarkah N and Adelia T 2019 Thesis examination timetabling using genetic algorithm *Int. Electron. Symp. Knowl. Creat. Intell. Comput. IES-KCIC 2018 - Proc.* 6–10

IOP Conf. Series: Materials Science and Engineering **796** (2020) 012034 doi:10.1088/1757-899X/796/1/012034

Properties of blended cement paste with diatomite from Aceh Province Indonesia

M Hasan[1]*, T Saidi[1]
[1]Dept. of Civil Engineering, Universitas Syiah Kuala, Banda Aceh, Indonesia

E-mail: muttaqin@unsyiah.ac.id

Abstract. This paper presents the properties of blended cement paste, i.e. water demand, setting time, flow, flexural strength and compressive strength with diatomite from Aceh Besar District, Aceh Province, Indonesia. Diatomite was calcined at the temperature of 500 °C for 4 hours, ground and sieved with sieve #200. The diatomite content used in this study was 0%, 10%, 20% and 30% of blended cement weight. The diatomite powder, cement, and water were mixed in a mixer to produce blended cement paste. EN and ASTM Standards were used for the test procedures. Water demand for normal consistency, setting time and flow was tested on freshly blended cement paste. Flexural and compression tests were performed on the 40 mm x 40 mm x 160 mm prism and 50 mm size cube specimens at the specimen age of 28 days. The test results show that the water demand increases with increasing the diatomite content. Initial setting time, flow and flexural strength decrease with increasing the diatomite contents. The use of 10 % and 20 % diatomite does not significantly alter the compressive strength.

1. Introduction

Indonesia is a country rich in mineral and energy resources, one of which is diatomite. Based on the data from the Aceh Province Mining and Energy Office the deposit of diatomite in Aceh Besar District, Aceh Province, Indonesia was quite high with an estimated 40,353,700 tons. Chemically, the main composition of diatomite is amorphous silica which reaches about 55-70%, depending on the local environment. The contents of silica compounds in diatomite vary widely, as well as the structure and are very much influenced by its origin [1]. The utilization of silica-rich material for partially cement replacement has been studied [2-3].

There have been some studies on using diatomite to produce blended cement. The water demand for blended cement with diatomite is higher than that of ordinary Portland cement [4-6]. Using calcareous diatomaceous rock from Zakinthos Island, the Ionian Sea, which contained mainly $CaCO_3$ and amorphous silica of biogenic origin with the form of opal-A in cement blended up to 35 % reduces the 28 days compressive strength by 24.14 %. However, blended cement containing 10 % of the diatomite develope the same compressive strength as corresponding Portland cement [4]. Diatomite rocks could be used for the production of blended cement with high strength. The higher the reactive silica content of diatomite, the higher the strength of blended cement. However, the type and morphology of the amorphous silica (diatome frustules) and the presence of friable, fine-grained biogenic calcite are additional parameters that need to be taken into consideration since they might also affect the strength of the resulting cement [5]. Blended diatomite cement with Blaine exceeding

[1] Corresponding author: muttaqin@unsyiah.ac.id

Content from this work may be used under the terms of the Creative Commons Attribution 3.0 licence. Any further distribution of this work must maintain attribution to the author(s) and the title of the work, journal citation and DOI.
Published under licence by IOP Publishing Ltd

5000 cm^2/gr, the compressive strength is higher than that ordinary Portland cement [6]. Using diatomite as cement replacement in the concrete mixture can increase the resistance of concrete to chloride and sulfate [7,8].

The aim of this study is to evaluate the water demand, setting time, flow, flexural strength and compressive strength of blended cement paste with diatomite originating from Aceh Besar District, Aceh Province, Indonesia.

2. Materials and Methods

The materials used in this research were Ordinary Portland Cement (OPC) and diatomite from Aceh Province, Indonesia. Diatomite was crushed and calcined at a temperature of 500 °C for 4 hours. The calcined diatomite was then ground and sieved with sieve #200 to have diatomite powder. The density and absorption of diatomite powder used in this study are 767 kg/m^3 and 6.54 %. The chemical composition of the diatomite powder is 58,87 % SiO_2; 12,2 % CaO; 2,25 % Fe_2O_3 and 0,39 % Al_2O_3. The diatomite was blended with OPC with diatomite content (w/w) of 0 %, 10 %, 20 % and 30 %.

The water demand and setting time – initial and final – of blended cement paste were determined according to EN 196-3 and ASTM C191-01a [9,10]. The cement and diatomite powder was mixed in the mixer. In order to determine the water demand, the water was added little by little while the mixer continues to rotate until normal consistency measured by the Vicat test was reached. The initial and final setting time was measured on the normal consistency blended cement paste using the Vicat test.

The flow of normal mortar was measured according to ASTM C1437-07 and ASTM C230/C230M-08 [11,12] using the standard flow table. The standard cone was put on the center of the table and the freshly blended cement paste was filled in the cone about 25 mm in thickness and tamped 25 times with the tamper. Then the second layer of blended cement paste was filled in the cone and again tamped 25 times until the cone was full. Later the cone was lifted away from the blended cement paste 1 minute after completing the mixing operating and immediately the table was dropped 25 times in 15 seconds. The flow of the mixture was measured in two perpendicular diameters.

The flexural and compressive strength measurement was conducted according to EN 196-1 and ASTM C109/C109M-02 [13,14]. The prims specimens with 40 x 40 x 160 mm^3 size (EN 196-1) and cube specimens with 50 x 50 x 50 mm^3 size (ASTM C109/C109M-02) were prepared by casting the blended cement paste in the mold. The mold was removed from the specimens at the specimen age of 24 hours. The specimens were then cured in the water until the age of the specimens of 28 days. The flexural test was conducted on the prism specimens. The specimens were supported by two simple supports and the one-point load was applied on the specimens until the specimens get failure as shown in Figure 1 [13]. The compression test was carried out on the remaining half of the flexural test specimens. Two steel plate with 40 mm x 40 mm size was placed on the bottom and top surface of the specimens and the compression load was applied on the specimens through those plates as shown in Figure 2 [13]. In addition, the compression test was also carried out on 50 x 50 x 50 mm^3 cube specimens [12]. The specimens were put on the universal testing machine and the compression load was applied until the specimens get failure.

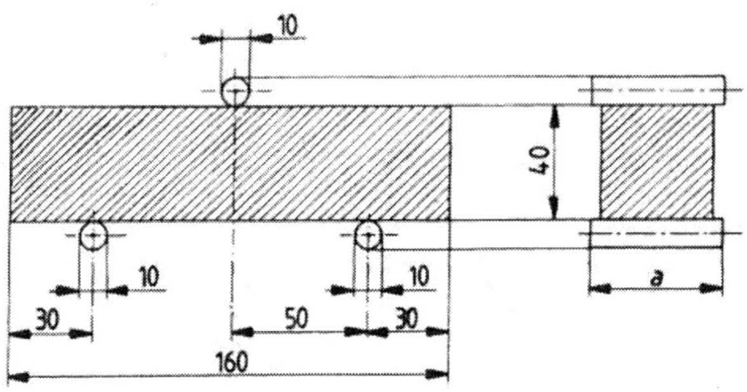

Figure 1. Flexural test on prism specimen

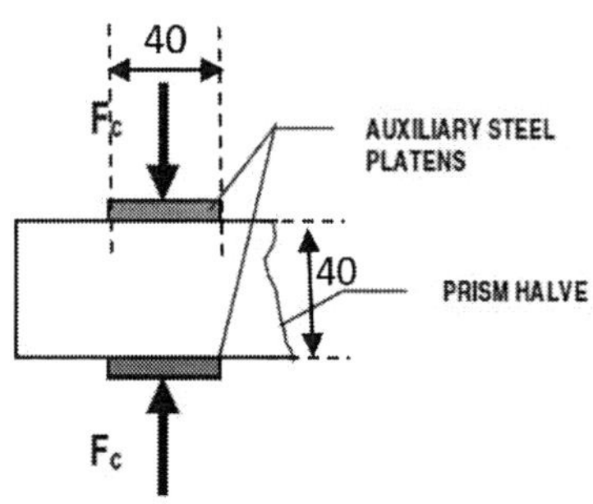

Figure 2. Compression test

3. Results

Figure 3 shows the water demand for blended cement paste as a function of diatomite content. The water demand is the quantity of water that is required in order to prepare a cement paste of normal consistency as specified in EN 196-3 [9]. From Figure 3, it can be seen that by increasing the diatomite content, the water demand also increases. The increasing water demand for blended cement paste is due to that the diatomite is an absorbent material and the finest grain size of diatomite. With the absorption of diatomite powder of 6.54 %, some mixing water is absorbed into diatomite powder while the water needs for the hydration process remain constant, hence the water demand for blended cement increases as increasing the diatomite powder content. By using 30 % diatomite, the water demand increases by 31.67 %. Same with the finding in this study, by using calcareous diatomite from Zakinthos Island, Ionian Sea, Greece, Kastis et al [4] reported that the water demand for blended cement is also higher than that for ordinary Portland cement. Water demand for blended cement can be expressed as follows:

$$a = a_o \exp(0{,}009\, d) \tag{1}$$

where: a = water demand for blended cement paste (% w/w), a_o = water demand for ordinary Portland cement paste (% w/w), d = diatomite content (% w/w).

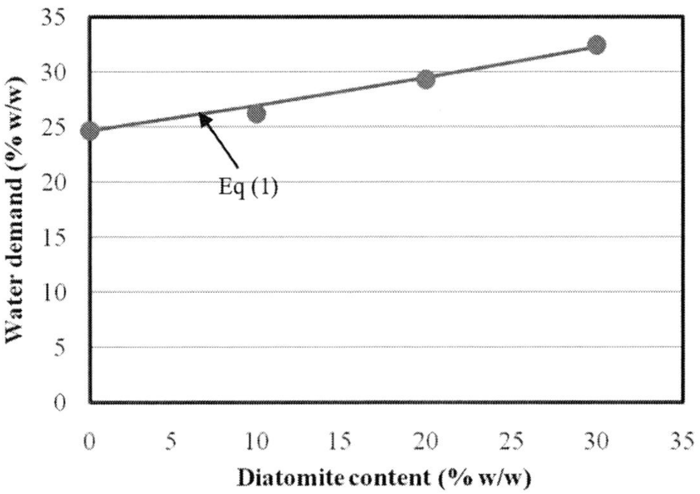

Figure 3. Water demand for blended cement

Figure 4 shows the initial and final setting time of blended cement paste. Initial setting time decreases with increasing the diatomite content. The final setting time of blended cement with 20 % and 30 % of diatomite content is almost the same as that of ordinary Portland cement. Blended cement with 10 % diatomite content has a lower final setting time. The flow of normal mortar of blended cement is shown in Figure 5. The flow of normal mortar decreases with the increase of diatomite content. This result confirms that the diatomite powder in blended cement absorbs some water into its particles.

Flexural strength of blended cement paste is shown in Figure 6. Blended cement paste has lower flexural strength than ordinary Portland cement paste. The lower flexural strength of blended diatomite cement paste leads to concrete structures composed of this type of cement that will experience crack at lower load [16].

The compressive strength of blended cement paste is shown in Figure 7. The blended cement paste has good compressive strength. Compare to ordinary Portland cement, the compressive strength of blended cement with 30 % diatomite reduces for only 9,26 %. This reduction of compressive strength of blended cement paste is related to the late reaction of SiO_2 in diatomite powder with $Ca(OH)_2$ resulted in cement hydration. This reaction to produce calcium silicate hydrates (CSH) usually occurs progressively after the age of cement paste of 56 days. More SiO_2 will react with $Ca(OH)_2$ after the cement paste age of 90 days makes blended cement paste denser, hence increases the compressive strength. Therefore, to obtain the contribution of SiO_2 in diatomite powder on the compressive strength of blended cement paste, it is necessary to test the compressive strength of blended cement paste at the age of 90 days. The study on this issue will be conducted in the next stage. The use of 10 % and 20 % diatomite does not significantly alter the compressive strength. The ratio between the compressive strength of blended cement paste to that of ordinary Portland cement is shown in Figure 8.

Using the formula proposed by Kastis et al [4], the relative compressive strength (RS) is the corrected value of blended diatomite cement paste compressive strength taking into account the pure cement content can be calculated. The relationship between the relative compressive strength of blended cement paste and the diatomite content is shown in Figure 9. Relative compressive strength indicates the effect of the addition of diatomite on the cement paste hydration process. From Figure 9, it can be seen that the relative compressive strength increases with increasing the diatomite content. This indicates that the addition of diatomite powder favors hydration and improves the development of the compressive strength of cement paste. This is attributed to the pozzolanic reaction of diatomite powder. With lower content of ordinary Portland cement in blended cement leads to less available $Ca(OH)_2$ in blended cement paste.

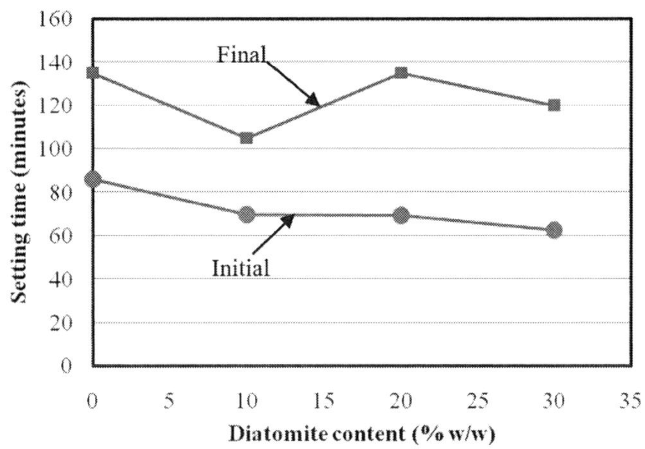

Figure 4. Setting time of blended cement

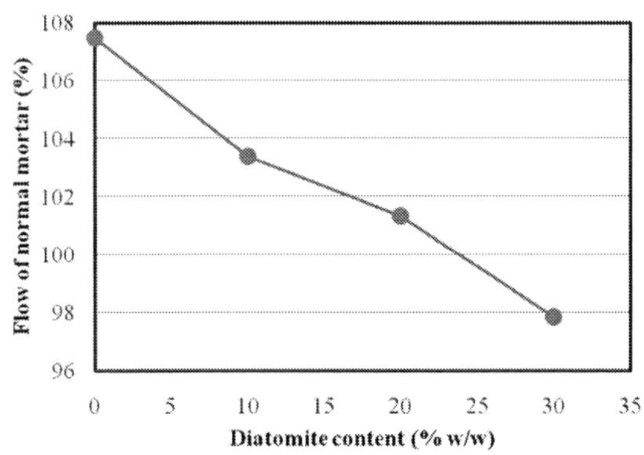

Figure 5. The flow of normal mortar

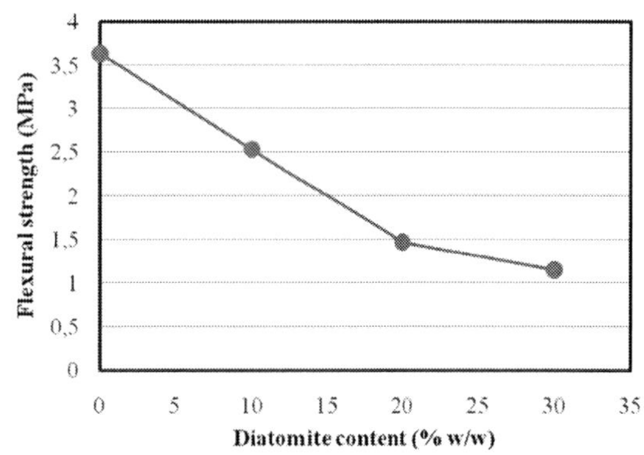

Figure 6. Flexural strength of blended cement paste

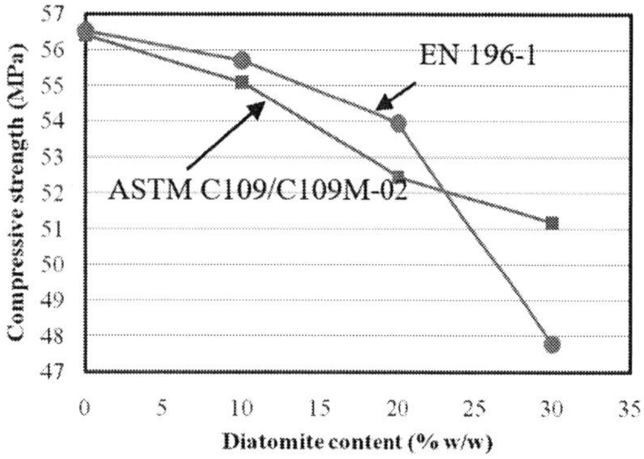

Figure 7. Compressive strength of blended cement paste

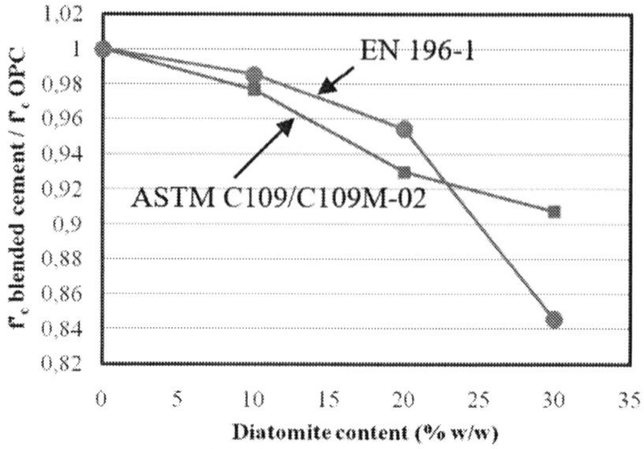

Figure 8. Reduction of compressive strength of blended cement paste

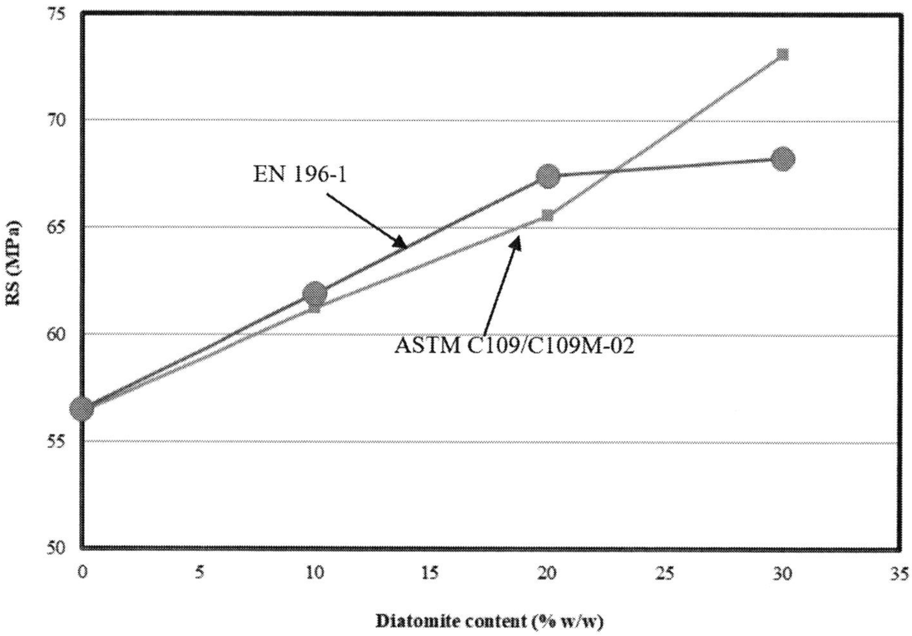

Figure 9. Relationship between relative compressive strength of blended cement paste and diatomite content

4. Conclusions

The addition of diatomite originating from Aceh Province, Indonesia to Portland cement results in a drastic increase in water demand. The initial setting time of blended cement increases as increasing the diatomite content. The flexural strength of blended cement decreases significantly with increasing of diatomite content. Blended cement having up 20 % diatomite content develop the almost same compressive strength as the ordinary Portland cement.

5. Acknowledgment

The authors wish to acknowledge the Directorate of Research and Community Service, the Ministry of Research, Technology and Higher Education of the Republic of Indonesia and LPPM of Syiah Kuala

University for providing research funding (Contract No. 12/UN11.2/PP/SP3/2019). The authors also thank Miss Sri Hayati, Mr. Mahlil and Mr. Razali for their assistance in experimental work.

6. References

[1] Khan S U 1980 *Pesticides in the Soil Environment,* Elsevier Scientific Publishing Co
[2] Aulia T B, Muttaqin, Afifuddin M, Zaki M, Merriza S 2018 *IOP Conf. Series: Mater. Sci. and Eng.* **434** 012199.
[3] Hamzani, Munirwansyah, Hasan M, Sugiarto S 2019 *IOP Conf. Series: Mater. Sci. and Eng.* **523** 012034
[4] Kastis D, Kakali G, Tsivilis S and Stamatakis M G 2006 *Cem. Concr. Res.* **36** 1821-1826
[5] Stamatakis M G, Fragoulis D, Csirik G, Bedelean I, Pedersen S 2003 *Cem. Concr. Compos.* **25** 177-184
[6] Fragoulis D, Stamatakis M G, Papageorgiou D, Chaniotakis E 2005 *Cem. Concr. Compos.* **27** 205-209
[7] Tagnit-Hamou A, Petrov N, Luke K 2003 *ACI Mater. J.* **100** 73-78
[8] Hasan M, Husin S, Nursaniah C 2016 *Key Engineering Materials* **711** 302-309
[9] EN 196-3:1995. Methods of Testing Cement – Part 1: Determination of Setting Time and Soundness.
[10] ASTM C191-01a, Standard Test Method for time of Setting of Hydraulic Cement by Vicat Needle
[11] ASTM C1473-07. Standard Test Method for Flow of Hydraulic Cement Mortar.
[12] ASTM C230/C230M-08. Standard Specification for Flow Table for Use in Tests of Hydraulic Cement
[13] EN 196-1:1995. Methods of Testing Cement – Part 1: Determination of Strength
[14] ASTM C109/C109M-02, Standard Test Method for Compressive Strength of Hydraulic Cement Mortars (Using 2-in. Or 50 mm Cube Specimens)
[15] Hasan M, Husaini, Abdullah N M 2019 *IOP Conf. Series: Mater. Sci. and Eng.* **523** 012037

IOP Conf. Series: Materials Science and Engineering **796** (2020) 012035 doi:10.1088/1757-899X/796/1/012035

Rapid material estimation for reinforced concrete construction utilizing a parametric tool for Aceh's post-disaster case

Atthaillah[1*], E Saputra[1], B Karsono[1], E Fahrizal[1], and C A Fithri[1]
[1]Department of Architecture Malikussaleh University, Jalan Sultanah Nahrasiyah, Lhokseumawe, Indonesia 24351

E-mail: atthaillah@unimal.ac.id

Abstract. This paper offered a new open-source parametric tool for reinforced-concrete construction estimation for post-disaster based on Aceh's case. The tool aimed at introducing and providing an innovative solution of estimation for building materials more dynamically and rapidly. The tool was initiated by the fact that the delay of relief shelter deployment in 2004 post-tsunami in Aceh of Indonesia was caused by the inability to provide a rapid calculation for building materials as a result of a massive request [1]. Further, the reinforced-concrete constructions had become a positive preference by the Acehnese [2]. In order to offer a new tool for rapid and dynamic estimation specifically for reinforced-concrete construction, the authors utilized visual scripting of Grasshopper with GHPython extension for optimization of the proposed tool. The tool introduced in this paper was the improvement of previously explored workflows by the authors and ended up with an innovative tool for modeling, visualizing, and estimating. Furthermore, the tool enhanced users' experiences to construct a workflow based on users' specific logic. The tool allowed alterations in the design without comprising time for estimation. In this way, the specific need of the victims in post-disaster was accommodated without obstruction to an estimation of building materials, thus, improved quality and management of the relief shelters for a post-disaster.

***Keywords**: Reinforced-concrete Construction, Relief Shelter, Parametric Estimation, Aceh Post-Disaster, GHPython, Grasshopper, Visual Scripting.*

1. Introduction

Aceh, an Indonesian province, was the second-worst affected by natural disasters. It was recorded more than 5 million people affected by the disaster from 2000 to 2017 [3]. The immediate need for the aftereffect of a disaster was the availability of shelter for the victims. a shelter was a very important entity that must be rapidly fulfilled for a disaster besides foods, water, and medicines. A good shelter ensured the victims appropriate safety and health. Therefore, the immediate availability of the shelter was a significant aspect in a post-disaster setting such as in Aceh.

Reinforced concrete construction had become a positive preference for post-disaster relief shelters in Aceh. This was confirmed by the author's survey conducted in 2017 for the type of relief shelters available for post-disaster in Banda Aceh, Pidie Jaya, North Aceh, and East Aceh. The main reason for the preference was its durability. Moreover, the role of government regulation, instructed by the Bureau of Reconstruction and Rehabilitation Aceh-Nias, was an important aspect in shaping reinforced-concrete construction for the shelter preferences [2]. For the 2004 post-tsunami, the type of construction with bricks infill as its wall was widely available after ten months of the disaster [1]. It was not an appropriate timeframe for relief shelters. One of the problems that liable for the delay of the relief shelter was the ability to quantify construction materials as the result of a massive request of

Content from this work may be used under the terms of the Creative Commons Attribution 3.0 licence. Any further distribution of this work must maintain attribution to the author(s) and the title of the work, journal citation and DOI.
Published under licence by IOP Publishing Ltd

the shelter [1]. The positive preference toward the reinforced-concrete shelter, in a post-disaster setting, was not supported by the ability for a rapid material estimation to fulfill the shelter demands.

Building material estimation was a crucial aspect for the success of construction, particularly, for an emergency case such as a post-disaster situation where rapidity was a requirement. Şener & Torus [4] advised rapidity as one important factor for post-disaster relief shelter. A parametric framework that puts attention to building estimation was showed in the work of Yeung & Harkins for Solomon Island's post-earthquake disaster [5]. However, they focused on timber-structured buildings. Furthermore, it was necessary to invent a better way in order to upgrade the current building estimation culture. Besides, this paper was highly related to the author's previous works [6] and [7] since it had significantly contributed and directed this study as discussed in this paper. In this study, we developed the previously explored workflows into a parametric tool.

This tool was aimed at the optimization of algorithms and allowed a more dynamic framework to be constructed. The tool allowed for future development in a more organized and simple way. Also, we extended the ability of our previous workflows, therefore, it was able to handle rather complex construction. These features were missing in our previous papers. In addition, we utilized GHPython to create a parametric tool for dynamic and rapid material estimation for reinforced-concrete shelters to suit a post-disaster condition such as in Aceh's post-disaster relief shelter. Moreover, the tool that we proposed was adaptable to every reinforced-concrete structure in the Indonesia-wide construction industry.

2. Methods

Our parametric framework was under the Grasshopper platform. Grasshopper defined as a visual node-based add-ons editor for Rhinoceros that was invented by David Rutten in 2007 initially known as explicit history [8]. Further, it visual-based scripting made it easy to be comprehended by people with no prior programming skills, particularly, for architect alike. Moreover, GHPython was utilized to develop the estimation tool, thus, it extended the Grasshopper components' capabilities. The improved capabilities were related to the ability of the component to model, organize the data and calculate the reinforced-concrete construction material. GHPython defined as an additional Grasshopper component that performed as an IronPython translator [9]. It allowed access for Rhinoceros or Grasshopper functionalities within GHPython component. In addition, we utilized the Grasshopper list component for calculation coefficients and yKTools for data streaming into a spreadsheet in our workflows.

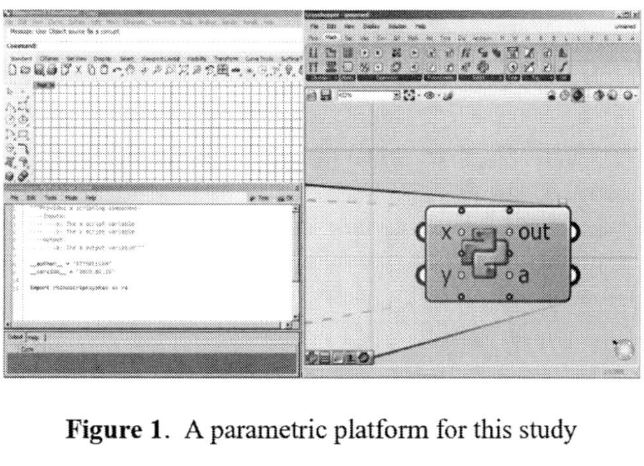

Figure 1. A parametric platform for this study

Figure 1 illustrated the parametric platform for this study. Rhinoceros on the upper left as the main platform for the entire system. Grasshopper on the left as an environment for GHPython. It showed the GHPython component on Grasshopper canvas with its editor was activated as revealed on the bottom left. In addition, Figure 1 illustrated that the GHPython components consisted of amendable input and output parameters. In this case, the authors put the description at the beginning of the GHPython script. All the descriptions for and inside the code were written in Indonesian. Both inputs and outputs took and produced objects, numbers or texts. It depended on the intended codes inside the GHPython editor. Further, both input and output descriptions were amendable for the clarity of either the input or the output. Besides, the icon for each new component was modifiable. In this way, components for our tool were created. Appreciatively, for the structure of the scripting inside GHPython was inspired by the Ladybug Tools [10].

The intended codes inside GHPyhton editor for reinforced-concrete construction estimation were written based on Indonesian Standards for reinforced-concrete estimation. Those standards were [11] for concrete, [12] for the wall, [13] for rendering and [14] for foundation works. Steel reinforcement estimation based on the formula (1) explained in [7]. Meanwhile, the roof coefficients were based on the rule of thumbs calculation for metal roofing in Indonesia. Thus, it required re-evaluation and re-validation processes for the coefficients, as authors had not discovered standard for metal roofing calculation.

3. Result

The tool produced in this paper was the result of the author's previous exploration of rapid estimation frameworks. To this stage, we had created twenty-seven components for the tool. The tool was installed by dragging the entire components into the Grasshopper's canvas. In order to create the effectiveness of our tool, we divided it into several categories as showed in figure 2. Firstly, the data consisted of components for sorting out the calculated data in certain order prior to it streamed into a spreadsheet. Secondly, material estimation contained components for material calculation based on SNI. Thirdly, coefficients for the calculation, which were created using Grasshopper's value, list components. Next, modeling, it involved components for modeling of the intended construction. The aimed of modeling components were for a more efficient of algorithms since some of Grasshopper's default components were coded into a single GHPython component. Finally, extra components for unforeseen functionalities that helped and might be developed in the future for the effectiveness of estimation.

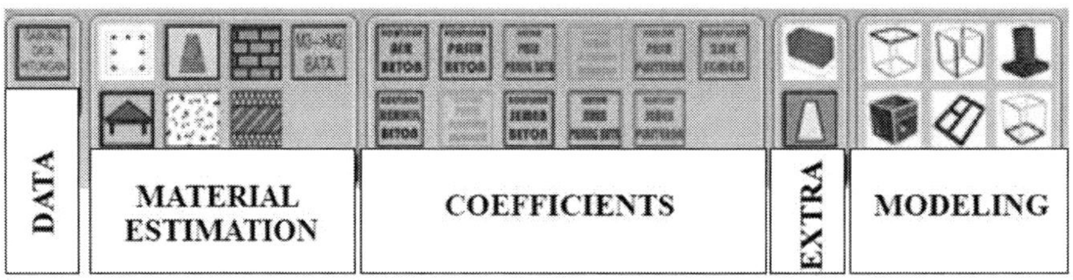

Figure 2. Parametric estimation tool

4. Discussion

From the above modeling category, the component for making a roof was missing. This component was in progress for its development. Up to this stage, the roof modeling relied on Grasshopper capability or Rhinoceros model as an input model for calculation for materials using the component presented in this paper. However, the simple ridge and gable roof systems were successfully modeled using Grasshopper algorithms. The roof modeling took inputs from a rectangle assigned to the perimeter of the building and positioned it in its elevation by taking an elevation input as a base from the height of the shelter. The authors aimed at the most optimum roof shape for the limited budget, as budget constraints were an important factor in determining the final shape of the relief shelters.

Furthermore, the components created above had allowed modeling and estimating being optimized and improved the adaptability of a workflow to users' specific senses. For example, to create a column in default Grasshopper, at least, it needed four components, while using GHPython it required one component. Furthermore, the workflow was able to be more dynamic, while it still maintained the rapidity of the calculation, since it open possibilities to adjust to users' specific logic of construction. For instance, if a user needed to directly utilize a 3D model from Rhinoceros, thus, the user was allowed to input the geometry into estimation components. Either way, if a user required inputting centerlines and points as inputs, also, it was allowable. This method needed a user to input to modeling components in out parametric estimation tool as demonstrated in figure 3. In addition, figure 3 exemplified the use of the tool to model and calculate for related construction materials. In this case, columns and foundations materials for concrete structure in Grasshopper canvas. The calculated data had not yet written into a spreadsheet.

By using a similar working method, the components presented in this study were allowable to be organized to users' intended logic for a more complex building system. For example, the visual algorithms in figure 4, is designed to input from centerlines drew in Rhinoceros. The optional inputs, for the workflow, were windows and doors locations. Either these modeled in Rhinoceros or using Grasshopper geometries was acceptable by the algorithm. The algorithm simultaneously performed modeling, visualization, and estimation for the shelter. The calculated data streamed in a real-time to a spreadsheet even the design of the shelter changed or altered to a specific request.

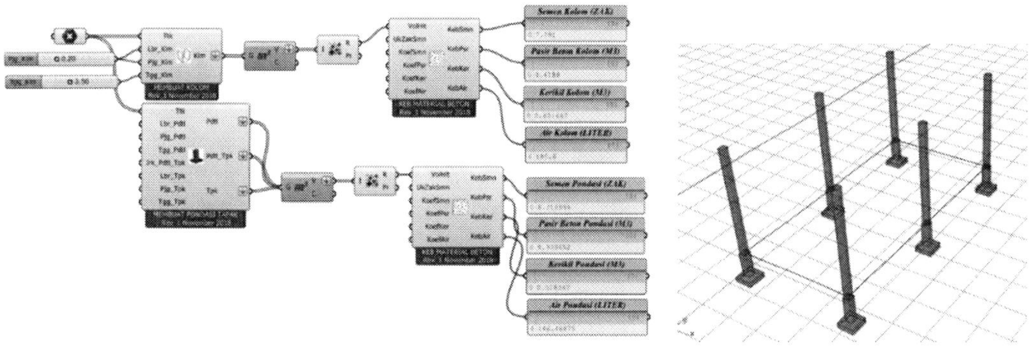

Figure 3. Workflow's algorithm using a parametric tool (left), Model produced in Rhinoceros (right)

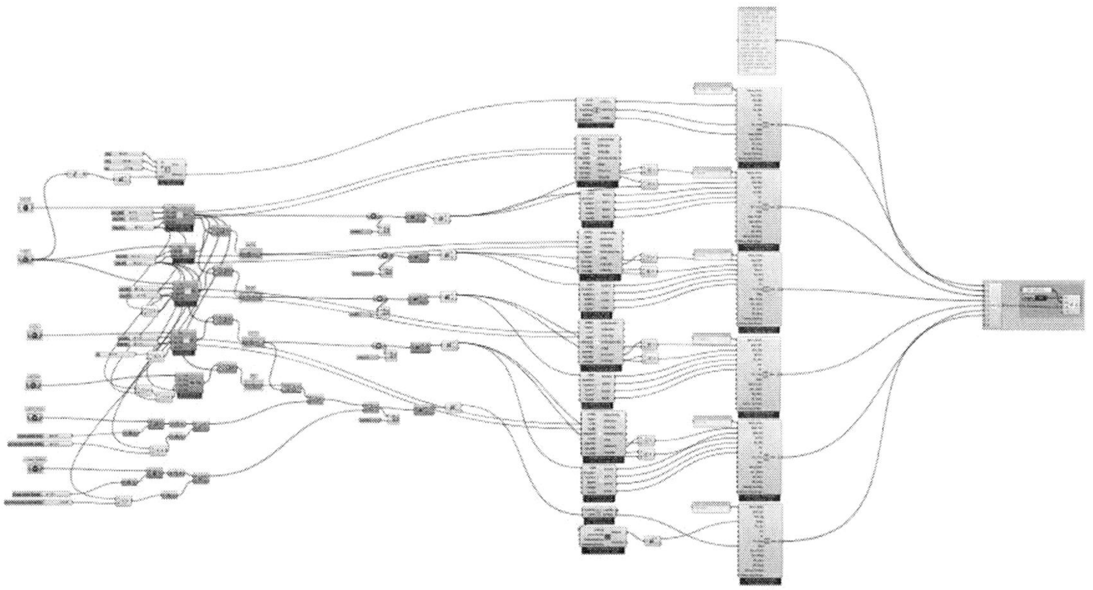

Figure 4. A parametric workflow using components invented from this study to perform a calculation for shelter materials

Once the algorithm was established, accordingly, to intended shelter design, it required only input processes while the modeling, visualization, and calculation for building materials were performed simultaneously. The authors discovered the process of inputting took one and a half minutes for a beginner. Obviously, this approach was solving the inability to quantify and provide information for shelter material requirements problematic as suggested by Ashmore et al. [1] for Aceh's post-disaster case. Thus, improved rapidity and this a positive situation as this important factor to make a success for the relief shelters in post-disaster as recommended by Şener & Torus [4]. The sample of a relative powerful algorithm to perform materials estimation was showed in figure 4. A completed material calculation was still missing in this paper; however, the most complicated çalculator for reinforcement

IOP Conf. Series: Materials Science and Engineering **796** (2020) 012035 doi:10.1088/1757-899X/796/1/012035

of concrete was successfully created. Alongside with future development of this parametric estimation tool, the rest of the easier and lighter components for modeling and calculation of the shelter construction materials will be available.

At the most right as presented in figure 4 were components yKTools. It allowed access for writing the data into a spreadsheet, in this case, Microsoft Excel. Using the components helped to stream the calculated data in a real-time into a spreadsheet without having to retype nor copy for the data. Also, along with some other Grasshopper components integrated into the workflow, it had proved the compatibility of the components presented in this paper with either other Grasshopper components or its add-ons. Therefore, the parametric estimation components presented in this paper had extended the capability of Grasshopper for a special task such as intended within this study. Figure 5 illustrated the modeling, visualization and calculation data written into a spreadsheet using the parametric estimation tool. The purpose of writing the data into a spreadsheet was considering that in the post-disaster settings people involved came from different backgrounds. Thus, it allowed for easier communication for relief shelter such as decision making for the shelter design based on a plotted budget. In addition, the roof algorithms were created Grasshopper un-optimized default components as explained earlier.

Figure 6 demonstrated the capability of the parametric components and workflow, as showed in figure 5, for shelter different configurations. The rapidity for modeling, visualization and estimation had allowed for comparison of a more suitable shelter configuration to adjust to a plotted budget. This was an important factor to assist decision making to select a suitable design for a budget scheme by a donor body and at the same time fulfilling victims' needs.

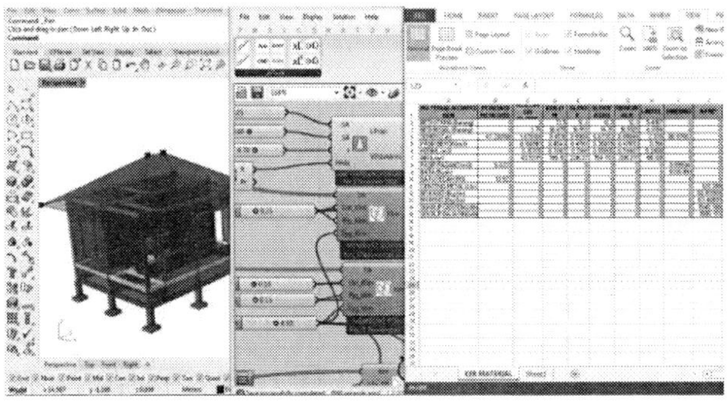

Figure 5. The parametric estimation tool in action. Rhinoceros in the left, Grasshopper in the middle and Microsoft Excel in the right.

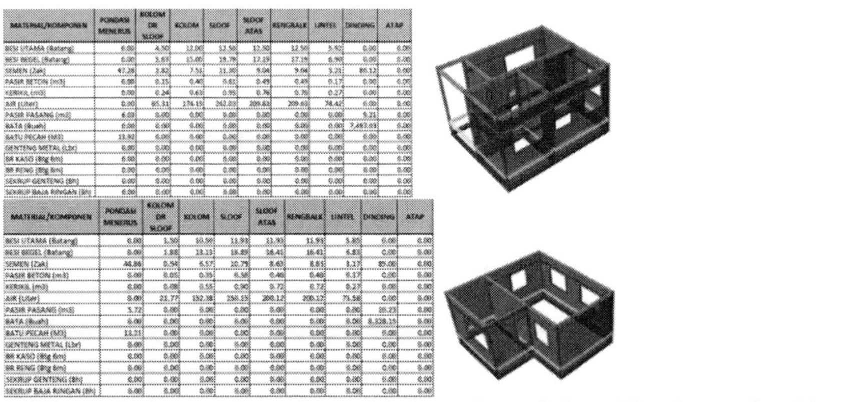

MATERIAL/KOMPONEN	PONDASI MENERUS	KOLOM DR SLOOF	KOLOM	SLOOF	SLOOF ATAS	RINGBALK	LINTEL	DINDING	ATAP
BESI UTAMA (Batang)	0.00	4.50	12.00	12.50	12.50	12.50	5.92	0.00	0.00
BESI BEGEL (Batang)	0.00	3.63	15.00	19.79	17.19	17.19	6.90	0.00	0.00
SEMEN (Zak)	47.28	2.82	7.51	11.30	9.04	9.04	3.21	86.12	0.00
PASIR BETON (m3)	0.00	0.15	0.40	0.61	0.49	0.49	0.17	0.00	0.00
KERIKIL (m3)	0.00	0.24	0.65	0.95	0.76	0.76	0.27	0.00	0.00
AIR (Liter)	0.00	85.31	174.15	262.03	209.63	209.63	74.42	0.00	0.00
PASIR PASANG (m3)	6.03	0.00	0.00	0.00	0.00	0.00	0.00	5.21	0.00
BATA (Buah)	0.00	0.00	0.00	0.00	0.00	0.00	0.00	7,487.03	0.00
BATU PECAH (M3)	13.92	0.00	0.00	0.00	0.00	0.00	0.00	0.00	0.00
GENTENG METAL (Lbr)	0.00	0.00	0.00	0.00	0.00	0.00	0.00	0.00	0.00
BR KASO (Btg 6m)	0.00	0.00	0.00	0.00	0.00	0.00	0.00	0.00	0.00
BR RENG (Btg 6m)	0.00	0.00	0.00	0.00	0.00	0.00	0.00	0.00	0.00
SEKRUP GENTENG (Bh)	0.00	0.00	0.00	0.00	0.00	0.00	0.00	0.00	0.00
SEKRUP BAJA RINGAN (Bh)	0.00	0.00	0.00	0.00	0.00	0.00	0.00	0.00	0.00

MATERIAL/KOMPONEN	PONDASI MENERUS	KOLOM DR SLOOF	KOLOM	SLOOF	SLOOF ATAS	RINGBALK	LINTEL	DINDING	ATAP
BESI UTAMA (Batang)	0.00	1.50	10.50	11.93	11.93	11.93	5.80	0.00	0.00
BESI BEGEL (Batang)	0.00	1.88	13.13	16.89	16.41	16.41	6.83	0.00	0.00
SEMEN (Zak)	46.86	0.94	6.57	10.79	8.63	8.63	3.17	85.00	0.00
PASIR BETON (m3)	0.00	0.05	0.35	0.58	0.46	0.46	0.17	0.00	0.00
KERIKIL (m3)	0.00	0.08	0.55	0.90	0.72	0.72	0.27	0.00	0.00
AIR (Liter)	0.00	21.77	152.38	250.15	200.12	200.12	73.58	0.00	0.00
PASIR PASANG (m3)	5.72	0.00	0.00	0.00	0.00	0.00	0.00	10.23	0.00
BATA (Buah)	0.00	0.00	0.00	0.00	0.00	0.00	0.00	8,328.15	0.00
BATU PECAH (M3)	11.21	0.00	0.00	0.00	0.00	0.00	0.00	0.00	0.00
GENTENG METAL (Lbr)	0.00	0.00	0.00	0.00	0.00	0.00	0.00	0.00	0.00
BR KASO (Btg 6m)	0.00	0.00	0.00	0.00	0.00	0.00	0.00	0.00	0.00
BR RENG (Btg 6m)	0.00	0.00	0.00	0.00	0.00	0.00	0.00	0.00	0.00
SEKRUP GENTENG (Bh)	0.00	0.00	0.00	0.00	0.00	0.00	0.00	0.00	0.00
SEKRUP BAJA RINGAN (Bh)	0.00	0.00	0.00	0.00	0.00	0.00	0.00	0.00	0.00

Figure 6. Sample of different shelter configurations with the estimated data. The data updated in real-time as the design changed.

In addition, figure 6 showed the calculated material for the rubble stone foundation. Yet, the foundation was not modeled and visualized. For this algorithm, the author used LTrap component to calculate for the construction material. This component had the capability to calculate accurate

IOP Conf. Series: Materials Science and Engineering **796** (2020) 012035 doi:10.1088/1757-899X/796/1/012035

information for the rubble stone foundation system without having to model the geometry of the foundation. However, for clarity and for the purpose of easier validation author recommended using the component for modeling the geometry. This was to show the capability of an extra component created in this study.

5. Conclusion

This study had resolved the slow material calculation method using a parametric estimation tool. This tool helped the estimation for reinforced concrete construction shelter more rapidly, accurately and flexibly. Using components introduced in this paper, it allowed the algorithms to be constructed more efficiently. Since it combined separated functionalities into one component using GHPython script. Furthermore, this improved validation of the algorithms more easily, if necessary, thus it reduced the risk of human error. Moreover, the components presented here enhanced the adaptability of workflows to a user-specific logics and needs. This opened an opportunity for the components to adjust for the estimation of more complex reinforced concrete buildings.

With the potentials explained above, the authors convinced, the parametric estimation tool was implementable in a post-disaster. The design alterations, it more likely to happen in a post-disaster as victims' requests, thus, it was not problematic. Obviously, this improved some obstructions explained by Ashmore et al. [1] in 2004 Aceh post-disaster. Further, rapidity as an important factor, as suggested by [4], was enhanced using the parametric estimation tool. The rapidity of information availability, for building materials, was one crucial factor that improved the time for a permanent relief shelter to be decided.

Despite the potentials explained earlier, there were limitations to this tool. Firstly, this work was limited to estimation for a core structural component of a reinforced concrete shelter. The development of the tool was allowable for estimation that was not explained in this paper. The improvements and developments were either for its functionality extensions or alteration for codes and coefficients, for example, the roof material calculator component. Next, the tool was tested and compatible with Rhinoceros 5 and Grasshopper 0.9.0076 above. The authors did not recommend it used in older versions. Also, the tool tab name within Grasshopper still used the first author's name, and this did not follow the common sense of Rhinoceros which used animal names for its add-ons. Apart from this, the tool will be released as an open-source tool. In this way, the authors believe that the development of this tool will be faster. Therefore, it will be useable immediately for implementation in a relevant post-disaster relief shelter.

6. References

[1] J. Ashmore, J. Fowler, J. Kennedy 2008 *IASC Emergency Shelter Cluster Shelter Projects* UN-HABITAT, Un-Specified.
[2] J. da Silva, 2010 *Practical Action Publishing*, Warwickshire.
[3] Badan Nasional dan Penanggulangan Bencana 2017 *Statistik Bencana* [Online]. Available: http://dibi.bnpb.go.id. [Accessed: 19-Jun-2017].
[4] S. M. Şener, B. Torus 2009 *Container Post Disaster Shelter-CPoDS.* 599–604.
[5] W. K. Yeung, J. Harkins 2010 *Int. J. Archit. Comput.* **9**, 17–31.
[6] Atthaillah, E. Saputra 2018 *J. Arsit.* **5**, 68.
[7] Atthaillah, E. Saputra, M. Iqbal 2018 *IOP Conf. Series: Materials Science and Engineering* ***352***, 012061.
[8] A. Tedeschi 2014 *AAD-Algorithms-Aided Design: Parametric strategies using Grasshopper*. Brienza: Le Penseur
[9] S. Davidson, 2018. *An Overview of the GhPython Component with Python*. [Online]. Available: https://developer.rhino3d.com/guides/rhinopython/ghpython-component/. [Accessed: 15-Jan-2019].
[10] M. S. Roudsari, M. Pak 2013 the *13th Conference of International Building Performance Simulation Association*, 3128–3135.
[11] A. SNI 7394 2008. Badan Standardisasi Nasional, Jakarta.
[12] A. SNI 6897 2008 Badan Standardisasi Nasional, Jakarta.
[13] A. SNI 2837 2008 Badan Standardisasi Nasional, Jakarta, 2008.
[14] A. SNI 2836, Badan Standardisasi Nasional, Jakarta, 2008.

Safety factors and stability analysis of escape hill critical high towards tsunami free-board and run-up level

Munirwansyah[1], M Irsyam[2], R P Munirwan[1*)], H Yunita[1], H Munirwan[3]

[1]Department of Civil Engineering, Faculty of Engineering, Universitas Syiah Kuala, Banda Aceh, Indonesia, 23111.
[2]Faculty of Civil and Environmental Engineering, Institut Teknologi Bandung, Bandung, Indonesia, 40132
[3]Department of Urban and Regional Planning, Institut Teknologi Sumatera, Bandar Lampung, Indonesia, 35365.

[*]E-mail: r.munirwan@unsyiah.ac.id

Abstract. Earthquake threats that had caused tsunami hazards with a lack of optimal efforts in the construction of evacuation sites, it is necessary to conduct studies on evacuation sites as escape hill. One method of overcoming building construction on the effect of freeboard and run-up level of vertical walls in one of tsunami evacuation models that need to be analyzed for the maximum height that is suitable against the lateral force of tsunami wave. This study's purpose is to calculate the stability of slope and embankment high on escape hill and calculate the safety factor of soil bearing capacity. The data obtained from the undisturbed sample are then taken to the laboratory for physical and mechanical properties analysis and slope failure of escape hill calculated using the finite element method. This study of maximum embankment height the escape hill for evacuation site, stability obtained at an altitude of 10.0 m and safety factors are 2.14 to 2.84 which is very safe. From the material set design that has been done, it is shown that the modeling of materials set with variations of soil cohesion value (c) and the value of the shear angle in the soil (φ) the safety factor of 2.412.

1. Introduction

Indonesia as an archipelago is geographically vulnerable to tsunami natural disasters. Most earthquakes epicenter on the bottom of the Indian Ocean and some of them caused large ocean waves (tsunami). One of the provinces of Aceh was Banda Aceh, including the tsunami-affected area on December 26, 2004, with the strength of the Mw 9.0 earthquake or Mw 9.3 [1], [2]. Given the various earthquake threats that have caused tsunami [3] and the lack of optimal efforts in the construction of evacuation sites in open areas [1], it is necessary to conduct studies on temporary evacuation sites, namely escape hills [4].

One way of overcoming building construction to the effect of the maximum height of the coming wave (freeboard) and when the creeping waves hit a structure crawling up the escape hill surface (run-up level) is the use of upright walls as can be seen in figure 1 (sketch of escape hill based on geotechnical consideration). The critical embankment height issue states that the soil which will be

[1] E-mail: r.munirwan@unsyiah.ac.id.

Content from this work may be used under the terms of the Creative Commons Attribution 3.0 licence. Any further distribution of this work must maintain attribution to the author(s) and the title of the work, journal citation and DOI.
Published under licence by IOP Publishing Ltd

built the construction above must have special treatment or must be engineered so that the soil bearing capacity is sufficient to the planned load. Safety risk analysis should be considered as suggested by [5] and [6] If the weight and height exceed the value of subgrade support, it is possible for the lateral collapse and movement to occur. Vertical wall or retaining wall serves to resist the amount of soil pressure due to lateral forces [4]. The escape hill design is an addition of a higher tsunami wave run-up level and freeboard height.

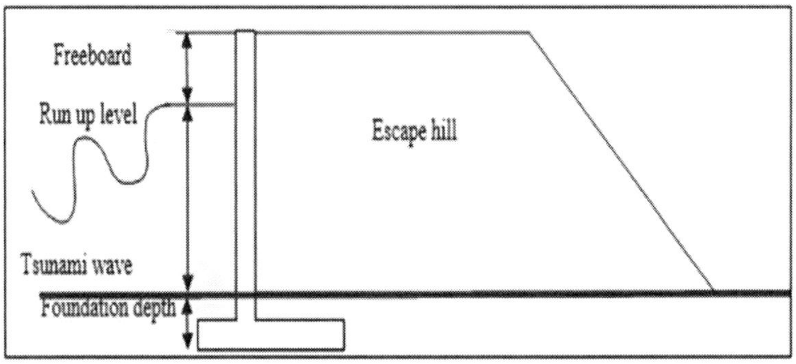

Figure 1. The height of the wall as escape hill construction on freeboard problems, run-up level and depth of the foundation

2. Methodology

The research data from this article are extended research of [4]. Based on the review and data obtained from the tsunami, the location of the review was carried out in Tibang Village, Syiah Kuala sub-district, the maximum run-up at the time of the 2004 tsunami reached 3.70 m. The built escape hill must represent the distance of the local population to evacuate within a sufficient radius or distance [7].

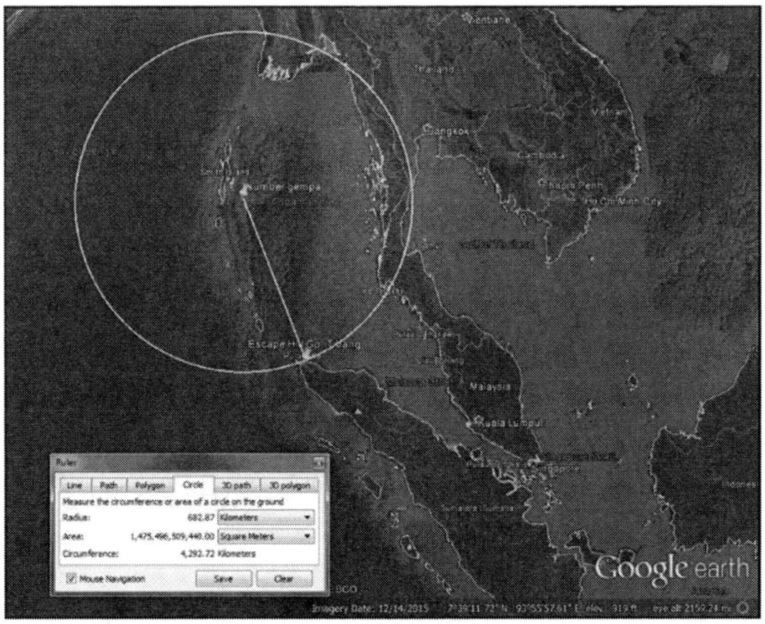

Figure 2. Map of Distance Measurement from the 2004 Tsunami Earthquake Epicentre to Escape Hill Location

To prevent the sliding and spreading around the embankment on subgrade under saturated conditions, it is necessary to know the collapse of the bearing capacity of the subgrade due to embankment [8], [9]. The easiest way to do is by calculating the height of the critical embankment (H_{crit}). Run-up levels, namely when waves come in against a structure, water carried by its momentum is pushed up and creeps up to the surface of the structure. The maximum height of the wave that comes is called freeboard.

Tabel 1. Soil parameter value of *Escape Hill design*

Parameter	Name	Subgrade			Landfill	Unit
		Sand	Sand	Sand	Sand	
		Silty	Silty	Silty		
		1m	2m	3 m	> 3m	
Type of behavior	Type	Undrained	Undrained	Undrained	Drained	
Dry soil weight	γ dry	8.90	10.10	10.40	16.10	kN/m_3
Wet soil wet	γ wet	22.84	22.76	22.96	19.24	kN/m_3
Horizontal permeability	kx	8.64E-03	8.64E-03	8.64E-03	4.53E-03	m/day
Vertical permeability	ky	8.64E-03	8.64E-03	8.64E-03	4.53E-03	m/day
Young's modulus	E ref	15000	10000	8000	10000	kN/m_2
Poisson's Ratio	υ	0.33	0.33	0.33	0.30	-
Cohesion	C ref	8.60	12.90	10.70	0.043	kN/m_2
Friction angle	ϕ	31.60	29.00	32.30	30.923	°
Dilatancy angle	Ψ	0	0	0	0	-

Figure 3. Tsunami height memorial poles 3.7 m run-up at X: 5°58'33.91"; Y: 95°35'06.37"

IOP Conf. Series: Materials Science and Engineering **796** (2020) 012036 doi:10.1088/1757-899X/796/1/012036

Soil parameter value from soil sampling in finite element analysis used is based on Tabel 1 and Tabel 2. The height of the maximum tsunami wave experienced was based on tsunami height memorial poles of 3.7 m run-up as can be seen in figure 3.

3. Result and Discussion

The estimated escape distance and evacuation time have been calculated. Based on the data of the December 26, 2004 tsunami earthquake, the distance of the epicenter to the escape hill location was 619.17 km with a tsunami wave speed of 500 Km/h up to 700 Km/h as shown in Figure 2. Moreover, the distance of the epicenter (Andaman-Nicobar Island) to the location of escape hill is 682.87 Km by Tsunami Wave speed 700Km/h is 58.53 minute.

Table 2. Soil parameter modeling

c variant modeling	ϕ variant modeling	Safety Factor (SF) with a heap height of 10 m	Safety Factor (SF) with a heap height of 12 m
c	ϕ	2.412	2.351
c + 25%	ϕ - 5%	2.546	2.484
c + 50%	ϕ - 10%	2.269	2.207
c + 75%	ϕ - 15%	2.140	2.075
c - 25%	ϕ + 5%	2.691	2.614
c - 50%	ϕ + 10%	2.834	2.517

The results of slope stability analysis by reviewing the strengthening conditions using a counterfort wall type of retaining wall were analyzed using the plaxis program, the slope angle used was the permissible angle of 15°. The following are the results of variations in soil parameter modeling in table 2 with heaps of 10 m and 12 m.

Table 2 shows the results of variations in soil parameter modeling, the smaller the cohesion value and the greater the value of the shear angle, the value of the safety factor increases compared to the high cohesion value and the small shear angle value. Based on these results, the highest value of safety factors is chosen. From the modeling that has been done, the most stable conditions are obtained in the modeling of c+50% and φ-10%. The result of the calculation derived from modeling with the heap height of the 12 meters is shown as in Figure 4.

In Figure 4, calculations are made by the plaxis program with variations in the modeling by increasing the cohesion value (c) and decreasing the shear angle value (φ) to obtain a condition that is safer than natural conditions. From the modeling that has been done, the most stable conditions are obtained in the modeling of c+50% dan φ-10%. The calculation results from modeling with a heap height of 12 meters.

4. Conclusion

Based on the calculation of slope stability using the counterfort wall type of retaining wall displayed by the plaxis program, it is safe, the safety factor obtained at the condition of the 12 m heap height is 2.207 meaning that the slope has met the appropriate safety number requirements. Firstly, the results of vertical wall stability analysis on the temporary tsunami evacuation model against freeboard and run-up level using counterfort wall type retaining wall can improve slope stability, the safety factor obtained at the condition of 12 m heap height is 2.207, in soil modeling with a variation of c+50% and φ-10%.

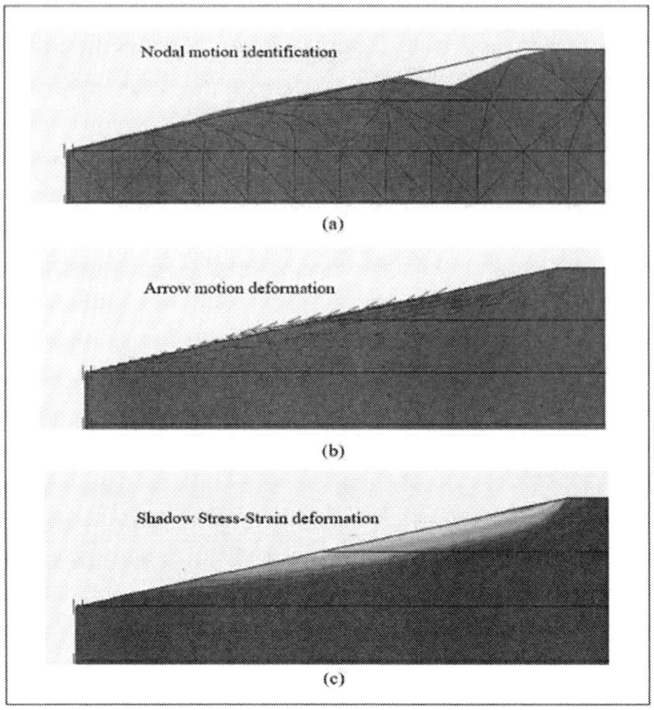

Figure 4. (a) Deformed mesh heap height of 12 m (c+50% dan φ-10%), (b) Arrow direction deformation and (c) Total Stress-strain displacement shading for heap height of 12 m (c+50% dan φ- 10%)

Secondly, from the modeling that has been done, it shows that the modeling of site materials with variations of c+50% and φ-10% with a safety factor of 2.207 produces the highest safety factor compared to variations c and φ; c+25% and φ-5%; c+50% and φ-10%; c+75% and φ-15%; c-50% and φ+10% with safety factors 2.351; 2.484; 2.207; 2.075; 2.517. Lastly, the values of soil parameters (material set) with variations of c = 0.032 kN/m^2 and φ = 32.469^o could be proposed to determine the Term of Reference (TOR) and Work Plan and Requirement in implementing this escape hill construction at this location.

5. Acknowledgments
The author gratefully acknowledges financial support from the Ministry of Research, Technology and Higher Education, Universitas Syiah Kuala Banda Aceh Indonesia, for the Profesor Research Funding, Grant No. 520/UN11/SPK/PNBP/2019, February 8, 2019.

6. References
[1] Munirwansyah, Munirwan R P, Yunita H, 2018 *International Journal on Advanced Science, Engineering and Information Technology.* **8**(3) 870-875.
[2] Munirwansyah, Munirwan R P, Sungkar M, Fachrurrazi, 2019 *AIP Conference Proceedings.* **2059**, 1.
[3] Fachrurrazi, Syamsidik, Al'ala M, Mahardi W 2017 *IOP Conf. Ser. Earth Environ. Sci.* **56** 012015
[4] Munirwan R P, Munirwansyah, 2019 *IOP Conference Series: Earth and Environmental Science* **273**, 1.

[5] Munirwansyah, Irsyam M, Munirwan R P, Yunita H, and Usrina M Z, 2018 *IOP Conference Series: Materials Science and Engineering* **352**, 1.
[6] Munirwansyah, Munirwan R P, Sungkar M, Melinda Z, 2019 *IOP Conference Series: Materials Science and Engineering* **523**.
[7] Jayakody R R J C, Amaratungga D, Haigh R, 2016 *the 6th International Conference on Building Resilience.* 1022-1030.
[8] Maula B H, Zhang L, 2011 *Procedia Eng.* **14** 559-66
[9] James M A, Reinoso E, Ordaz M, Garnica B H, Silva R, Mendoza E, Rodriguez J C, 2016 *Ocean and Coastal Management*, 119, 68-75.

IOP Conf. Series: Materials Science and Engineering **796** (2020) 012037 doi:10.1088/1757-899X/796/1/012037

Modeling of groundwater level fluctuation in the tropical peatland area of Riau, Indonesia

H Pratama[1]*, S Sutikno[1,2*], M Yusa[1]
[1]Civil Engineering Department, University of Riau, Pekanbaru, Indonesia 28293
[2]Centre for Disaster Studies, University of Riau, Pekanbaru, Indonesia 28293

E-mail: damai.ria@bppt.go.id

Abstract. The groundwater level (GWL) is a key parameter for assessing the level of fire risk in the peatland so that if the predicted fluctuations of GWL in the next few days or weeks can be predicted; the risk of peat fire can also be estimated. The purpose of this study is to develop a hydrological model using regression analysis that can be used to predict GWL in the peatland. The data used for modeling were historically recorded of rainfall and GWL fluctuations from SESAME equipment in Dompas village, Riau, Indonesia. Regression analysis was carried out using four data length scenarios, such as one-month, two-months, three-months, and six-months to find out which time period could represent hydrological conditions in the field. The results showed that the regression analysis using three-months and six-months data represented the best results of the GWL prediction with a correlation coefficient of 0.95. However, the prediction using one-month and two-months data represented reasonable prediction results with a correlation coefficient of 0.86 and 0.89 respectively. Judging from observed and predicted GWL that was always in the lower position than 0.4 m depth, this area was always in the high risk of peat fire throughout the year.

1. Introduction

Peatlands are a kind of soil that is formed from accumulated organic matter deposits. These deposits are formed naturally from weathering vegetation that grows above the ground in a period of hundreds of years. The decomposition process of materials is hampered due to anaerobic and wet conditions, this causes peat soils to be found in swamps. Peat has unique physical characteristics, which are able to absorb water very much. The pristine peatlands consist of almost 90% water and the remaining 10% is residual decomposed plant material [1], therefore peat in natural conditions (undisturbed) will always be wet and humid. Therefore, the hydrological process is very important in the management of peatlands [2]. In the dry season, the peat soil functions as a water-retaining layer and releases water slowly. Peat water appears brownish-black, because of the long-term submergence of various organic materials on peatlands [3].

Almost every year, the haze disaster because of peatland fire occurs in Indonesia, especially in Kalimantan and Sumatera island which causes huge economic losses and harmful public health problems. The fires also cause a decrease of peat soil thickness of 10-15 cm and resulted in 100% mortality of soil flora and fauna [4]. Unlike the rapid occurrence of earthquake disasters [5], peatland fires tend to occur slowly which usually begins with limited rainfall events, so that actually there is still sufficient time to prepare an emergency response. The peatland fires are very difficult to be extinguished because the fires penetrate to the below-ground of peatland. The historical evidence shows that only rainfall can stop the peat fire completely [6], [7]. To anticipate the possibility of disaster recurring, prevention efforts are needed. Based on this historical experience, it is necessary to have a device that can be used as a basis for

Content from this work may be used under the terms of the Creative Commons Attribution 3.0 licence. Any further distribution of this work must maintain attribution to the author(s) and the title of the work, journal citation and DOI.
Published under licence by IOP Publishing Ltd

IOP Conf. Series: Materials Science and Engineering **796** (2020) 012037 doi:10.1088/1757-899X/796/1/012037

early warning of the peatland fires risk, so that efforts for fire prevention can be prepared as early as possible.

The GWL is a key parameter for assessing the level of fire risk in peatlands [8], [9], so that if the GWL fluctuations in the next few days or weeks can be predicted, then the risk of fire also can be estimated. Tropical peatlands will be burnt easily if they are dry with a depth of GWL more than 0.4 m [10]. To be able to find out the level of risk of drought and fires on peatlands, the Government of the Republic of Indonesia through the Peat Restoration Agency (BRG) monitors GWL in real-time using the SESAME (Sensory Data Transmission Service Assisted by Midori Engineering) tool. The SESAME tool was installed in several peatland areas to monitor GWL, temperature and rainfall data.

The purpose of this study is to develop a hydrological model using regression analysis that can be used to predict GWL fluctuation in peatlands so that it can be used to predict the level of risk of land fires as a basis for early warning. The SESAME tool in Dompas village, Bengkalis Regency, Riau Province was picked up as a study site for developing the model. Historically, the village has severe peat fire damage in 2013, 2014 and 2015. The results of this research are presented in this paper.

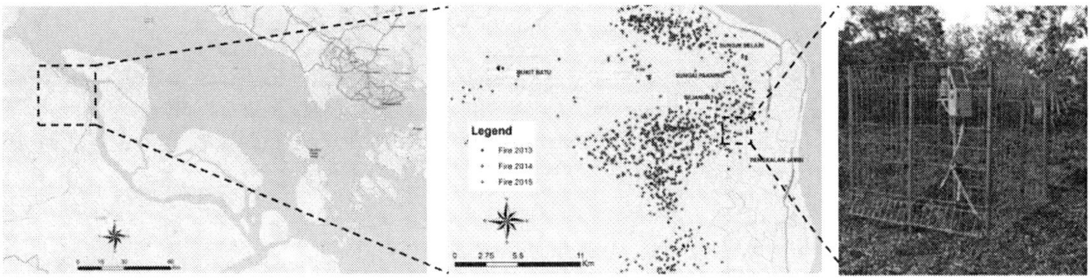

(a) (b) (c)

Figure 1. Study area at Dompas village, Bengkalis Regency (a), the fire history in 2013, 2014, 2015 (b), and the SESAME tool at Dompas village (c).

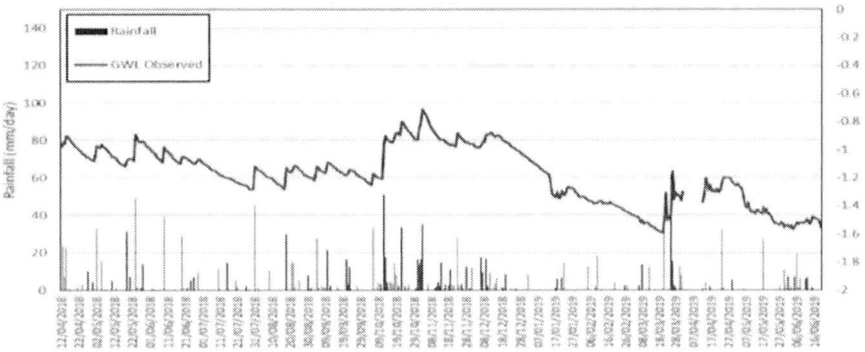

Figure 2. The historical data of GWL fluctuation and rainfall from the SESAME at Dompas village, Bengkalis Regency

2. Method

2.1. Material Study Area and Data Acquisitions

This research was investigated at Dompas village, Bengkalis Regency, Riau province, Indonesia which is located in Sungai Rokan-Sungai Siak Kecil Peatland Hydrological Unit (PHU) (See Figure 1a). The Dompas village and many villages in Bengkalis have severe peat fire in 2013, 2014 and 2015. The historical fire spots at Dompas village in 2013, 2014 and 20154 are presented in Fig. 1b. A SESAME tool has been installed by Peat Restoration Agency, Republic of Indonesian (BRG-RI) at Dompas village to monitor the GWL fluctuation for a fire risk alert in this area Since April 2018 (Fig. 1c). The recorded data from this tool is presented in Fig. 2.

2.2. Basic Concept for Modeling

The GWL information in the peatlands is very important for the management and prevention of peatland fires, therefore prediction of GWL in peatlands is very important to do. Spatial and temporal groundwater prediction can be done both by modeling and mapping GWL [11] and by processing remote sensing data [8]. However, the results of these predictions and modeling are highly dependent on existing satellite data globally with a detailed resolution.

The GWL prediction in the peatlands can be carried out by mathematical modeling and a statistical approach based on the concept of water balance in peatlands as presented in Figure 3 [12]. Based on the basic principle of water balance for modeling tropical peatland as presented in Fig.3, the daily fluctuation of GWL is estimated using the difference between an increase of GWL due to the daily rainfall and a decrease of GWL due to daily evapotranspiration. It can be formulated mathematically as follows:

$$dW = dW_{rain} - dW_{loss} \tag{1}$$

where, dW is the daily GWL change, dW_{rain} is the increases of daily GWL because of rainfall, and dW_{loss} is a decrease in daily GWL because of evapotranspiration and runoff. The dW_{loss} and dW_{rain} parameters are a correlation which is a function of each evapotranspiration and rain. The dW_{loss} and dW_{rain} parameters represent the response of GWL fluctuation because of rainfall and losses in the peatland hydrological cycles. Thus, the GWL in the next day can be estimated with the following formulation.

$$W_{n+1} = W_n + dW_{rain} - dW_{loss} \tag{2}$$

where, W_{n+1} is the estimated of GWL in the next day and W_n is the GWL in the previous day.

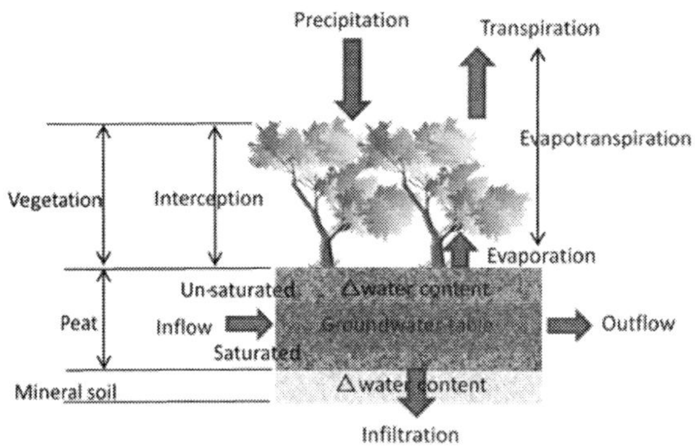

(Source: Takahashi, 2017)

Figure 3. The basic principle of water balance for modeling tropical peatland

2.3. Modeling Procedures

Using the historical data of GWL and rainfall, the response of GWL fluctuation was recorded statistically. The dW_{rain} which is defined as the correlation between the raised of GWL and the amount of rainfall was developed. However, the dW_{loss} is defined as the correlation between daily GWL loss and GWL distance from ground surface. The dW_{rain} and dW_{loss} parameters were analyzed using regression analysis to understand the trend and the parameter determination. The determination coefficient (R^2) is a statistical measure to describe how close the data are to the fitted regression line. It is the percentage of the response variable variation that is explained by a linear model. The higher the determination coefficient, the better the model fits to the data. The value of the determination coefficient will have between 0 and 1. The regression analysis was carried out for both parameters using four data length scenarios, such as 1-month,

2-months, 3-months, and 6-months to find out which time period could represent the best hydrological conditions like in the field.

The GWL for the next day can be predicted using eq. 2 using the regression equations of dW_{rain} and dW_{loss} from the regression analysis. The predicted GWL was compared with the observed GWL to know the correlation between them. The correlation coefficient is a statistical measure that calculates the strength of the relationship between those variables. The values range between -1.0 and 1.0. A calculated number is greater than 1.0 or less than -1.0 means that there was an error in the correlation measurement. A correlation of -1.0 shows a perfect negative correlation, while a correlation of 1.0 shows a perfect positive correlation. A correlation of 0.0 shows no relationship between the movement of the two variables [13]. The strength of the relationship varies in degree based on the value of the correlation coefficient. Experts do not consider correlations significant until the value surpasses at least 0.8. However, a correlation coefficient with an absolute value of 0.9 or greater would represent a very strong relationship.

3. Result and Discussion

3.1. Modeling using one-month data

The regression analysis of dW_{rain} and dW_{loss} using 1-month data of May 2018 is presented in Fig. 4a and Fig. 4b respectively. There were six rainfall events during May 2018. The data in April 2018 was not used in this modeling because there were only two days of rainfall events. The determination coefficients of the dW_{rain} and dW_{loss} regression analysis were 0.7548 and 0.1504 respectively. It means that the rainfall has a response about 75.48% to the raised of GWL, and the GWL distance from the ground surface has a response about 15.04% to the daily water loss of GWL.

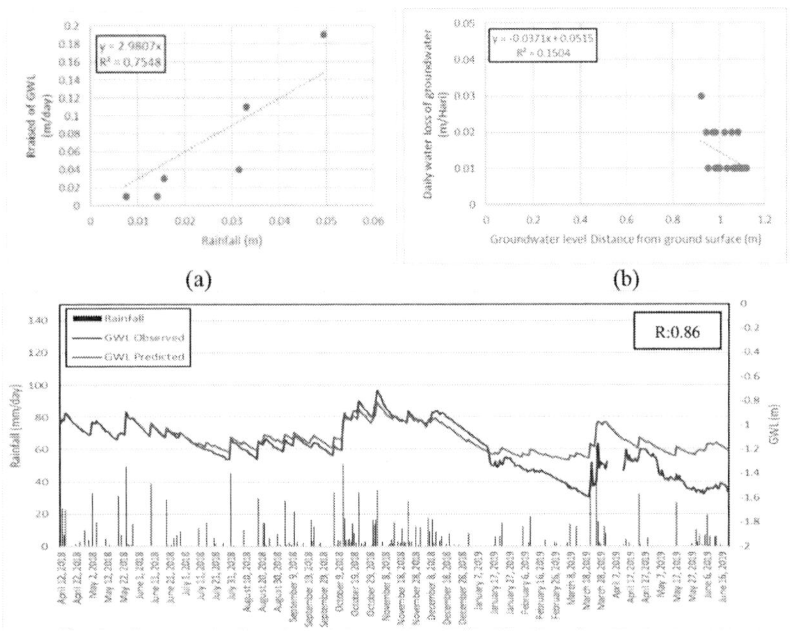

(c)

Figure 4. The modeling process of dW_{rain} (a), dW_{loss} (b) and comparison of predicted GWL with observed GWL (c) using one-month data (May 2018)

Using rainfall data input and the GWL distance from the ground surface as well as the equation resulted from regression analysis of dW_{rain} and dW_{loss}, the GWL for the next day was predicted using eq.2. The result of the GWL prediction using those data and equations was presented in Fig. 4c. The figure shows that the predicted GWL has a good pattern with the observed GWL from June 2018 to January 2019. The correlation coefficient between the predicted GWL and observed GWL was about 0.86. This means that the predicted GWL has a significant correlation with the observed GWL. It was an acceptable correlation coefficient for the modeling. However, the observed GWL data from March to June 2019 showed the strange fluctuation, and there were some missing data. This condition could be because of the disturbance on the data logger equipment in the site area.

3.2. Modeling using two-months, three-months, and six-months data

The regression analysis of dW_{rain} and dW_{loss} using two-month data from May to June 2018 is presented in Fig. 5a and Fig. 5b respectively. With one-month additional data (three rainfall additional data), the determination coefficients of the dW_{rain} and dW_{loss} regression analysis became 0.7446 and 0.1071 respectively. Although the determination coefficients using two-month data regression analysis were smaller than that of one-month data analysis, the correlation coefficient between the predicted GWL and observed GWL was stronger which was about 0.89 (Fig. 5c).

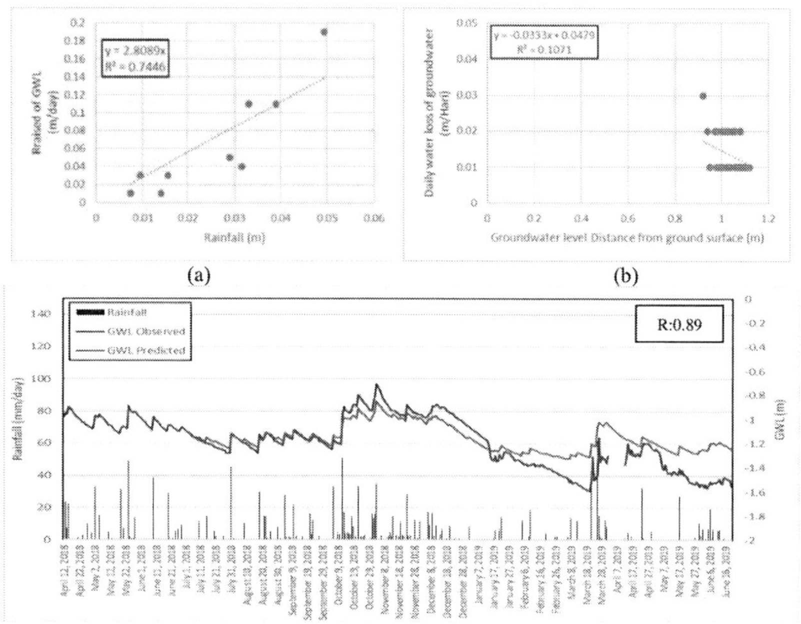

Figure 5. The modeling process of dW_{rain} (a), dW_{loss} (b) and comparison of predicted GWL with observed GWL (c) using two-months data (May-June 2018)

The regression analysis of dW_{rain} and dW_{loss} using three-month data from May to July 2018 yielded the determination coefficients of 0.7791 and 0.111 respectively as presented in Fig. 6a and Fig. 6c. This determination coefficient of dW_{rain} was higher than that of one-month and two-months data modeling. However, the determination coefficient of dW_{loss} was lower than that of one-month data modeling and was higher than that of two-months data modeling. The prediction of GWL using three-months data modeling represented a very strong relationship with the observed GWL data with the correlation coefficient of 0.95 as presented in Fig. 5c. The prediction of GWL using six-months data modeling also represented a very strong relationship with the observed GWL data with the correlation coefficient of 0.95 as presented in Fig. 7c. In the regression modeling process, the determination coefficient of dW_{loss} was the highest with was about 0.1672 as presented in Fig. 7b. However, the determination coefficient of dW_{rain} was slightly lower than that of three-months data modeling and was higher than that of one-month and two-months data modeling. The determination coefficient of dW_{rain} using six-months data modeling was about 0.7709 as presented in Fig. 7a.

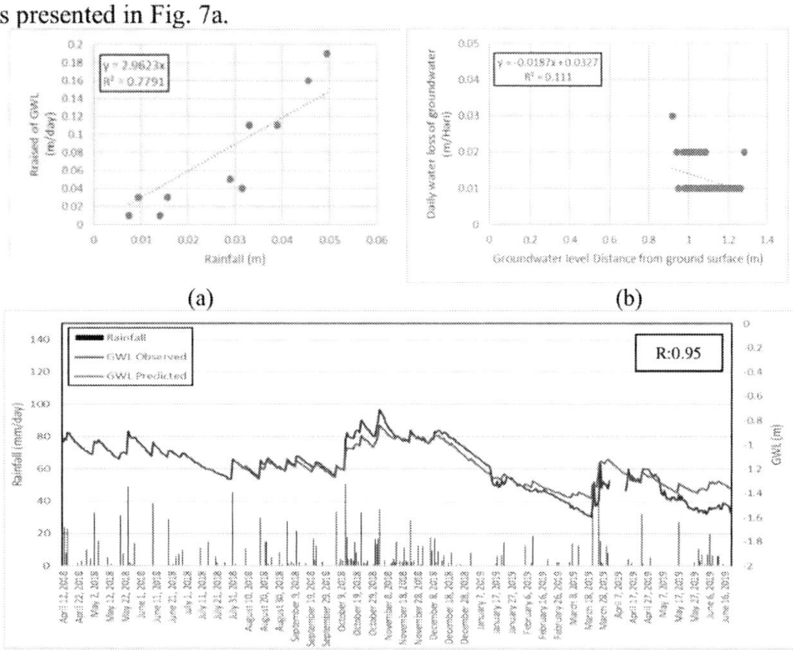

(c)

Figure 6. The modeling process of dW_{rain} (a), dW_{loss} (b) and comparison of predicted GWL with observed GWL (c) using tree-months data (May-July)

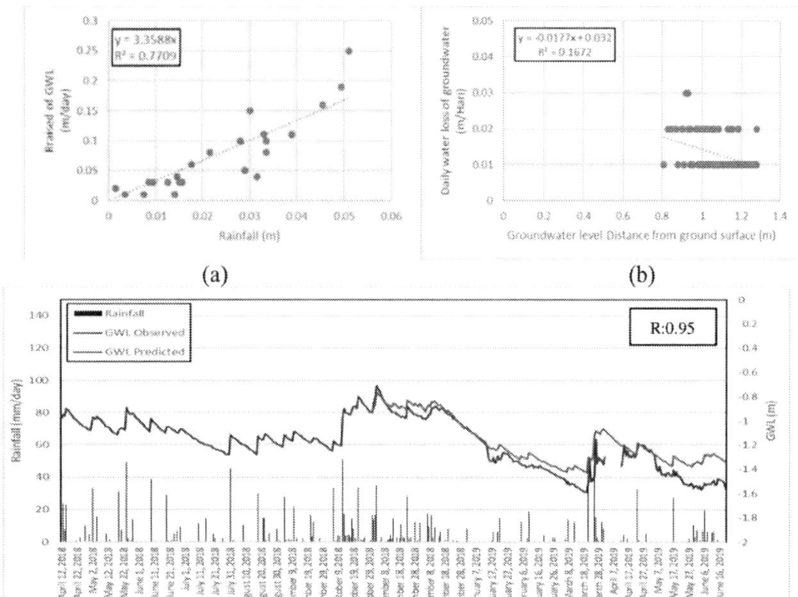

(c)

Figure 7. The modeling process of dW_{rain} (a), dW_{loss} (b) and comparison of predicted GWL with observed GWL (c) using six-months data (May-October)

From the above discussion, it was clear that the more data used for modeling, the better of the GWL prediction results were obtained. It was because the model could represent the response of GWL fluctuation in a wide range of hydrological conditions in both dry and rainy season very well. However, using one-month data only for modeling was sufficient for GWL prediction with acceptable results, if there was enough number of the rainfall event and the GWL fluctuation could represent well the response as the impact of the rainfall. This information is useful if there is a limited number of data for the prediction of GWL in the future.

Judging from all observed and predicted GWL that the GWL is always in a lower position than 0.4 m depth in Dompas village, this area is always in the high risk of peat fire throughout the year. To mitigate this condition, it is very important to manage water resources properly by a rewetting approach using canal blocking to raised GWL in the peatland [14].

4. Conclusion

This research developed a hydrological model using regression analysis that can be used to predict GWL in the peatland so that it can be used to predict the level of risk of peat fires as a basis for early warning. The GWL is a key parameter for assessing the level of fire risk in the peatland. Regression analysis was carried out using four data length scenarios, such as one-month, two-months, three-months, and six-months to find out which time period could represent hydrological conditions in the field. The results showed that the regression analysis with a data length of three-months and six-months represented the best results of the GWL prediction with a correlation coefficient value of 0.95. However, the prediction of GWL using regression analysis with data lengths of one-month and two-months represented a reasonable prediction result with a correlation coefficient value of 0.86 and 0.89 respectively. This means that the GWL can be predicted even only use one-month data if the response of GWL because of rainfall represents the hydrological process in the peatland along the year very well.

Judging from all observed and predicted GWL that the GWL is always in a lower position than 0.4 m depth in Dompas village, this area is always in the high risk of peat fire throughout the year. To mitigate this condition, it is very important to manage water resources properly by a rewetting approach using canal blocking to raised GWL in the peatland.

5. Acknowledgement

The authors would like to thank the Directorate General of Higher Education (DIKTI) - Ministry of Research, Technology and Higher Education for its research grants program.

6. References

[1] J. Jaenicke, H. Wösten, A. Budiman, F. Siegert 2010 *Mitig. Adapt. Strateg. Glob. Chang.* **15** (3), 23–239.
[2] J. Labadz *et al.*, *Rev. Lit. Arts Am.* **52**, 2010.
[3] A. Dohong 2017 Bolstering Peatlands Restoration in Indonesia through 3Rs Approach in *Developing International Collaborations to Address Fire and Other Conservation Issues in Central Kalimantan, Indonesia.*
[4] B. Wasis, E. I. Putra, B. H. Saharjo 2019. *B IOD I V E R S I T A S*, **20**, 1770–1775.
[5] R. Suryanita, Mardiyono, A. Adnan 2017 *Telkomnika (Telecommunication Comput. Electron. Control,* 15(4), 1830–1840.
[6] A. Sandhyavitri, M. A. Perdana, S. Sutikno, F. H. Widodo, *IOP Conf. Ser. Mater. Sci. Eng.*, **309**, 012016, 2018.
[7] V. Huijnen *et al.* 2016 *Sci. Rep.*, **6**, 1–8.
[8] M. M. Rahman, G. J. McDermid, M. Strack, J. Lovitt, 2017 *Remote Sens.* **9**, 10, p. 1057.
[9] C. E. Ballard, N. McIntyre, H. S. Wheater, J. Holden, Z. E. Wallage, 2011 *J. Hydrol.* **407**, 81–93.
[10] G. E. . B. Susilo, K. . Yamamoto, T. . Imai, Y. . Ishii, H. . Fukami, M. . Sekine, 2013 *Hydrol. Sci.*

*J.*58(3), 539–548.
[11] A. Sulaiman, K. Kimura, K. Takano, 2017 the *1st International Round Table Peatland Restoration.*
[12] H. Takahashi, 2017. *Monitoring and forecast of groundwater level in a tropical peatland and a rice field - Central Kalimantan, West Kalimantan, Jambi, Riau and Central Jawa Provinces*, Indonesia.
[13] S. M. Ross, 2010. CHAPTER 12 - Linear Regression, S. M. B. T.-I. S. (Third E. Ross, Ed. Boston: Academic Press.
[14] S. Sutikno *et al.*2019 *MATEC Web Conf,* **6003**, 276, 1–7.

Structural style of the North Sumatra basin, offshore Aceh

Muchlis[1,2,*], Chris Elders[1]
[1]School of Earth and Planetary Sciences, Curtin University
[2]Departments of Earth Science and Engineering, Syiah Kuala University

E-mail: Muchlis@postgrad.curtin.edu.au

Abstract. The Offshore North Sumatra Basin is an underexplored basin. Only a limited amount of exploration has occurred in the past resulting in the discovery of one economic field that is currently producing, a number of uneconomic discoveries and some unsuccessful wells. This is reflected in limited geological studies of the area. This paper demonstrates the importance of N-S trending faults associated with Basin formation in the Oligocene, continued fault activity in the late Oligocene and both compressional and extensional reactivation in the Miocene.

1. Introduction

The North Sumatra Basin (NSB) is located in the western part of Indonesia, part of Aceh and offshore North Sumatra. The basin is bordered by Mergui Ridge to the west and Malacca Platform to the east. To the south, it is bounded by the Barisan Mountains. To the north, it extends into Thailand offshore, where it is known as the Mergui Basin (Figure1).

Exploration in offshore Aceh began in the 1970s. Most wells drilled targets on the crest of horst structures. So far, only one field is in operation [1]. Other fields have not been developed because of economic issues. Some wells did not encounter hydrocarbon.

The offshore NSB needs to be further studied to understand its petroleum potential. This paper aims to introduce and discuss the structural evolution of the offshore area of Aceh Province, where fewer structures have been drilled than in the onshore area.

2. Regional Geology

2.1. Tectonic Setting

The North Sumatra Basin is a Paleogene-Neogene basin with a complex tectonic history located in a back-arc setting relative to the India-Sumatra subduction trench [2]. The initiation of the India-Sumatra subduction began in Cretaceous [3]. The rate of subduction in Sumatra is between 6 to 7 cm/year [4]. The activity of the plates is indicated by the myriad of seismic shocks that occur along the trench.

The Barisan Mountains separate the North Sumatra Basin from the India-Sumatra trench. They are cut by the Sumatra Fault System (SFS). The dextral Sumatra dextral strike-slip Fault is associated with the oblique subduction of the Indian oceanic plate beneath Sumatra. Several authors stated that the initiation of the Sumatra Fault System and spreading in the Andaman Sea occurred at the same time during the mid-Miocene [5]. Uplift of the Barisan Mountains occurred in the Pliocene [6]. This

Content from this work may be used under the terms of the Creative Commons Attribution 3.0 licence. Any further distribution of this work must maintain attribution to the author(s) and the title of the work, journal citation and DOI.
Published under licence by IOP Publishing Ltd

illustrates the importance of understanding the interaction between extension, compression and strike-slip tectonics in the evolution of the North Sumatra Basin.

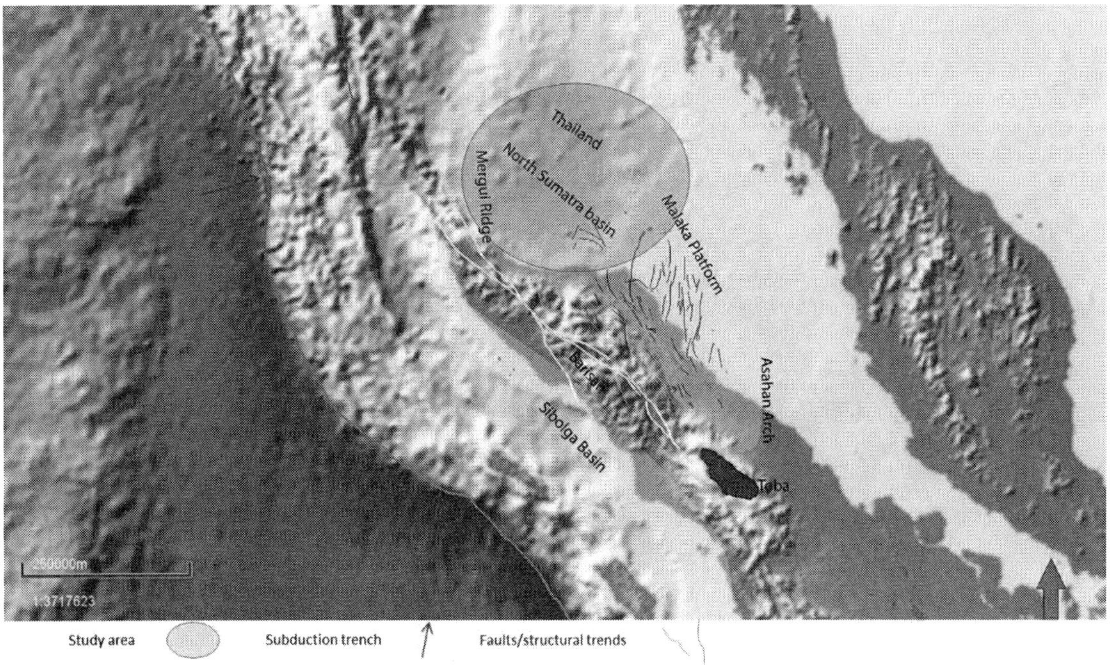

Figure 1. Location of North Sumatra Basin and its regional geology

2.2 *Regional Stratigraphy*

The North Sumatra Basin contains sedimentary rocks from late Eocene to Recent age. However, most exploration wells only penetrated the crest of horst blocks and found Oligocene rocks deposited in the basement. Thus, the age of the sedimentary sequences that filled the graben is less well constrained. A summary of the stratigraphy of NSB is shown by figure 2 and is described as the following section.

The bottom part is the basement. Most of the wells drilled into basement rock in the NSB encountered metamorphic rocks of Cretaceous age. According to Cameron et al. (1980) and Barber (2000), the Cretaceous basement is categorized as the Woyla Group which outcrops along the western margin of the Barisan Mountains. Then followed by Meucampli/Tampur Formation (Late Eocene). Tampur Formation primarily consists of carbonate sediment and dolomites with cherts nodules. The Tampur Formation was deposited under sub-littoral to marine conditions [7], in the eastern part of basin high near the Asahan arch. Meanwhile, the Meucampli Formation consists of clastic sediments that outcrop in the Northwest of Aceh. Next is Parapat Formation (Early Oligocene). The Parapat Formation consists of conglomerate and sandstone. The sands came from local contributions from the horst blocks. The sediments were deposited in fluvial, lacustrine and shallow marine environments (Meckel, 2013). After that is Bampoo Fm (Late Oligocene). The Bampo Formation primarily consists of poorly laminated, pyritic carbonaceous shale, with a minor amount of carbonaceous material. These sediments were deposited during the transgression and deepening of the graben. The Bampo Formation is the primary source rock that generated gas and is reported to be the source for the giant onshore Arun field. In addition, this formation could be a candidate for unconventional gas plays in the future. The Bampoo Formation is followed by Belumai/Peutu Formation (Early Miocene). Sandstones are dominant in the Peutu Formation. Cameron et al. (1980) defined the sediments as the Belumai Formation, consisting of fine- to medium-grained sandstones. Carbonate reefs of the Peutu

Formation are major producers of condensates and host the Arun field (Mobil). Then, Baong Formation is deposited in Middle Miocene. The Baong Formation (Middle Miocene) primarily contains massive dark grey, variably calcareous shale, rich in foraminifera, which indicates a neritic marine depositional environment. Then, it is followed by Keutapang Formation (Late Miocene). The Keutapang Formation comprises of fine-grain sandstones and shales with a thin layer of limestones [9]. Seurula Formation is on top of Keutapang Formation. The Seurula Formation contains coarse-grained sandstone and shales [9]. Finally, Julu Rayeuk Formation is on top of Seurula Formation. This formation is dominated by sandstone sedimentary rocks which vary from fine to coarser grain from the base to the top of the sequence [9].

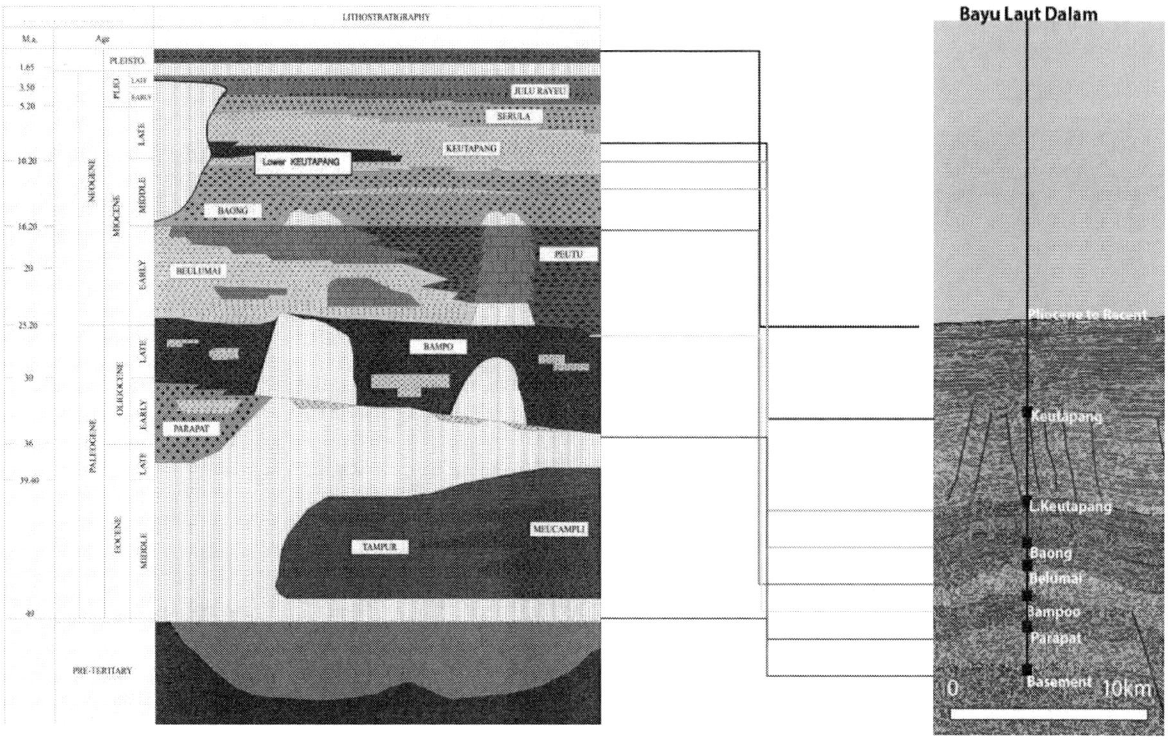

Figure 2. Left is the general Stratigraphy chart of the North Sumatra Basin [10]. Right is top formations from the Bayu Laut Dalam well which are tied to the seismic.

3. Data Set

More than 60 lines of 2D seismic data from offshore NSB are used to build a structure map. The length of the line varies from 50 km to 180 km. The spacing line is 10kms. The record length of each seismic line varies from 4000 milliseconds to 6000 milliseconds (ms) two-way time (TWT). Most lines are oriented East-West and North-South. NE-SW and NW-SE oriented lines are also present. The resolution of the image varies from poor to moderate. The seismic amplitudes are heterogeneous varying between different surveys. The stratigraphy and geological structures of the offshore NSB were interpreted and mapped using Petrel. There is one well, Bayu Laut Dalam (BLD) which tied into the seismic and controlled the stratigraphic interpretation (figure 2). The well was spudded on a basement high (horst) and contained recent to pre-Paleogene sediment. Eight horizons were interpreted from the pre-Paleogene basement to younger sediments of Pliocene to recent. They are; Basement, Parapat, Bampoo, Belumai, Baong, Lower Keutapang, Keutapang and Pliocene to younger.

4. Major Sequences

The interpreted horizons are categorized as pre-rift, early syn-rift, late syn-rift, and post-rift, syn-inversion and post inversion (see Figure 3 & 4). Sedimentary rocks that filled at each stage represent the dominant environmental and the basin setting at that time.

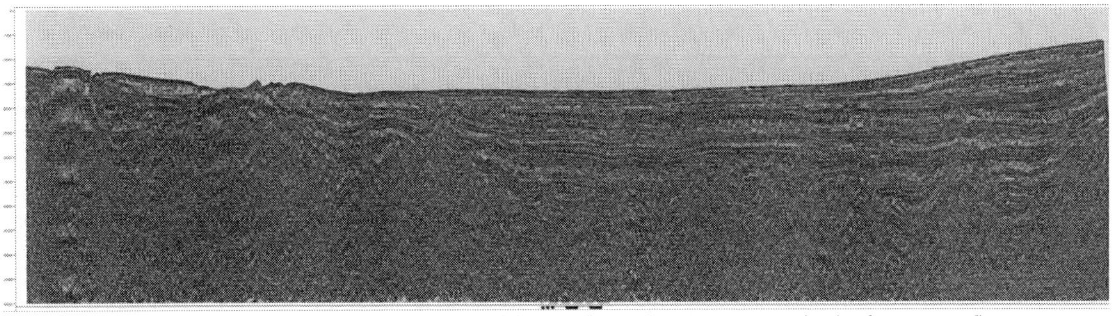

Figure 3. The seismic line from offshore of North Sumatra Basin (uninterpreted).

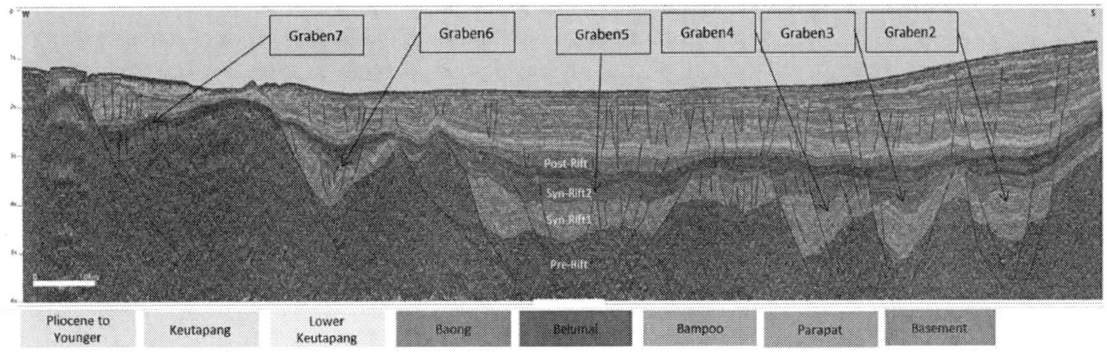

Figure 4. Seismic Interpretation from offshore of North Sumatra Basin. This is the longest seismic line (180km) showing the regions of the offshore North Sumatra Basin. This section shows the structural deformation in every sequence and also the distribution of sediments in the basin

4.1 Pre-rift (Basement)

The basement is characterized by a chaotic and dimmed reflector. There are several major faults that cut through the basement. First, extensional faults that shape horst and grabens (Figure 4). The displacement varies from 900 to 1450 meters. Dip angles are variable.

4.2 Early and Late Syn-Rift, Early to Late Oligocene (Parapat and Bampoo Formation)

This unit is characterized by parallel and continuous reflectors. In the upper part, it shows a high amplitude, but the lower part shows medium to low amplitude. Early syn-rift is marked by thicker Parapat sediments toward the major faults within the grabens than on the horst. The late syn-rift is shown by the thicker Bampoo unit toward the major faults.

4.3 Post-Rift, Early to Middle Miocene (Belumai and Bampoo Formation)

This unit is characterized by moderate to high amplitude, continuous, and parallel reflectors. During the deposition of this unit, there is less evidence of the deformation. The thickness of this sequence varies from thinning at the edges and thicker in the depocenter.

4.4 Syn-Inversion, Late Miocene (Keutapang)

The syn-compressional sequence is described by a thinning sequence deposited on to the anticline (very symmetrical) crest. Within the Keutapang, there are onlap features cutting the anticline indicating the beginning of compression. The seismic is characterized by parallel and continuous reflectors. Populations of minor faults are present within this sequence.

4.5 Post-Inversion (Late Miocene to Recent)

The seismic characteristic of this unit is high amplitude, continuous and parallel reflectors. However, in the lower slope area at the east, chaotic and transparent reflectors are identified. There are few faults, although one major fault reactivated old structures during the Pliocene. It is located in the eastern area of the Basin margin.

5. Structural style and timing

North-South trending faults define a series of elongated horsts and grabens. Faults show variable dip and angle both to be East and West. Then, several minor faults are present within the graben that terminate at the top of Oligocene time. Those, plus the syn-rift characteristics of Parapat formation show that extension occurred during Oligocene. The fact that the timing of the extension is unclear as the older sediments within the grabens have not been penetrated. During the late Oligocene, the basin experienced extension for the second time, which caused some Major faults reactivated. It is shown by the thickening of Bampoo formation within the grabens towards the faults. During this time, the fault may increase the angle of rotation. It recreates the large displacement between the upthrown and downthrown blocks. In some parts, the vertical displacement may exceed 1400 meters. The magnitude of the grabens varies from 35 km to 10 km wide. The length also ranged from 70 km to 80 km. Inversion structure present in the west part. This structure is defined by a symmetrical anticline above the footwall. This structure formed during the late Miocene that we indicated by the thinning sequence within the Keutapang Formation. Some minor faults also present during this age. During Pliocene, older structures were reactivated by major normal faults. The reactivation is defined by different thicknesses of the unit in the lower and upper parts that cut by the fault. The fault terminates within the Pliocene sediment unit.

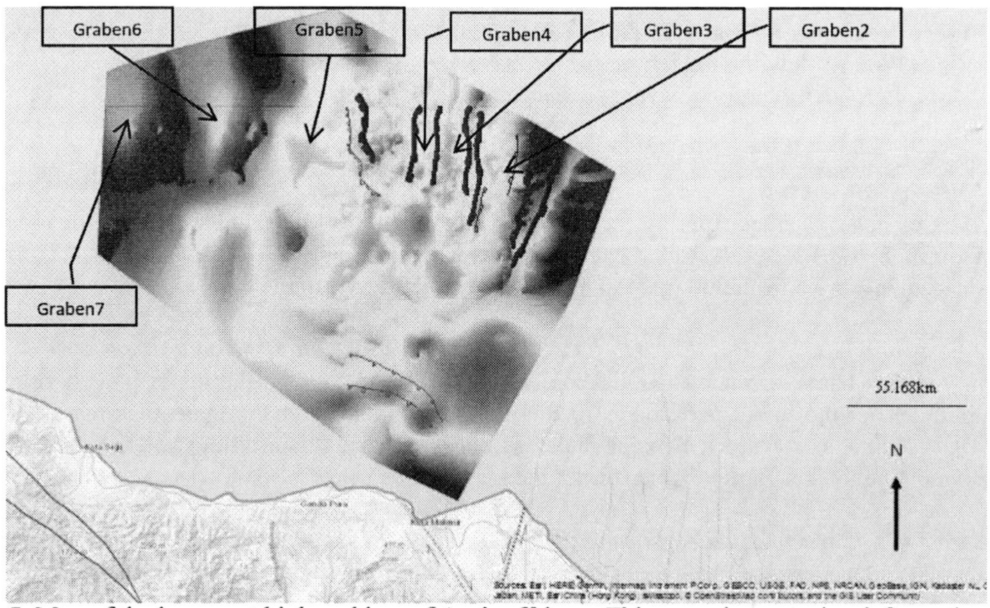

Figure 5. Map of the basement high and low of Aceh offshore. This map shows major deformations that occurred during Paleogene-Neogene. This map was done by Arc-Map software, whereas the input from interpreted seismic horizons using Petrel.

6. Conclusion

The interpretation of 2D seismic lines has revealed the major structures in the offshore Aceh of North Sumatra Basin. Mostly, those structures have oblique N-S orientation. These major faults controlled the horsts and grabens development during early and late Oligocene under the extensional regime.
Then, fault activity continued both compressional and extensional reactivation in the Miocene and the Pliocene. Offshore Aceh still attracts many companies to explore for oil and gas. As a result, the data of this basin will increase and hopefully can be opened to academia in Aceh. By doing so, it is expected that this work can be updated with details and more precise.

References

[1] McArthur A, Helm, R. 1982 *Indonesia Petroleum Association.* **11**, 127-146
[2] Hall R. 2012 *Tectonophysics.* **570**, 1-41.
[3] Morley C 2002 *Tectonophysics.* **347**, 189-215
[4] Hamilton W 1973 *United States Geological Survey Professional Paper.* **1078**
[5] Curray J, Moore D, Lawver L, Emmel, F, Raitt R, Henry M, Kieckhefer R 1979 *American Association of Petroleum Geologists Memoirs.* **29**,189-198.
[6] De Smet M, Barber A 2005 *Geological Society London Memoirs.* **31**, 86-97.
[7] Cameron N R, Clarke M, Aldiss D, Aspden J, Djunuddin A 1980 *Indonesia Petroleum Association.* **9**, 149-187
[8] Barber A 2000 *Journal of Asian Earth Sciences.* **18**, 713-738
[9] Mulhadiono P H, Soedaldjo P 1978 *Indonesia Petroleum Association.* **7**, 107-132
[10]Anderson B, Bon J, Wahono H 1993 *Indonesia Petroleum Association.* **22**, 169-189

Application of precast foamed concrete panels for the structural deck of green roof system

A Munir[1*], Muslimsyah[1], M Afifuddin[2], Mahlil[2]
[1] Department of Architecture and Planning, Syiah Kuala University, Banda Aceh, Indonesia
[2] Department of Civil Engineering, Syiah Kuala University, Banda Aceh, Indonesia

E-mail: munirsyadi@unsyiah.ac.id

Abstract. Rapid development in urban areas has caused many environmental problems, one of which is the phenomenon of urban heat island (UHI) which results in excessive energy usage. The building sector is the most responsible for energy use, which accounts for 40% of total energy consumption. A green roof is a sustainable solution to solve the problems related to urban heat islands. To reduce more structural self-weight, lightweight foamed concrete could be applied as a roof construction material. This paper presents experimental results on the structural behavior of a U-shape precast panel made of lightweight foamed. Three foamed concrete panels in the U-shaped form having different width dimensions of 200mm, 400mm and 600mm with a clear span of 1800 mm were cast and tested until failure under a double mid-span concentrated load. The panels were considered as a simple U-shaped beam filled with a growing medium for a green roof system. This result shows a potential use of foamed concrete for roof deck structure integrated with a green roof system. The U-shaped panel with a width of 400mm is the optimal form of the three tested specimens for the structural roof deck of the green roof system.

1. Introduction

Building construction in urban areas causes many environmental problems, including the urban heat island phenomenon. The temperature of cities continues to increase because of the heat island phenomenon and the climatic change. Using green roofs is an enormous potential for the urban environment in urban heat island mitigation [1,2] and shows an increase in the thermal performance of buildings in tropical regions [3,4,5].

The existence of plants on the building roof provides many ecological and economic benefits, including stormwater management, energy conservation, mitigating the effects of urban heat islands and extending the life of roofing materials, and providing aesthetically better environment [6]. More than an aesthetic function, green roofs give a greater advantage to buildings by reducing heat flow through evapotranspiration, shading the roof, and increasing roof insulation and thermal mass, thereby reducing energy requirements for cooling systems [7].

Construction and maintenance costs are the reasons green roofs have not been widely applied. The main disadvantages of the green roof are cost and weight. The green roofs add more weight to the structural load, which leads to changes in the structural design of columns, beams, and slabs, resulting in a more expensive structural cost [8,9]. Green roofs add 60 - 150 kg/m^2 (extensive) and 180 – 500

Content from this work may be used under the terms of the Creative Commons Attribution 3.0 licence. Any further distribution of this work must maintain attribution to the author(s) and the title of the work, journal citation and DOI.
Published under licence by IOP Publishing Ltd

kg/m^2 (intensive) for the roof construction load with the green roof medium of 60-200 mm for extensive and 150 – 400mm for intensive [10]. This is a major disadvantage when applied to earthquake-prone areas like Indonesia. Using extensive types of green roofs is more likely to be applied in Indonesia given the lighter structural load and simpler application. Reducing the self-weight of the roof by using lightweight foamed concrete is an alternative to reduce the risk of building damage due to earthquake disasters.

Lightweight concrete has been used widely today because it has many advantages, including low density, low thermal conductivity, high flowability, and self-compacting. The density of foamed concrete ranges from 300 to 1800 kg/m^3, which is significantly lower than that of normal concrete (2400 kg/m^3). Cement, water, and foam compose foamed concrete [11,12,13]. With the density ranges that can be made, foamed concrete can be applied for both structural and non-structural construction. Due to distinctive properties of foamed concrete including density reduction, low thermal conductivity, high flowability, and self-compacting concrete, and given the ease to produce and its relative cost-effectiveness, foamed concrete has found applications in many civil and structural engineering areas [14].

One example application of lightweight concrete is as a structural panel wall in the form of precast. Many researches have focused on the use of lightweight foamed concrete for structural elements because it has the advantage of being lighter than conventional concrete. Using foam concrete panels for walls of buildings is commonly applied, but for the roof is still limited. This research initiates the use of precast foamed concrete panels for roofs by maximizing its benefits of being lightweight material with high heat resistance and the ability to reduce the heat flow to the room beneath by absorbing solar thermal radiation by the green roofs system. This combination is expected to reduce energy usage in buildings to mitigate the effects of urban heat islands in urban areas.

This study focusses on the structural behavior of foamed concrete panels when applied to a green roof system. The aim of this research is to investigate the flexural behaviors of a U-shaped lightweight foamed concrete panel. Three shapes of concrete panels were cast with a foamed concrete density of 1400 kg/m^3 and reinforced by 8mm rebars and 1mm galvanized wire mesh. The behavior of the test object against loading is also analyzed based on crack patterns and failure modes. This paper discusses structural aspects in terms of the ability of foamed concrete panels to carry roof loads by adding green roof systems.

2. Experimental Program

2.1. Material

Materials used for the concrete mix are Portland cement (PC) Type-I, foam agent, and water. The physical and chemical properties of PC were not examined in this study since the material has been certified for Indonesian Standard (SNI) 03-2847-2002. However, we checked the physical condition of the packaging and the cement grains, which are still in good condition. The drinking water from the city supply was used in the experiments. The foam controls the density of concrete by incorporating pre-formed foam into a fresh concrete mix to obtain a target density of 1400 kg/m^3 of fresh concrete. The ratio of a foaming agent to water is 1:30 by volume.

The foam generator produced foam from the foaming agent and a certain quantity of foam was poured to the concrete mix shortly after preparation to obtain the target density. To improve workability, we added a small amount of super-plasticizer to the concrete mix, which does not influence the characteristics of foamed concrete. The composition of foamed concrete mix calculated on the basis of specific gravity (SG 1.4) and water to cement ratio (w/c 0.4) comprises 1000 kg/m^3 of PC, 400 kg/m^3 of water, and 282.5 liter/m^3 of foam. Six cylindrical specimens (Ø15-30) were cast to analyze the concrete characteristics in terms of their specific gravity and compressive strength. The samples were cured by wet covering (moist curing) method in a shaded area for 28 days before the test.

2.2. Precast panel specimens

Specimens for precast foamed concrete roof panels are U-shaped beams prepared with 3 different widths, namely 200mm, 400mm and 600mm with 40mm of thickness. The depth space for the green roof medium is 150mm and the total length of the beam is 2000mm. The panels use Ø8 mm steel and galvanized welded wire mesh (wire diameter 1.0 mm, opening square 1 inch) for reinforcement. Figure 1 shows the details of the panels with reinforcement bars and dimensions of the beams.

The specimens were cast based on the proportion of the mix design. The PC and water were weighed and mixed in a concrete mixer until it was mixed uniformly, then added the foam into the wet mix and controlled for the target density of 1400 kg/m^3. The fresh concrete mix is then poured into the mold that has been prepared by the dimensions specified. We open the mold after 7 days of the concrete age and then cured by wet covering fabric (moist covering method) exposed to shaded open air before the flexural testing schedules.

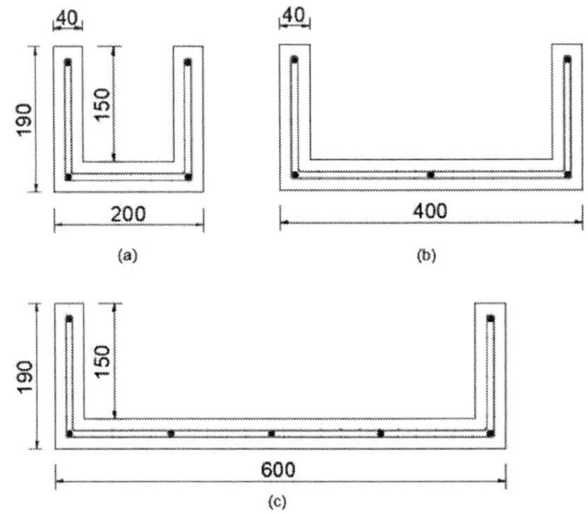

Figure 1. Cross-sections of precast foamed concrete panels
with different width (unit in mm).

2.3. Compression Test

The compressive strength test was conducted using a universal compression test machine with a constant load rate of 0.1 kN/s after 28 days of age. The test was performed under ASTM Standard C234. An axial compressive load with a specified rate of loading was applied to Ø15cm-50 cm until failure occurred. The compressive strength was obtained based on the average crushing strength of three crushed cylindrical specimens.

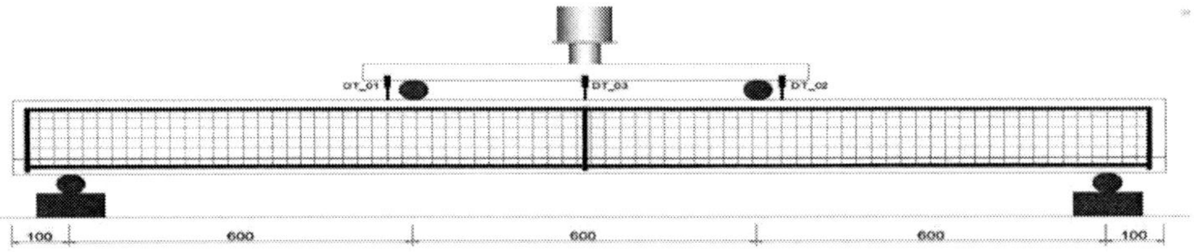

Figure 2. Test specimen set-up.

2.4. *Full-Scale Flexural Test*

All beams were tested to investigate the flexural capacity under pure bending by using *four*-point loading tests. Four-point symmetrical loading tests were adopted for the beam specimens under ASTM D6272-17. The dimension and the details of the specimens are shown in Figure 2. All beams were loaded constantly with a loading speed of 0.1 kN/s using a hydraulic press with an automatic recording of the applied force until failure.

To determine deflections, linear displacement transducers (LDT) set at five points along the specimen length (at supports, at loading points and in the middle of the span). Two LDTs at the loading positions and two LDTs at support positions were installed to ensure that both loading and support points were balanced during the test and one LDT was placed under the mid-span of the beam to obtain the maximum deflection of the beams. Concrete strain gauges were installed in the tension bar (bottom bar reinforcement) at the mid-span of the beams. The strain gauge and five LDTs that attached to the tested beam were connected to a data logger to obtain concrete strains as well as deflection at distinct levels of loading respectively throughout the testing period. During the test, photos were taken to observe cracking behavior and failure mode for the beams for visual analysis of strains and displacements, especially crack pattern and propagation.

3. Result and Discussion

3.1. *Compressive strength*

Consistency and stability checks are necessary to ensure that the mix is considerably stable. The stable mixture is achieved when the ratio of fresh density to chosen density and the ratio of fresh density to hardened density are kept to unity [15]. Table 1 shows the test result of cylindrical samples for the compressive strength and stability of the lightweight foamed concrete at 28 days of age. The average density of all samples is 1392 kg/m^3, only 0.5% less than the designated density of 1400 kg/m^3. This result shows a stable condition of foam with a ratio of 1.01 on average compared to the density of fresh foamed concrete. The average compressive strength is about 12 MPa, which is right for its density to density 1400 kg/m^3 as summarized from various literature [16]. The mixtures got a compressive strength lower than 17 MPa, which is the minimum requirement compressive strength for structural used concrete. However, the roof panels can be considered as a simple beam structure that mostly works on flexural load only.

Table 1. Compressive test results of foamed concrete cylindrical samples

Cylindrical specimen	Specific gravity [a] (Kg/m^3)	Maximum load (Ton)	Compressive strength (MPa)	Stability [b]
Sample-1	1373.21	20.00	11.10	1.02
Sample-2	1361.89	22.00	12.21	1.03
Sample-3	1384.53	21.00	11.65	1.01
Sample-4	1450.55	24.00	13.32	0.97
Sample-5	1384.53	22.00	12.21	1.01
Sample-6	1399.62	21.00	11.65	1.00
Average	1392.39	21.67	12.02	1.01
SD	31.16	1.37	0.76	0.02

[a] based on the dry weight of cylinder samples at age 28 days
[b] ratio of fresh to harden foamed concrete

3.2. *Deflections*

Figure 3 illustrates the relationship between the load given to the deflection that occurs in the U-shaped beam specimen. The evaluation of the flexural behavior was made by recording failure load and analysis of deformations during the test. Table 2 shows a summary of the achieved results for all beams. The Table includes the ultimate load, mid-span deflection at the ultimate load, experimental ultimate moment, and predicted moment for the green roof load on the day the experiment.

Table 2. Flexural test results for maximum load capacity, deflection, and estimated moment value for the extensive green roof.

Specimen	Pmax (tf)	Defection (mm)	Mmax[a] (tf.m)	Mmax, green roof[b] (t.m)
U20	2.89	21.14	0.87	0.053
U40	4.2	22.25	1.26	0.066
U60	3.49	14.63	1.05	0.079

[a] Maximum bending moment calculated based on Pmax test results

[b] Maximum bending moment estimated based on maximum green roof load (60 - 150 kg/m^2) for extensive green roof [10] and 100 kg at mid-span for live-load for a worker.

(a)

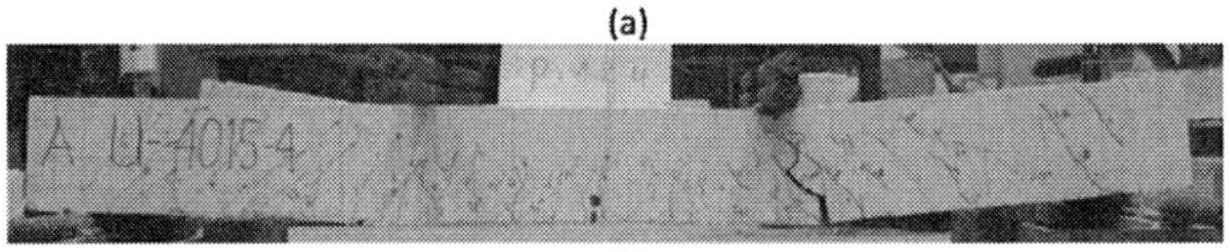

(b)

(c)

Figure 3. Test specimen cracking condition after maximum loading, (a) U-20, width = 200 mm, (a) U-40, width = 400 mm, (a) U-60, width = 600 mm.

3.3. Ultimate moment capacity and cracking pattern

Based on the test results on all precast specimens, it shows that there are different behaviors for each test object. This result concludes that all beams fail in bending during the loading process. Figure 4 illustrates the load-deflection relationship of the specimens. The highest ultimate load and the ultimate moment is obtained for the U-40 specimen, namely 4.2 tf and 1.26 tf.m, respectively. Increasing the width of the panel cross-section does not increase the ultimate moment, thus reducing the ultimate load. This weakening can be explained based on the cracking and failure patterns that occurred during the test.

The failure of U-60 is different from that for U-20 and U-40, as shown in Figure 4. The initial cracking and failure of the specimens U-20 and U40 occurred at the bottom part of the specimens (tension zone) at the loading points. As for the U-60, the bending failure occurs at the top side of the specimen (compression zone). It can be explained that the failure occurred due to the increase of the cross-section are and reinforcement in the tension zone, while in the compressive zone there was no difference with the U-20 and U-40. This causes an increase in the moment of resistance in the tension zone, resulting in weakening in the compressive zone. From these results, it can be concluded that increasing the size of the precast panel does not improve moment capacity. Based on this test, the optimum panel width is 400mm.

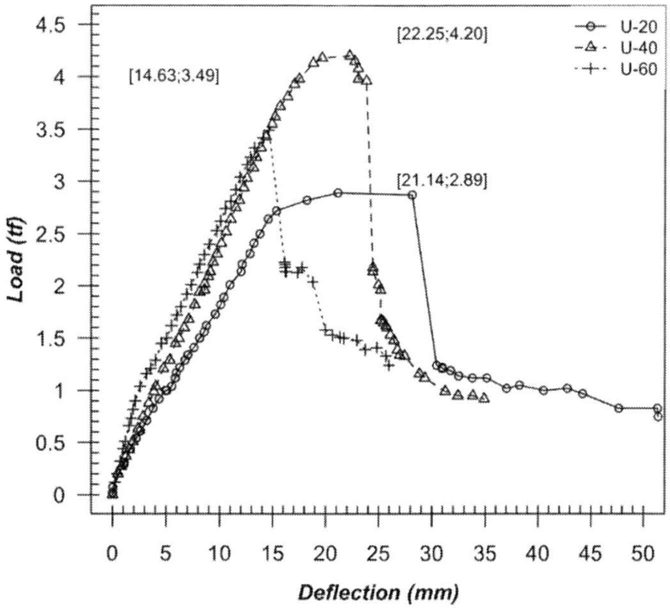

Figure 4. Deflection at the mid-span recorded during the test for each specimen.

3.4. Green roof application

The test results of precast foamed concrete panel specimens have shown the performance of foamed concrete for roof construction, which also functions as a green roof system. In this study, the roof panel is treated as a simple beam on two supports with a 1800mm of span. Based on the loading scheme provided, the strength of the beam can be seen from the value of the ultimate moment for U-40 is equal to 1.26 tm when the panel failure due to maximum loading. Table 2 shows the comparison of the ultimate moments and maximum moment for all panel specimens with the green roof load assumed to be 60 - 150 kg/m^2 for an extensive green roof system [10]. The panels also add the live load that works on the roof at 100 kg (the maximum moment at mid-span). The maximum moment with a green roof load that occurs in each panel (Table 2) is still much smaller than the ultimate moment. The U-40 panel that has the best performance shows that the maximum moment of the green roof load is only 5% of the ultimate moment. With this assumption, using a 4000mm span still produces a maximum moment of 0.22 tm (10% ultimate moment) for the U-40 panels. However, this assumption is still general, which requires further study of the behavior of U-shaped panels from various dimensions and other aspects.

Another application form of the panel as a green roof is by treating the panel as a roof cover that is placed on top of the girder with a shorter span (for example 1000mm), as on conventional roof construction. The panels only function as a container for the green roof and transfer the load to the girder construction underneath. Thus, the thickness of the panel can be reduced so that its weight also decreases, which then makes the installation process easier.

This research has shown the prospect of using integrated precast foamed concrete panels as a green roof system as an initiation of the use of green roofs for areas prone to earthquake hazards. However, further investigation is still needed about the behavior of foam concrete to loading for various combinations of dimensions of the U-shaped precast panel. The use of panels as a roof also still requires practical applications for the systems such as inter-panel connection systems, waterproofing systems, green roof drainage systems, and various other technical aspects.

4. Conclusions
Applications of lightweight foamed concrete as green roof construction to reduce structural self-weight have been discussed in this paper. This paper presents experimental results on the structural behavior of a U-shape precast panel made of lightweight foamed. Three foamed concrete panels in the U-shaped form having different width dimensions of 200mm, 400mm and 600mm were tested until failure under a double mid-span concentrated load. The panels were considered as a simple U-shaped beam filled with a growing medium for a green roof system. The U-shaped panel with a width of 400 mm is the best form of the three tested specimens for the structural roof deck of the green roof system. The maximum moment calculated based on a green roof load is lower than the ultimate moment based on test results. This result shows a potential use of foamed concrete for roof deck structure integrated with a green roof system. Further study requires practical applications of systems such as inter-panel connection systems, waterproofing systems, green roof drainage systems, and various other technical aspects.

5. Acknowledgment
The authors would like to thank the University of Syiah Kuala and the Directorate General of Higher Education, the Ministry of National Education, Indonesia for the financial support of this research.

6. References
[1] Santamouris M 2014 *Sol Energy* **103** 682–703.
[2] Raji B, Tenpierik MJ, Van Den Dobbelsteen A 2015 *Renew Sustain Energy Rev* **45** 610–23.
[3] Jim CY 2012 *Landsc Ecol Eng* **8**(2) 173–87.
[4] Hashemi SSG, Mahmud H Bin, Ashraf MA 2015 *Renew Sustain Energy Rev* **52** 669–79.
[5] Chua KJ, Chou SK 2010 *Energy* **35**(2):667–78.
[6] Getter KL, Rowe DB 2006 *HortScience* **41**(5) 1276–85.
[7] Blank L, Vasl A, Levy S, Grant G, Kadas G, Dafni A, et al. 2013 *Build Environ* **66** 23–28.
[8] Bianchini F, Haque ABMR, Hewage K, Alam MS 2016 *Earthquakes Struct* **11**(2) 265–80.
[9] Bianchini F, Hewage K 2012 *Build Environ* **48**(1) 57–65.
[10] Besir AB, Cuce E 2018 *Renew Sustain Energy Rev* 82 915–39.
[11] Ramamurthy K, Kunhanandan Nambiar EK, Indu Siva Ranjani G 2013 *Cem Concr Compos* **31**(6) 388–96.
[12] Chica L, Alzate A 2019 *Constr Build Mater* **200** 637–47.
[13] Mohd Sari KA, Mohammed Sani AR 2017 *MATEC Web Conf.* **97**.
[14] Amran YHM, Farzadnia N, Abang Ali AA *Constr Build Mater* **101**(3) 990–1005.
[15] Kunhanandan Nambiar EK, Ramamurthy K 2008 *J Mater Civ Eng* **20**(2) 111–7.
[16] Abd AM, Abd SM 2017 Case *Stud Constr Mater* **6** 8–15.

Response structure analysis of prestressed box girder concrete bridge due to earthquake loads

M Miranda[1], R Suryanita[1]*, E Yuniarto[1]
[1]Civil Engineering Department, Faculty of Engineering, University of Riau, Pekanbaru, Indonesia

E-mail: reni.suryanita@eng.unri.ac.id

Abstract. The damage of the bridge structure that occurs in Indonesia is almost close to the number of bridge construction. The damage condition is also related to the influence of Indonesia on active tectonics and volcanic pathways which cause damage to the bridge structures. The analysis of the structural response of the prestressed box girder concrete bridge aims to identify the value of the displacement structure on the bridge pier against the earthquake load. Meanwhile, the structure responses numerical analysis of the prestressed box girder concrete bridge using MIDAS CIVIL software. The analysis of structure response prestressed box girder concrete-bridge structure is carried out with three soil conditions, that is hard soil, medium soil, and soft soil. The results of the study, the displacement value on the pier bridge on X direction for hard soil conditions is 1.26 cm, medium soil is 1.53 cm, and soft soil is 2.02 cm. Thus, it can be concluded that the value of the bridge displacement due to the planned earthquake load on hard soil conditions is 21.4% smaller than that in the medium-soil condition and 60.32% smaller than the soft soil.

1. Introduction

According to the Statistics Information Book [1], in 2016, Indonesia recorded has 18,014 bridge units, but around 61.59% of bridges in Indonesia were not in good condition and damaged. Indonesia is one of the most seismically active countries on earth. Indonesia occupies a very active tectonic zone because the three large plates of the world and other small plates meet each other in the territory of Indonesia. The existence of interactions between these plates places the territory of Indonesia as a region that is very prone to earthquakes [2].

The pier system on the bridge can be a single or compound column or can be a full wall. In the SNI 2833-2016 regulation of bridge design against earthquake loads, the difference in pier systems is in the modification factor of the response of earthquake values (R) that can be deployed by each bridge structure system or subsystem [3].

An earthquake is a disaster that can be harmful to the community, such as financial loss and also a risk of death. Pekanbaru is a city located in the middle of Sumatera Island. Even though the city of Pekanbaru is a city that rarely occurs earthquake, but Pekanbaru has ever felt the impact of the big earthquake that occurred in West Sumatera in September 2009 [4].

In the western part of Indonesia, the process of subducting the Indo-Australian Plate into the Eurasian Plate in western Sumatra resulted in the occurrence of earthquakes with magnitudes 8 or 9 that are very vulnerable. For example, in 2004, the earthquake in Banda Aceh City with magnitude 9.2 caused damage to 120 bridges with a category of severe damage [5], [6].

[2] E-mail: reni.suryanita@eng.unri.ac.id

Content from this work may be used under the terms of the Creative Commons Attribution 3.0 licence. Any further distribution of this work must maintain attribution to the author(s) and the title of the work, journal citation and DOI.
Published under licence by IOP Publishing Ltd

IOP Conf. Series: Materials Science and Engineering **796** (2020) 012040 doi:10.1088/1757-899X/796/1/012040

The purpose of this study was to analyze the earthquake acceleration performance with three soil conditions, namely hard, medium and soft soil on the bridge pier. The choice of analysis on the pier section of the bridge is based on the fact that the failure of the structure on the bridge pier will fail the overall bridge structure. This analysis uses nonlinear static, which is using MIDAS CIVIL software so that the planned bridge has a high earthquake resistance and is more economical.

2. Literature Review

In this analysis, the bridge girder used is a box-type girder type which is designed for prestressed concrete that has a long span, which is 200 meters. According to [7], [8], prestressed concrete is reinforced concrete which has been given low stress to reduce the potential tensile stress in the concrete due to workload. According to [9],[10], loading is the magnitude of the load used in calculating the structure so that it does not experience destruction during the life of the bridge. The loads that must be analyzed in a bridge plan are divided into three groups, namely permanent load, traffic load, and earthquake load.

The permanent load consists of the weight and additional dead load.

1. Self-weight is the weight of materials and the parts of a bridge consisting of structural elements such as girder, floor plate, and diaphragm.
2. An additional dead load is the weight of all materials which give rise to a load on the bridge consisting of non-structural elements. The additional dead loads are asphalt, sidewalks, curbs, and safety fences.

In this study, earthquake analysis was divided into three land categories, namely the categories of hard, medium and soft soil types based on PGA values from the Pekanbaru area around 0.214 g. Based on SNI 2833-2016, the spectrum response is a graph that states the relationship between the period of vibrating structure (T) and the maximum structural response when experiencing existing earthquake vibrations. For practical needs, the spectrum response is simplified in graphical form as can be seen in Figure 1.

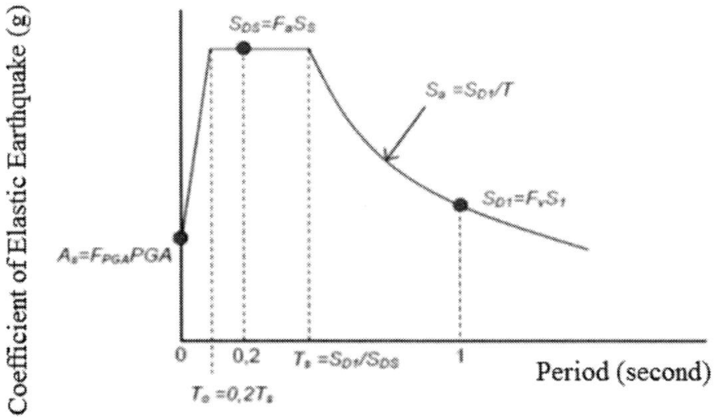

Figure 1. Typical Form of Response to the Earthquake Spectrum

The requirement for the number of variations is regulated in SNI 2833: 2016 where the number of variations must be sufficient to get the participation of combined mass variations of at least 90% of the

IOP Conf. Series: Materials Science and Engineering **796** (2020) 012040 doi:10.1088/1757-899X/796/1/012040

actual mass in each orthogonal horizontal direction of the response reviewed by the model. If it is not sufficient, then the amount of variety must be increased to meet the requirements for mass participation.

3. Research Methodology

Bridge structure modeling refers to the case study of the construction of the SEI SIAK II (MYC) prestressed concrete girder bridge on the Pekanbaru-Duri Cross Road located in the Rumbai District area, Pekanbaru City.

The bridge structure data analyzed are as follows:

1. Type of girder	= box girder	7. Width of lane	= 5 meters
2. The quality of concrete girder	= fc'45	8. Number of girder	= 51 pieces
3. Total length	= 200 meters	9. Girder dimensions	= varies
4. Total width	= 10.6 meters	10. Floor plate thickness	= 0.25 meters
5. Number of spans	= 1 span	11. Pier depth	= 5 meters
6. Total lane	= 2 lanes		

Mechanical Properties of Structural Materials

Structural Concrete with characteristics:

1. Concrete compressive strength, fc '= 45 Mpa
2. Modulus of Elasticity, Ec = 4700√ (fc ') = 4700 √45 = 31528,55848 Mpa

Tendon with characteristics:

1. Steel Structure, BJ 37
2. Melting Point 240 Mpa
3. Modulus of Elasticity, E = 200000 Mpa
4. Types of strands: uncoated 7 wire super strands ASTM A-416 grade 270
5. Diameter strands = 0, 01270 m

The number of strands is used as follows:

1. ns1 = 11 tendons, 19 strands / tendons = 209 strands with tendon sheaths = 85 mm
2. ns2 = 11 tendons, 19 strands / tendon = 209 strands with tendon sheath = 85 mm

Loading Bridge Structure

Permanent Load

1. Weight of box girder (QMS)

a. Box girder weight per m (QMS)
= Average × γ
= 9.224 × 25.5
= 235,212 kN / m

b. Maximum shear force (VMS)
= 10/8 × QMS × L
= 10/8 × 235,212 × 200
= 14974.84 KN

2. *Additional dead load (MA)*

a. Weight (bitumen + rainwater) = 13,384 kN / m

3. Traffic Load

Line load on the girder:

QTD = q × S × ƔTD
= 9.0 × 4.8 × 1.8
= 48.6 kN / m

PTD = (1 + DLA) × p × S × DTD
= (1 + 0.3) × 49.0 × 4.8 × 1.8

= 343.98 kN

Maximum shear force:
VTD = (1/2 × QTD × L) + PTD = 5203.98 kN

Earthquake Load

For Hard Soil
$AS = FPGA \times PGA$= 1.187 x 0.213g = 0.253g
$SMS = Fa \times Ss$ = 1.2 x 0.433g = 0.520g
$SM1 = Fv \times S1$ =1.526 x 0.274g = 0,418g
SDS = 2/3 x SMS = 2/3 x 0.520 = 0.346 g
SD1 = 2/3 x SM1 = 2/3 x 0.418 = 0.279 g

For Medium Soil
$AS = FPGA \times PGA$ = 1.374 x 0.213g =0.293g
$SMS = Fa \times Ss$ = 1.453 x 0.433g = 0.629g
$SM1 = Fv \times S1$ = 1.851 x 0.507g = 0.507g
SDS = 2/3 x SMS = 2/3 x 0.629=0.419 g
SD1 = 2/3 x SM1 = 2/3 x 0.507=0.338 g

For Soft Soil.
$AS = FPGA \times PGA$ = 1.636 x 0.213g = 0.348g
$SMS = Fa \times Ss$ = 1.913 x 0.433g = 0.828g
$SM1 = Fv \times S1$ = 2.903 x 0.274g = 0.795g
SDS = 2/3 x SMS = 2/3 x 0.828=0.552 g
SD1 = 2/3 x SM1 = 2/3 x 0.795=0.530 g

MTD = (1/8 × QTD × L2) + (1/4 × PTD × L)
= 260199 kNm

Calculates the parameters T0 and Ts
For Hard Soil
$T0$ = 0.2 x SD1/SDS
= 0.2 x 0.279/0.346 = 0.161 s
Ts = SD1/SDS
= 0.279/0.346 = 0.805 s

For Medium Soil
$T0$ = 0.2 x SD1/SDS
= 0.2 x 0.338/0.419 = 0.161 s
Ts = SD1/SDS
= 0.338/0.419 = 0.806 s

For Soft Soil
$T0$ = 0.2 x SD1/SDS
= 0.2 x 0.530/0.552 = 0.192 s
Ts = SD1/SDS
= 0.530/0.552 = 0.960 s

4. Results and Discussion

4.1 Analysis of Spectrum Response Variety of Bridges (Capital Analysis)

The natural period and natural frequency of the bridge model must consider the percentage of the mass participation factor (capital mass participation factor). Based on SNI 2833: 2016, the amount of diversity must be taken in such a way as to produce cumulative mass participation of more than 90%. Analysis of variance was carried out up to 88th variance with cumulative mass participation reaching 100% and 100%, for X and Y directions. Because cumulative mass participation had exceeded 90%, the SNI 2833: 2016 requirement to continue the earthquake spectrum response analysis had fulfilled. The cumulative mass participation of the bridge model in this study can be seen in Figure 2.

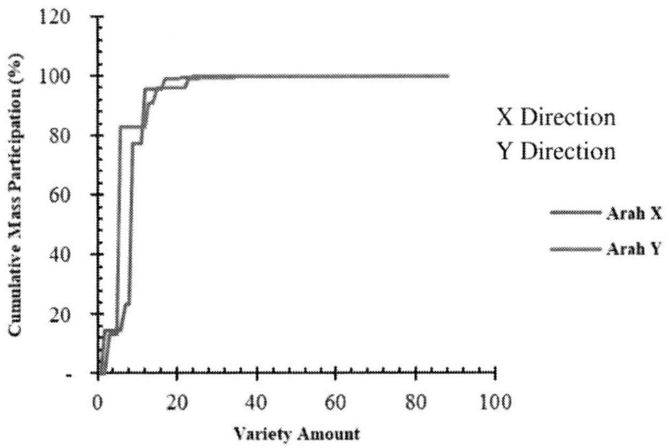

Figure 2. Cumulative Variety Mass Participation

The free vibration period of structure (T) is a product of structural mass and structural rigidity. The first three forms of vibrating structures are shown in Figure 3 The period in the first range of 0.38374 seconds is called the fundamental period of vibration because it has the highest period value.

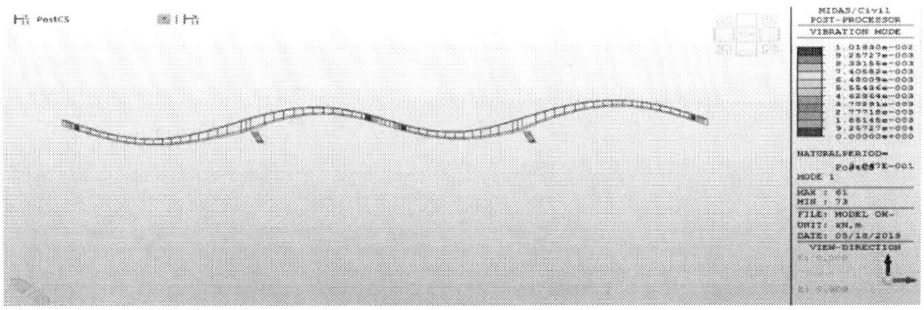

MODE I
Period = 0,383734 seconds
Frequency = 2,605972 Hz

Figure 3. The fundamental period of Structural Vibration

4.2 Response Analysis of Earthquake Structures of Bridge Structures

This section contains an explanation of the results of the spectrum response analysis. The observation point on the bridge reviewed is located on the bridge pier which can be seen in Figure 4. The spectrum response analysis in this study was conducted to obtain the structural response results on the box girder prestressed concrete bridges in the form of displacements.

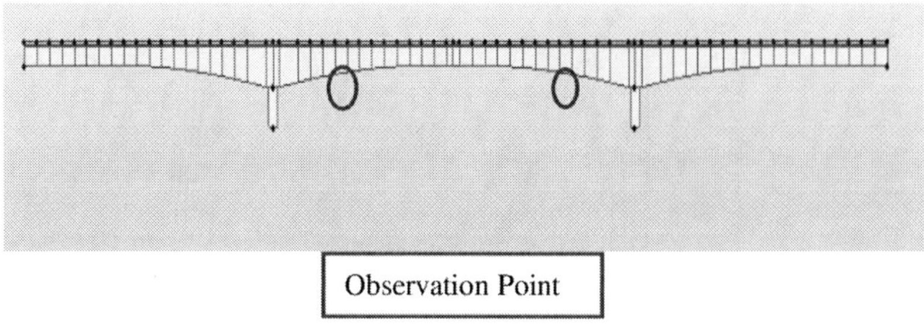

Figure 4 Observation Point

The response structure reviewed is located on the bridge pier. The result of the structural response from the MIDAS CIVIL software for hard soil, as shown in Tabel 1 and Figure 5.

Table 1 Displacement Value of the Right Pier and Left Pier with Hard Soil Conditions.

Pier Height (m)	Displacement Value (m)
5,0	0,877832
4,4	0,658374
3,8	0,658374
3,1	0,438916
2,5	0,438916
1,9	0,438916
1,3	0,219458
0,6	0,219458
0,0	0,000000

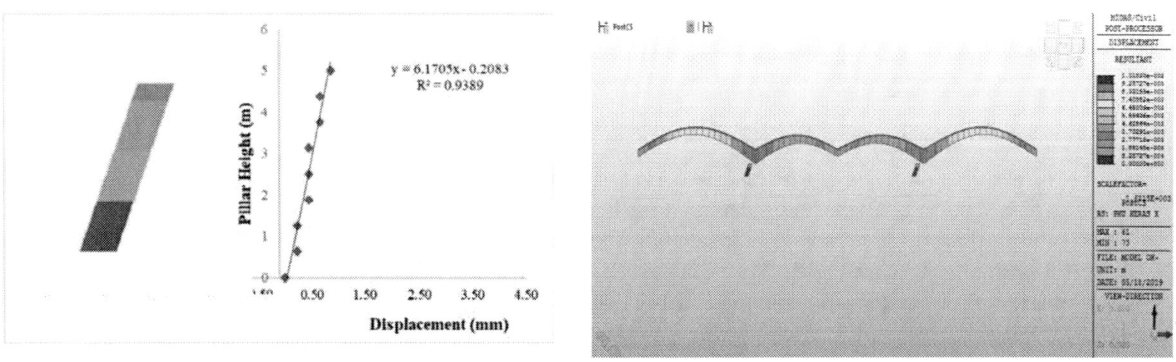

Figure 5 The Value of Displacement Graph and Contour for Hard Soil Conditions

IOP Conf. Series: Materials Science and Engineering **796** (2020) 012040 doi:10.1088/1757-899X/796/1/012040

From the graph, it can be seen that the displacement of the bridge pier structure with hard soil conditions increases at the top of the structure, whereas displacement value for Medium soil and soft soil can be shown in Table 2 and Table 3. The description of the displacement pattern for each type of soil in a sequence is shown in the Figure 7.

Table 2 Displacement Value of X Direction in the Right Pillar and Left Pillar with Medium Soil Conditions On the Pillar of the Bridge

Pier Height (m)	Displacement Value (m)
5,0	1,06330
4,4	0,79748
3,8	0,79748
3,1	0,53165
2,5	0,53165
1,9	0,53165
1,3	0,26583
0,6	0,26583
0,0	0,00000

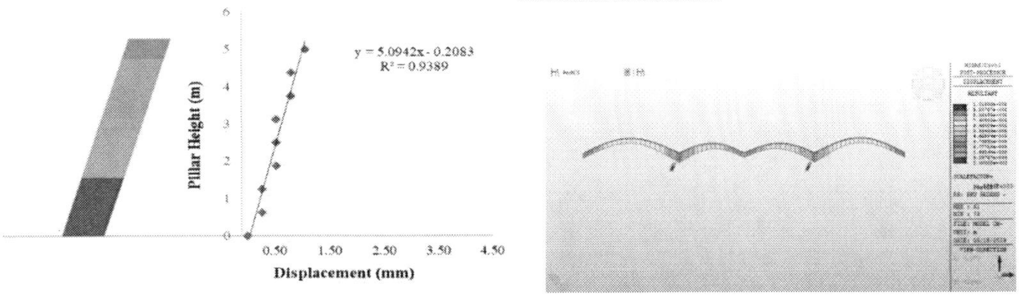

Figure 6 The Value of Displacement Graph and Contour for Medium Soil Conditions in X Direction

Table 3 Displacement Value of the Right Pillar and Left Pillar with Soft Soil Conditions

Pier Height (m)	Displacement Value (m)
5,0	1,39570
4,4	1,04678
3,8	1,04678
3,1	0,69875
2,5	0,69875
1,9	0,69875
1,3	0,34893
0,6	0,34893
0,0	0,00000

IOP Conf. Series: Materials Science and Engineering **796** (2020) 012040 doi:10.1088/1757-899X/796/1/012040

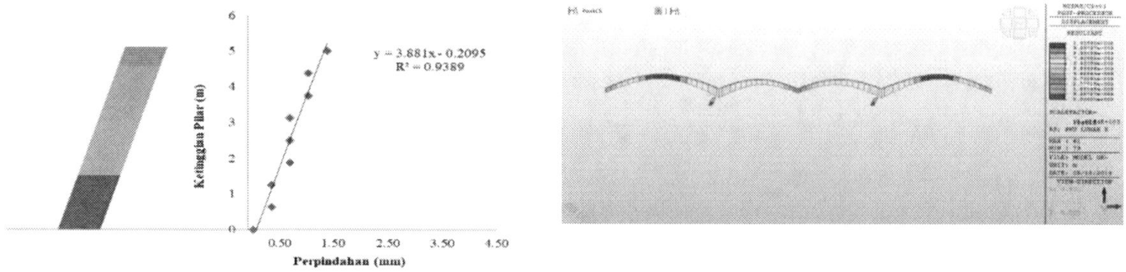

Figure 7. The Value of Displacement Graph and Contour for Soft Soil Conditions

Table 4. Maximum Displacement Summary Value for Multiple Soil Types

No.	Location	Type of soil	Max Displacement (mm)
1.		Hard	0,8778
2.	Pekanbaru	Medium	1,0633
3.		Soft	1,3957

The maximum of displacement for three types of soil can be shown in Figure 8.

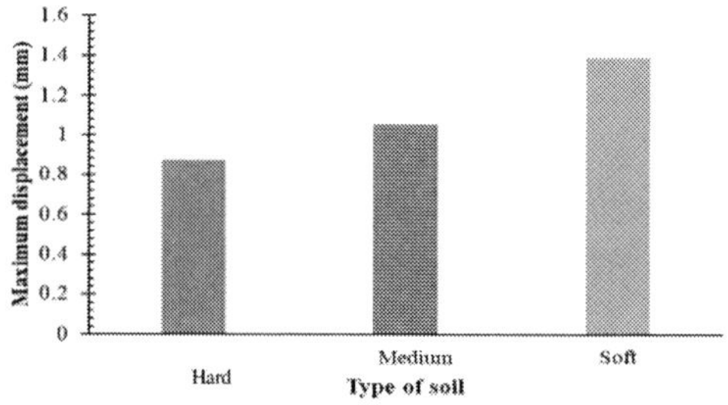

Figure 8. Maximum Displacement Value Graph on The Bridge Pier

5. Conclusions

Based on the results of the study can conclude several things as follows:

1. The pattern of maximum displacement value follows the pattern of peak surface soil spectrum (SDS); the higher the SDS for a type of soil, causes the value of displacement will also be higher. The bridge structure model has a relatively low vibration period (<0.3837 seconds). Meanwhile, the bridge structure is quite rigid in resisting dynamic loads from the earthquake load of the city of Pekanbaru.
2. The results of the structural girder prestressed concrete bridge structural response responses based on the Indonesian Earthquake Spectrum Response to SNI 2833: 2016, show that the structural response in the city of Pekanbaru in the longitudinal direction on the pillar section of the largest box girder prestressed concrete bridge occurs in the condition of soft soil with a displacement value of 1.40 mm. Meanwhile, the structural response in the city of Pekanbaru the smallest longitudinal direction occurred in hard soil conditions with a displacement value of 0.88 mm, while for medium conditions the displacement value of 1.06 mm was obtained. Thus, it can be concluded that the value of the bridge displacement due to earthquake load planned for the transverse direction in hard

IOP Conf. Series: Materials Science and Engineering **796** (2020) 012040 doi:10.1088/1757-899X/796/1/012040

soil conditions is 16.98% smaller than in medium soil conditions and 59.10% smaller than in soft soils.

6. References

[1]. Adi Prasetya, W., Setya Budi, G. 2017 *Perhitungan Jembatan Layang (Flyover) Dengan Tipe Box Girder Beton Prategang (Prestressed Concrete) Untuk Pertemuanjalan Mayor Alianyang Dan Jalan Soekarno-Hatta Kabupaten Kubu Raya*, 169–191.
[2]. Kementrian Pekerjaan Umum Dan Perumahan Rakyat 1992. *Bridge Design Manual.*
[3]. Kementrian Pekerjaan Umum Dan Perumahan Rakyat 2017 *Buku Informasi Statistik.*
[4]. Kementrian Pekerjaan Umum Dan Perumahan Rakyat 2017 *Buku Peta Gempa.*
[5]. Kurniawandy A., Nakazawa S., Hendry A., Ridwan, Firdaus R 2017 *AIP Conference Proceedings* **1892**,1.
[6]. Manalip, H., Handono, B. D. 2018 *Perencanaan Balok Girder Profil I Pada Jembatan Prestressed Dengan Variasi Bentang*, **6(2),** 67–74.
[7]. Nasional, B. S. 2016 *Pembebanan Untuk Jembatan.*
[8]. SNI 2833-2016. 2016. *Perencanaan Jembatan Terhadap Beban Gempa.*
[9]. Suryanita, R., Djauhari, Z. 2016. *Berdasarkan Spektrum Gempa Wilayah Sumatera*, **15(1),** 18–24.
[10]. Wijaya, A., Suryanita, R., Djauhari, Z. 2016. *Jom Fteknik*, **3(1)**, 1–15.

IOP Conf. Series: Materials Science and Engineering **796** (2020) 012041 doi:10.1088/1757-899X/796/1/012041

Fire risk analysis based on groundwater level in *rewetting* peatland, Sungaitohor village, kepulauan Meranti district, Riau province

SH Silviana[1*], BH Saharjo[2], S Sutikno[3]
[1]Faculty of Forestry, Graduate School of Tropical Silviculture IPB-University, Bogor, Indonesia
[2]Faculty of Forestry, IPB-University, Bogor, Indonesia
[3]Civil Engineering Department, University of Riau, Pekanbaru, Indonesia 28293

[*]E-mail: sintaharyatis@gmail.com

Abstract. The hydrological condition of peatlands is one of the important factors affecting peatland's vulnerability to fire. The groundwater level (GWL) of peatlands or so-called groundwater depths is an indicator in determining the hydrological balance of peatlands. In normal conditions, peatlands are not prone to fire because they are able to hold and absorb water. Drainage on a large scale on peatlands causes the peat water level to drop so that the peat experiences water loss and is prone to fire. The disruption of the hydrological balance of peatlands can be seen in the wet month, when the water level is not much different from the conditions on a dry moon, which is still below the surface, causing peat to become flammable. The focus of this study was to determine the level of risk of fire vulnerability based on groundwater levels in peatlands, which was carried out restoration efforts, in the form of *rewetting*. This study, as a piece of early warning information on forest fires on GWL measurements of >40 cm, especially in the peatlands of Sungaitohor Village, Tebingtinggi Timur District, Kepulauan Meranti Regency, Riau Province. Groundwater level measurements were carried out directly in the field by installing 66 monitoring well points placed on different land uses (33 in unburned areas and 33 in burnt areas) during February - June 2019. The results showed that this region, in general, had a high level of risk of forest and land fires. The highest level of fire risk based on GWL >40 cm (danger category) was 99.63% in March. The high value of GWS >40 cm in this region made very susceptible to forest fires. In addition, extreme weather worsened the condition of peatlands into drought and was prone to fire.

1. Introduction

The increase in forest and land fires in Indonesia in the past few decades has been an unresolved problem. Various parties felt an interest in tackling the problem of forest and land fires in Indonesia because of its very large impact. Fire affects the loss of biodiversity, global warming, health, and the loss of economic opportunities for society. One of the causes of forest and land fires is deforestation, especially its conversion to Industrial Plantation Forest [1]. Fire events can be related to activities such as land clearing [2]. The land conversion followed by the construction of canals and drainage on a large scale causes a decrease in the water table, so that it is vulnerable to fire, especially in the dry season [3].

Content from this work may be used under the terms of the Creative Commons Attribution 3.0 licence. Any further distribution of this work must maintain attribution to the author(s) and the title of the work, journal citation and DOI.
Published under licence by IOP Publishing Ltd

Forest and land fires often occur on peatlands, peat drought caused by the presence of canals has an impact on its vulnerability to fire. In the early to mid-2014, there had been a large fire in the area in Sungaitohor Village, Kepulauan Meranti Regency, Riau Province. Peatlands fires as a result of very dry peatlands conditions due to the sequestration of the area from existing canals. Therefore, it is said that canal construction is the initial cause of peatlands drying out, making it susceptible to fire supported by a long dry season.

This recurrent fire is a problem that must be resolved by the government. Various initiatives are carried out by villagers who also work together with other agencies both government and NGOs. The Government of Indonesia issued Government Regulation (PP) No. 71/2014 in an effort to prevent peatlands damage. Land use and management of peatlands for protection must maintain the hydrological function of peatlands. Based on PP No. 71/2014, that peatlands are said to be damaged if in the protected function drainage networks are made, whereas in peatlands the cultivation function is declared damaged if the groundwater level (GWL) is more than 40 cm below the surface. In this case, the government has obligatory to maintain the GWL at the level of 40 cm so that the peatlands remain at a safe level. In 2015, after a year of the fire, efforts were made to avoid fires in this area by building a canal blocking. Peatland was degraded because the canal opening could be attempted to be re-wetted by closing the canal (*canal blocking*) [4],[5]. The condition of degraded peat swamp forests caused by fires can be improved with restoration efforts, namely by restoring ecosystems back to their natural structure and function [6]. One of the restoration efforts is carried out by rewetting through canal blocking. Canal blocking either in the whole or in part canal is expected to increase the groundwater level. Efforts to rewetting peatlands (rewetting) with canal block construction have been and are still being conducted in Sungaitohor Village, Tebingtinggi Timur District, Meranti Islands Regency, Riau Province (Meranti Island Peat Hydrology Unit). *Rewetting* will maintain the groundwater level at the desired condition so that the peat will always be moist. Research on fire risk analysis based on the water level in restored peatlands is conducted with the rewetting method. Rewetting efforts in Indonesia and tropical peat are still very few and are in its initial stages [7]. Therefore, it is necessary to study the condition of the peat water level in the face of peat water as information in the early warning of forest and land fires.

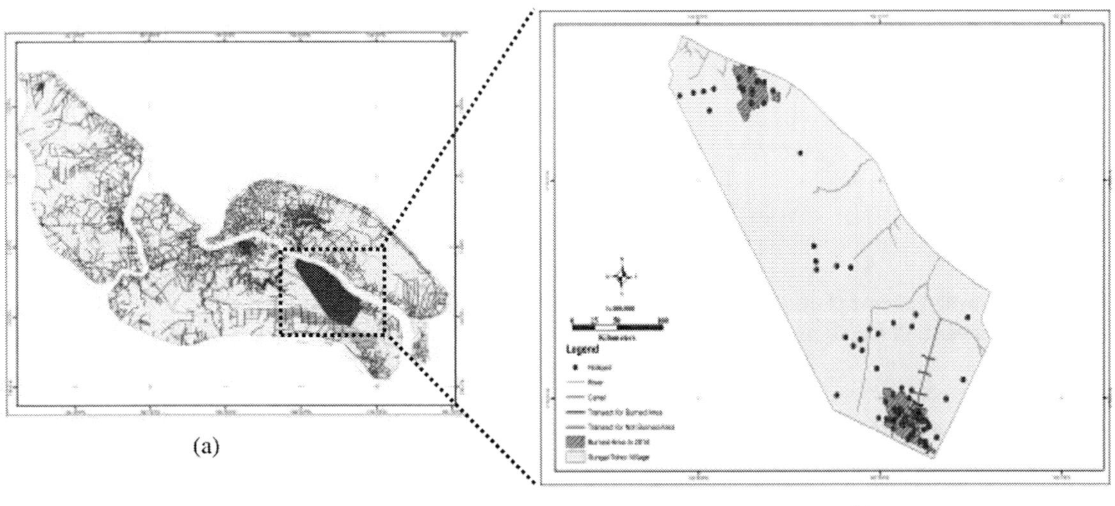

(a) (b)

Figure 1. (a) Map of research location (b) Monitoring point of groundwater level in unburned and burned land and hotspot distribution in 2014.

2 Methods

2.1 Place and time of research

The study was conducted in Sungaitohor Village, Tebingtinggi Timur District, Kepulauan Meranti District, Riau Province (Fig 1). Data collection and measurement carried out on peatlands post hydrological restoration in the form of rewetting with the construction of canal blocking. Measurement of groundwater level was carried out in different land uses on unburned and burned land. The unburned land was in sago plantations, rubber plantations, secondary forests. Measurement of groundwater level data was carried out in February - June 2019. Measurement of the groundwater level was conducted every 2 times a month.

2.2 Research design

Retrieval of groundwater level data was carried out on a canal flow using repetition on unburned and burned land. The total number of monitoring wells was 66 monitoring well points, which were divided into 33 points each on burned and unburned land (Table 1).

Table 1. Design of the research

Parameter	Land Use		
	Unburned Peatland		
	Sago Plantation	Rubber plantation	Secondary Forest
Number of sampling location	1	1	1
Number of monitoring well/side location	11	11	11
Parameter	Burned Peatland		
	Burned 1	Burned 2	Burned 3
Number of sampling location	1	1	1
Number of monitoring well/side location	11	11	11

2.3 Tools

The tools used in this study were as follows:

1. Monitoring wells were used for the following purposes:
   - For monitor peat groundwater level, using a float and put a meter pipe on it. The scale of water level reading could be directly read above ground level.
2. Sensory Data Transmission Service Assisted by Midori Engineering (SESAME) which was a field of near-real-time monitoring of field data transmission hydrological conditions located in the area of this study.
   - For measurement of groundwater level in real-time
   - For rainfall calculation

2.4 Materials

The monitor well model was made of a 2.5-inch piezometer pipe with a float-equipped with a groundwater level gauge. In addition, monitoring wells were equipped with upper and lower pipe caps, collars (wooden plates 20 cm x 20 cm) buoys, measuring pipes, meters (Fig 2).

Figure 2. Groundwater level monitoring model

2.5 Research procedure

Measurement of groundwater level was conducted based on Government Regulation (PP) No. 71/2014 in an effort to prevent damage to peatlands. The level of risk of fire on peatlands referred to the minimum threshold of GWL of 40 cm and then divided into 3 classifications (Fig 3), namely:

1. Safe 0 – 25 cm
2. Fire alert 26 – 40 cm
3. Danger > 40 cm

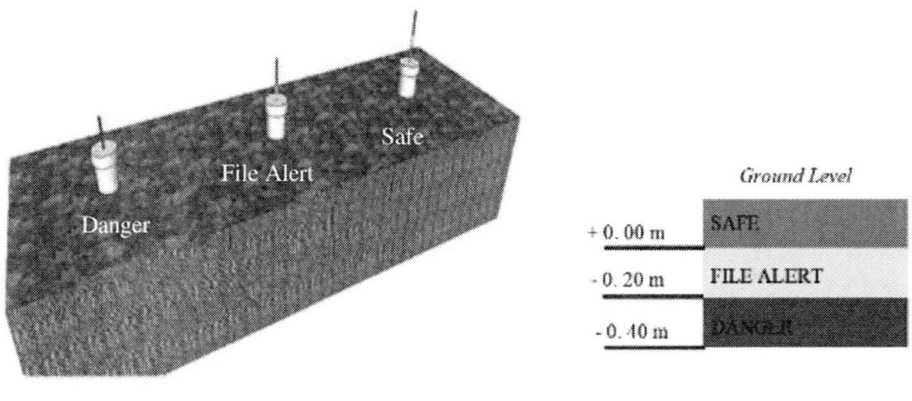

Figure 3. (a) Category of fire risk level on peatlands based on groundwater level (b) Groundwater level index based on color (Source Image: Sutikno et al. 2018) [8].

The research was carried out through several stages including data collection, data processing, and data analysis presented in the research flowchart (Fig 4).

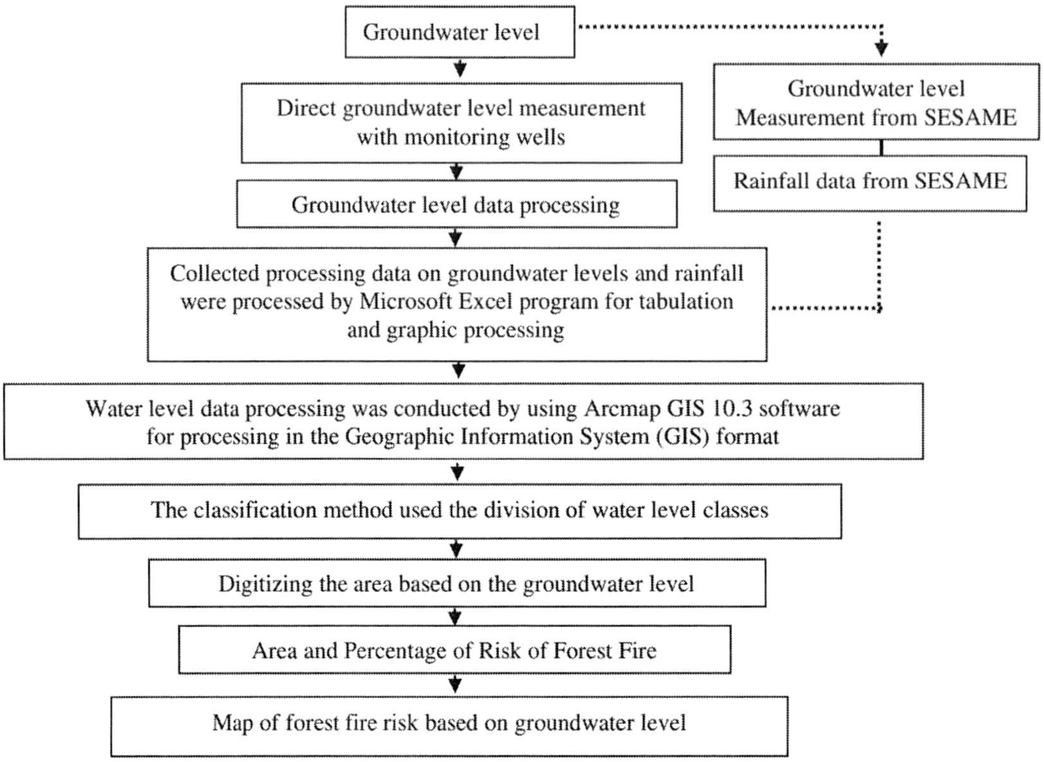

Figure 4. Research flow diagram

3. Results and Discussion

3.1 Dynamics of Groundwater Level

GWL fluctuation calculation is done directly in the field, where GWL fluctuation measurements are carried out in February 2019 - June 2019. Measurements are made manually once every two weeks. This is, according to the Minister of Environment and Forestry Regulation (Permen KLHK) No.15/2017 that manual measurement of GWL is done at least once every two weeks, while automatic measurements are carried out at least once a day. Knowledge about the dynamics of GWL in peatlands is absolutely necessary to know the condition of the land or for further management. Field measurements show that the dynamics of groundwater levels vary greatly on unburnt peatland: sago plantations (Fig. 5a), rubber plantations (Fig. 5b), secondary forests (Fig. 5c), and burnt peatland: burnt peatland 1 (Fig. 5d), burnt peatland 2 (Fig. 5e), burnt peatland 3 (Fig. 5f).

The frequency of water level in the land is very related to the decomposition of peat composting materials, its cover and hydrological conditions. In addition, external factors such as climate change, rainfall, and sunlight intensity are influential. The existence of this relationship will be very important as an early warning system in fire information. Dry peatland will act as a material that is ready to be burned or burned so that the area is prone to fire. The availability of water in a study area is described by the volume of rainfall, while the potential for temporal distribution of water is expressed by the number of rainy days. The results of rainfall calculations at SESAME during the months of February 2019 - June

2019 found that every month there was an increase in the amount of rainfall. The amount of rainfall in February was 16 mm / month, March 39.50 mm/month April 119.50 mm/month, May 138.50 mm/month and June 130.11 mm/month.

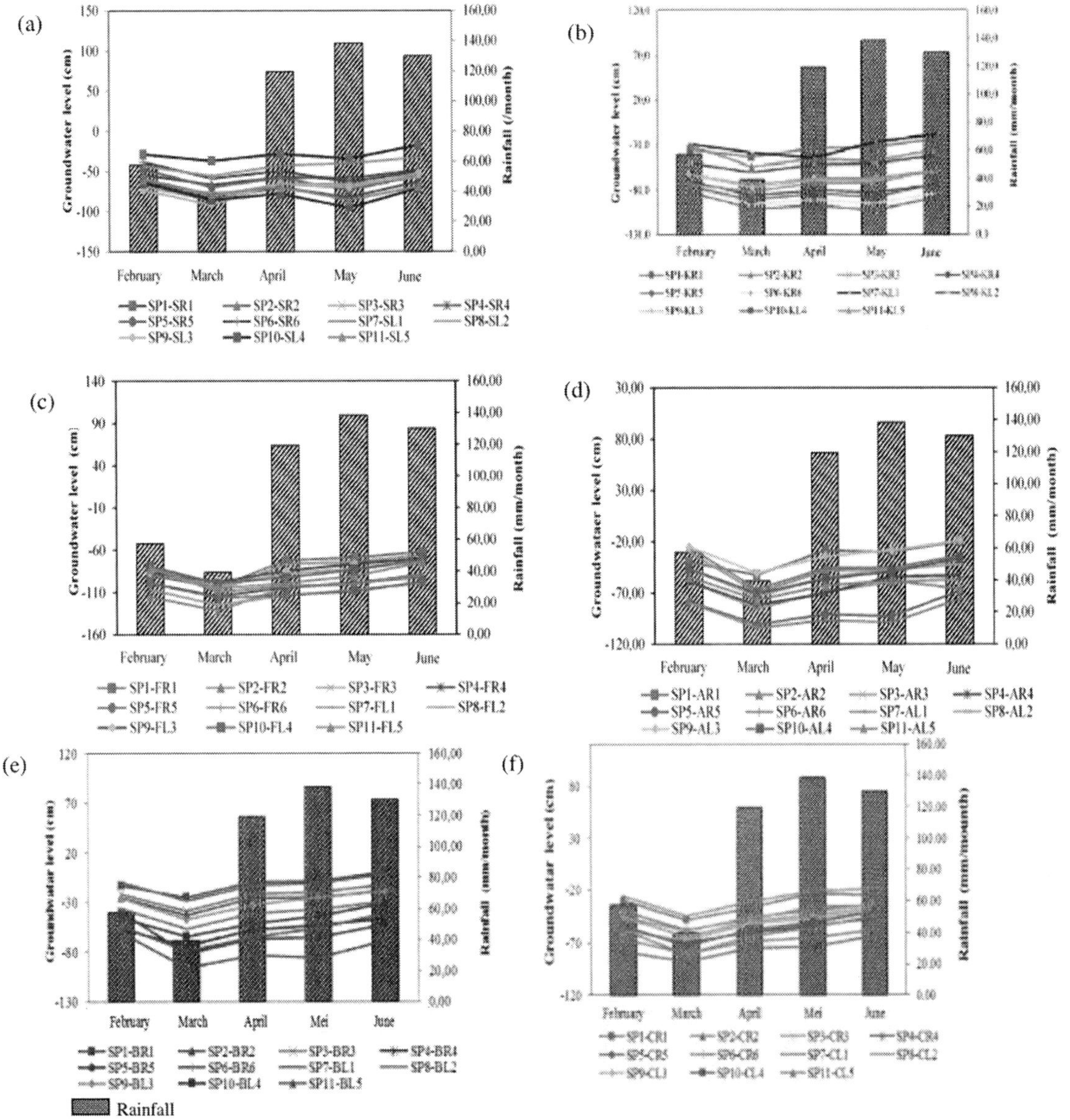

Figure 5. Groundwater level fluctuation and rainfall (a) Sago Plantation; (b) Rubber Plantation; (c) Secondary Forest; (d) Burnt Peatland 1; (e) Burnt Peatland 2; (f) Burnt Peatland 3

Information
SP1-SP11: Deep well 1 - Deep well 11; SR1-SR6: Sago Right 1 – Sago Right 6; SL1-SL5: Sago Left 1 – Sago Left 5; KR1-KR6: Rubber Right 1- Rubber Right 6; KL1-KL5: Rubber Left 1 - Rubber Left 5; FR1-FR6: Secondary Forest 1 – Secondary

Forest 6; FL1-FL5: Secondary Forest 1 – Secondary Forest 5; AR1-AR6: Burnt area Right 1 – Burnt area Right 6; AL1-AL5: Burnt area Left 1- Burnt area Left 5; BR1-BR6: Burnt area Right 1 – Burnt area Right 6; BL1-BL5: Burnt area Left 1- Burnt area Left 5; CR1-AR6: Burnt area Right 1 – Burnt area Right 6; CL1-AL5: Burnt area Left 1- Burnt area Left 5
Right: Right side of the canal; Left: Left side of the canal

GWL is influenced by rainfall as a source of water availability. Low rainfall intensity in March resulted in a reduction in GWL. If we see the correlation between GWL and rainfall, there is a relationship, but rainfall does not directly increase GWL. The increase in GWL on peat soils that have high drought must be followed by high rainfall intensity as well. Thus, the greater the rainfall that occurs, the greater the increase in GWL. The inherent nature of peatlands makes peat very sensitive to environmental changes. Sarwono [9] said that the inherent nature of tropical peat is irreversible drying and subsidence. When drying occurs in this case the low rainfall, which is a factor in the availability of water in peatlands causes peat colloids to be damaged so that changing peat to dry is not back. In addition, the level of soil saturation is a factor preventing water entry. If the soil has a high level of drought, then the water supply which must fill the pore space of the soil must also be high. This can happen because rainwater has to fill a lot of empty pore space in peat soil. When the soil is saturated, even though the rainfall is categorized as small to moderate, it is able to raise the GWL to exceed the soil surface. Firmansyah and Mockhtar [10] said that peatlands generally experience drought in the surface layer during the dry season, this causes a decrease in GWL on peatlands so it is vulnerable to fire hazards.

Weather greatly affects the level of vulnerability to fire dangers, dry hot weather will cause fires. The nature of the weather affects the occurrence of fire [11]. Fire weather is defined as weather conditions that affect the start of a fire. In February the rainfall is at the lowest among other months with 3 days of rainfall, while on other days there is no rainfall or 0 mm/day. The duration of heat intensity and solar radiation this month also causes the air temperature to increase. Areas with high temperatures will cause the fuel to dry out quickly and facilitate fires [12]. Air temperature is an important weather factor that causes fires [13]. Constant air temperature is a factor that influences the temperature of the fuel and the ease of the fuel to burn. [14] In the morning with a fairly low temperature of around 20oC coupled with low wind speeds keep the fire from developing so it is concentrated at one point. While during the day with a temperature of 30 - 35°C, while the water content of the fuel is quite low (<30%) making the combustion process take place quickly and the form of the fire was not a single point, but changing because of the influence of the wind. In the following month, there was an increase in rainfall. High rainfall will affect the groundwater level so that soil moisture is also high and the condition of vegetation which is a fuel in the forest as well If the fuel moisture is high then it will be difficult to fire [12]. Months with little rainfall, the drought index is quite high conversely, in the months with high rainfall, the drought index is low, even reaching zero. This shows that rainfall affects the water content of fuel [15].

3.2 Risk of Fire in Peatlands Based on Groundwater Level

The results of the analysis of the risk of forest fires based on the highest groundwater level or at danger level (GWL> 40 cm) occurred in March - April (Table 2). Danger category in March reached 99.63% and April reached 97.58%. The high level of this category was caused by very extreme weather factors in this area. The absence of rain made the peatland dry. In March, this risk was influenced by the factor of low rainfall in the previous month, February. In February, the rainfall was only around 16 mm/month - 33 mm/month with the number of rainy days 3-7 days a month. Besides the low rainfall that occurred in the previous month, this month was also followed by a low amount of rainfall, so that the GWL category of >40 cm was the highest. This resulted in an effect on the level of fire risk in April, which was also high even though the level of rainfall had started to be high. The dry season occurring in the previous month resulted in a very large shrinkage of the groundwater level and made peatlands lose a lot of water. Drought-affected peat results in reduced soil moisture content. If a long drought occurs continuously results in total drought in this area and the peat cannot be re-wet or undergoes an irreversible drying process. In addition, the ability of peat will change from very high in absorbing water (hydrophilic) into rejecting water (hydrophobic). [14] Decreased peat's ability to absorb water is related to the availability of hydrophilic compounds in peat materials, namely carboxylates and OH^- Phenolates. Therefore, peat which is in dry condition causes the ability to absorb water to not work.

Table 2. Recapitulation of fire risk percentage based on groundwater level

Measurement Month	Groundwater level category (cm)	Total area (m^2)	Fire Risk (%)
February	0 - 25	51138.38	1.35%
	- 40	284588.94	7.50%
	>40	3456639.09	91.15%
March	0 - 25	0.00	0.00%
	26 - 40	14125.94	0.37%
	>40	3778240.47	99.63%
April	0 - 25	26393.41	0.70%
	26 - 40	65384.39	1.72%
	>40	3700588.61	97.58%
May	0 - 25	48589.59	1.28%
	26 - 40	133902.13	3.53%
	>40	3609874.69	95.19%
June	0 - 25	105194.15	2.77%
	26 - 40	517825.70	13.65%
	>40	3169346.56	83.57%

The dry season that occurs not only causes a decrease in the surface of the groundwater, but it also causes the peat to burn more easily. In addition, the existence of making drainage or trench that is too wide has an impact on the lack of ability of peat to store water and the process of peat decomposition. Peat turns into available fuel because the organic matter in the peatland has dried up. When there is a fire igniting above ground level in this area (land burning), it allows fires to take place quickly. Because peat as a fuel has been in a dry condition. [2] The lower the water content or the lower wetness level, the more fragile the peat is. The level of fire risk is shown in (figure 6). February (Fig. 6 a), March (Fig. 6b), April (Fig. 6c), May (Fig. 6 d), June (Fig. 6 e).

Measurement of water level data in the field can also be used as an indicator of the level of fire risk in the dry season (dry month). Results of groundwater level measurements obtained in general have a high level of risk of forest fires. The difference in rainfall intensity that occurs is not much different among months that have high or low rainfall, where in general rainfall remains below the surface of the land, causing peatlands to become flammable. Data management in May and June reveals a decrease in the danger category (GWL >40cm) by 18.45%. This is caused by the rainfall this month is quite high since April - May ranging between 119.5 mm/month and 130.1 mm/month. An increase in rainfall makes the water source on this peatland. Good condition on peatlands is <40 cm because in this condition peat is difficult to burn. Water level will affect the level of peat water, while rainfall will affect the water level of peatlands.

IOP Conf. Series: Materials Science and Engineering **796** (2020) 012041 doi:10.1088/1757-899X/796/1/012041

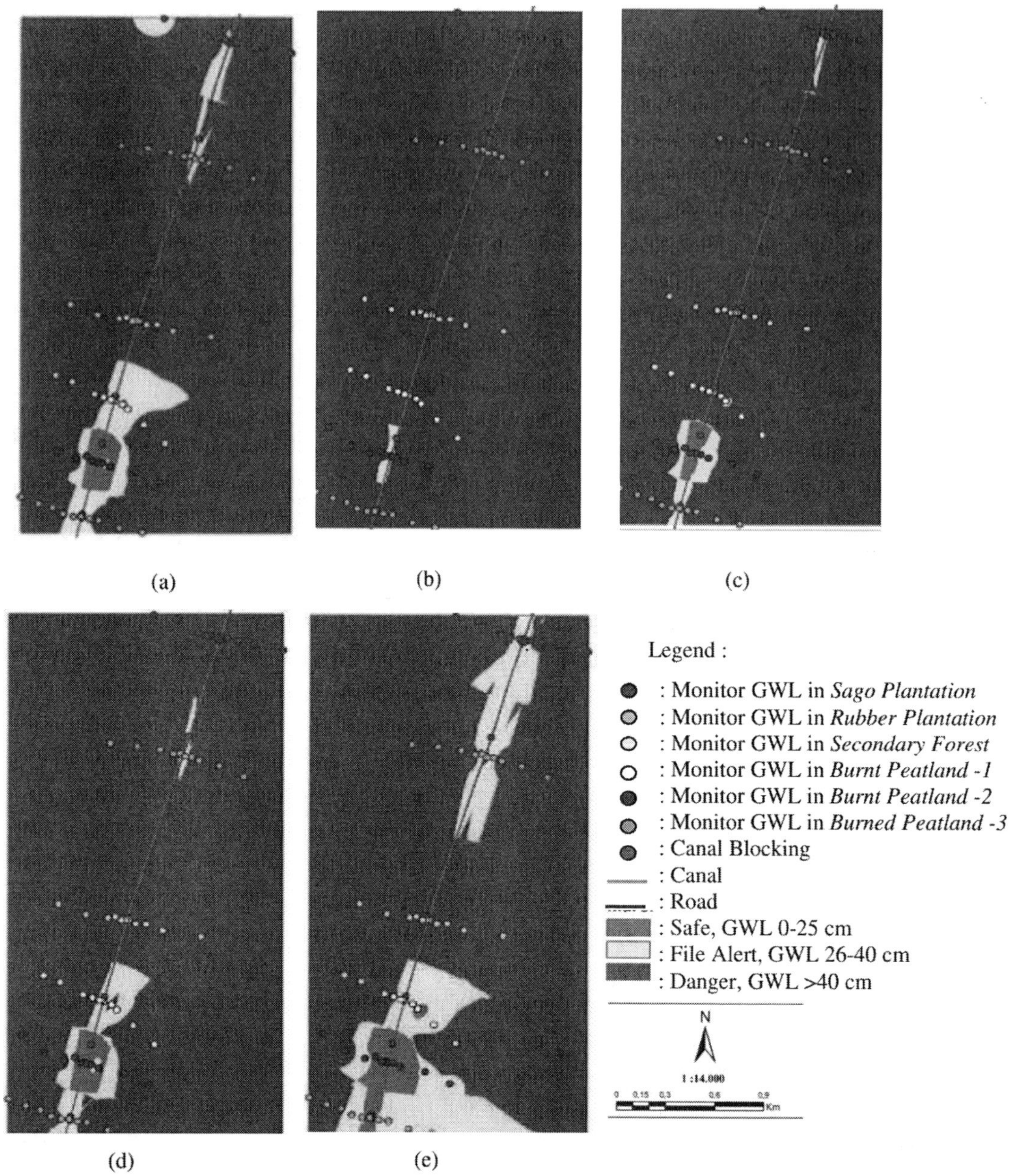

Figure 6. Fire risk map based on groundwater level in Sungaitohor Village, Kepulauan Meranti 2019 (a). February; (b) March, (c); April; (c) May; (d) June.

4. Conclusion

Groundwater level (>40 cm) can be used as an early warning system for the risk of forest and land fire dangers. The highest level of fire risk based on GWL of >40 cm (danger category) is 99.63% in March. In the dry season, the rainfall was lower, so that peat water is very dry resulting in higher fire risk. Peatland fires are preceded by low water content in peatlands. In this situation, peatlands are very dry and prone to burning.

5. Acknowledgment

The author would like to thank the Ministry of Research, Technology, and Higher Education of the Republic of Indonesia for supporting the fund of this research through the PMDSU Scholarship.

6. References

[1] Najiyati S L L, Muslihat, Suryadipura 2005 *Bogor: Wetlands International-Indonesia Programme and Wildlife Habitat in Canada.*
[2] Syaufina L 2008 *Kebakaran Hutan Dan Lahan Di Indonesia.* Malang: PT. Bayu Media Publishing.
[3] Verwer C P, van der Meer, Jnabuurs G 2008. *Review of carbon flux estimates and other greenhouse gas emissions from oil palm cultivation on tropical peatlandsIdentifying the gaps in knowledge.* Alterrarapport 1731. Alterra. Wageningen. 44.
[4] Suryadiputra I N N, Dohong A, Waspodo R S B, Muslihat L, Lubis I R., Hasudungan 2005 *A guide to the blocking of canals and ditched in conjunction with the community. Bogor: Wetlands International Indonesia Programme and Wildlife Habitat Canada.*
[5] Ritzema H, Limin S, Kusin K, Jauhiainen J, Wösten H 2014 *Catena.* **114**, 11-20.
[6] Page S E, Rieley J O, Banks C J 2011 *Global Change Biology.* **17(2)**, 798-818.
[7] Giessen W, Sari E N N. 2018 *Tropical Peatland Restoration Report: Indonesian Case.* Millennium Challenge Account – Indonesia. 82p.
[8] Sutikno S, Nasrul B, Gunawan H, Jayadi R, Rinaldi, Saputra E, Yamamoto K *MATEC Web of Conferences.* **276**
[9] Fuller M 1991 *Forest Fire: an Introduction to Wildland Fire Behaviour, Management, Fire Fighting and Prevention.* Wiley Nature Edition. New York: JohnWiley and Sons, Inc.
[10] Direktorat Jendral Perlindungan Hutan dan Pelestarian Alam. 1994a. *Surat Keputusan Dirjen PHPA Nomor 244/Kpts/DJ VI/1994 Tentang Petunjuk Teknis Pemadaman Kebakaran Hutan.* Departemen Kehutanan. Jakarta.
[11] Young R A, Giese RL 1991 *Introduction to Forest Fire.* John Wiley and Sons Inc. Toronto Canada.
[12] Saharjo B H 1997 *Mengapa Hutan dan Lahan Terbakar.* Harian Republika.
[13] Syaufina L 2014 Peran strategis sektor pertanian dalam pengendalian kebakaran lahan gambut. *Risalah Kebijakan Pertanian dan Lingkungan I* : 35-39
[14] Sabiham S 2000 J. *Tanah Tropika.* **11**, 21-30.
[15] Saharjo B H, Syaufina 2015 *Kebakaran Hutan dan Lahan Gambut.* Center for International Forestry Research. Bogor. Indonesia.

IOP Conf. Series: Materials Science and Engineering **796** (2020) 012042 doi:10.1088/1757-899X/796/1/012042

Application of paper disc velocimeter for groundwater flow measurement in tropical peatland

S Sutikno[1,2*], RD Ramawilis[1], FA Hutapea[1], Rinaldi[1], K Yamamoto[3]
[1]Civil Engineering Department, University of Riau, Pekanbaru, Indonesia 28293
[2]Centre for Disaster Studies, University of Riau, Pekanbaru, Indonesia 28293
[3]Civil Engineering Department, Yamaguchi University, Japan

*E-mail: ssutiknoyk@yahoo.com

Abstract. This paper presents the application of paper disc velocimeter (PDV) for groundwater flow measurement in tropical peatland. The PDV is a simple and low-cost method used without electricity that is developed by Yamaguchi University to measure groundwater flow. This method was applied to understand the groundwater hydraulic in the tropical peatland as the impact of rewetting efforts using canal blocking. A study site at Lukun village, Kepulauan Meranti Regency, Riau in Pulau Tebing Tinggi peatland hydrological unit (PHU) was picked up for the application of this method. Five deep wells in each transect of three transects totally were prepared for the measurement using PDV. The Transect-1 and Transect-2 were located at the upstream side of canal block with a distance of about 400 m and 100 m respectively. The Transect-3 was located on the downstream side of the canal block with a distance of about 100 m. The results showed that the groundwater velocity in the upstream side of canal block tends to lower than that in the downstream side. This was because the groundwater level in the upstream was maintained in a stable condition due to the canal blocking. This finding showed the impact of canal blocking on the rewetting effort in the tropical peatland.

1. Introduction

Peat is a soil that was formed from the decomposition of vegetations in the lowlands which is always high saturated condition which makes the peat contain more than 75% organic content [1]. Peat is an organic material that is formed naturally from the remnants of plants that are imperfectly decomposed with a thickness of more than 50 cm and accumulates in the swamp. The land is specially made by the remnants of vegetation that accumulates over a long period and makes peat soils [2]. The unique physical characteristics of peat that need to be understood include maturity, color, specific gravity, porosity, irreversible dryness, subsidence, and flammability. In terms of maturity, peat varies in maturity because it is formed from different materials, environmental conditions, and time. Peatland is very vulnerable against fire in the dry season and it is very difficult to be stopped due to the underground fire [3].

Riau Province holds the largest area of tropical peatland in Indonesia. Unfortunately, forest fires occur almost every year, especially during the dry season in Riau peatland. In the dry season, the peatland should be as a water-retaining layer functions and releases water slowly [4]. In natural conditions, peatlands are not flammable because of their sponge-like nature, which absorbs and retains water optimally so that in the rainy and dry seasons there is no difference in extreme conditions. Peatland fire is a fire that occurs in the peat area caused by human actions or due to natural factors such as drought or dry season. One of the efforts to prevent the problem of drought and forest fires is by rewetting activities using canal blocking [5]. Generally, the goal of rewetting degraded peatland is to restore the hydrological function of peat which increases the wetness and peat moisture. The basic principle of canal blocking is to hold and to store water as long as possible within the peatland hydrological unit area.

The groundwater flows around canal blocks should be analyzed to understand the effect of canal blocking in the peatland. The analysis was carried out using a paper disc velocimeter (PDV) device that was installed at the monitoring deep wells around the canal block. The PDV is a simple and low-cost method used without electricity that is developed by Yamaguchi University to measure groundwater flow [6]. Based on the laboratory

Content from this work may be used under the terms of the Creative Commons Attribution 3.0 licence. Any further distribution of this work must maintain attribution to the author(s) and the title of the work, journal citation and DOI.
Published under licence by IOP Publishing Ltd

test, the PDV can be used to measure the groundwater velocity with a range of 0.03 cm/min – 0.3 cm/min and also the velocity direction with the accuracy of about 33% [7]. This method was applied to understand the groundwater hydraulic in the tropical peatland as the impact of rewetting efforts using canal blocking. A study site at Lukun village, Kepulauan Meranti Regency, Riau, Indonesia in Pulau Tebing Tinggi peatland hydrological unit (PHU) was picked up for the application of this method. Five deep wells in each transect of three transects totally were prepared for the measurement using PDV. The Transect-1 and Transect-2 were located at the upstream side of canal block with a distance of about 400 m and 100 m respectively. The Transect-3 was located on the downstream side of the canal block with a distance of about 100 m.

2. Method

2.1. Study Area

The study was carried out at a research site which is located in Pulau Tebing Tinggi peatland hydrological unit (PHU), Kepulauan Meranti Regency, Riau Province, Indonesia as presented in Fig. 1a. The Pulau Tebing Tinggi PHU is one of the most priority areas for peatland restoration by the Indonesian government [8]. Based on the topography map, Lukun Village generally has a lowland and peatland area, with an area of 154.6 km^2. (see Figure 1b).

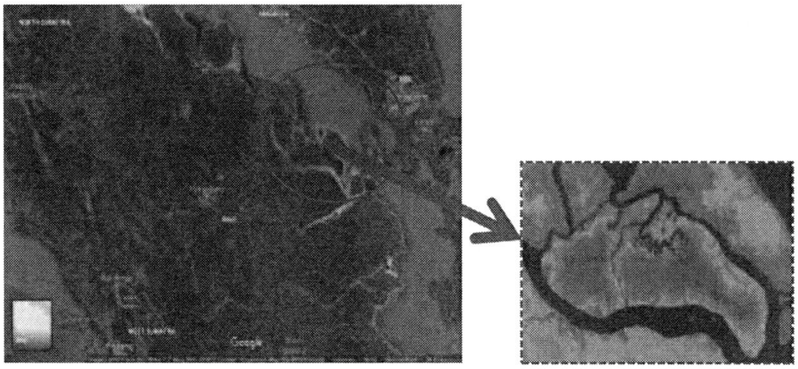

(a) (b)

Figure 1. Lukun Village, Tebing Tinggi District, Kepulauan Meranti Regency, Riau Province (a), the topographic map of Pulau Tebing Tinggi peatland hydrological unit (b)

2.2. Experiment set up

Together with the measurement of groundwater flow, the other data were used for analysis the flow characteristics due to canal blocking in this study were elevation data and groundwater level data. The measurement of the groundwater level was carried out using PDV by submerging the sensor for 60 minutes in each deep well around the canal block (Fig. 2a). The deep wells were made using PVC pipe (Polyvinyl Chloride) with a diameter of 2 inches and a height of 2 m as presented in Fig. 2b. The sensor of PDV consists of a paper disc covered with some layers of sponges, as presented in Fig. 1c.

The measurements were made on 3 transects, namely Transect-1 in upstream, Transect-2 in middle, and Transect-3 in downstream as presented in Fig. 3. The deep wells in the first transect are 1-1, 1-2, 1-3, 1-4, 1-5, the deep wells in the second transect are 2-1, 2-2, 2-3, 2 -4, 2-5, and the deep wells in the third transect are 3-1, 3-2, 3-3, 3-4, 3-5. The distance of each monitoring deep well is 1 m, 51 m, 101 m, 201 m, and 301 m from the canal. The cross-section of each transect is presented in Fig. 4.

The PDV tool uses ink spread on paper to represent the speed and direction of groundwater flow. Figure 5 shows the paper disc before and after used for measurement, and after ImageJ analysis. After submerging the PDV sensor for 60 minutes, the dotted ink in the center of paper disc will be dispersed because of the groundwater flow. There are two methods to interpret the measurement result, such as using naked eye analysis and using image analysis. The naked eye analysis measures the length of the ink tailing (α) in the paper disc, and the velocity magnitude can be calculated using Eq. 1 [6]. If the result is smaller than 0, then there is no flow.

$$v = \frac{\alpha - 0{,}194}{19{,}765} \quad (1)$$

Image analysis using ImageJ application is used to delineate the ink dispersion so that the direction and the length of the tailing can be observed. Based on this, the velocity magnitude and the direction can be justified. The results of the image analysis on the paper disc using ImageJ application are XM tailing, XM Dot and YM tailing, YM Dot as presented in Fig. 6. The velocity magnitude using this method can be calculated using Eq. 2 [6].

$$\text{Velocity 60 min} = \frac{1}{10{,}761} x0{,}1x((\text{XM dot} - \text{XM Tailing})^2 + (\text{YM dot} - \text{YM Tailing})^2)^{0.5} \quad (2)$$

(b) (a) (c)

Figure 2. Method of groundwater flow measurement in the peatland (a), using deep well (b), and PDV sensor (c).

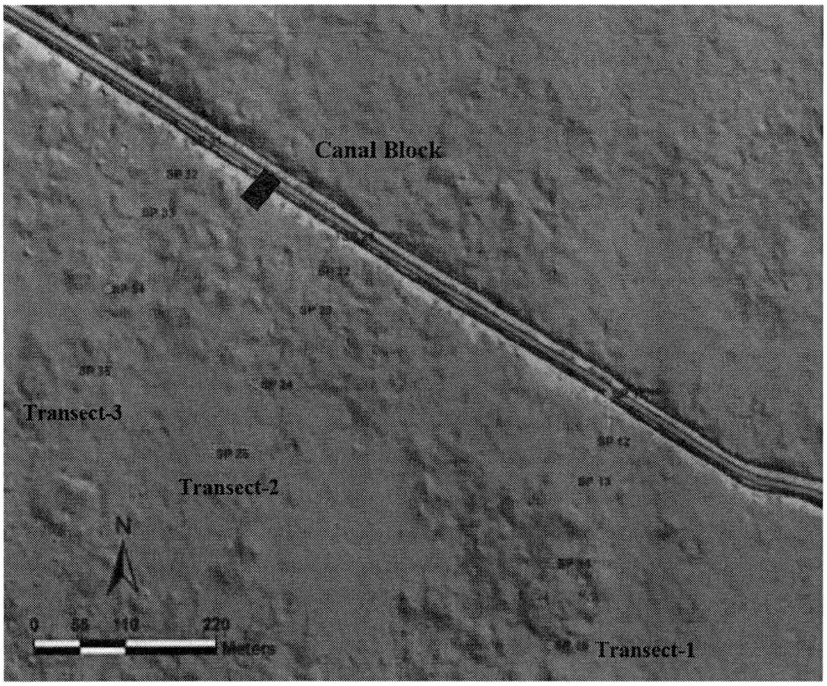

Figure 3. The configuration of deep wells in each transect and canal blocks.

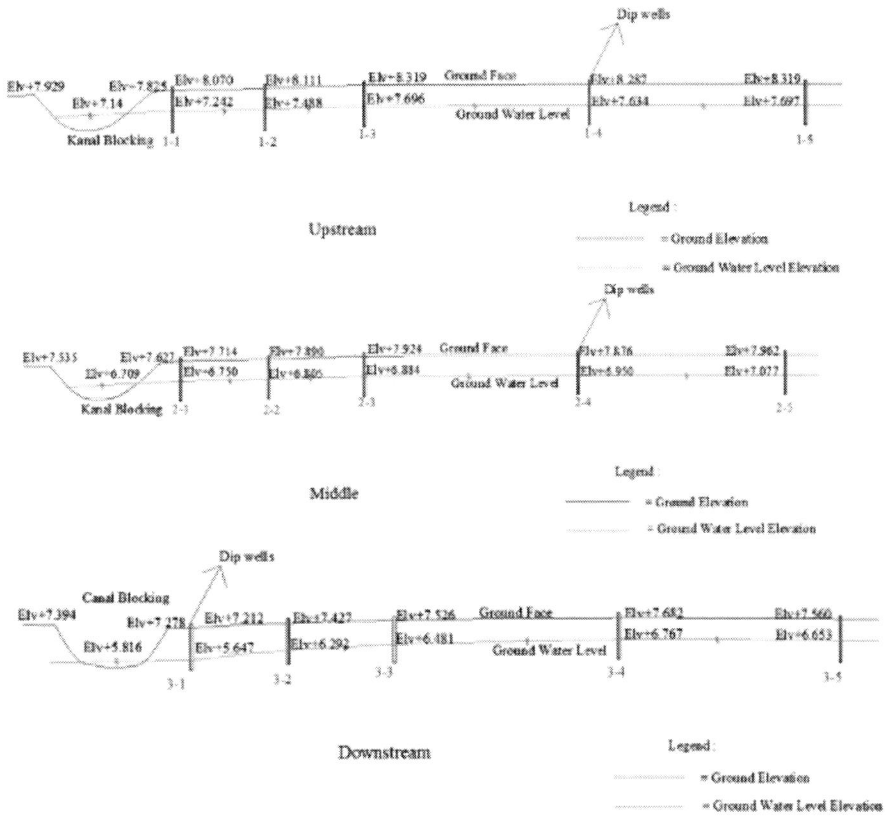

Figure 4. Cross-section of each transect

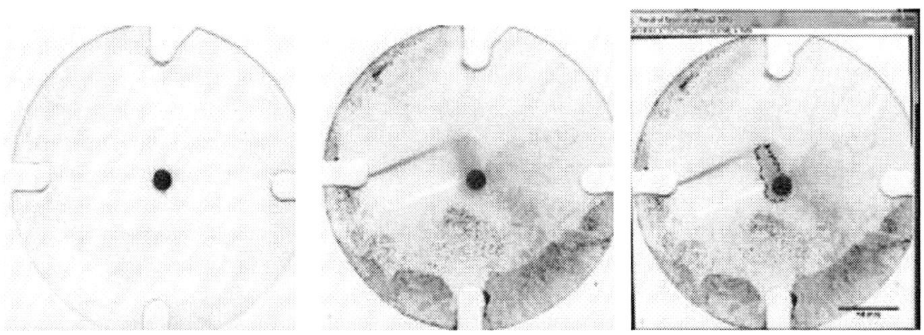

Figure 5. The paper disc before used, after used and after analyzed using ImageJ

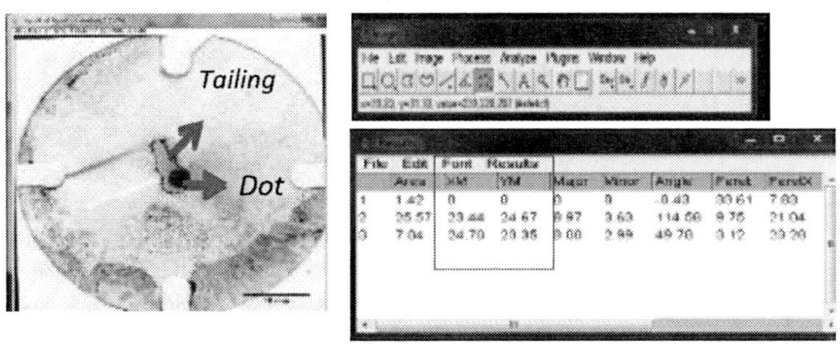

Figure 6. Velocity analysis of paper disc tailing using ImageJ application

3. Result and Discussion

3.1. The result from ImageJ and naked eye analysis

Figure 7 shows the comparison of velocity results using ImageJ and naked eye analysis on each deep well. The velocity magnitudes tend to have a similar trend between the analysis using the naked eye and using ImageJ analysis. Both methods show that the velocity in the Transect-1 and Transect-2 have a lower velocity and in the Transect-3 have a higher velocity. Some points have the same velocity magnitude as the result of the analysis using the naked eye and ImageJ. However, some points have different velocity magnitude. This is because of the ink can be finely identified using ImageJ analysis, so that the ink tailing can be delineated accurately. The naked eye analysis has a high potential for human error.

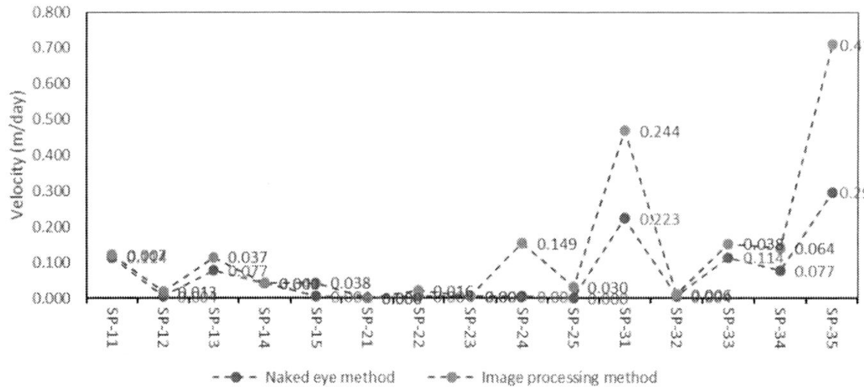

Figure 7. The comparison of velocity result using ImageJ and naked eye analysis

3.2. Groundwater flow around canal block

Measurement of the groundwater flow around the canal block was carried out in order to understand the impact of canal blocking for rewetting effort in tropical peatland. The map of velocity magnitude and direction as a result of measurement from 15 deep wells are presented in Fig.8. The map shows that the velocity in Transect-3 has a higher magnitude than that in Transect-2 and Transect-1 (also see Fig. 7). This phenomenon was caused by canal blocking. Canal blocking made the water level in the upstream side of the canal increase and the downstream side was not impacted. There was a different water elevation (head) between the upstream and downstream sides of the canal block because of canal blocking. The increasing water level in the upstream side of the canal block penetrated the peatland area and made the groundwater level higher. This made groundwater level in the peatland area of upstream of canal block higher than that in the peatland area of downstream of canal block, as presented in Fig. 9. As a result, the head in the Transect-3 was higher that caused the groundwater velocity higher than before canal blocking. However, the groundwater flow in the upstream side of the canal block tends to lower than that on the downstream side. This was because the groundwater level was maintained in a stable condition due to the canal blocking. This finding showed the effectiveness of canal blocking on the rewetting effort in the tropical peatland.

Theoretically, the groundwater flows from high to low groundwater levels. The measurement using PDV also can identify the direction of groundwater flow. The maps of the direction of groundwater flow as a result of PDV measurement are presented in Fig. 8 and Fig. 9. The flow direction showed a good agreement with the theory that the flow should be from high to low groundwater level. Most of the measurement at the deep well in Transect-3 showed the flow direction to the downstream which has the lower groundwater level and high head due to canal blocking. However, the groundwater flow direction in the Transect-1 and Transect-2 were not in a similar pattern but were still follow the basic theory of flow. This was because of the groundwater level in the upstream side of canal block were in a stable condition due to the canal blocking. There was no significant difference in the groundwater level in the upstream peatland of canal block. However, the flow should be going to the peatland area from the canal if the water level in the canal starts to increase. This condition usually occurs at the beginning of the rainy season.

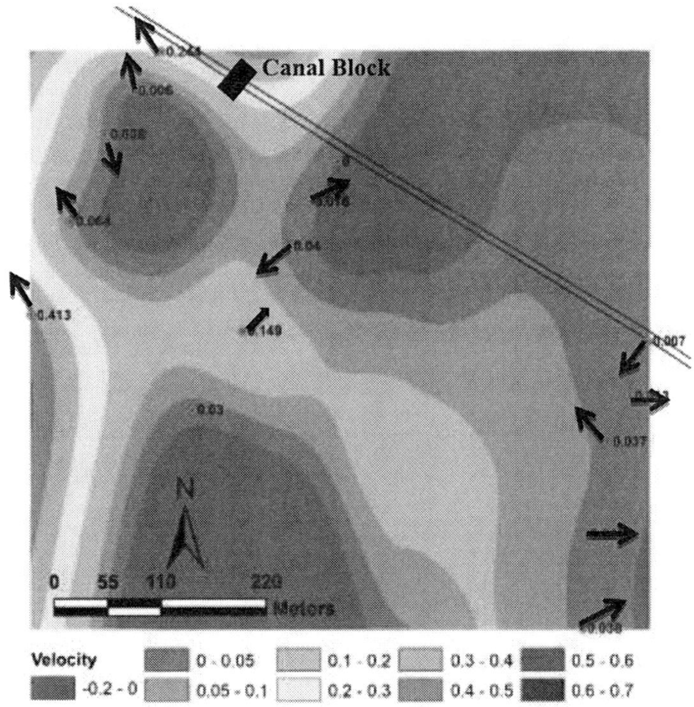

Figure 8. The map of groundwater velocity magnitude and direction

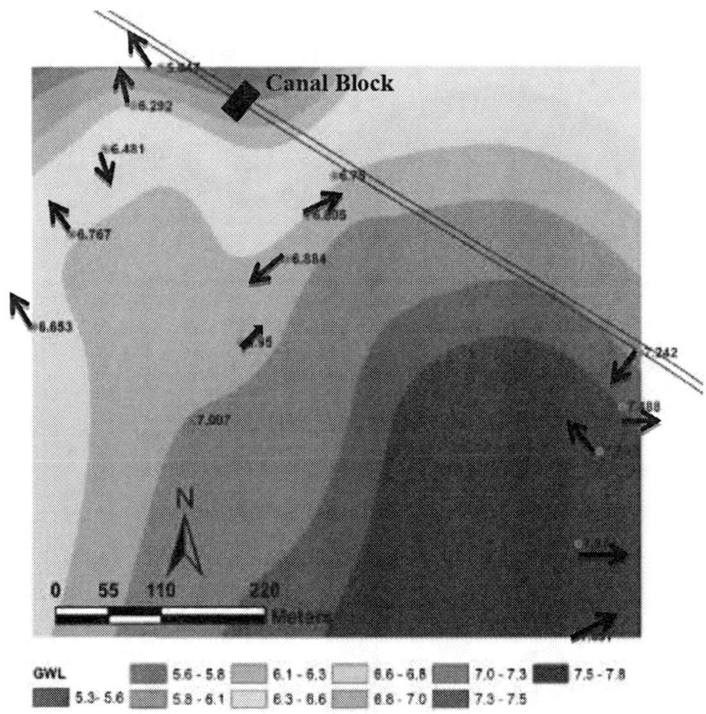

Figure 9. The map of groundwater level elevation and the flow direction

4. Result and Discussion

This paper presents the application of paper disc velocimeter (PDV) for groundwater flow measurement in tropical peatland. The PDV is a simple and low-cost method used without electricity that is developed by Yamaguchi University to measure groundwater flow. This method was applied to understand the groundwater hydraulic in the tropical peatland as the impact of rewetting efforts using canal blocking. The results showed that the groundwater flow in the upstream side of canal block tends to lower than that in the downstream side. This was because the groundwater level was maintained in a stable condition due to the canal blocking. However, there was a large head difference between upstream and downstream of the canal block that made the Transect-3 had a higher velocity. This finding showed the effectiveness of canal blocking on the rewetting effort in the tropical peatland.

5. References

[1] N. E. Mochtar, F. E. Yulianto, T. R. S. 2014 *Pengaruh Usia Stabilisasi pada Tanah Gambut Berserat yang Distabilisasi dengan Campuran CaCO3 dan Pozolan,* Surabaya.

[2] P. D. T. Supriatna *et al. 2008 Rencana Tindak Pengelolaan Lahan Gambut Berkelanjutan*, **39(5)**. Jakarta: Departemen Dalam Negeri.

[3] A. Sandhyavitri, M. A. Perdana, S. Sutikno, F. H. Widodo. 2018 *IOP Conf. Series: Materials Science and Engineering*, 0–9.

[4] A. Dohong *et al.* 2017 *Modul pelatihan Pembangunan Infrastruktur Pembasahan Gambut Sekat Kanal Berbasis Masyarakat*, Jakarta: Kedeputian Bidang Konstruksi, Operasi dan Pemeliharaan Badan Restorasi Gambut (BRG), Republik Indonesia, 2017.

[5] S. Sutikno *et al.* 2009 *MATEC Web Conf.*, **6003**(276), 1–7.

[6] K. Yamamoto, F. Ono, K. Tsushima, M. Ohishi, A. Kanno, M. Sekine. 2014 *J. Japan Soc. Civ. Eng. Ser. G Environmental Res.* **70**, p. III_355-III_363.

[7] F. Ono, K. Yamamoto, T. Akashi, A. Kanno, M. Sekine. 2016 *J. Japan Soc. Civ. Eng. Ser. C (geosph. Eng.)* **72(4)**, 377-384.

[8] Peraturan Presiden Republik Indonesia Nomor 1 tahun 2016 *tentang Badan Restorasi Gambut*. 1–16.

IOP Conf. Series: Materials Science and Engineering **796** (2020) 012043 doi:10.1088/1757-899X/796/1/012043

Preparation of magnetic chitosan microspheres using iron sand particles prepared by ball milling method

Rahmi[1]*, Lelifajri[1], and ASW Ayu[1]
[1]Chemistry Department, Faculty of Mathematics and Natural Sciences, Universitas Syiah Kuala, Banda Aceh, Indonesia

*email: rahmi@fmipa.unsyiah.ac.id

Abstract. Iron sand particles prepared by ball milling method have been used in the magnetic chitosan microspheres preparation. FTIR, XRD and SEM analyses were performed to characterize the magnetic chitosan microspheres. SEM images and FTIR spectra confirmed the formation of magnetic chitosan microspheres. The patterns of XRD showed the presence of iron sand particles reduced the crystallinity of chitosan. The adsorption experiments were conducted for mercury removal from aqueous solutions. AAS was performed to determine the mercury concentrations in the solutions after the adsorption process. The results showed the adsorption capacity (Q) of the magnetic chitosan microspheres was not significantly different from magnetic chitosan microspheres prepared using iron oxide.

1. Introduction

Chitosan is the second abundant polymer in nature after cellulose. Chitosan is one of the attractive polymers in the adsorption process due to its high adsorption capacity. It contains amine and hydroxyl groups on the chain backbone that can form complex with heavy metal ions. However, the use of chitosan in the powder form in the adsorption process will be difficult to be separated from the solution after the adsorption process.

Some studies have been conducted to modify chitosan such as magnetic chitosan microspheres preparation by using iron oxide. The magnetic chitosan microspheres can be removed easily from the solution using a magnet after the adsorption process, without centrifugation or filtration [1]. Most studies reported the iron oxide used in the magnetic chitosan microspheres preparation was commercial iron oxide that will produce high-cost adsorbent [2-5]. While it is known that iron sand contains a high content of iron oxide. Iron sand is easily found and abundant in the environment especially in Aceh, Indonesia [6]. Therefore, in order to get low-cost adsorbent, iron sand can be used as an iron oxide source for magnetic chitosan microspheres preparation. In our previous work, iron oxide nanoparticles have been prepared from iron sand [7-10]. The iron oxide nanoparticles were used in the magnetic chitosan microspheres preparation. However, in order to obtain the iron oxide nanoparticles, some chemicals were used that produce chemical wastes. Therefore, in this study, we used iron sand nanoparticles obtained by the ball milling method which is a green method without using any chemicals.

In this work, magnetic chitosan microspheres preparation was conducted using iron sand particles obtained by ball milling method. The magnetic chitosan microspheres were characterized by FTIR, XRD and SEM analysis. The adsorption capacity of magnetic chitosan microspheres was evaluated for mercury ions and the result was compared with the magnetic chitosan microspheres prepared using iron oxide.

Content from this work may be used under the terms of the Creative Commons Attribution 3.0 licence. Any further distribution of this work must maintain attribution to the author(s) and the title of the work, journal citation and DOI.
Published under licence by IOP Publishing Ltd

2. Methods

2.1. Preparation of iron sand particles

Iron sand was collected using a magnet, washed with hot distilled water several times and dried at 70°C overnight. And then, the iron sand was powdered using a ball mill for 20 hours.

2.2. Preparation of magnetic chitosan microspheres

Chitosan was placed in an Erlenmeyer flask containing acetic acid (2%, 20 mL) and stirred for 2 hours to form a chitosan solution. Iron sand particles were dispersed in the chitosan solution and stirred for an hour. The mixture was inserted into the syringe and dripped gradually into the solution of NaOH (3 M). The obtained magnetic chitosan microspheres were filtered and washed with distilled water until neutral pH was reached. Finally, it was dried at 40°C for 7 hours.

2.3. Adsorption

Magnetic chitosan microspheres (0.1 g) were added in a beaker containing mercury solution (300 ppm, 10 mL). The mixture was shaken (135 rpm) at room temperature for 80 minutes. The magnetic chitosan microspheres were separated from the solution by a magnet after the adsorption process. AAS analysis was performed to determine the mercury concentration after adsorption. The adsorption capacity (Q) was calculated using the equation:

$$Q = \frac{(C_i - C_e)\,V}{W} \qquad (1)$$

where Ce is the mercury concentration at equilibrium and Ci is the initial concentration of mercury. V is the volume of the mercury solution and W is the mass of the adsorbent.

2.4. Instrumentations

The concentration of mercury metal ions was determined by Atomic Absorption Spectrometer (Shimadzu AA-6300). FTIR spectra were analyzed by Shimadzu FTIR-Prestige 21 Series Fourier Transform Infrared Spectrometer; wave number from 4500-400 cm^{-1}. The XRD patterns were analyzed by Shimadzu XRD-7000 Series X-Ray Diffractometer which operated at 40 kV and 30 mA produced $Cu_{k\alpha}$ with $\lambda = 0.154$ nm at a distance of $2\theta = 10°\text{-}80°$ using the step size of 0.02°/minute. JSM-6510A/JSM-6510LA (Analytical/Analytical low vacuum SEM) was performed to obtain the micrographs at 100× and 30,000× magnifications.

3. Results and Discussion

3.1. FTIR

In order to ensure the interaction between components of the microspheres and to observe the functional groups of the materials, the FTIR spectroscopic analysis was performed. The results of FTIR analysis of iron sand, chitosan and magnetic chitosan microspheres were shown in Figure 1.

Figure 1a shows the FTIR spectrum of iron sand, where the existence of the Fe-O vibration is observed at wave number 582.50 cm^{-1}. The FTIR spectrum of chitosan is shown in Figure 1b, where it shows the absorption band of OH stretching vibrations at wave number 3446.79 cm^{-1}, aliphatic -CH vibration at wave number 2922.16 cm^{-1} and $-CH_2$ bending vibrations at wave number 2879.72 cm^{-1}. The absorption band of amide vibration is observed at wave number 1654.92 cm^{-1}. Figure 1c shows the spectrum of magnetic chitosan, where the absorption band of OH vibration appears at a lower wavenumber (3427.51 cm^{-1}) than the absorption band of OH vibration of chitosan. The absorption band of amide vibration also shifted to lower wavenumber (1633.71 cm^{-1}). It was due to the interactions between chitosan and iron sand such as hydrogen bonding and electrostatic force. The absorption band of Fe-O vibration of magnetic chitosan appears at a lower wavenumber (576.72 cm^{-1}) than iron sand. It confirmed the interaction between a component in the magnetic chitosan and the formation of magnetic chitosan.

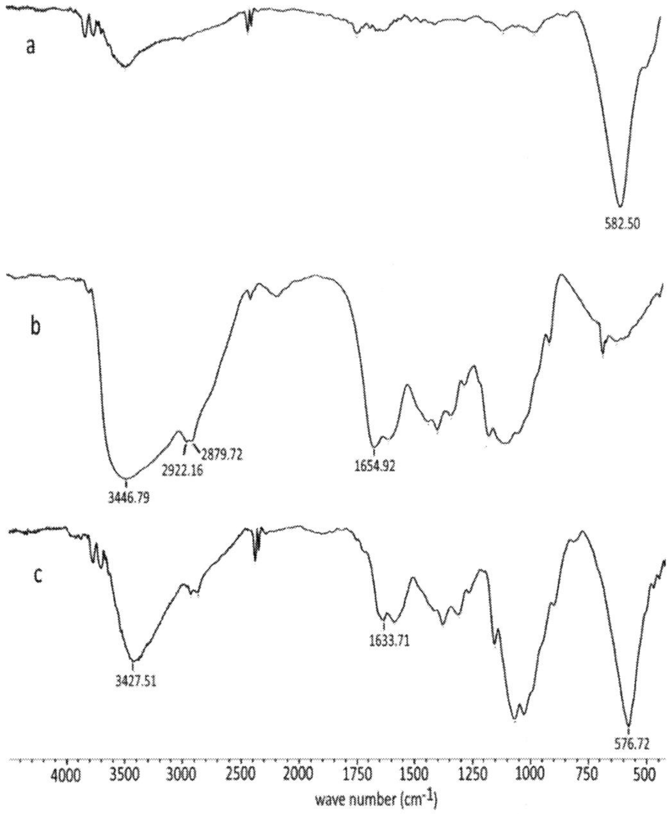

Figure 1. FTIR spectra of iron sand particles (a), chitosan (b) and chitosan magnetic microsphere (c)

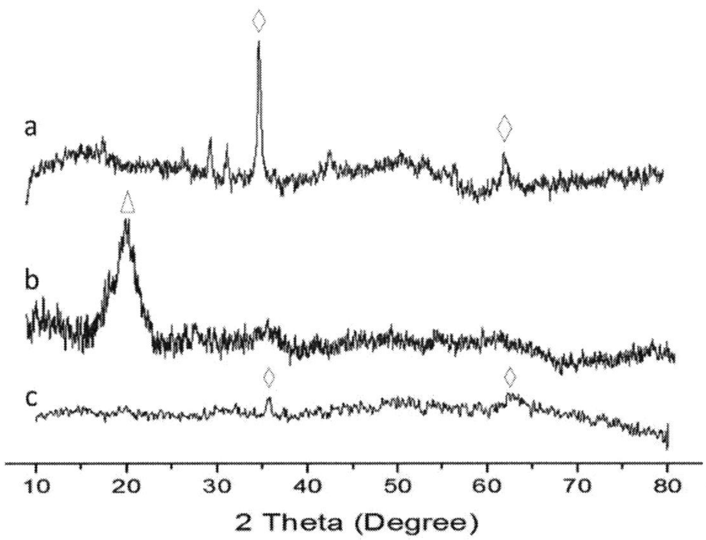

Figure 2. XRD patterns of iron sand particles (a), chitosan (b) and magnetic chitosan microsphere (c)

3.2. XRD

The structural features of materials were studied by XRD analysis. The XRD pattern of iron sand particles prepared by the ball mill method is shown in Figure 2a. It exhibits typical peaks of Fe_3O_4 at $2\theta = 35.43°$ and $62.56°$ because Fe_3O_4 is the main component of iron sand. Figure 2b is the XRD pattern of chitosan that shows semicrystalline materials. It exhibits a typical peak of chitosan at $2\theta = 20.79°$ [11]. Figure 2c shows the XRD pattern of magnetic chitosan microspheres, where chitosan loses its crystallinity. This is because the structure of chitosan is more amorphous due to less organized. During the adsorption process, mercury has high accessibility to bind to the active side of an amorphous polymer.

Typical peaks of Fe_3O_4 are also found in the magnetic chitosan diffractogram at $2\theta = 35.64°$ and $62.48°$. It confirmed that the iron sand main composition was not changed during modification and the iron sand particles distributed on the surface of chitosan polymer.

3.3. SEM

The SEM images of iron sand particles prepared by the ball mill method and the magnetic chitosan microspheres are shown in Figure 3. Figure 3a is an SEM image of iron sand at magnification 30,000×. It shows irregular shapes of particles with various particle sizes from micro to nanosized particles.

Figure 3b is an SEM image of magnetic chitosan microspheres at magnification 100×. It shows a microsphere shape of chitosan magnetic, where iron sand particles were covered by chitosan polymer. The formation of a microsphere was performed at an alkaline solution because chitosan is a pH-responsive polymer where chitosan dissolves at low pH (acid solution) and shrinks at high pH (alkaline solution). Based on the SEM images, the formation of magnetic chitosan microspheres was successfully performed.

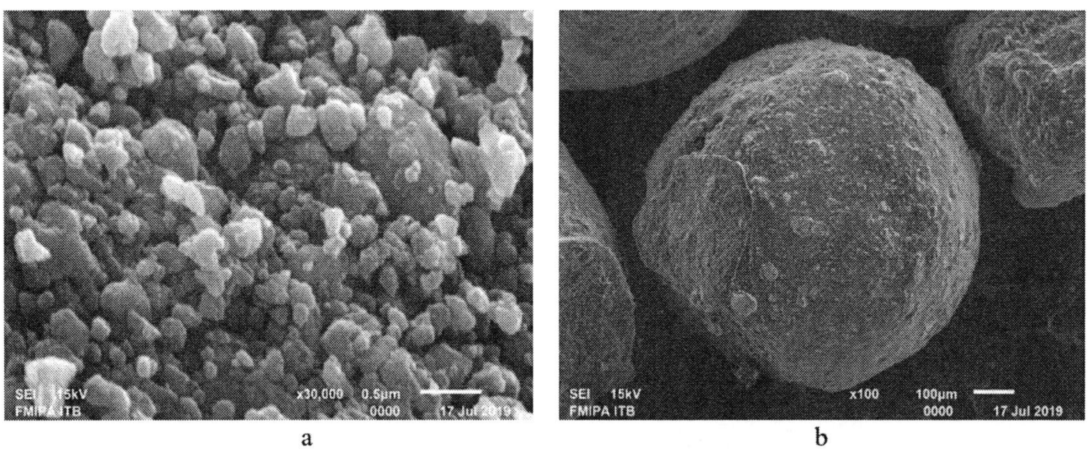

Figure 3. SEM analysis of iron sand particles prepared by ball mill method at magnification 30,000× (a) and magnetic chitosan microspheres at magnification 100×

3.4. Adsorption Study

Adsorption experiments were performed to investigate the adsorption capacity of magnetic chitosan microspheres for mercury ions and the results are shown in Figure 4. It shows that the magnetic chitosan microspheres prepared using iron sand particles have lower adsorption capacity than the magnetic chitosan microspheres prepared using iron oxide. It was due to the iron sand particles prepared by the ball mill method still contained impurities and the particle size was relatively larger than iron oxide. However, the difference in their adsorption capacities was not significantly. The adsorption capacity of the magnetic chitosan microspheres prepared using iron oxide particles and magnetic chitosan microspheres prepared using iron sand particles was 29.94 mg/g and 28.62 mg/g, respectively, where the difference was only 4.52%.

These results indicate that the preparation of magnetic chitosan microspheres using iron sand particles obtained by the ball mill method was a relatively green method and low cost because the preparation of iron sand particles without chemicals and the adsorption capacity was not significantly different with magnetic chitosan microspheres prepared using iron oxide.

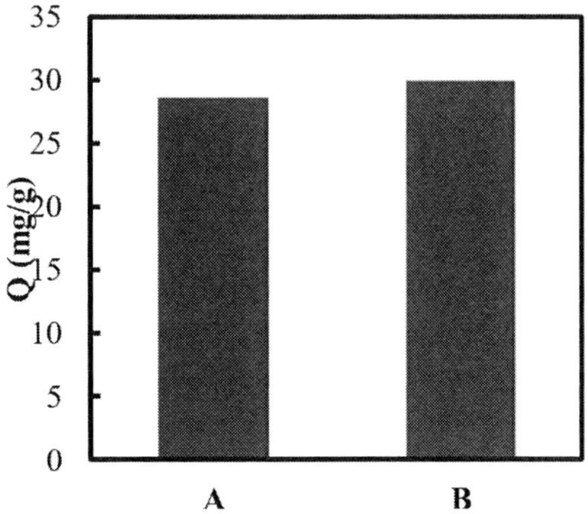

Figure 4. The adsorption capacity of magnetic chitosan microspheres prepared using iron sand particles (A) and magnetic chitosan microspheres prepared using iron oxide (B)

4. Conclusions

Iron sand particles prepared through the mechanical milling process were used in magnetic chitosan microspheres preparation. The characterizations confirmed the formation of magnetic chitosan microspheres. The adsorption capacity of magnetic chitosan microspheres prepared using iron sand particles was not significant with the adsorption capacity of magnetic chitosan microspheres prepared using iron oxide particles. This study recommends the low-cost and green adsorbent for mercury.

References

[1] El-Reash Y G A 2016 *Journal of Environmental Chemical Engineering*,**4** 3835-3847
[2] Alizadeh B, Delvanaz M, Shakeri A 2018 *Carbohydrate Polymers* **182** 675-683
[3] Zhao W, Huang X, Wang Y, Sun S, Zhao C 2016 *Carbohydrate Polymers* **150** 201-208
[4] Xu B, Zheng H, Zhou H, Wang Y, Luo K, Zhao C, Peng Y and Zheng X 2018 *Journal of Molecular Liquids* **256** 424-432
[5] Ma H, Pu S, You Y, Zhu R, Zinhenko A and Chu W 2018 *Chemical Engineering Journal* **345** 556-565
[6] Jalil Z, Sari E N, and Handoko E 2014 *Journal of Applied Physics* **4(1)** 110–14
[7] Rahmi, Fathurrahmi, Lelifajri, and PurnamaWati F 2019 *Heliyon* **5** e01731.
[8] Rahmi, Ishmaturrahmi, Mustafa I 2019 *Microchemical Journal* **144** 397-402.
[9] Rahmi, Fathurrahmi, Lelifajri, Purnamawati F, Sembiring R 2019 *IOP Conf. Ser.: Earth Environ. Sci.* **276** 012004.
[10] Ishmaturrahmi I, Rahmi R, Mustafa I 2019 *IOP Conf. Ser.: Mater. Sci. Eng.* **523** 012020
[11] Mahdavinia G R, Soleymani M, Etemadi H, Sabri M 2018 Journal of Biological Macromolecules **107** 719-729

Slope stability analysis due to extreme precipitation

H Gunawan[1], N Al-Huda[1], M Sungkar[1], A Yulianur[1], and B Setiawan[2,*]
[1] Jurusan Teknik Sipil, Fakultas Teknik, Universitas Syiah Kuala, Indonesia
[2] Program Studi Teknik Geologi, Fakultas Teknik, Universitas Syiah Kuala, Indonesia

E-mail: bambang.setiawan@unsyiah.ac.id

Abstract. Slope stability analysis due to extreme weather, i.e. precipitation is an interesting topic to investigate in the high rainfall region. This paper aims to investigate the effect of extreme rainfall on slope stability. A study case at a location inside the highly vulnerable formation in Aceh-Indonesia to failure, which is Tutut Formation, was carried out. Soil samples at the investigated site were collected and tested for further examination. The geometry of the landslide location was measured using Theodolite. Both the soil and surface geometry data were used to develop a computer model from which the slope failure was justified. A failure mechanism due to extreme precipitation is suggested. This finding is consistent with the field observations.

1. Introduction

Landslide is a natural hazard that often poses threats to both the society and economy ([1]; [2]). As one of the natural hazards, this landslide causes infrastructure and property damages, casualties, and severe injuries. This hazard may be attributed to many destabilizing factors [3]. Most of the landslides are caused by natural dynamic processes [3], including significant rainfall seasonal variation [4]; as this rainfall variation causes enormous impacts on soil characteristics and phreatic groundwater conditions [2]. In many cases of a landslide, extreme or prolonged precipitation is the main trigger of the slope instability.

In general concept, the incidence of the landslide is mainly controlled by outside (external) and/or inside (internal) factors ([5]). The external factors consist of the increasing of load due to a wide range of activities such as loads by earth dumping, slope cutting or excavation, vibration (including machine, turbine and seismic vibrations), loads associated with surface and groundwater flow, and rainfall ([6]; [7]; [8]; [9]; [10]). The internal factors are mainly associated with the characteristics of the soil and rocks. These internal factors include the geological conditions (rock types, stratigraphy, geological structures), landforms (slope), soil type, soil thickness, physical and mechanical properties of soil/rock, and pore water pressure ([11]; [12]; [13]; [14]). These many factors as aforementioned above suggest the complexity of the landslide phenomenon.

This paper aims to investigate slope failure due to extreme precipitation. Recently, extreme precipitation is documented as one of the main causes of natural and engineered earth structures instability ([15]; [16]; [17]). Furthermore, this study is also assessing the damage of road infrastructure at the study location to provide information to relevant stakeholders. The results of this investigation can be used for slope stabilization efforts by relevant stakeholders.

[2,*] Corresponding author: bambang.setiawan@unsyiah.ac.id

Content from this work may be used under the terms of the Creative Commons Attribution 3.0 licence. Any further distribution of this work must maintain attribution to the author(s) and the title of the work, journal citation and DOI.
Published under licence by IOP Publishing Ltd

Figure 1. A landslide along the main Meulaboh-Geumpang Pidie road [1]

2. Study site

The study site is in the Aceh Singkil district. Figure 2 shows the simplified geological map of the Aceh Singkil district. There are 7 lithology types at Aceh Singkil district, which are Kluet Formation, Sibolga Formation, Barus Formation, Trumon Volcanic Formation, Toba Tuff, Tutut Formation, and Alluvium. The last two lithology types are very dominant in the Aceh Singkil district [18].

It is estimated that there are two rock formations namely Tutut and Alluvium Formations at the study case site. The Tutut Formation consists of conglomerate sandstone, a little siltstone, and mudstone. Tutut Formation is compact to friable. The red mark in Figure 2 is the point of the measured location of this study. The morphology of the study case site, Lae Pinang village, Singkohor sub-district, Aceh Singkil district, is steep- to gently-slope (Figure 3a).

3. Methods

3.1. Data acquisition

The data field acquisition was carried out in early 2019. The weather was generally calm with no strong winds or rain during the data collection. This filed data collection was using several supporting tools such as a theodolite, measurement tools, camera, survey board, tubes for collecting an undisturbed soil sample. Secondary data were also collected during the field trip. Secondly, the process of collecting field data including the two undisturbed samples (UD#1 and UD#2), visual description, and geometric measurement for modeling was carried simultaneously (see Figure 3b). All the field data is important for modeling slope stability.

3.2. Data analysis

Slope stability analysis is performed to assess the stability of a slope. A limit equilibrium condition is used in this analysis. Currently, this limit equilibrium method is mostly used in the slope stability analysis as it involves to most simplicity and accuracy. Generally, the method must search for critical surfaces by using geometry. A slope is stable if the safety factor, computed along any potential sliding surface, is larger than 1. The smallest value of the safety factor is representing the global stability condition of the slope.

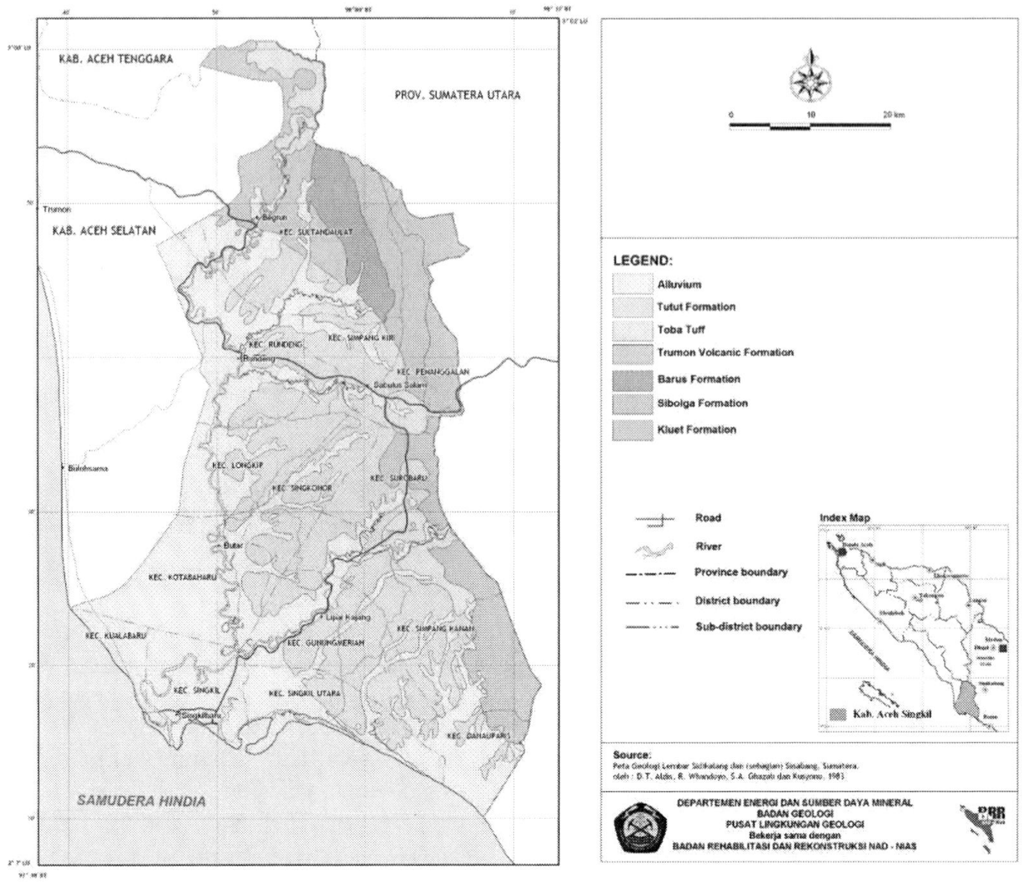

Figure 2. Simplified geological map of Aceh Singkil district [18]

(a) (b)

Figure 3. (a) Morphology of the study site, and (b) visual appearance of the soil.

3.3. Developed model
The developed model is based on field geometry data and laboratory test results. Two soil types with different physical and mechanical characteristics are used in the developed models (see Figures 4a and

4b). Then, three different conditions are applied to the two developed models. The three conditions include: 1) the condition when the soil is quite dry, 2) the condition when the soil is wet/moist, and 3) the condition when the soil is fully saturated with water. Details of the used parameters in the development of the models are shown in Table 1.

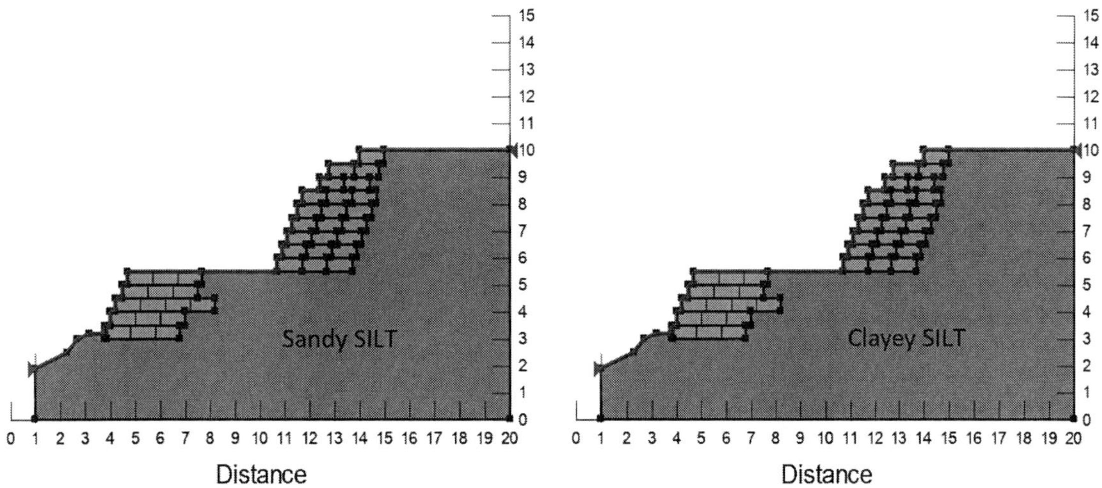

(a) (b)

Figure 4. Developed models (a) for Sandy SILT, and (b) Clayey SILT

Table 1. Description of the model parameters

Parameter	Model A	Model B	Remarks
Soil type	Sandy SILT	Clayey SILT	Visual
Soil classification	CLML	MH	USCS
Cohesion (kg/cm^2)	0.0221	0.0038	-
Internal angle of friction (degree)	20.54	21.59	-

4. Results and discussion

As aforementioned, the results of the field measurement and laboratory testing data are for developing slope models. An analysis was carried in three different situations, namely: (1) the condition when the soil is dry, (2) the condition at the time of the soil is moist, and (3) the condition of the soil is fully saturated with water. Furthermore, two different models of Model A (based on UD#1) and Model B (based on UD#2) were used in the analysis. The results of the simulation are shown in Figures 5 to Figure 7.

4.1. Results

A summary of the simulation results is detailed in Table 2. The results show that only in dry conditions the both Model A and Model B have more than 1 (one) safety factor (SF) that can be interpreted as 'SAFE' or collapse will not occur in this condition. However, when the ground soil is wet or partly saturated from 1-meter depth, the emerging SF is below 1 (one), which is indicating 'NO SAFE' condition or collapse will occur.

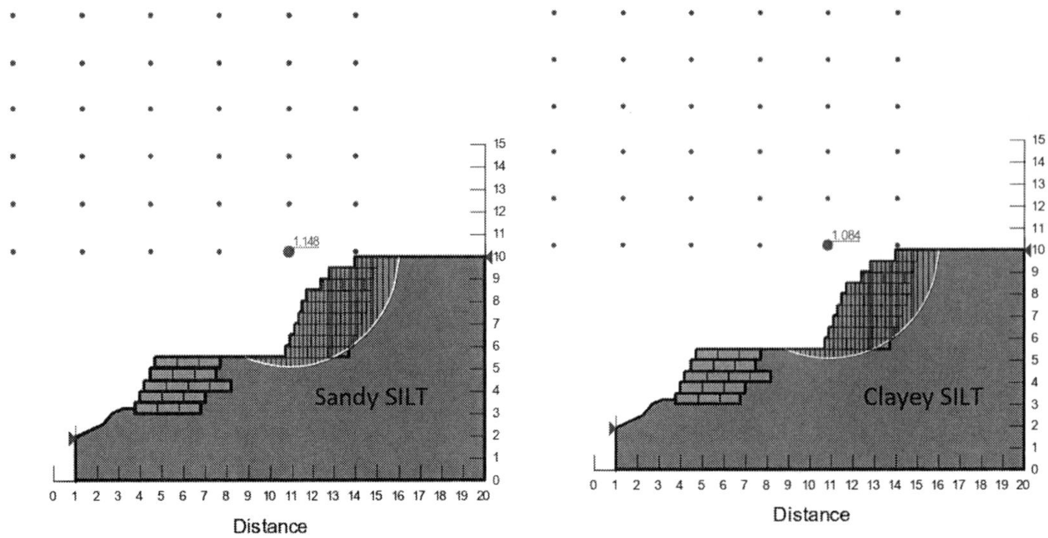

(a) (b)

Figure 5. Simulated model when the soil moisture is dry for (a) model A, and (b) model B

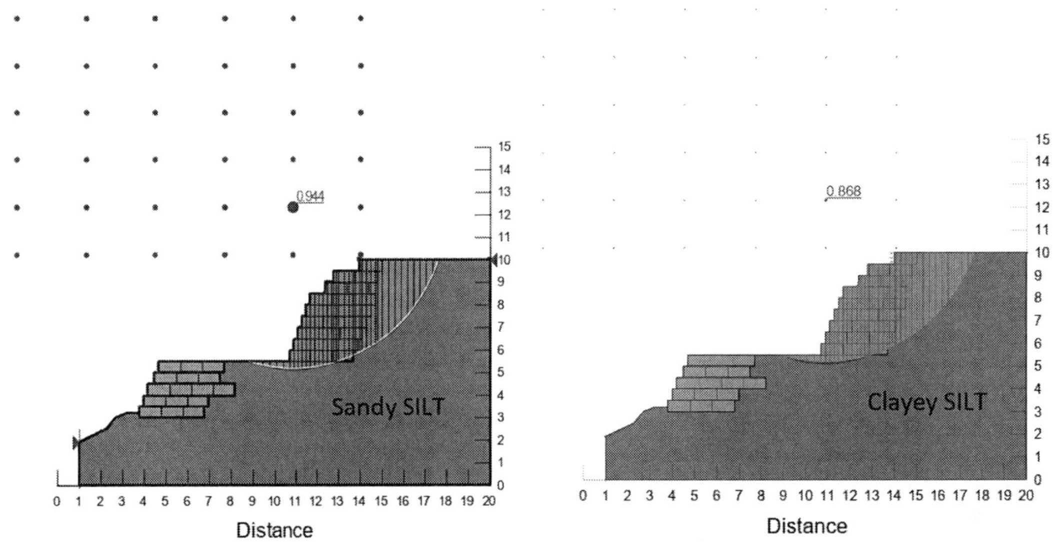

(a) (b)

Figure 6. Simulated model when the soil moisture is wet/moist for (a) model A, and (b) model B

Table 2. A summary of the analysis results

Soil moisture condition	Model A	Model B	Remarks
Dry	SAFE	SAFE	SAFE is SF>1
Wet/moist	NOT SAFE	NOT SAFE	NOT SAFE is SF<1
Fully saturated	NOT SAFE	NOT SAFE	

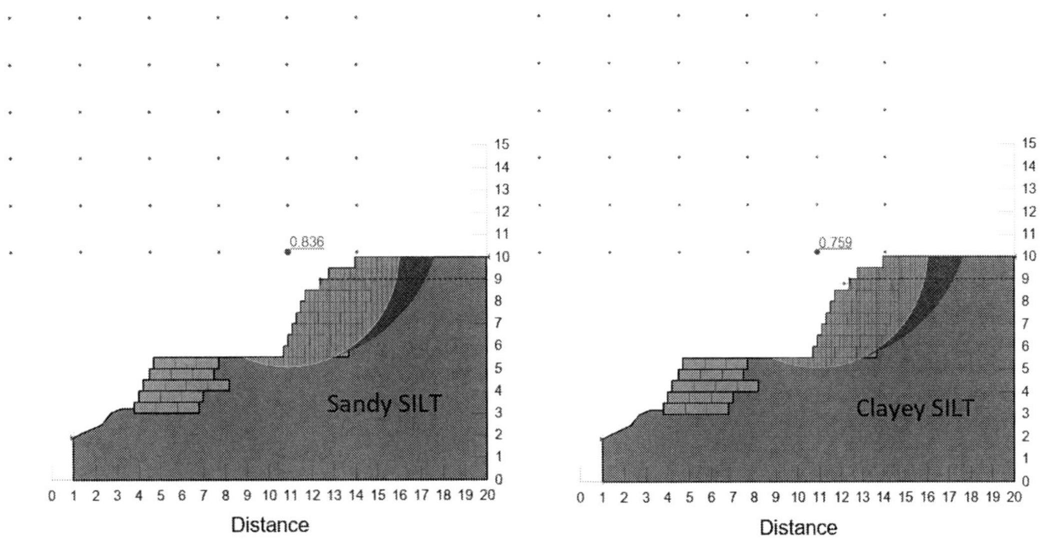

Figure 7. Simulated model when the soil is fully saturated with water(a) Model A, and (b) Model B

Figure 5 displays the analysis of slope stability when the soil is dry. The critical safety factor in this condition is more than 1, which suggests a stable slope. Figure 6 illustrates the simulation when the soil is moist. The simulation suggests a safety factor less than 1 for both models A and model B. Figure 7 demonstrates the impact of high rain intensity on the SF. As can be seen, the SF reaches the lowest value when the soil is fully saturated.

4.2. Discussion

A homogenous sandy SILT and clayey SILT soil slopes (Figure 4) are used to illustrate the impact of the precipitation on the behavior of a natural slope. The model includes a 10m high slope with ~40° to ~70° inclination. This typical slope geometry is reported in the literature for landslide modeling [19]. The analysis implements the classic effective shear strength parameters (i.e., cohesion and angle of friction). In general, Figures 5 to 7 demonstrate the decrease in SF under the projected precipitation. As can be seen in Figure 5, the slope is stable under the initial conditions. Once the soil moisture is increasing until moist/wet due to heavy rainfall, localized tension (represented by the red zones in Figures 6) appears in the middle slope boundary, which suggests that the slope's boundary approaches the limit state to failure. As the projected precipitation persists up to the fully saturated soil, the slope's boundary is overestimated the limit state, which is indicative of active sliding. This can be seen in the upper slope boundary of Figure 7. Such a significant surge in the pore-water pressure causes a substantial reduction in the soil shear strength and increment of the overburden load. This failure mechanism is consistent with the field observations. Extreme rainfall was occurring at the study site in November to the beginning of December 2017 as suggested by several official statements of government agencies and institutions located around the scene. Furthermore, a similar slope failure mechanism was suggested by Cruden and Varnes [20]. A further slope stability assessment using the passive seismic method [21] is an interesting subject to explore.

5. Conclusion

Generally, an extreme rainfall intensity can expose a natural slope, engineered slope or earth retaining structure to significant increases in pore pressures and reduce in bearing capacity. These changes can lead to slope destabilization. This study aims to raise awareness regarding the impacts of extreme rain intensity on slope stability. A case study was carried out. The presented models show substantial impacts of extreme precipitation on the slope stability. It is imperative to consider the extreme precipitation to ensure the stability of geotechnical structures.

6. Acknowledgment

We acknowledge the help and assistance given by Universitas Syiah Kuala for the study under Lektor Research Scheme Year 2019 of Universitas Syiah Kuala (Contract No. 522/UN11/SPK/PNPB/2019 on date 8 February 2019). Also, the authors are grateful to the Fakultas Teknik Universitas Syiah Kuala for their support.

7. References

[1] Serambi News. 2017. *Meulaboh-Geumpang Lumpuh*. Accessed on 27 August 2019 https://aceh.tribunnews.com/2017/05/12/meulaboh-geumpang-lumpuh
[2] Intrieri, E., Gigli, G., Mugnai, F., Fanti, R., Casagli, N. 2012 *Engineering Geology*. 147-148: 124-136.
[3] Leshchinsky, B., Vahedifard, F., Koo, H.B., Kim, S.H. 2015 *Landslides*. **12(5)**, 997–1005.
[4] Lu, N., Sener-Kaya, B., Wayllace, A., Godt, J.W. 2012 *Water Resources Research*. **48(9)**,W09524.
[5] Zuoan, W., Shihai, L., Wang, J. G., Ling, W. 2006 *Engineering Geology*. **84**, 1–11.
[6] Binet, S., Jomard, H., Lebourg, T., Guglielmi, Y., Tric, E., Bertrand, C., Mudry, J. 2007 *Hydrological Processes*. **21(25)**, 3463-3472.
[7] Huang, A-B., Lee, J-T., Ho, Y-T., Chiu, Y-F., Cheng, S-Y. 2012 *Soils and Foundation*. **52 (4)**, 737-747.
[8] Piccarreta, M., Capolongo, D., Boenzi, F. 2004 *International Journal of Climatology*. **24**, 907–922.
[9] Lazzari, M., Piccarreta, M., Capolongo, D. 2011 *Proceedings of the Second World Landslide Forum*. **2**, 671–678.
[10] Robinson, J. D., Vahedifard, F., AghaKouchak, A. 2016 *Canadian Geotechnical Journal*. **54**, 117–127.
[11] He, Y., Liao, C., Xu, P., Zhang, B. 2005 *Journal of Mountain Science*. **2(1)**, 42-49.
[12] Askari, F., Farzaneh, O., 2008 *International Journal of Civil Engineering*. **6 (1)**, 10-23.
[13] Perrone, A., Vasallo, R., Lapenna, V., Maio, C. D. 2008 *Journal of Geophysics and Engineering*. **5(3)**, 323-337.
[14] Gunadi, D. S. A., Jaya, I. N. S., Tjahjono, B. 2017 *Indonesian Journal of Electrical Engineering and Computer Science*. **5(1)**, 139-146.
[15] NRC. 2008. *Transportation Research Board, Division on Earth and Life Studies, National Research Council of the National Academies Washington*, D.C. ISBN: 978-0-309-11306-9.
[16] NRC. 2013. *Committee on Understanding and Monitoring Abrupt Climate Change and Its Impacts; Board on Atmospheric Sciences and Climate; Division on Earth and Life Studies; National Research Council of the National Academies*, Washington, D.C. ISBN:978-0-309-28773-9.
[17] Sorooshian, S., AghaKouchak, A., Arkin, P., Eylander, J., Foufoula-Georgiou, E., Harmon, R., Skofronick-Jackson, G. 2011 *Bulletin of the American Meteorological Society*. **92(10)**.
[18] Pusat Lingkungan Geologi. 2006 *Departemen Energi dan Sumber Daya Mineral bekerja sama dengan Badan Rehabilitasi dan Rekonstruksi NAD-NIAS* (BRR).
[19] Lu, N., Sener-Kaya, B., Wayllace, A., Godt, J.W. 2012 *Water Resources Research*. **48(9)**, W09524.
[20] Cruden, D.M., Varnes, D.J. 1996 *Transportation Research Board Special Report 247*. Edited by A.K. Turner and R.L. Schuster, National Academies Press, Washington, D.C. pp. 36–75.
[21] Setiawan, B., Jaksa, M., Griffith, M., Love, D. 2018 *Data in Brief*. **20**, 735-747.

IOP Conf. Series: Materials Science and Engineering **796** (2020) 012045 doi:10.1088/1757-899X/796/1/012045

Chemical and microbiological characteristics of Pacik Kule, a traditional fermented carp from Southeast Aceh

M Muzaifa[1], Martunis[1], C Nilda[1*], Nurhayati[1]
[1]Agricultural Product Technology Department, Agriculture Faculty, Universitas Syiah Kuala, Kopelma Darussalam - Banda Aceh 23111, Indonesia

*E-mail: cutnilda@unsyiah.ac.id

Abstract. *Pacik kule* is a dish made from carp (Cyprinus carpio) which is commonly found in Southeast Aceh as traditional food. Before being processed into *pacik kule*, carps are fermented for 1 to 2 days at room temperature using salt or without salt. This study aims to analyze the effect of salt addition and fermentation time on the chemical and microbial characteristics of *pacik kule*. This research used factorial completely randomized design consisting of 2 factors. The first factor was the fermentation time which consists of 2 levels, namely 1 day (K1) fermentation, and 2 days (K2) fermentation. The second factor is the addition of salt consisting of 2 levels, which is S1 (without salt) and (S2) with salt (1.5%). The analysis conducted is a chemical analysis and microbiological analysis. The results showed that the length of fermentation affected the total lactic acid bacteria but did not affect other parameters such as protein content, total volatile bases, and pH. The addition of salt affected the water content and total lactic acid bacteria but did not affect protein content, pH, and total volatile bases (TVB). The interaction between length of fermentation and salt addition affected water content, protein content, total volatile bases (TVB), and total lactic acid bacteria. Based on the results of chemical and microbiological analysis, fermentation for 1 day with the addition of salt (1.5%) was the best treatment with the highest protein content (11.93%), total lactic acid bacteria (9.4 log CFU/g), water content the lowest (17.41%) and the lowest total volatile base (13.63 mgN / 100g).

1. Introduction

Indonesia has a variety of cultures, traditions, and characteristics, especially the characteristics of traditional foods that are diverse from various regions. One of these traditional food is processed fermented fish. Fish is an animal that is consumed by many people and available at an affordable price. Fermented fish products in Indonesia have different forms, raw materials, and types of fermentation and generally still use a spontaneous process that is fermentation using salt [1, 2]. According to [3], fermented food products usually contain higher nutritional value than the original ingredients. This result can be caused by microbes in fermentation products that can break down complex components in food into simpler ingredients, easier to digest and also microbes can synthesize some vitamins.

Fish fermentation products that are well-known in Indonesia include *Picungan*, a traditional fermented fish product that is processed using *picungan* seeds (*Pangium edule reinw*) originating from Banten [4], *Terasi* which is a traditional fish fermentation product found in Cirebon [5], *Pakasam* originating from South Kalimantan, made from fish marinated through salt fermentation process [6], and *Wadi* which is a fish fermentation product originating from Central Kalimantan [7]. In Aceh, fish fermentation products that have been studied are *bude* and *belacan depik*. *Bude* is one of the fermented fish products which is

Content from this work may be used under the terms of the Creative Commons Attribution 3.0 licence. Any further distribution of this work must maintain attribution to the author(s) and the title of the work, journal citation and DOI.
Published under licence by IOP Publishing Ltd

processed using anchovy (*Stolephorus* sp) and salt as raw materials. It has a rather thick liquid form, that can be consumed as side dishes in a raw or processed condition [1]. *Belacan depik* is one of the fermented fish products in the form of pasta, which is made from several raw materials including *depik fish* (*Rasboraawarensis*), laos, lemongrass, turmeric, and salt, which can be consumed as seasonings in the processed form [8].

One of the other fermented fish products that have not been studied is *pacik kule*. Based on the results of a field survey (2019), *pacik kule* is a dish made from carp (*Cyprinus carpio*) which is widely found in Southeast Aceh as traditional food. Before being processed into *pacik kule*, caps are fermented for 1 and 2 days at room temperature using salt or without salt. The variation of salt used in fermentation is around 1% to 2%. *Pacik kule* is taken from a combination of Alas and Gayo language where *pacik* means rotten and *kule* means tiger. *Pacik kule* means processed fish that are not good or rotten and then reused to becomes a traditional food that can be consumed. The word *pacik kule* comes from the story of a tiger that has a habit of spoiling food before eating. That is the beginning of the story of the process of *pacik kule* name.

Some changes occur during the fermentation process such as in the aroma, color, and texture. These changes were indicated by the smell of acid, the color begins to turn pale gray, and the texture begins to soften. According to [9], the principle of preservation in fish fermentation products is caused by several factors consisting of a decrease in water activity by salt and a decrease in pH due to the formation of acid by microbes. Fermentation aims to select spoilage and pathogenic microbes with the result that the microbes contained in fish during fermentation are microbes that are resistant to salt conditions and can play a role during fermentation.

The presence of salt can affect the types of microorganisms that play a role during the fermentation process. Different salt concentrations can affect microbial growth and safety of fermented products. Optimum growth of lactic acid bacteria is depending on salt concentrations that are not higher than 6% to 7% for *plaa-som* fermentation that occurs for 4-7 days [10]. *Plaa-som* is a fermented fish product that is generally made by Thai people, which is traditionally made using raw materials from freshwater fish or sea fish. Salt fermentation can also cause changes in the chemical and microbiological of fish meat. Therefore, this study aims to determine the changes in chemical and microbiological of *pacik kule* during fermentation with the addition of salt.

2. Materials and Method

2.1 Materials

The main raw materials used in this study are carp and salt. *Pacik kule* was made based on procedures carried out by the people of Southeast Aceh. The chemicals used in the analysis were aquades, de Man Rogosa Sharpe Agar (MRSA), natrium agar (NA), Pepton, NaOH, HCl, HgO, K_2SO_4, H_2SO_4, HBO_3, H_3BO_3, NaOH-Na_2S2O_3, K_2CO_3, TCA, methyl red, methylene blue, phenopthalein indicator, and alcohol. The tools used in the process of making *pacik kule* are plastic jars, knives, and digital scales. The tools used to analyze were petri dishes, pipette drops, spatulas, incubators, test tubes, measuring cups, erlenmeyers, analytical scales, ovens, desiccators, burettes, destruction, kieltec system, pH meters, colony counters, filter paper, titration, furnaces, blender, outer chamber, digital refractometer, measuring flask, kjeldhal flask, condenser, bunsen, and laminar flow cabinet.

2.2 Experimental Design

This study used a completely randomized design with two independent variables. The first variable was the length time of fermentation being used (K1= 1 day and K2 = 2 days) whilst the second variable was the salt addition (S1=without salt and S2= addition 1.5% salt. Then all the treatment combinations were done in three repetitions, resulting the total experiment units of 12 samples. Analysis of Variance (ANOVA) and Duncan Multiple Range Test (DMRT) was used for statistical analysis.

2.3 Research Procedure

The process of fermented *pacik kule* was performed with common procedures carried out by the people of Southeast Aceh, with variations in salt concentration and different length of fermentation. The first step of the procedure was removing the gills, bones, head, and fins of carp. Furthermore, the fish is washed, then drained, weighed and added with salt according to the treatment (without salt and salt addition (1.5%). Finally, at around 1500 grams of fish meat is put in a tightly closed jar and fermented at room temperature according to the treatment (1 day and 2 days). The analysis conducted in this study was the water content [11], protein content with Kjedahl method [11], pH [12], total volatile bases (TVB), and total lactic acid bacteria [13].

3. Result and discussion

3.1 Moisture Content

The value of moisture content ranged from 17.41% to 23.25% with an average of around 20.80%. Analysis of variance shows that the interaction length of fermentation and the addition of salt have a significant effect ($P \leq 0.05$) on the moisture content of *pacik kule* (Figure 1). Figure 1 shows the comparison between treatments without salt has a higher water content value (23.25%) compared to the treatment of adding salt (1.5%) around 17.41%. The function of salt in the fermentation process aims to reduce the water content in the material so that the moisture content in food will be reduced and can inhibit bacterial growth. According to [14], a decrease in water content occurs due to the salting process carried out in the fermentation process, because salt can attract water molecules that exist in fish tissue.

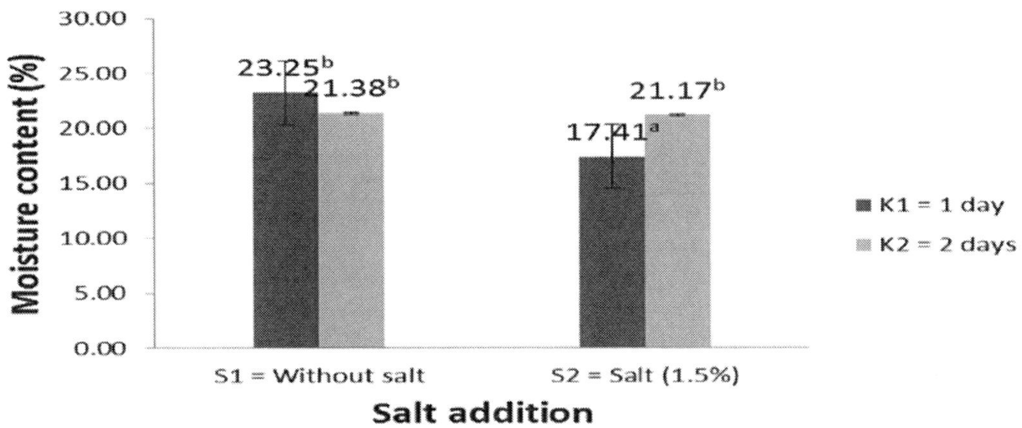

Figure 1. Interaction of length of fermentation and salt addition towards moisture content of *pacik kule* at $DMRT_{0.05}$ level 1 = 3.31, level 2 = 3.44, level 3 = 3.52, and coefficient of variants = 8.44% (Values followed by the same letter indicate no significant difference)

3.2 Protein content

Protein content ranged from 9.18% to 11.93% with an average of 10.38%. Based on the results of ANOVA, the interaction period of fermentation and addition of salt (KS) has a significant effect ($P \leq 0.05$). Based on Figure 2, it can be seen that the protein content of *pacik kule* in the first day of fermentation is around 11.25% compared to the second-day fermentation which is around 9.17%, it is assumed that the longer the fermentation time, the lower the value of the protein produced. According to [15] decreasing protein value during fermentation occurs due to protein hydrolysis to amino acids and peptides, and then amino acids will be further decomposed into other components that play a role in the formation of product flavors. The increase in the value of protein content in salt addition (1.5%) is around 11.93%, due to the low water content in this treatment. The previous study by [16] claims that water content is inversely proportional to protein content. This result is in accordance with the research by [17], that the higher the water content produced from food, the lower the protein will be due to myogen and protein dissolved in water.

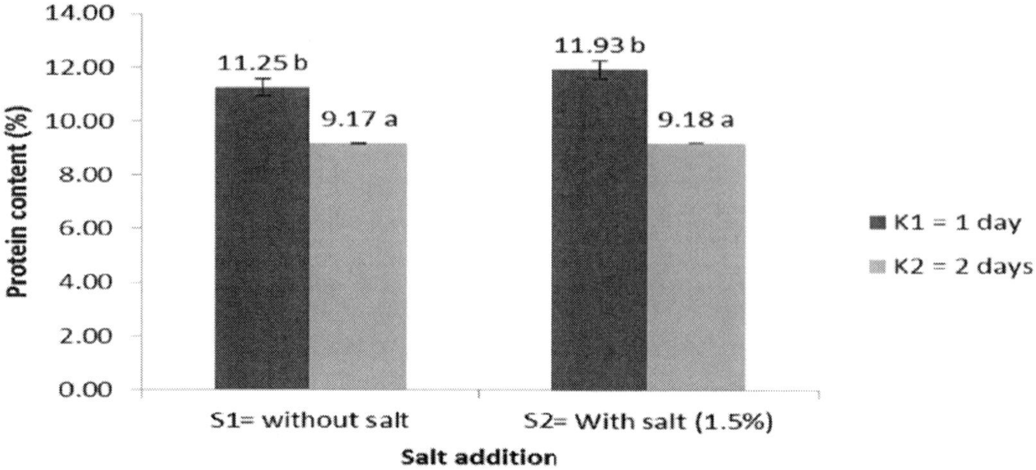

Figure 2. Interaction of length of fermentation and salt addition towards protein content of *pacik kule* at $DMRT_{0.05}$ level 1 = 3.07, level 2 = 3.19, level 3 = 3.27, and coefficient of variants = 15.7% (Values followed by the same letter indicate no significant difference)

3.3 Total Volatile Bases (TVB)

TVB value is a parameter used to notice the freshness of fish and has an important meaning in the process of deterioration of fish quality. TVB values of *pacik kule* ranged from 13.63 to 32.68 mg N /100g with an average value of 23.77 mg N/100g. Analysis of variance shows the interaction length of fermentation and salt addition has a very significant effect ($P \leq 0.01$) on the value of TVB *pacik kule*. Figure 3 shows that the total volatile bases of *pacik kule* which is fermented for 2 days without salt are higher (32.68 mgN/100g) compared to salt addition fermentation (1.5%) for 2 days with a value of 24.65 mgN/100g. The low value of TVB in the addition of salt treatment (1.5), presumably due to the addition of salt inhibits decay during the fermentation process. The previous study by [18] explained that the *rusip* fermentation added with palm sugar has a lower TVB value. The presence of the lactic acid bacteria in *rusip* fermented can suppress the growth of TVB-forming microbes.

The high salt content can be used as a preservative and can inhibit the growth of microorganisms, such as decomposing microorganisms that will produce alkaline compounds (TVB). The longer fermentation time usually caused the TVB detected levels higher. This result can be caused by the accumulation of metabolites of microorganisms including TVB in food products that indicate the occurrence of decay [19].

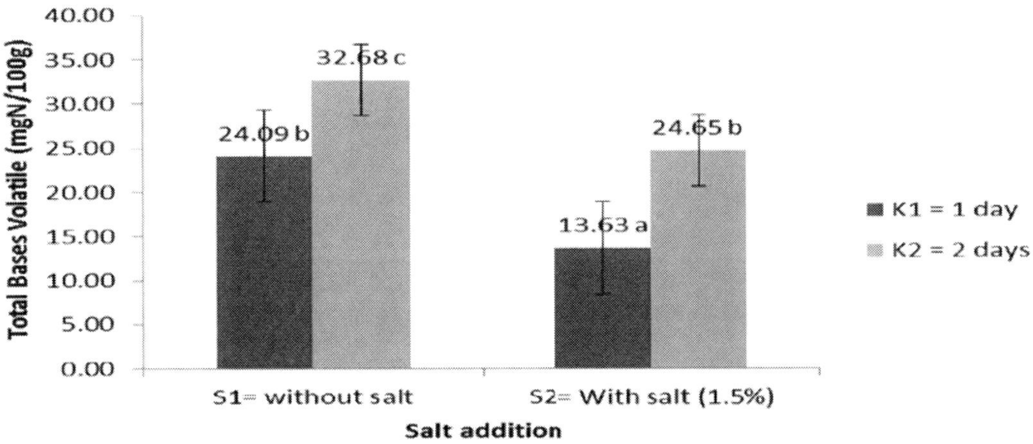

Figure 3. Interaction of length of fermentation and salt addition towards total volatile bases of *pacik kule* at $DMRT_{0.05}$ level 1 =8.86, level 2 = 9.21, level 3 =9.44, and coefficient of variants = 19.8% (Values followed by the same letter indicate no significant difference)

3.4 pH

The pH value is the main indicator that can determine the success of the fermentation process. The pH value of *pacik kule* during the fermentation process changed from initially 6.42 to 6.07. In this study, the results of variance for the pH of *pacik kule* showed that the influence of fermentation time (K), the addition of salt (S), and interaction time of fermentation and addition of salt had no significant effect ($P \geq 0.05$).

Previous research by [20] mentioned that the pH value of fresh fish ranged from 5.80 to 6.94, while during the fermentation process it ranged from 6.60 to 5.20 so that it can be seen that the pH obtained in this study was not much different from the pH from [20]. Furthermore, a study by [21] explained the longer the fermentation process, the pH value will decrease. At the time of the fermentation process, glucose will be converted to eventually become lactic acid, thereby increasing levels of lactic acid and causing pH values to decrease. Changes in pH that occur in the fish meat are caused by the process of autolysis and bacterial attack [22].

3.5 Total Lactic Acid Bacteria (LAB)

Lactic acid bacteria are a class of Gram-positive bacteria, which do not form spores in their life processes. These bacteria are common found in fermented foods to control fermentation. They contribute to the sensory and nutritional characteristics of fermented products [23-25]. Lactic acid bacteria can break down carbohydrates (sugar) and produce lactic acid. The results of the total value of lactic acid bacteria ranged from 5.54 log CFU/g to 9.10 log CFU/g, with an average value of 7.96 log CFU/g. Analysis of variance showed that the length of fermentation (K) and the addition of salt (S) had a very significant effect ($P \leq 0.01$) on the total lactic acid bacteria *pacik kule*.

Figure 4 shows the number of lactic acid bacteria on two days fermentation and without salt treatment had the lowest number of total lactic acid bacteria (5.54 log CFU/g) compared to other treatments. This result is suspected because lactic acid bacteria quickly adapt to the environmental conditions of growth on the first day, and the trend on the second day was decline because nutrition has started to decrease. The previous study by [26] explained that the number of initial microbial cells will accelerate the adaptation phase and during the fermentation process, resulting in an increase in the total number of microbes. The high amount of lactic acid bacteria at the beginning of making *rusip* causes the breakdown of sugar into lactic acid is greater, and as the result, microbes that cannot stand with acid will be dying.

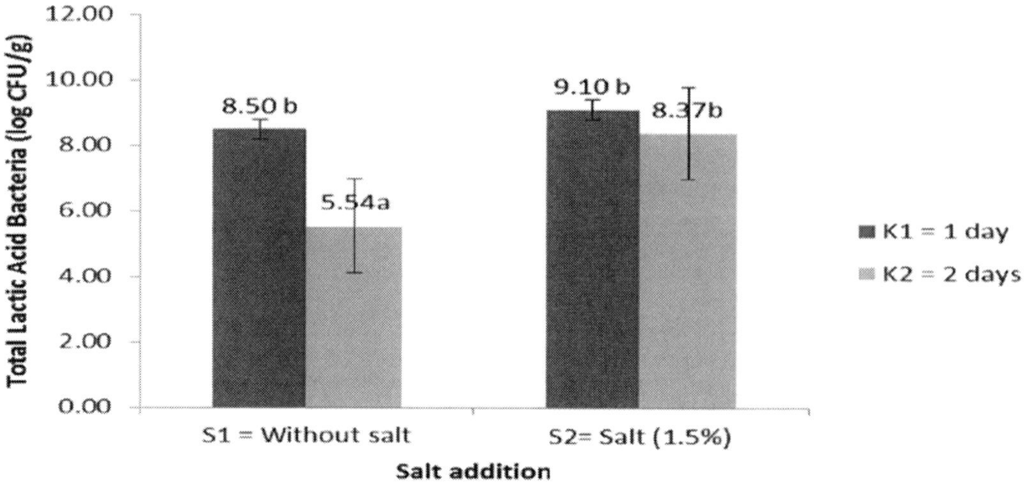

Figure 4. Interaction of length of fermentation and salt addition towards total volatile bases of *pacik kule* at $DMRT_{0.05}$ level 1 =2.85, level 2 = 2.96, level 3 =3.04 and coefficient of variants = 17.24% (Values followed by the same letter indicate no significant difference)

Furthermore, the increasing number of lactic acid bacteria to the treatment of adding salt can be caused by the presence of a salt solution that can stimulate the growth of lactic acid bacteria [27]. Research conducted by [28] explains that the higher the amount of glucose concentration used in the fermentation process, the higher the amount of lactic acid and halophilic bacteria produced. The use of salt and a tightly closed container in the process of making *belacan depik* affects the growth of lactic acid bacteria [8].

4. Conclusion

The longer the fermentation time, resulting in the lower value of the moisture content, protein content, and total lactic acid bacteria. The higher the salt concentration is used, the higher the protein content and total lactic acid bacteria. The best treatment was obtained from treatment with one-day fermentation and with the addition of salt (1.5%) due to the highest protein content of 11.93%, the highest total lactic acid bacteria 9.4 log CFU/g, the lowest amount of water content (17.41%) and the lowest total volatile base (13.63 mgN/100g).

5.References

[1] Muzaifa, M. and N. Arpi 2015 *Jurusan Teknologi Hasil Pertanian*. Fakultas Pertanian
[2] Amri, M 2007 *Jurnal Ilmu-Ilmu Pertanian Indonesia.* **9,** 71-76.
[3] Berlian, Z., Syarifah, Imamul, Huda 2016 *Jurnal Biota.* **2,**162-157.
[4] Heruwati, E. S, Widyasari, H. E., Haluan, J 2007 *Jurnal Pascapanen dan Bioteknologi Kelautan dan Perikanan.* **2,** 9-18.
[5] Romadhon, Rianingsih, L., Aanggo Apri Dwi 2018 *Jurnal Pengolahan Hasil Perikanan Indonesia.* **21**, 68-76.
[6] Hestiyana 2017 *Jurnal Totobuang*, **5**, 255-269.
[7] Rizky, M. Y., Fitri, R.D., Hastuti, U.S., Prabaningtyas, S 2017 *Jurnal Bionature.* **18**, 87-98.
[8] Muzaifa, M 2015 *Jurnal Sagu* **14**, 19-22.
[9] Candra, J.I 2006 *Skripsi*, Fakultas Perikanan dan Ilmu Kelautan Institut Pertanian Bogor.
[10] Desniar, Poernomo., D, Wijayatur, W 2009 *Jurnal Pengolahan Hasil Perikanan Indonesia.* **12**, 73-87.
[11] Association of Official Analytical Chemist (AOAC). 2015. *Official Method of Analysis*. 16th edition. Washington

[12] Association of Official Analytical Chemist (AOAC). 1995. *Official Method of Analysis*. 18th edition.Washington
[13] Fardiaz, S. 1989. *Mikrobiologi Pangan*. PAU Pangan dan Gizi. IPB.
[14] Mumtianah, O. N., Kusdiyantini, E., Budiharjo, A 2014 *Jurnal Biologi*. **3**, 20-30.
[15] Adawyah, M.P. 2006. *Pengolahan dan Pengawetan Ikan*. PT. Bumi Askara, Jakarta.
[16] Suyatno, Sari. N.I., Loekman, S 2015 *Jurnal JOM* **3**,1-8.
[17] Hadiwiyoto, S. 1993. *Teknologi Pengolahan Hasil Perikanan Jilid 1*. Liberty, Yogyajakarta.
[18] Koeseoemawardani, D., dan Yuliana, N. 2009. *Jurnal Sains dan Teknologi Indonesia*. **11**, 205-211.
[19] Dissaraphong S., S. Benjakul and W., Visessanguan 2006 *Bioresource Technol.* **10**, 77-88.
[20] Desniar, Iriani, S., Retno, S.S 2002 *Jurnal Pengolahan Hasil Perikanan Indonesia*. **15**, 232-239.
[21] Asma., Nisa, K., Wardani, A.K. 2016. *Jurnal Pangan dan Agroindustri*. **4**, 367-376.
[22] Suptijah, P., Yayandi, Gushagia., Dadi Rochmadi Sukarsa 2008 Fakultas Perikanan dan Ilmu Kelautan. IPB : Bogor.
[23] Mayo B, Aleksandrak-Piekarczyk T, Fernendez M, Kowalczyk M, Alvarez-Martin P, Bardowski J, 2010. Updates in the Metabolism of Lactic Acid Bacteria, pp. 3-33. In Mozzi F, Raya RR and Vignolo G (Eds.) Biotechnology of Lactic Acid Bacteria Novel Application, Wiley-Blackwell, Iowa, USA.
[24] De Vuyst, L., G. Vrancken, F. Ravyts, T. Rimaux and S. Weckx 2009 *Food Microbiol.* **26**, 666–675
[25] Felis GE, Dellaglio F 2007 *Microbiol.* **8**, 44–61.
[26] Koeseoemawardani, D., Rian, S., Tauhid, M 2013 *Jurnal Aritech*. **33**, 265-271.
[27] Anwar L. O., L Hardjito, dan Desniar 2014 *Jurnal pengolahan hasil perikanan Indonesia* **17**, 254-262.
[28] Killinc, B., S. Cakli, S., Tolasa and T. Dincer 2006 *European Food Research Technology*. **222**, 604-613.

IOP Conf. Series: Materials Science and Engineering **796** (2020) 012046 doi:10.1088/1757-899X/796/1/012046

Spray drying of mangosteen (*Garcinia mangostana* L.) juice and analysis of antioxidant activity and total phenolic content

DM Setyawati[1]*, E Mardliyati[1], S Rahman[1], and AF Wijaya[1]
[1]Pusat Teknologi Farmasi dan Medika, Badan Pengkajian dan Penerapan Teknologi, Jakarta, Indonesia

E-mail: damai.ria@bppt.go.id

Abstract. Mangosteen contains natural antioxidant compounds such as polyphenolic compounds. It is most fresh consumed and can not be stored for long especially after separated from its pericarp. One alternative to extend the shelf life is by drying the mangosteen fruit. Spray drying was investigated to this study to produce mangosteen juice powder which was examined for antioxidant activity and total phenolic content. Mangosteen juice powder was dried with the addition of maltodextrin in a ratio of mangosteen: maltodextrin of 10:90; 30:70 and 50:50, which mangosteen was based on total solid value. Spray drying was carried out in 2 inlet temperatures condition, 140°C and 160°C, to observe the effect of temperature on antioxidant activity and total phenolic content. These 2 inlet temperatures of spray dryer did not affect the physical appearance, colour, taste, and aroma of mangosteen powder. Increment of spray dryer temperature did not show significant differences in total phenolic content and IC50 of antioxidant activity, except IC50 in the ratio of 50:50. At the ratio 50:50, the increment of spray dryer temperature decreases the antioxidant activity. Therefore, spray drying is a prospective drying option for mangosteen juice at the appropriate addition of maltodextrin and temperature process. It should maintain the physical and chemical properties of mangosteen and finally support ease of storage and shipping.

1. Introduction

Mangosteen (*Garcinia mangostana* L.) is a tropical fruit and widely cultivated in Indonesia. It contains natural antioxidant compounds such as phenolic acids, anthocyanins, and xanthone derivatives [1] (Xie et al. 2015). Xanthone is largely responsible for its bioactivity. It is a polyphenolic group compound and abundance can be found in mangosteen [2]. Mangosteen pericarp has been used as a medicine against inflammatory and diarrhea and treatment of skin infection and wound. Meanwhile, mangosteen fruit aril was consumed fresh or juice processed [3][4]. Mangosteen juice, combined with other fruit or vegetable which posses antioxidant, is now widely available in the market as a dietary supplement [1].

Drying of juice could extend shelf-life and add the value of mangosteen products. The drying process will remove the water and prevent the growth of fungi and microbial [3]. Powdered juice should be a good alternative for healthy ingredient formulated food [5]. The drying method should be considered so that phenolic contents are maintained. Encapsulation of phenolic content, extracted from fruits or foods, should be a good alternative for maintaining the stability of these compounds. Encapsulation prevents the compounds from light, enzyme, moisture and oxygen exposure [6].

One of the most widely used encapsulation techniques is spray drying, which provides rapid evaporation of water and maintains the low temperature in the particles. Spray drying method suitable

Content from this work may be used under the terms of the Creative Commons Attribution 3.0 licence. Any further distribution of this work must maintain attribution to the author(s) and the title of the work, journal citation and DOI.
Published under licence by IOP Publishing Ltd

for the drying of heat-sensitive compounds such as polyphenolic compounds [6]. In conclusion, the spray dry method is capable of drying large quantities of liquid rapidly, minimal negative impact on the product, without damaging the surrounding environment, lowest capital and operating costs [7][8]. Drying agent addition such as maltodextrin helps to dry and prevent sticking on the dryer wall. The properties of additives for spray drying in use is critical, as it has an influence on the process parameter and physico-chemical properties of spray drying [9]. Maltodextrin is inexpensive, greater availability, widespread use in foods, possesses low viscosity at high solid content, shown able to reduce product's hygroscopicity and increases product yield [5][7].

Encapsulation of procyanidins from grape seed extract using spray drying method was carried out by mixing the core substances and wall material (gum arabic and maltodextrin) by the ratio of 30:70. The result showed that spray dry processing did not change the procyanidin [10]. Fang and Bhandari (2011) reported that spray drying using maltodextrin as the carrier could maintain the total phenolic content and anthocyanin in bayberry juice. They also reported that the inlet temperature of 150°C was suitable for spray drying of the heat-sensitive material [11]. Fazaeli *et al.* (2012) reported that a higher concentration of maltodextrin and increment of spray drying inlet temperature will increase the final yield of black mulberry juice powder [12]. The objective of the present study was to evaluate the effect of maltodextrin addition to mangosteen juice on total phenolic content and antioxidant activity. Evaluation on organoleptic, total phenolic content and antioxidant activity was measured on mangosteen juice powder after spray drying process on inlet temperature of 140°C and 160°C, and mangosteen juice: maltodextrin ratio of 10:90; 30:70 and 50:50.

2. Method

2.1. Materials

Mangosteen fruit is harvested from Purwakarta, West Java. Materials of 2,2-dihenyl picrylydrazil (DPPH), Folin Ciocalteau reagent, ascorbic acid, gallic acid were purchased from Sigma Co (St. Louis, MO, USA). Sodium carbonate, citric acid, methanol, ethanol were purchased from Merck (Germany). Maltodextrin DE 18-20 was purchased from Zhuceng Dongxiao Biotechnology Co., Ltd. (China). Instruments used for this research were spray drying (Lab Plant, UK Ltd., UK), spectrophotometer (Heλios α, Thermo Spectronic, USA), refractometer pocket (Atago, Atago Co. Ltd., Japan), and magnetic stirrer (Heidolph, Heidolph Instruments GmbH. & Co., Germany).

2.2. Mangosteen juice preparation

Juice preparation was carried out based on Setyawati *et al* (2019) method by steaming pretreatment. Mangosteen fruit aril was separated from its pericarp and washed with water. Fruit aril was separated from the rotten and yellow ones. Fruit aril was steamed on boiling water for 15 minutes and homogenized in a filtered-blender to obtained mangosteen juice. Mangosteen juice was stored at -20°C (freezer) until the next testing [13].

2.3. Spray drying of mangosteen juice

Production of mangosteen juice powder was carried out using a lab-scale spray dryer instrument (Lab Plant UK Ltd.). Mangosteen juice was mixed with maltodextrin DE 18-20, which ratios of mangosteen juice and maltodextrin were 10:90; 30:70 and 50:50. Mangosteen juice was weighed based on total soluble solid (TS) value, which the parts of 10, 30 and 50 of mangosteen juice were counted as 100% TS value. TS values of mangosteen juice and maltodextrin mixture were measured. The mixture was divided into 2 parts for drying at 2 different inlet temperature of the spray drying process, those were 140°C and 160°C. The flow rate was 3 ml per minute. Sample powder was stored in a desiccator at 25°C for further analysis [3].

2.4. Determination of antioxidant activity by DPPH radical scavenging

Antioxidant activity can be measured by the DPPH radical scavenging activity method based on Ghafar *et al.* (2010) method with some modification. A 50 µL of sample mixed with 1 mL 0.04 M DPPH in 80% ethanol, shaken and left to stand at room temperature of 25°C for 30 minutes in a dark room. Ascorbic acid was used as a positive control at a concentration of 5, 10, 25, 50 and 75 µg/mL. The negative control contained only DPPH solution, while the blank contained 80% ethanol. DPPH scavenging activity was measured using a spectrophotometer at 517 nm. The scavenging effect was determined by the ratio of DPPH absorption decrease against the absorption of DPPH solution (negative control) using the equation (1).

$$Scavenging(\%) = \frac{Abs_{(control)} - Abs_{(sample)}}{Abs_{(control)}} \times 100 \qquad (1)$$

Inhibition concentration 50 (IC_{50}) value was obtained from the plotted graph of scavenging against the concentration of the sample (mg/ mL) [14].

2.5. Determination of total phenolic content

The total phenolics were determined by using Folin-Ciocalteu method. The sample was prepared by diluted mangosteen aril juice by 10-fold dilution in distilled water. 200 µL of the sample was mixed with 750 µL of Folin Ciocalteau reagent. After 5 minutes, 750 µL of 6% sodium carbonate solution was added to the mixture and stand for 60 minutes at room temperature of 25°C. The absorbance was measured by UV-Vis spectrophotometer at 725 nm. The standard curve of gallic acid was prepared by the same procedure. The maximum concentration of gallic acid of 100 µg/mL was prepared with methanol solvent, from which a serial dilution was made to give concentrations of 70, 50, 30, and 20 µg/mL. Total phenolic content was expressed as µg gallic acid equivalent (GAE) per gram sample [14].

2.6. Statistical analysis

Total phenolic content and antioxidant activity of each spray drying sample powder were compared to each other using Student's t-test.

3. Results and Discussion

Keeping the fruit fresh is the way to maintain its nutritional value but most storage techniques require low temperature (Sagar & Suresh. 2010). Mangosteen can not be stored for a long period and fruit aril will be browning after it is separated from its pericarp. Pretreatment of mangosteen juice preparation by steamed in boiling water for 15 minutes was carried out to prevent the juice from browning and maintain total phenolic content and antioxidant activity (Setyawati *et al.* 2019). Mangosteen pericarp extract industry will utilize the pericarp and unfortunately, discard the mangosteen fruit aril. Drying of mangosteen fruit aril is a suitable alternative for post-harvest management and to enhance storage stability, minimize packaging requirements and reduce transport weight (Sagar & Suresh. 2010).

Spray drying of mangosteen fruit aril was performed at 2 operating temperatures and mangosteen juice - maltodextrin ratio to determine the effect on physical properties, total phenolic content and antioxidant activity. Maltodextrin is the most commercially available additives used in spray drying. The properties of additives for spray drying in use is critical, as it influences the process parameter and physico-chemical properties of spray drying (Lee, Taip & Abdullah. 2018). Physical quality parameters associated with dried fruit products include colour, visual appeal, the shape of product and flavour should be observed (Sagar & Suresh. 2010).

Physically, mangosteen powder dried by spray drying at inlet temperature 140°C and 160°C had a similar appearance, as well as the taste and smell (Figure 1). While different ratios gave different colours, tastes, and smells. The higher part of mangosteen juice, the browner the colour, the sourer and the more mangosteen typical smell (Table 1). The ratio of 50:50 had better physical characteristics of powder and better sour taste like fresh consumed mangosteen. Finally, the spray drying technique is able to maintain the sweet and sour taste of mangosteen.

Figure 1. The physical appearance of mangosteen powder at an inlet temperature of spray drying of 140°C and 160°C, and the ratio of mangosteen juice: maltodextrin 10:90; 30:70 and 50:50

Table 1. Physical properties of mangosteen powder

The ratio of mangosteen juice : maltodextrin	The inlet temperature of spray drying	
	140°C	160°C
10:90	Fine powder, slightly brownish-white, slightly sticky, rather sweet, less sour taste	Fine powder, slightly brownish-white, slightly sticky, rather sweet, less sour taste
30:70	Fine powder, brownish-white, slightly sticky, rather sweet, a little sour taste and the typical smell of mangosteen	Fine powder, brownish-white, slightly sticky, rather sweet, a little sour taste and the typical smell of mangosteen
50:50	Fine powder, white brown, slightly sticky, rather sweet, sour taste, the typical smell of mangosteen	Fine powder, white brown, slightly sticky, rather sweet, sour taste, the typical smell of mangosteen

The hydroxyl groups of phenols or polyphenols in plant extract will react specifically with the yellow Folin-Ciocalteu reagent to form a blue chromophore complex. The more phenolic compounds or polyphenols, the higher the intensity of blue produced [15]. Total phenolic contents of the mangosteen powder ratio of 10:90 between inlet temperature of spray dryer of 140°C and 160°C were not significantly different (significant level $P > 0.01$), as well as the ratio of 30:70 and 50:50. It showed that the spray dryer inlet temperature between 140°C and 160°C did not affect the total phenolic content of mangosteen powder (Table 2). The condition of the inlet temperature of the spray dryer should be considered, although a short time process in the spray dryer was applied. The increment of heating temperature decreases total phenolic and antioxidant activity [16], but Fang and Bhandari (2011) reported that spray drying can preserve total phenolic and anthocyanin content higher than 91%. They also suggested that the inlet temperature of 150°C and the outlet temperature of 80°C is suitable for spray drying heat-sensitive material [11].

Total phenolic contents of mangosteen powder ratio of 30:70 after spray-dried at an inlet temperature of 140°C and 160°C were equal to mangosteen juice that is 2.35 mg GAE/g. Mangosteen juice and mangosteen powder in this research contained a higher amount of total phenolic contents than grape (1.58 mg GAE/g), black plum (1.435 mg GAE/g) and cherries (1.054 mg GAE/g) [17].

Table 2. Total phenolic content of mangosteen powder

The ratio of mangosteen juice : maltodextrin	Total phenolic content (mg GAE/g)	
	The inlet temperature of spray drying 140°C	The inlet temperature of spray drying 160°C
10:90	1.02 ± 0.1	0.83 ± 0.07
30:70	2.35 ± 0.13	2.4 ± 0.13
50:50	5.56 ± 0.24	5.41 ± 0.04

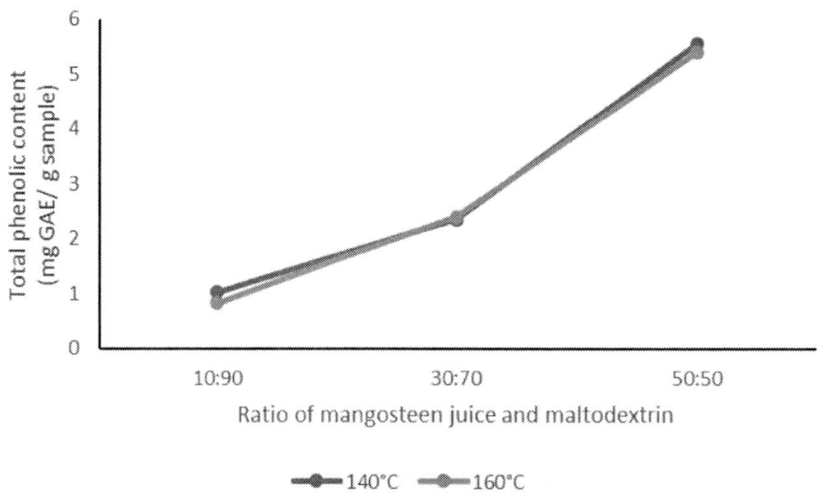

Figure 2. Total phenolic content of mangosteen powder at an inlet temperature of spray drying of 140°C and 160°C, and the ratio of mangosteen juice: maltodextrin 10:90; 30:70 and 50:50

DPPH is an oxidizing material which can be reacted with the antioxidant compound. DPPH reduction will occur and change the violet color of DPPH to a pale yellow color. So, high scavenging activity of sample means a high reduction of DPPH [18]. The IC_{50} value of the mangosteen powder ratio of 10:90 between inlet temperature of spray dryer of 140°C and 160°C was not significantly different (significant level $P > 0.01$), as well as the ratio of 30:70. However, the ratio of 50:50 between the inlet temperature of the spray dryer of 140°C and 160°C was significantly different (significant level $P > 0.01$). Antioxidant activity at the ratio of 50:50 showed that increment of inlet temperature (160°C) decrease antioxidant capacity (increase IC_{50} antioxidant activity) (Table 3). It is correlated with Chen and Lin (2007), where the increment of heating temperature decreases total phenolic and antioxidant activity. But, at the ratio of 10:90 and 30:70, the increment of inlet temperature did not affect the antioxidant capacity. It may be possible due to the higher part of maltodextrin (70% and 90%) can encapsulate heat-sensitive compound better. The higher part of maltodextrin can preserve the antioxidative compound even when exposed to a higher temperature.

Table 3. Radical scavenging activity of mangosteen powder

The ratio of mangosteen juice : maltodextrin	Inhibition concentration 50% (IC_{50}) (mg/mL)	
	The inlet temperature of spray drying 140°C	The inlet temperature of spray drying 160°C
10:90	552.43 ± 32.01	529.52 ± 34.17
30:70	146.27 ± 4.23	137.43 ± 2.85
50:50	37.50 ± 3.13	57.96 ± 2.16

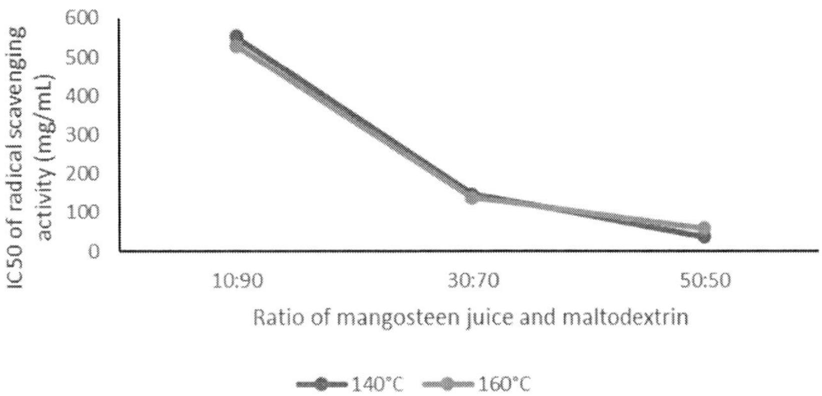

Figure 3. IC_{50} of radical scavenging activity of mangosteen powder at an inlet temperature of spray drying of 140°C and 160°C, and the ratio of mangosteen juice: maltodextrin 10:90; 30:70 and 50:50

4. Conclusion

Spray drying process and addition of maltodextrin as drying agent were suitable for drying mangosteen juice which contains polyphenolic compounds. Spray drying capable to maintain the original taste and sour of mangosteen fruit and total phenolic content and antioxidant activity. While maltodextrin capable to encapsulate heat-sensitive compounds such as phenolic compounds from heat transfer of spray dryer.

5. Acknowledgements

This research was supported by the Insinas research grant from the Ministry of Research, Technology, and Higher Education of the Republic of Indonesia.

6. References

[1] Xie Z, Sintara M, Chang T, Ou B 2015 *Food Sciences and Nutrition.* **3,** 32-38.
[2] Minh NH 2015 *International Journal of Pure & Applied Bioscience.* **3,** 297-304.
[3] Chaovanalikit A, Mingmuang A, KItbunleuwit T, Choldumrongkool N, Sondee J,Chupratum S 2012 *Int Food Res Journal.* **19,**1047-1053
[4] Chaverri PJ, Rodriguez NC, Ibarra MO, Rojas JMP 2008 *Food and Chemical Toxicology.* **46,** 3227-3239

[5] Moser P, Souza RTD, Telis VRN 2017 *Journal of Food Processing and Preservation.* **41,** el2852
[6] Rigon RT, Norena CPZ 2016 *J Food Sci Technol.* **53,** 1515-1524
[7] Looi YF, Ong SP, Julkifle A, Alias MS 2019 *J Food Process Preserv.* **43**: el3915
[8] Sagar VR, Suresh KP 2010 *Journal of Food Sciences and Technology.* **47**:15-26
[9] Lee JKM, Taip FS, Abdullah Z 2018 *Food Research.* **2,** 486-499
[10] Zhang L, Mou D, Du Y 2007 *Journal of the Science of Food and Agriculture.* **87,** 2191-2197
[11] Fang Z, Bhandari B 2011 *Food Chemistry.* **129**, 1139-1147
[12] Fazaeli M, Djoemah ZE, Ashtari Ak, Omid M 2012 *Food & Bioproducts Processing.* **90,** 667-675
[13] Setyawati DR, Mardliyati E, Putri RA, Kamila M, Rifada MA, Meliana Y, Fernando 2019 *International Journal of Engineering Research and Application.* **9,** 49-52
[14] Ghafar MFA, Prasad KN, Ismail A 2010. *African J Biotechnol.* **9,** 326-330
[15] Blainski A, LOpes GC, De Mello JCP 2013 *Molecules.* **18**, 6852-6965
[16] Chen YT, Lin KW 2007 *Food Chemistry.* **101,** 955-963
[17] Karakaya S, El SN, Tas AA 2001 *International Journal of Food Sciences and Nutrition.* **52,** 501-508
[18] Aksoy L, Kolay E, Ağılönü Y, Aslan Z, Kargıoğlu M 2013 *Journal Biol Sci.* **20**, 235-239.

IOP Conf. Series: Materials Science and Engineering **796** (2020) 012047 doi:10.1088/1757-899X/796/1/012047

Endophytic Bacteria isolated from the leaf of Langusei (*Ficus minahassae* Tesym. & De Vr.) and their antibacterial activities

T E Tallei[1,*], Y T Linelejan[1], S D Umboh[1], A A Adam[2], Muslem[3], and R Idroes[3,4]

[1]Department of Biology, Faculty of Mathematics and Natural Sciences, Sam Ratulangi University, Manado, Indonesia
[2]Dentistry Study Program, Faculty of Medicine, Sam Ratulangi University, Manado, Indonesia
[3]Department of Chemistry, Faculty of Mathematics and Natural Sciences, Syiah Kuala University, Banda Aceh, Indonesia
[4]Department of Pharmacy, Faculty of Mathematics and Natural Sciences, Syiah Kuala University, Banda Aceh, Indonesia

Corresponding author: trina_tallei@unsrat.ac.id

Abstract. Endophytic bacteria colonizing the internal tissues of plants are being studied for its potential beneficial properties including antibacterial activity. The aims of this study were to identify the endophytic bacteria residing in the leaf of *Ficus minahassae*, a native plant of North Sulawesi, as well as their antibacterial activities against laboratory-adapted reference bacterial strains, *Staphylococcus aureus*, and *Escherichia coli*. The isolate identification was conducted using the 16S rRNA gene marker and the antibacterial activity test was done using a well diffusion method. A yellow-pigmented (designated as YL1) and a beige-pigmented (designated as YL2) bacteria were isolated from the internal tissue of langusei. Using the BLAST search, the isolate YL1 had a 100% identity with *Brachybacterium muris*, while YL2 has 99% identity with *Pseudacidovorax intermedius*. Ez-Taxon database confirmed that YL1 was *B. muris* with 99.81% similarity, while YL2 was *P. intermedius* with 99.80% similarity. Isolate YL2 was able to inhibit the growth of *S. aureus* but not *E. coli*. Isolate YL1 did not inhibit both reference bacteria. This study represents the first investigation of endophytic bacteria isolated from langusei leaf and their antibacterial activity.

1. Introduction

In recent years, the mining of microbial resources contained in plant tissue has begun to receive much attention. These endophytic bacteria colonize the intercellular spaces and vascular systems of the plants. These microbes have been studied for various purposes. Compounds produced by endophytes have many benefits, such as antibacterials. The active compound of plant endophytic microbes often has a stronger activity than that produced by their host plants. The abilities of the active compounds of plant endophytic bacteria still need to be explored to discover new bacterial strains that have the potential to be further explored [1].

Several studies have reported that endophytic bacteria can be found in ficus plants, for example in *Ficus carica* [2] and *F. racemosa* [3]. One of the ficus plants native to North Sulawesi is *F. minahassae*. The local people names this plant as langusei. This plant is found in the primary forest, especially along the river, to a height of about 135 meters above the sea. Dichloromethane extract of the leaves contained 2-hydroxyethyl benzoate, phytyl fatty acid ester, squalene, and β-sitosterol [4]. The increasing emergence of bacterial resistance creates demands in focusing on new and more potent antimicrobial sources. Therefore, this study was aimed to identify endophytic bacteria in *F. minahassae* and to assay their antibacterial activity.

Content from this work may be used under the terms of the Creative Commons Attribution 3.0 licence. Any further distribution of this work must maintain attribution to the author(s) and the title of the work, journal citation and DOI.
Published under licence by IOP Publishing Ltd

2. Materials and Methods

2.1. Sample Preparation

The leaf of langusei was obtained from the Batukota river bank in Manado. The leaf was surface sterilized by incubation in 5% NaOCl for 10 min, followed by in 70% alcohol for 2 min, and rinsed three times in sterile dH_2O. To check the sterilisation procedure, 100 ul aliquot of dH_20 taken from the rinsing of the sterilized leaf was spread over the nutrient agar (NA) media, incubated at 37°C for 2 days, and examined for the presence or absence of the growth of microorganisms.

2.2. Isolation of the Endophytes

The plant leaf was acceptically ground in a sterile mortar and pestle until producing juice. One hundred microlitters of the juice were spread over the NA media. The plates were incubated at 37°C for 2 to 4 days. The growing bacterial isolates were grouped based on their morphology and characteristics of their colonies such as size, shape, and colour. Each separated colony was picked using Oose and streaked on NA media, incubated at 37°C for 2 days. Each separately grown colony was considered as a pure isolate to be used for further examination.

2.3 Identification of the Endophytes

The pure selected isolates were identified using Gram staining followed by identification based on the 16S rRNA gene. The DNA genome was isolated using the Plant Genomic DNA mini kit (Geheaid). MyTaq HS Red Mix (Bioline) was used for PCR amplification, with 10 μM of each BKXF and BKXR primer. The PCR was performed using the following condition: initial denaturation at 95°C for 3 min followed by 35 cycles of denaturation of 95°C for 30 sec, annealing at 50°C for 30 sec, elongation at 72°C for 90 sec, and final elongation at 72°C for 1 min. The PCR products were sent to 1st BASE Malaysia for sequencing. The aligned contiguous consensus sequences were used for the homology search by BLAST software (http://blast.ncbi.nlm.nih.gov) algorithm at the National Center for Biotechnology Information (NCBI). The identification was further confirmed using the Ez-Taxon database (https://www.ezbiocloud.net/)[5].

2.4 Phylogenetic Analysis

The phylogenetic analysis was performed using the stepwise procedure described previously [6]. The chromatograms obtained from 1st BASE Malaysia were processed using Geneious 10.1.3. The primer sequences were removed by trimming approximately 50 nucleotides at the beginning of DNA sequences. Errors of the reading of the nucleotides were corrected accordingly. By using the pairwise alignment of forward and reverse sequences, the consensus sequences were generated. All similar sequences obtained from Genebank were aligned using multiple sequence alignment with hierarchical clustering [7], which is available online at (http://multalin.toulouse.inra.fr/multalin/), and trimmed accordingly to obtain the core sequences of 16S rRNA. By using the Neighbor Joining method integrated into MEGA v10.0.4 [8], the phylogenetic tree was reconstructed.

2.5 Antibacterial Test of the Isolates

The antimicrobial activity of the isolates was determined using the agar well-diffusion method. Nutrient agar media was poured into plates equipped with stainless-steel cylinders. After the media were solidified, the cylinders were removed leaving a 5 mm diameter of wells. Two hundred microliters of an overnight culture of indicator strains (*Escherichia coli* (Migula) Castellani and Chalmers (ATCC® 25922™) and *Staphylococcus aureus* subsp. *aureus* Rosenbach (ATCC® 25923™)) containing approximately 10^6 cells/ml were used for this purpose. Bacterial isolates were grown overnight, heat-killed by incubating at 80°C for 1 hour, and the supernatant was collected by centrifuging at 6000 rpm for 5 min. One hundred microlitters of each isolate was added into each well. Inhibitory zone of the isolates was checked after 24 hours incubation at 37°C. Fifty μg/ml 1-cyclopropyl-6-fluoro-1,4-dihydro-4-oxo-7-(1-piperazinyl)-3-quinolinecarboxylic acid was used as positive control and sterile dH_2O as the negative control.

3. Results and Discussion

3.1. Endophytic Bacterial Isolation

Of several endophytic bacteria that were successfully isolated and purified from the surface-disinfected leaf of langusei, two isolates (YL1 and YL2) were further investigated. Gram staining indicated that YL1 was a positive-Gram bacterium. The isolate was yellow-pigmented on nutrient agar (Fig. 1a). YL2 was a negative-Gram bacterium and beige-pigmented on nutrient agar (Fig. 1b).

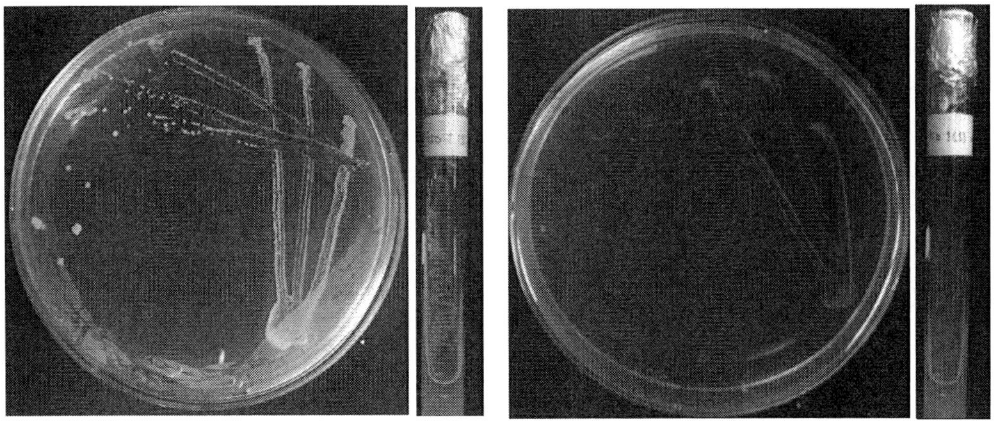

a b

Figure 1. a) Colonies of isolate YL1 and b) colonies of isolate YL2 on nutrient agar media

3.2. Isolates Identification

Identification of the isolates was conducted using 16S rRNA gene marker analysis. Using the BLAST search to compare the obtained sequence with reference sequences at GenBank, it was confirmed that YL1 (GenBank accession number MK764941) had 100% identity with *Brachybacterium muris* and YL2 (GenBank accession number MK764961) had 99% identity with *Pseudacidovorax intermedius*. Using the Ez-Taxon database, YL1 was confirmed as *B. muris* with 99.81% similarity with *B. muris* strain C3H-21, and YL2 was confirmed as *P. intermedius* with 99.80% similarity with *P. intermedius* strain DSM 21352. *Brachybacterium muris* is an actinobacteria while *P. intermedius* belongs to β-proteobacteria. *Brachybacterium* sp. has been isolated from tea (*Camellia sinensis*) [9], fresh tissue of banana (*Musa paradisiaca*) [10], rice (*Oryza sativa*) leaf [11], vine (*Vitis* sp.) stems [12], dragon blood tree (*Dracaena cochinchinensis*) [13], and arils of durian (*Durio zibethinus*) [14].

Other endophytic Brachybacteria among others are *B. paraconglomeratum* isolated from roots of safed musli (*Chlorophytum borivilianum*) [15], *B. hainanense* isolated from noni (*Morinda citrifolia* L.) branch [16], and *B. faecium* isolated from Mexican bamboo (*Polygonum cuspidatum*) [17]. Some of Brachybacteria exhibited antimicrobial activity [18 – 20].

Pseudacidovorax intermedius has been isolated from soil sample [21], surface sediments of the South China Sea [22], rice seeds [23], *Vetiver zizanioides* - a perennial grass of the Poaceae family [24], hanging root of benjamin fig (*F. benjamina*) [25], and date palm roots [26]. Other *Pseudacidovorax* sp. has been isolated from growing shoot tips of banana (*Musa* sp.) [10], potato (*Solanum tuberosum*) [27], and rice stem [28].

To the best of our knowledge, there has been no report on the isolation of endophytes from langusei, therefor this finding is the first report of the isolation of *P. intermedius* and *B. muris* from fig leaf. Some endophytic bacteria from fig among others are *Klebsiella oxytoca*, *Pseudomonas* sp. and *Pantoea* sp, isolated from fig (*F. carica*) [29], *Psedomonas aeruginosa*, isolated from red-stem fig (*F. variegata*)

[30], *Bacillus subtilis*, *B. amyloliquefaciens*, and *B. amyloliquefaciens*, isolated aerial roots of bajan fig (*F. benghalensis*) [25], and agrobacteria which were detected in many healthy benjamin fig [31].

3.3. Phylogenetic Analysis

The construction of a phylogenetic tree through the Neighbor joining method (Fig. 2) demonstrates the differences between taxa of the endophytes found in the leaf of *F. minahassae*. *Brachybacterium muris* belong to Actinobacteria and *P. intermedius* belongs to β-Proteobacteria. Endophytic bacteria of fig tree reported previously were from α-Proteobacteria, γ-Proteobacteria, and Firmicutes. Most of the endophytic bacteria of rice were α-Proteobacteria (46%) and Actinomycetales (16%) [11]. Bacterial endophytes are polyphyletic, belonging to a broad range of taxa, including Actinobacteria, Firmicutes, α-Proteobacteria, β-Proteobacteria, and γ-Proteobacteria [32]. Of these, *Pseudomonas* sp. has frequently found as dominant bacterial endophytes [33].

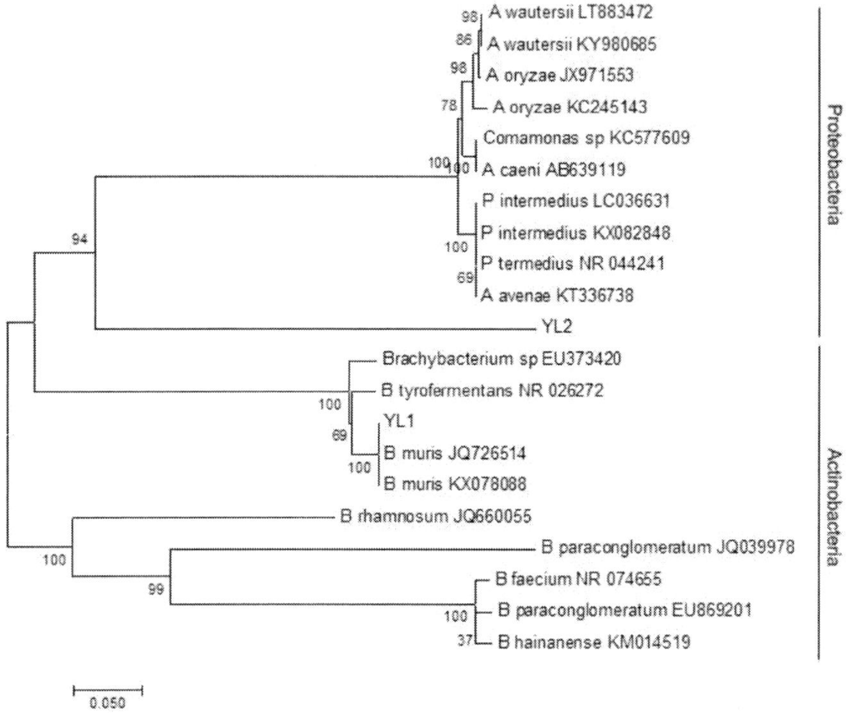

Figure 2. Phylogenetic tree constructed using the Neighbor Joining method based on the sequences core region of 16s rRNA of endophytic bacteria obtained from the leaf of langusei. The complement sequences were obtained from GenBank as indicated by their accession numbers. YL1 is *B. muris* and YL2 is *P. intermedius*.

3.4. Antimicrobial Test

The antibacterial activities of B. muris YL1 and P. intermedius YL2 were examined using the well diffusion method. B. muris did not inhibit the growth of S. aureus nor E. coli. P. intermedius displayed anti-bacterial activity against S. aureus (Fig 3a) but not against E. coli (Fig 3b). The diameter of the inhibition zone was 18.75 ± 0.096. The similar finding showed that *Brachybacterium* sp. isolated from *D. cochinchinensis* did not inhibit bacteria including *S. aureus*, *E. coli*, *B. subtilis*, *K. pneumoniae*, and fungi including *A. niger* and *A. fumigatus*, but inhibited plant fungus *Gibberella fujikuroi* [13].

The current study suggests that endophytic bacteria of langusei are a source of potential novel antibacterial. Endophytic bacteria from *F. variegata* produced antibacterial compounds [30]. *Streptomyces* sp. isolated from leaves of *F. racemosa* showed an antibacterial activity [3]. *Pseudacidovorax intermedius* was confirmed as nitrogen-fixing bacterium [21, 22, 24]. Nitrogen-fixing nitrogenase (nifH) homolog gene was also found in *Pseudacidovorax* sp. isolated from potato [27]. Nitrogenase is an enzyme used by some organisms to fix atmospheric nitrogen gas (N_2).

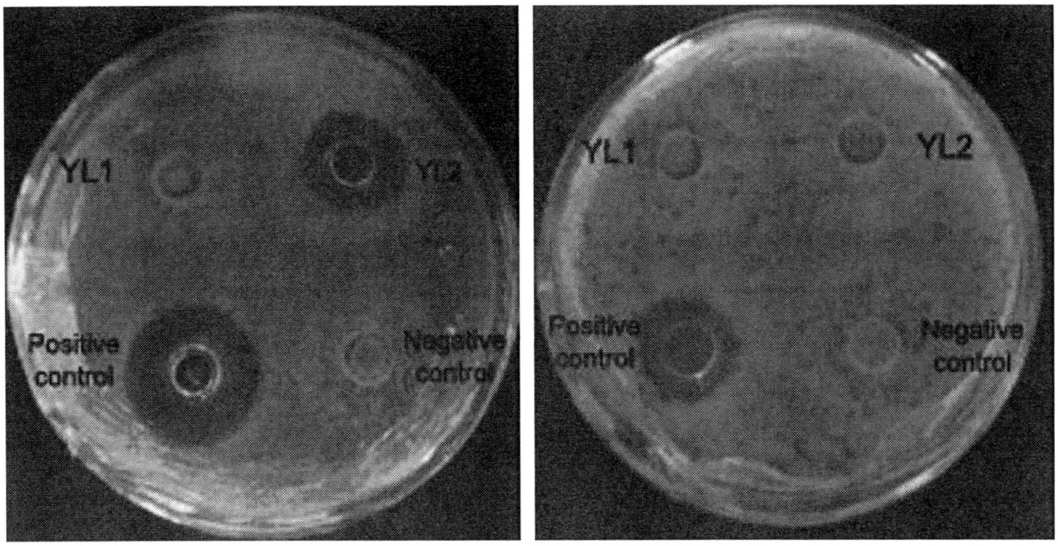

(a) (b)

Figure 3. Antibacterial activity test of *B. muris* YL1 and *P. intermedius* YL2 against *S. aureus* (a) and *E. coli* (b); + = positive control containing 50ug/ml 1-cyclopropyl-6-fluoro-1,4-dihydro-4-oxo-7-(1-piperazinyl)-3-quinolinecarboxylic acid; - = negative control (sterile dH_20).

Identification using NMR, dichloromethane extract of the leaf of *F. minahassae* contains 2-hydroxyethyl benzoate, phytyl fatty acid ester, squalene, and β-sitosterol [4]. These compounds are probably induced by endophytic bacteria living in the intracelullar tissues of the leaves. Polyethylene terephthalate (PET) hydrolase is produced by bacteria hydrolase PET to release among others is 2-hydroxyethyl benzoate (HEB) [34]. Phytyl esters, an intracellular wax esters, are also found in bacteria and generally considered to be energy storage components [35]. This compound has an antimicrobial property [36]. Squalene hopene cyclases have been found in bacteria [37]. Squalene-hopene cyclase (SHC) catalyzes the conversion of acyclic squalene into the pentacyclic hopene and hopanol in prokaryotes. Squalene has antimicrobial properties [38, 39]. β-sitosterol, one of the severol phytosterols, also has antibacterial activity [40]. This compound has been isolated from endophytic fungi of Amazon plants [41] and the Chinese medicinal plant *Arisaema erubescens* [42]. It may also be produced by endophytic bacteria.

4. Conclusion

Two endophytic bacteria isolated from *F. minahassae* were confirmed as *B. muris* (YL1) and *P. intermedius* (YL2) based on molecular identification using 16S rRNA gene marker. Among the two, only *P. intermedius* showed activity against the growth of *S. aureus*. Further studies are required to reveal the potential of these endophytes as antibacterial agents.

5. Conflict of Interest

There is no conflict of interest.

IOP Conf. Series: Materials Science and Engineering **796** (2020) 012047 doi:10.1088/1757-899X/796/1/012047

6. References

[1] Muzzamal H, Sarwar R, Sajid I, Hasnain S 2011 *Pak. J. Zool.* **44,** 249-257.
[2] Ma YM, Liang XA, Zhang HC, Liu R 2016 *J. Agric. Food Chem.* **64(19)**, 3789-93.
[3] Smita V, Dipak V 2015 *Int. J. Res. Biosci.* **4(4),** 68-74.
[4] Apostol PG, De Los Reyes MM, Mandia EH, Shen CC, Ragasa CY 2016 *Der Pharma Chemica* **8(20),** 220-223.
[5] Yoon SH, Ha SM, Kwon S, Lim J, Kim Y, Seo H, Chun J *Int. J. Syst. Evol. Microbiol.* **67,** 1613-1617.
[6] Tallei TE, Kolondam BJ 2015 *Hayati J. Biosci.* **22(1),** 41-47.
[7] Corpet F 1988 *Nucl. Acids Res.* **16(22),** 10881-10890.
[8] Kumar S, Stecher G, Li M, Knyaz C, Tamura K 2018 *Mol. Biol. Evol.* **35,** 1547-1549.
[9] Das PP, Medhi T 2018 *Annals Plant Sci.* **7(4),** 2119-2125.
[10] Thomas P, Soly TA 2009 *Microb. Ecol.* **58(4),** 952-64.
[11] Ferrando F, Mañay JF, Scavino AF 2012 *FEMS Microbiol. Ecol.* **80,** 696-708.
[12] Andreolli M, Lampis S, Zapparoli G, Angelini E, Vallini G *Microbiol. Res.* **183,** 42-52.
[13] Salam N, Khieu TN, Liu MJ, Vu TT, Son CK, Quach NT, Phi QT, Prabhu M, Rao N, Fontana A, Sarter S and Li WJ 2017 *BioMed. Res. Int.* **2017,** 1308563.
[14] Suhandono S and Utari IB 2014 *Microbiol. Indonesia* **8(4)** ,161-169.
[15] Barnawal D, Bharti N, Tripathi A, Shanker S, Chandan P, Chanotiya S and Kalra A 2016 *J. Plant Growth Regul.* **35,** 553.
[16] Liu Y, Zhai L, Yao S, Cao Y, Cao Y, Zhang X, Su J, Ge Y, Zhao R, Cheng C 2015 *Int. J. Syst. Evol. Microb.* **65,** 4196-4201.
[17] Sun H, He Y, Xiao Q, Ye R, Tian Y 2013 *AJMR African J. Microb. Res.* **7(16),** 1496-1504.
[18] Claverías FP, Undabarrena A, González M, Seeger M, Cámara B 2015 *Front Microbiol.* **6,** 737.
[19] Radjasa OK 2007 *J Coastal Dev.* **10(3),** 143-151.
[20] Leiva S, Alvarado P, Huang Y, Wang J, Garrido I *2015 FEMS Microbiol Lett.* **362(24),** 206.
[21] Kämpfer P, Thummes K, Chu HI, Tan CC, Arun AB, Chen WM, Lai WA, Shen FT, Rekha PD, Young CC 2008 *Int. J. Syst. Evol. Microbiol.* **58(Pt 2),** 491-5.
[22] Zhang LH, Chen SF 2012 *World J. Microbiol. Biotechnol.* **28(9),** 2839-47.
[23] Midha S, Bansal K, Sharma S, Kumar N, Patil PP, Chaudhry V, Patil PB 2016 *Front. Microbiol.* **6,** 1551.
[24] Zhao X, Javed CH, He Y, Zhang Z, Peng G, Tan Z *2009 Acta Microbiologica Sinica* **49(11),** 1430-1437
[25] Pathak K, Keharia H, Kharkwal AC 2009 *Symbiotic Fungi, Soil Biology* vol 18, eds Varma and A.C. Kharkwal (Berlin Heidelberg: Springer-Verlag).
[26] Yaish MW, Al-Harrasi I, Alansari AS, Al-Yahyai R, Glick BR 2016 *Int. Microb.* **19(3),** 143-155.
[27] Pageni BB, Lupwayi NZ, Akter Z, Larney FJ, Kawchuk LM, Gan YT 2014 *Canadian J. Plant Sci.* **94(5),** 835-844.
[28] Raweekul W, Wuttitummaporn S, Sodchuen W, Kittiwongwattana C 2016 *Thammasat Int. J. Sci. Tech.* **21(1),** 8-17.
[29] Alvarado-Marchena L, Smith-Durán A, Alvarado-Ulloa C, Chacón-Cerdas R, Flores-Mora D, Flores-Mora D 2016 ARPN *American J. Agric. Biol. Sci.* **11(7),** 290-297.

[30] Leonita S, Bintang M, Pasaribu FH 2016 *Current Biochem.* **2(3),** 116-128.
[31] Zoina A, Raio A, Peluso R, Spasiano A 2008 *Plant Pathol.* **50(5),** 620-627.
[32] Miliute I, Buzaite O, Baniulis D and Stanys V 2015 *Zemdirbyste-Agriculture* **102(4)**, 465-478.
[33] Pratiwi RH, Hidayat I, Hanafi M, Mangunwardoyo W 2016 *Malaysian J. Microb.* **12(4),** 300-307.
[34] Acton A 2013 *Bacillus - Advances in Research and Application* (Atlanta USA: Scolarly Editions)
[35] Alvarez HM, Pucci OH, Steinbüchel A 1997 *Appl. Microbiol. Biotechnol.* **47,** 132-139.
[36] Ragasa CY, Hofileña J, Rideout JA 2004 *Philippine J. Sci.* **133(1)** 1-5.
[37] Nair IM, Jayachandran K 2017 *Int. J. Health & Life-Sciences* **3(2),** 161-177 doi:10.20319/lijhls.2017.32.161177
[38] Popa O, Băbeanu NE, Popa I, Niţă S. Dinu-Pârvu CE 2015 *BioMed Res. Int.* Article ID 367202.
[39] Kopicová Z, Vavreinová S 2007 *Czech J. Food Sci.* **25,** 195-201.
[40] Ododo MM, Choudhury MK, Dekebo AH 2016 *SpringerPlus* **5(1),** 1210.
[41] Carvalho JM, da Paixão LKO, Dolabela MF, Marinho PSB, Marinho AMR 2016 *Acta Amaz.* **46(1),** 69.
[42] Wang LW, Xu BG, Wang JY, Su ZZ, Lin FC, Zhang CL, Kubicek CP 2012 *App. Microbiol. Biotechnol.* **93(3),** 1231-9.

IOP Conf. Series: Materials Science and Engineering **796** (2020) 012048 doi:10.1088/1757-899X/796/1/012048

$K_2O/NbOPO_4$ solid acid catalyst for biodiesel production

N Nurhazanah[1], Y Sy[1], H Husin[2*], CM Rosnelly[2], A Maulana[2]
[1]Graduate School of Chemical Engineering, Universitas Syiah Kuala, Banda Aceh, Indonesia 23111
[2]Department of Chemical Engineering, Universitas Syiah Kuala, Banda Aceh, Indonesia, 23111

E-mail: husni_husin@che.unsyiah.ac.id

Abstract. Biodiesel production from palm oil and alcohol has been carried out using a $K_2O/NbOPO_4$ solid acid catalyst. $NbOPO_4$ catalyst was synthesized through the sol-gel method, followed by drying and calcination at a temperature of 800°C for 6 hours. The catalyst was further treated using K_2CO_3 with various concentrations of 15, 25, 35, 45 and 55% through the impregnation technique. Morphology and composition of the catalyst were evaluated by SEM, EDX and XRD instruments. The results showed that the catalyst has non homogeneous particles with size ranging between 1.86 μm to 25 μm, accompanied by the agglomeration. The results of the analysis also confirmed that the prepared catalyst contains potassium, niobium, phosphorus and oxygen with a composition of 10.15; 17.67; 6.11 and 41.83%, respectively. The catalytic activity was tested in the transesterification reaction for 3 hours, at a temperature of 65°C with the molar ratio of palm oil to alcohol of 12:1 and catalyst loading of 6%. The highest yield of 98.87% was obtained on the process using the catalyst impregnated with 35% K_2CO_3. The overall results show that there is a very promising potential of NbOPO4 catalysts for biodiesel production.

1. Introduction

The demand for fuel sourced from non-renewable fuels is currently increasing. This is caused by the increased utilization of fuel in daily life with the increasing growth of population [1, 2]. The impact caused by the continuous use of fuel sourced from fossils, especially in the field of transportation, induces concern about environmental and health damage [3]. The study on the utilization of vegetable materials as a substitute for fossil-sourced fuels continues to be explored. Of many renewable sources, palm oil can be an alternative material to produce renewable, sustainable and biodegradable biofuel [4-6].

There are several methods can be used to convert vegetable oils into biofuels such as transesterification, fermentation, thermal cracking, catalytic cracking, and hydro-processing. Each method has advantages and disadvantages [7]. The transesterification is a method commonly used to react triglycerides with alcohol to produce methyl esters and by-products in the form of glycerol. In general, biodiesel production uses a homogeneous base catalyst such as K or Na hydroxide in a transesterification reaction. However, the separation of homogeneous catalysts used and purification of the generated biodiesel is difficult due to saponification and emulsification [8-10]. In addition, the glycerol by-product can be easily contaminated by K or Na salts hence making it not possible to be

* To whom any correspondence should be addressed.

Content from this work may be used under the terms of the Creative Commons Attribution 3.0 licence. Any further distribution of this work must maintain attribution to the author(s) and the title of the work, journal citation and DOI.
Published under licence by IOP Publishing Ltd

IOP Conf. Series: Materials Science and Engineering **796** (2020) 012048 doi:10.1088/1757-899X/796/1/012048

directly used as a commodity [11]. The alternative approach of using heterogeneous catalysts continues to be studied to overcome the above problems. Some researchers have studied the characteristics of heterogeneous solid acid catalysts and it is reported that heterogeneous catalysts have good thermal stability, selectivity, catalyst performance and environmentally friendly [12-16]. Ion exchange organic resins, zeolite, and ZrO_2 have been used as solid acid catalysts [17]. However, some of these catalysts have limitations, such as low thermal stability, mass transfer resistance (diffusion limits which reduce reaction rates) or loss of acidic active sites in polar media [18].

The niobium metal transition has been reported as an interesting catalyst for various reactions [19-21]. Niobium as the main catalyst component has proven a very promising performance as a heterogeneous catalyst. The catalytic activity of commercial niobium to has been carried out by reacting diluted H_2SO_4 or H_3PO_4 solutions to obtain strong acid active sites. Phosphate compounds in niobic acid will form niobium phosphate which acts as an active catalytic site. The glassy morphology of $NbOPO_4$ can expose the Nb^{5+} Lewis acid site at low temperatures, which is strong with a high tendency to absorb oxygen from triglycerides [22]. There was a report that potassium salts (such as K_2CO_3) as support and have shown high catalytic activity in transesterification reactions. In this study, $NbOPO_4$ catalyst was synthesized and modified with potassium (K) for application in biodiesel production through transesterification reactions. The effect of potassium loading was studied in detail on the characteristics of the catalyst, as well as the yield and quality of biodiesel.

2. Experimental

2.1. Material

The main material used for the synthesis of the solid acid catalyst was $NbCl_2$. The citric acid (Merck, Germany) was used as a solvent. Phosphoric acid (85% pure H_3PO_4), Ammonia (25% NH_3 solution), and methanol (99.8%) were all supplied from Merck, Germany. Palm oil obtained from the PT. Socfindo Nagan Raya Aceh District, Indonesia was used as source of biodisel.

2.2. Preparation of $K_2O/NbOPO_4$

The synthesis of $NbOPO_4$ was carried out by following the procedure of a number of references from previous studies [1, 15, 18, 19]. $NbCl_2$ was dissolved in a citric acid solution dropwise with continuous stirring for 1 hour. After that, H_2PO_4 and NH_3 5% were added as pH regulators. The solution was heated at a temperature of 70 - 80°C, stirring continued until a clear solution was obtained. The process was stopped after the gel was formed and then put in the oven at 80°C for 24 hours to remove the water content. Following that, the formed gel was put into the furnace at 800°C for 6 hours to stabilize the crystalline phase. Introduction of K_2CO_3 onto the $NbOPO_4$ catalyst was conducted by wet impregnation process. K_2CO_3 with varied concentration (15%, 25%, 35%, 45%, and 55%) was impregnated on $NbOPO_4$ for 6 hours, then put into the furnace for 4 hours at 200°C.

2.3. Catalyst characterization

To study the characteristics of the prepared catalysts, the analysis was done through several techniques. Composition analysis was carried out using an X-ray diffractometer (XRD) instrument which was recorded using Cu Kα radiation on a D/max-2500 diffractometer, the diffractogram reading was carried out at an angle of 2θ. The morphology and structure were studied by scanning electron microscopy (SEM). Each element of the catalyst has a unique atomic structure and was identified by the spectrum of electromagnetic emissions using EDX.

2.4. Vegetable oil characterization

The suitability of vegetable oil as a fuel for diesel engines depends on its physical chemistry characteristics. Therefore, several tests were carried out, such as evaluation on the kinematic viscosity which greatly influences the friction and flow rate of biodiesel at a certain temperature. In addition, the density and fatty acid composition were also determined.

2.5. Transesterification reaction

Production of biodiesel (methyl ester) was conducted through the transesterification process using a three-neck round-bottom flask reactor equipped with stirring motor and thermometer. Palm oil and methanol were weighed with methanol to oil weight ratio of 12:1, it was then put into the flask with $K_2O/NbOPO_4$ catalyst as much as 2% (of oil weight). The transesterification process was carried out for 3 hours with continuous stirring at a constant temperature of 65°C. The reactor used was equipped with a condenser to avoid the evaporation of methanol during the transesterification process. After completion, the separation was carried out using filter paper and a separating funnel. During separation, the used catalyst was attached to the filter paper. The separated oil was left for 48 hours until 2 layers of liquid were formed, namely the upper layer in the form of methyl ester/biodiesel and the bottom layer in the form of glycerol. The washing process was then carried out on biodiesel using hot water (50°C) [22]. During washing, emulsions occurred between biodiesel and water. To achieve perfect separation, the unreacted methanol was removed and the remaining residual impurity/glycerol emulsion solution was left for 24 hours. The biodiesel yield was calculated based on Equation (1).

$$\% \, Yield = \frac{Mass \ of \ biodiesel}{Mass \ of \ oil} \ x \ 100 \ \% \tag{1}$$

3. Results and discussion

3.1. Characterization of the catalyst by XRD

The crystalline phase of the synthesized catalyst was identified by XRD analysis and results are given in Figure 1. Figure 1 shows that there are two broad peaks of 2Θ angle readings which are in the range 15–40° and 40–60°. The broad pattern (d = large) and low intensity indicate the amorphous character of $NbOPO_4$ [11, 19, 20]. There is no detectable crystalline phase from $NbOPO_4$ and/or Nb_2O_5 in the sample. Moreover, based on the diffractogram pattern there is no K diffraction peak observed which is presumably because the K content in the catalyst is in the size of very small nanoparticles and had spread evenly [11, 19]. A similar finding was also reported for the characteristics of the Ru metal transition on the $NbOPO_4$ solid catalyst used for cellulose conversion to sorbitol [19].

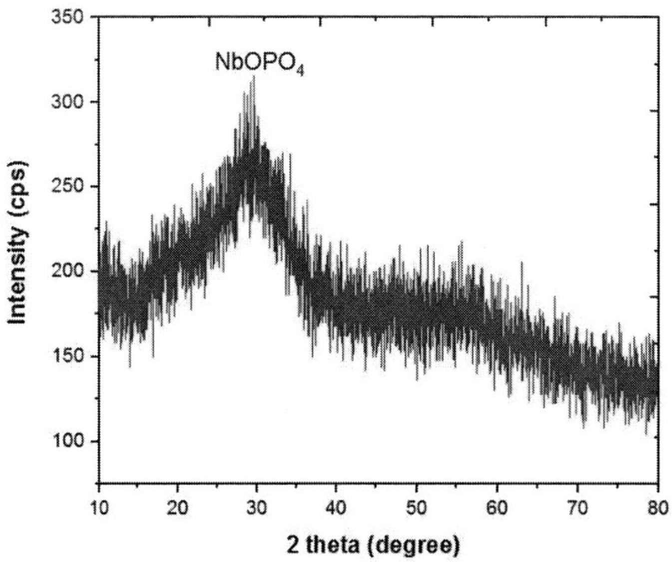

Figure 1. XRD diffractogram of $K_2O/NbOPO_4$ catalyst

3.2. Characterization of $K_2O/NbOPO_4$ catalyst by SEM -EDX

To determine the morphology and particle size of the synthesized $K_2O/NbOPO_4$ catalyst, SEM analysis was carried out. Figure 2 presents the results of the SEM microphotograph of $K_2O/NbOPO_4$ catalysts recorded at 5000x magnification, the image shows the presence of micro grains of different magnitudes and macro pores between these micro grains. The porous structure on the micrometric scale is illustrated in Figure 2.

Figure 2. SEM image of synthesized $K_2O/NbOPO_4$ catalyst with 35% K_2CO_3

It can be seen that the catalyst has the nonhomogeneous particles which size ranging from 1.86 - 25 µm accompanied by the phenomenon of agglomeration. It was found that morphological and mesopore structures were more or less dependent on potassium loading on the surface of $NbOPO_4$ [5]. EDX is an analysis technique to identify the elements contained in a sample. The impregnation of potassium onto $NbOPO_4$ solid acid catalyst was successful, as shown in Table 1. As depicted from Table 1, the catalyst consists of oxygen, phosphorus, potassium and niobium. The quantity of potassium and niobium were obtained of 0,91% and niobium 9,41%, respectively.

Table 1. Elements contained in the catalyst

Element	Weight (%)	Atom (%)
Oxygen	52.24	80.54
Phosphorus	11.33	9.14
Potassium	1.42	0.91
Niobium	35	9.41
Total	100	100

3.3. Effects of K_2CO_3 and catalyst loadings on the biodiesel yield

The effect of K_2CO_3 concentration on the performance of $K_2O/NbOPO_4$ catalysts was evaluated in the transesterification process of palm oil into biodiesel. The yield of biodiesel obtained regarding the concentration of K_2CO_3 used for impregnation process are shown in Figure 3. It can be seen that yield of biodiesel proportional to the loading of potassium. For example, yield of biodiesel reached of 92.89% and 94% at the potassium loading of 15 and 25% K_2CO_3, respectively. Catalysts which impregnated with 35% K_2CO_3 showed the best performance indicated by a high generated yield of 98.87%. It is believed that the distribution of potassium on the surface of $NbOPO_4$ acts as a buffer for the active phase. However, an increase in K_2CO_3 loading up to 55% causes a decrease in catalyst performance as seen from the reduced yield of biodiesel by 2% compared to the results using 35% $K_2O/NbOPO_4$. The effect of adding K_2CO_3 above 35% results in poor or too thick K scattering, causing distribution problems on the $NbOPO_4$, this can cause the closure of active sites [21, 22]. Other researchers mentioned that the excessive presence of active phases on the surface of the catalyst is not desirable [12]. In other words, biodiesel conversion is reduced due to decreased catalytic performance.

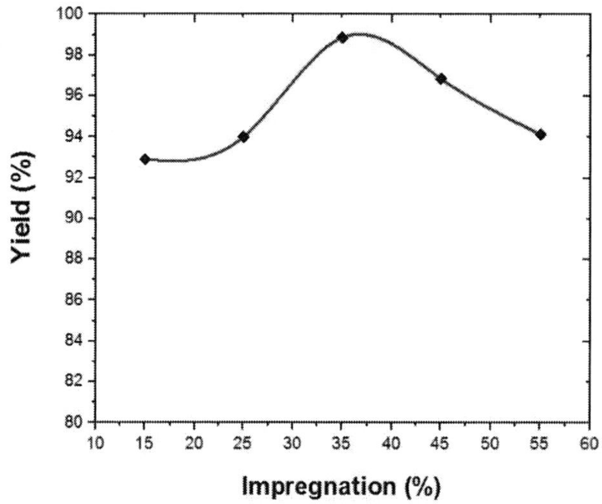

Figure 3. Effect of % K_2CO_3 on biodiesel yield (%)

3.4. Characteristic of biodiesel

Several tests were carried out to determine the characteristics of the produced biodiesel. Table 2. shows the results obtained in the physicochemical characterization of the palm oil biodiesel produced in this study. The physicochemical characteristics of biodiesel were only analyzed on that sample produced from the transesterification process using $K_2O/NbOPO_4$ catalyst with 35% K_2 loading as it showcased the highest yield. The quality of biodiesel in terms of density and viscosity are given in Table 2. It can be explained that the density, and kinematic viscosity are in accordance with the specifications of the Indonesian National Standard (SNI) which emphasized that the catalyst synthesized in this study has the potential to be applied in the production of palm oil into biodiesel.

Table 2. Characteristics of biodiesel produced in this study

Test Parameters	Unit	Research Results	SNI
Kinematic Viscosity	mm^2/s (cst)	4,72 – 5,55	2,3 – 6,0
Density	Kg/m^3	860 - 890	850 - 890

Figure 4 depicts the chromatogram profile from GCMs analysis of biodiesel produced in the transesterification process using 35% $K_2O/NbOPO_4$ catalyst. In Fig. 4, there is the appearance of some peaks that represent the fatty acid components contained in biodiesel. In detail, the data on fatty acid contents of biodiesel can be seen in Table 3. It can be seen that the produced biodiesel mainly consists of Palmitic acid, Linoleic acid, and Oleic acid. Figure 4 also shows that peaks 1 and 2 with an area of 48815069 representing palmitic acid appear at the retention time of 20-21 with the highest composition of 56.92%. Furthermore, peak 3 represents lenoleic acid and and peaks 4.5 are marking the presence of oleic acid with a composition of 3.52% and 39.57% respectively.

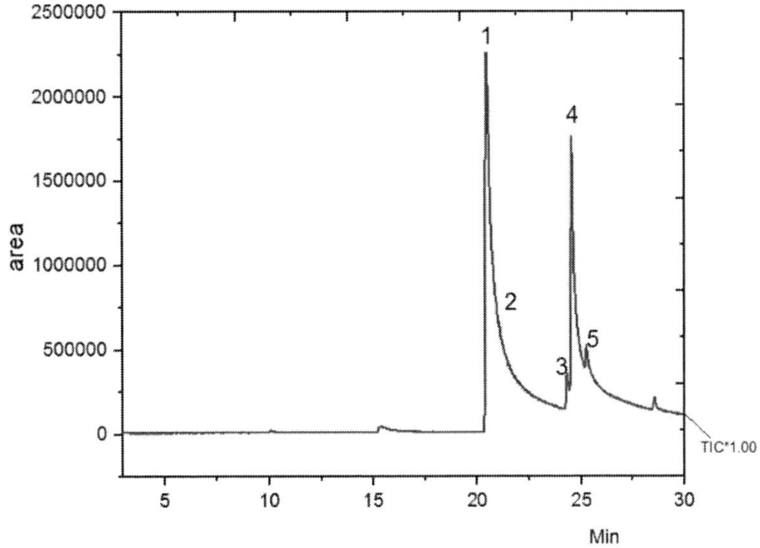

Figure 4. Chromatogram profile of biodiesel

Table 3. Composition (%) of biodiesel

Peak	Fatty Acid	Name	Composition (%)
1,2	Palmitic Acid	Hexadecanoic acid	56.92
3	Linoleic Acid	Octadecanoic acid	3.52
4,5	Oleic Acid	Octadecanoic acid	39.57

4. Conclusion

This study reports the loading effect of K_2 impregnation on the characteristics and performance of $K_2O/NbOPO_4$ as a heterogeneous solid acid catalyst for the biodiesel production. XRD analysis shows an amorphous character of mesoporous niobium phosphate catalyst. The results of SEM analysis stated that the size of the $K_2O/NbOPO_4$ catalyst particles is not homogeneous accompanied by the phenomenon of agglomeration. EDX analysis shows that the $K_2O/NbOPO_4$ catalyst is composed of potassium and niobium with a percent atom is 0,91% and 9,41%, respectively. From the results obtained, the loading of K_2 has a significant effect on the performance of $NbOPO_4$ catalysts in the process of transesterification of palm oil into biodiesel. The highest biodiesel yield was obtained as much as 98.87% in the reaction using the $NbOPO_4$ catalyst which impregnated with 35% of K_2CO_3. The produced biodiesel has qualities that are in accordance with SNI. According to the obtained results, this developed $K_2O/NbOPO_4$ catalyst can be considered as an attractive alternative to heterogeneous catalysts for biofuel production.

Acknowledgment

We greatly acknowledge the funding support from the Ministry of Research, Technology and Higher Education under the Master Thesis Research Grant.

References

[1] Husin H, Adisalamun, Sy Y, Asnawi T M and Hasfita F 2017 *AIP Conference Proceedings* **1788** 030073.
[2] Husin H, Mahidin M, Yunardi Y and Hafita F 2015 , *Key Engineering Materials* **659** 231-6, .
[3] Husin H, Pontas K, Yunardi, Adisalamun, Alam P N and Hasfita F 2017, *ASEAN Journal of Chemical Engineering* **17** 27-36
[4] Glisic S B, Pajnik J M and Orlović A M 2016 , *Applied Energy* **170** 176-85
[5] dos Santos L K, Hatanaka R R, de Oliveira J E and Flumignan D L 2019 , *Renewable Energy* **130** 633-40
[6] Nongbe M C, Ekou T, Ekou L, Yao K B, Le Grognec E and Felpin F-X 2017, *Renewable Energy* **106** 135-41
[7] Afshar Taromi A and Kaliaguine S 2018 , *Applied Catalysis A: General* **558** 140-9
[8] Syamsuddin Y and Husin H 2010 , *Jurnal Rekayasa Kimia & Lingkungan* **7**
[9] Husin H, Hasfita F and Rinaldi W 2013, *Proceedings of The Annual International Conference, Syiah Kuala University-Life Sciences & Engineering Chapter,* pp 14-8
[10] Zuhra Z, Husin H, Hasfita F and Rinaldi W 2015 , *Agritech* **35** 69-77
[11] Boonyuen S, Smith S M, Malaithong M, Prokaew A, Cherdhirunkorn B and Luengnaruemitchai A 2018 , *Journal of Cleaner Production* **177** 925-9
[12] da Conceição L R V, Carneiro L M, Rivaldi J D and de Castro H F 2016 , *Industrial Crops and Products* **89** 416-24
[13] Di Serio M, Mallardo S, Carotenuto G, Tesser R and Santacesaria E 2012 , *Catalysis Today* **195** 54-8
[14] Fitriana N, Husin H, Yanti D, Pontas K, Alam P N, Ridho M and Iskandar 2018, *IOP Conference Series: Materials Science and Engineering* **334** 012011
[15] Naveenkumar R and Baskar G 2019, *Bioresource Technology* **280** 493-6
[16] Nofiarli, Kasim A and Nurdin H 2013, *The Biodiesel characteristic of kepayang oil (Pangium edule Reinw)* vol 8
[17] Malhotra R and Ali A 2019, *Renewable Energy* **133** 606-19
[18] Hussain Z and Kumar R 2018, *International Journal of Green Energy* **15** 629-40
[19] Huang Y-T, Cheng R, Zhai P, Lee H, Chang Y-H and Feng S-P 2017, *Electrochimica Acta* **236** 131-9
[20] Siritanaratkul B, Maeda K, Hisatomi T and Domen K 2011, *ChemSusChem* **4** 74-8
[21] Yan J, Wu G, Guan N and Li L 2016 *Appl. Catal. B-Environ.* **152** 280–8
[22] Husin H, Marwan and Mahidin 2011 *Jurnal Reaktor* **13** 254-61

IOP Conf. Series: Materials Science and Engineering **796** (2020) 012049 doi:10.1088/1757-899X/796/1/012049

Fabrication of Activated Charcoal from Coconut Shell Combined with Coal Fly Ash from PLTU Nagan Raya for Adsorption of Methylene Blue

Mariana[1]*, F Mulana[1], Fauzi M Djuned[1], M Fadli[2], and M Meilian[2]
[1]Department of Chemical Engineering, Syiah Kuala University, Banda Aceh, Indonesia
[2]Undergraduate Student of Department of Chemical Engineering, Syiah Kuala University, Banda Aceh, Indonesia

E-mail: mariana_hasyim@yahoo.com

Abstract. Coconut shell is one of the abundant agricultural solid wastes in Aceh Province. Meanwhile, a solid waste of coal fly ash is a by-product from PLTU Nagan Raya. Both coconut shell and coal fly ash have a great potential to be utilized as adsorbent. The aim of study is to prepare and characterize the adsorbent made from combination of coconut shell activated charcoal and coal fly ash. The prepared adsorbent was characterized using Scanning Electron Microscopy (SEM) to study the morphological structure, Fourier Transform Infra-Red Spectrophotometer (FTIR) to detect the functional group and X-Ray Diffraction (XRD) to determine the crystalline structure. Performance of adsorbent was carried out by adsorption process of methylene blue solution on adsorbent. The characterization results showed that all four types of the prepared adsorbents met the requirements of water content quality and iodine adsorption according to SNI 06-3730-1995 concerning to activated charcoal. It was obtained that adsorbent with the best performance was coconut shell activated charcoal (without addition coal fly ash) having the adsorption capacity to methylene blue of 45.36 mg/g on adsorption contact time of 140 minutes.

1. Introduction

Aceh Province is one of the largest producers of coconut shell waste in Indonesia. The existance of coconut shell waste can be minimized by utilizing those agricultural by-product materials as raw material to produce activated charcoal. In addition to reducing waste, producing activated charcoal from coconut shell is a very cost-effective approach compared to the production of activated carbon from coal as a more expensive raw material. It was reported that activated carbon from coconut shell has a high density and purity and harder carbon properties and more resistant to attrition [1]. In addition, coconut shell activated carbon is reported to have an amorphous form, which capable of absorbing various kinds of gases, vapors and colloidal solids [2-4]. In 2015, Roesniani et al [5] used coconut shell waste as a material to produce activated charcoal adsorbent and successfully adsorbed ammonia by 34.87%. Besides source of coconut

[1]To whom any correspondence should be addressed.

Content from this work may be used under the terms of the Creative Commons Attribution 3.0 licence. Any further distribution of this work must maintain attribution to the author(s) and the title of the work, journal citation and DOI.
Published under licence by IOP Publishing Ltd

IOP Conf. Series: Materials Science and Engineering **796** (2020) 012049 doi:10.1088/1757-899X/796/1/012049

waste, Aceh Province especially in the Nagan Raya Region also generates coal fly ash (FA) waste due to the presence of electrical steam power plant (PLTU Nagan Raya) in surrounding area. Coal fly ash is a by-product as well as the main solid waste produced from the coal combustion process in the industry. If not handled properly, the presence of coal fly ash can damage the land and pollute the environment [6]. Application of coal fly ash as one of the materials for water and wastewater processing can offer great advantages economically and environmentally. Zhao et al [7] modified fly ash with TiO_2 and used it as an adsorbent to remove ammonia. They succeeded in removing ammonia by 79% for 3 hours of contact time [8].

In this study, coconut shell and coal fly ash were mixed to produce a new combined adsorbent. The combined adsorbent with the best characteristics was obtained by studying the appropriate combination ratio of both coconut shell and coal fly ash. The performance of the adsorbent was tested by the adsorption process of adsorbent in methylene blue solution

2. Experimental

2.1 Materials

The used materials were coconut shell, coal fly ash, 0.1 M of H_2SO_4 solution (Merck KGaA), methylene blue (Merck), 0.1 N of iodine solution (Merck), 1% starch indicator, 0.1 N of sodium thiosulfate (Merck) and distilled water.

2.2 Tools

The tools used in this study were Fourier Transform InfraRed Spectroscopy (FTIR) (Agilent Technologies Cary 630 FTIR), Scanning Electron Microscopy (SEM) (JEOL-JSM 6510 LA), X-Ray Diffraction (XRD) (Pan Analytical, Expert Pro), hot plate and stirrer, muffle furnace (Isuzu model EPTR-26R), Spectrophotometer UV-Visible (Shimadzu 1800), digital scales (Mettler PE 300l), mechanical sieves 100-200 mesh (TEST-SIEVE 410-69) and desiccator.

2.3 Preparation of adsorbents

Coconut shell was put into the furnace and then carbonized. Carbonization was carried out at temperature of 400° C for 2 hours [9]. The obtained coconut shell charcoal was then cooled for 24 hours, then milled and sieved using a 100-200 mesh sieving [10, 11]. The same procedure was performed on coal fly ash.

Combination of activated charcoals from coconut shell (TK) and coal fly ash (FA) was prepared with 4 (four) variation mass ratio of 100:0; 70:30; 30:70 and 0:100. A total of 50 grams of combined coconut shell charcoal : coal fly ash were soaked in 500 ml of sulfuric acid solution of 0.1 M for 24 hours then drained. After that, it was filtered and rinsed with distilled water until the pH was close to neutral. It was then placed into a muffle furnace at a temperature of 800 °C for 2 hours [12]. Once finished, the adsorbents were stored in the desiccator.

2.4 Characterizations of Adsorbents

The prepared adsorbents were characterized in terms of their morphological structure, chemical functional groups and crystalline structure using Scanning Electron Microscopy (SEM), Fourier Transform Infra-Red Spectrophotometer (FTIR) and X-Ray Diffractometer (XRD), respectively.

2.4.1 Water content

The adsorbent as much as 2 grams was put into the porcelain dish that has previously been weighed. The dish was placed into an oven at 105°C for 3 hours then cooled in a desiccator and weighed again.

$$\text{Water Content } (\%) = \frac{a - b}{a} \text{ x } 100\% \quad (1)$$

a and b are respective weights of adsorbent before and after drying (g)

2.4.2 Ash content

A total of 2 grams adsorbent was put into a porcelain dish and furnaced at 600 °C for 6 hours. Once done, the adsorbent was cooled in a desiccator. The mass of adsorbent before and after furnaced were weighed to determine the ash content.

$$\text{Ash Content } (\%) = \frac{a}{b} \text{ x } 100\% \quad (2)$$

a is the weight of ash (g) and b is the initial weight of adsorbent (g)

2.4.3 Adsorption of Iodine

The 0.15 grams of adsorbent was put into a dark and sealed beaker. Then, 50 mL of 0.1 N of iodine solution was put in the beaker then shaken for 15 minutes then filtered. A total of 10 mL of filtrate was titrated with 0.1 N of sodium thiosulfate solution. When the yellow color of the solution was seen almost disappeared, 1% starch indicator was added. Titration was continued until the blue color was gone.

$$\text{Adsorption of Iodine} \left(\frac{mg}{g}\right) = \frac{\left\{10 - \left(\frac{N \text{x} V}{0,1}\right)\right\}}{S} \text{ x } 12,69 \text{ x } 2,5 \quad (3)$$

Where, V is the needed volume sodium thiosulfate solution (V), N is the concentration of sodium thiosulfate solution solution (N), S is the weight of carbon (g), and 12,69 is a constant representing the iodine amount corresponding to 1 ml of 0.1 N sodium thiosulfate solution.

2.5 *Adsorption of Methylene Blue*

A total of 0.1 gram of adsorbent with a combination ratio of 100:0; 70:30; 30:70 and 0:100 was put into an erlenmeyer containing 50 ml of 100 ppm methylene blue solution. Following that, the erlenmeyer was closed and stirred at 100 rpm. The methylene blue solution in the erlenmeyer was then filtered. The concentration of methylene blue in the filtrate was determined by the curve of methylene blue absorbance.

$$\text{Removal } (\%) = \frac{\text{Initial conc. of Adsorbate - Final conc. of Adsorbate}}{\text{Initial conc.}} \text{ x } 100\% \quad (4)$$

$$\text{Adsorption Capacity} = \frac{\text{Initial conc. of Adsorbate - Final conc. of Adsorbate}}{\text{Adsorbent amount.}} \text{ x V of sol} \quad (5)$$

3. Results and Discussions

Characterization results of combined activated charcoal from coconut shell and coal fly ash is given in Table 1. The characterization of water content was treated by oven at 105°C for 4 hours. Meanwhile, the characterization of ash content was treated by furnace at 600°C for 3 hours. Table 1 shows the analysis results of the combined adsorbent, indicating that the water content and adsorption capacity of iodine have met the activated charcoal quality standard based on SNI No. 06-3730-1995. However, the ash content in the samples with ratio TK:FA of 70:30; TK: FA of 30:70 and TK: FA of 0: 100 have not complied to the SNI quality. This is because the standard quality in terms of ash content for combined activated charcoal from coconut shell : coal fly ash adsorbent is not yet available in the SNI record. In addition, coal fly ash is an ash, therefore, the higher the composition ratio of fly ash in the combination, the greater the ash content will be. Though the ash content is high, all adsorbents still showed a good iodine adsorption performance. Amongst all, activated charcoal from coconut shell (without addition of coal fly ash) showed the best iodine adsorption ability of 1988.1 mg/g.

Table 1. Characterization results of combined activated charcoal from coconut shell and coal fly ash

Parameter	SNI Standard Quality No. 06-3730-1995	Adsorbent	Analysis Results
Water content	Max. 15%	TK:FA 100:0	11.55 %
		TK:FA 70:30	8.81 %
		TK:FA 30:70	0.93 %
		TK:FA 0:100	0.12 %
Ash content	Max. 10%	TK:FA 100:0	0.72 %
		TK:FA 95:5	9%
		TK:FA 90:10	18%
		TK:FA 70:30	50.25 %
		TK:FA 30:70	96.46 %
		TK:FA 0:100	99.83 %
Iodine absorbancy	Min. 750 mg/g	TK:FA 100:0	1988,1 mg/g
		TK:FA 70:30	1692 mg/g
		TK:FA 30:70	1692 mg/g
		TK:FA 0:100	1607,4 mg/g

Note: TK for coconut shell and FA for coal fly ash

Performance on additional samples was also carried out by using two adsorbents with combination ratios of 90:10 and 95:5. The ash content in these additional samples was obtained as much as 18% and 9%, respectively. These results confirmed that the lesser the content of the fly ash in the combined adsorbent, the lower ash content. The ratio TK:FA of 95:5 adsorbent has an ash content which is in accordance with SNI No. 06-3730-1995.

3.1 Scanning Electron Microscope (SEM) analysis

The Scanning Electron Microscope (SEM) analysis was carried out to observe the surface morphology of the prepared adsorbents. The results of SEM imaging can be seen in Figure 1. Based on Figure 1, there are significant morphological differences on the surface of the combined adsorbents. The adsorbent with ratio TK:FA of 100:0 (no addition of fly ash) has larger pores with impurities seen attaching on the surface. Meanwhile, those that combined with fly ash tend to have smaller pores that attributed to the presence of fly ash.

3.2 Fourier Transform Infra Red (FTIR) Analysis

The characterization using a Fourier Transform Infra Red (FTIR) is aimed to observe the functional groups existing in adsorbents prepared from the activated charcoal of coconut shell : coal fly ash. The results of FTIR analysis can be seen in Figure 2.

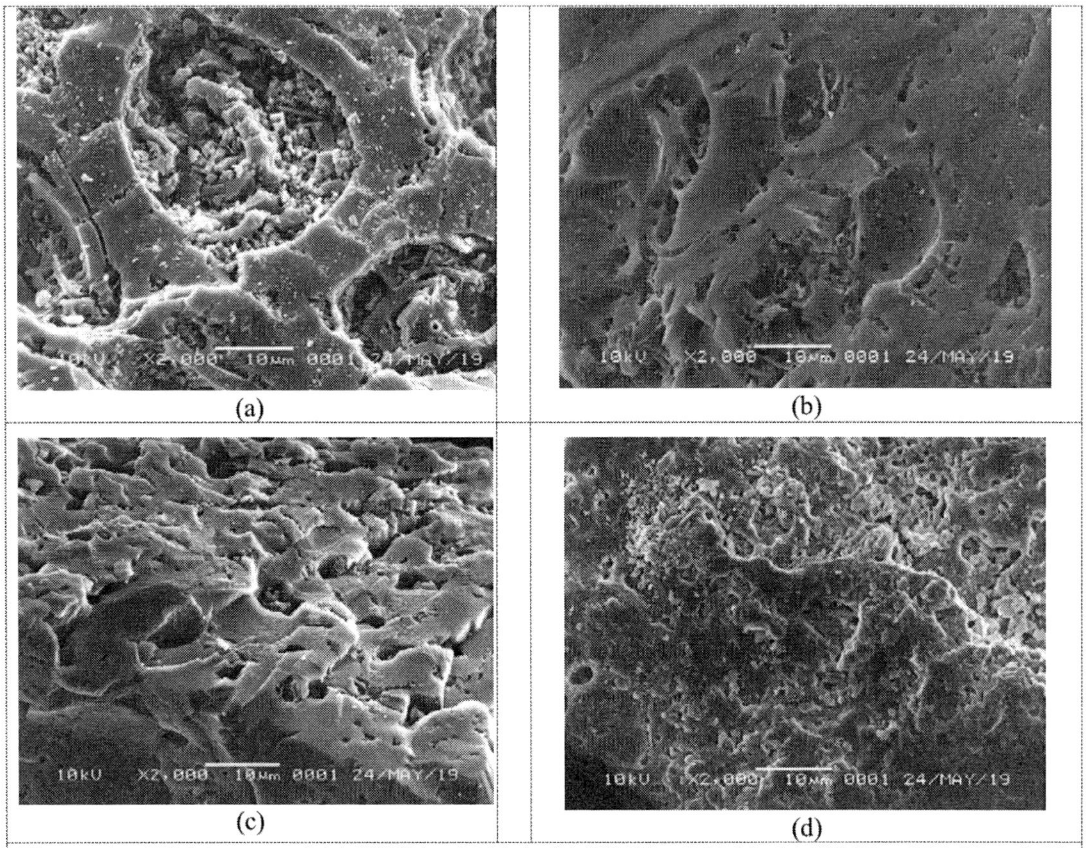

Figure 1. SEM images of (a) TK:FA of 100:0, (b) TK:FA of 70:30, (c) TK:FA of 30:70 and (d) TK:FA of 0:100 adsorbent at mag. of 2000x

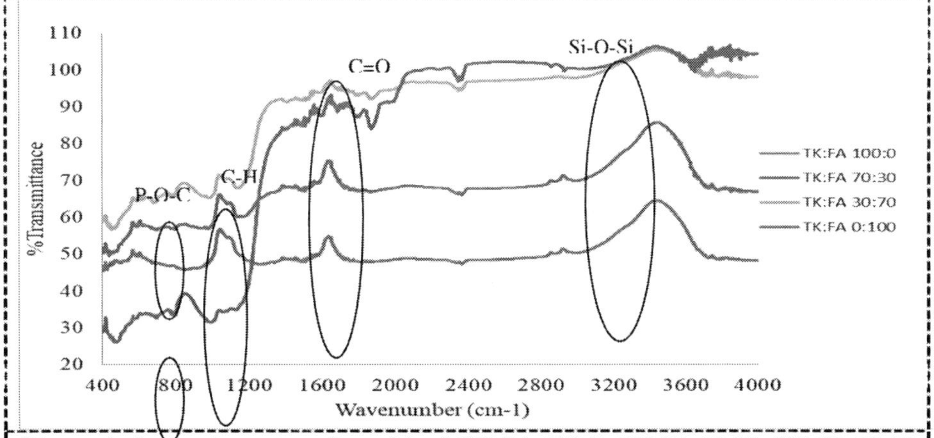

Figure 2. Infrared spectra of combined TK:FA 100:0; TK:FA 70:30; TK:FA 30:70 dan TK:FA 0:100 adsorbents

Based on Table 2, it can be seen that the absorption of the C-H, C = O and Si-O-Si groups dominate the four combined adsorbent samples. From the results obtained there was a decrease in intensity of absorption peak for samples with increased fly ash addition. This shows that the addition of fly ash

IOP Conf. Series: Materials Science and Engineering **796** (2020) 012049 doi:10.1088/1757-899X/796/1/012049

to coconut shell activated charcoal affects absorption intensity in wavelength regions and results in changes in functional group structure [14]. In addition, there is a formation of the P-O-C functional group formed on the coal fly ash combined adsorbent due to the presence of P element from fly ash. The absorption band at the wave numbers 3074.53 and 3653.18 cm-1 which is the H-O-H stretch vibration of a water molecule that has a weak hydrogen bond with the Si-O surface on the adsorbent with ratio of TK: FA = 0 : 100

Table 2. Resulting FTIR readings of coconut shell combined with coal fly ash activated carbon sorbent

Bond	*Wavenumbers range* (cm^{-1})				
	Reference [13]	**TK:FA of 100:0**	**TK:FA of 70:30**	**TK:FA of 70:30**	**TK:FA of 0:100**
P-O-C	1050-950	-	977.91	977.91	995.27
C–H	3000-2700	2758.21; 2872.01 and 2985.81	2870.08 and 2985.81	2792.93 and 2987.74	2856.58 and 2927.94
C=O	1900-1650	1726.29 and 1759.08	1726.29; 1759.08 and 1784.15	1683.86 and 1784.15	1683.86; 1784.15 and 1791.87
Si-O-Si	1130-1000	1083.99	1083.99	1062.78 and 1082.07	1058.92; 1080.14 and 1128.36
Si-OH	3700-3200	-	-	-	3074,53 dan 3653,18
Si-H	2250-2100	-	-	2133,27	-
Al-O	570-420	426.27 dan 493.78	437.84; 455.2 dan 476.42	433.98 dan 464.84	432.05 dan 478.35
Fe-O	≈567	588.29	588.29	588.29	580.57

3.3. X-Ray Diffraction (XRD) Analysis

The X-Ray Diffraction (XRD) characterization of the coconut shell combined coal fly ash adsorbent was carried out at an angle of 2θ. The results of the XRD analysis can be seen in Figure 3. Figure 3 presents the results of X-ray diffraction analysis of coconut shell combined with fly ash sample. Activated carbon shows a highly irregular microcrystalline structure in which graphite micro crystals are randomly oriented [15].

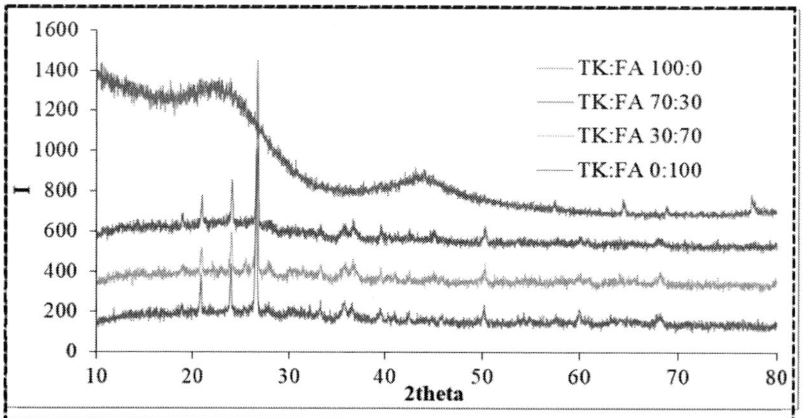

Figure 3. XRD pattern of coconut shell combined coal fly ash activated charcoal adsorbents

As in the TK: FA 100: 0 adsorbent contains a slight sharp peak that indicates the presence of SiO2, namely in the region 2θ = 12.97 o (d = 6.8242 Å). The slight sharp peak in adsorbent of TK:FA = 100: 0 indicates that this is an amorphous structure [16].

In each adsorbent, some of peaks are detected at a 2θ area of 12.97° (d = 6,8242) (TK:FA of 100:0); 26.82 ° (d = 3.3241 Å) (TK:FA of 70:30); 26.73° (d = 3.3357 Å) (TK:FA of 30:70) dan 26.63 ° (d = 3.3473 Å) (TK:FA of 0:100) indicating the presence of SiO_2. The emergence of sharp peaks indicating the better alignment of the layers that is the characteristic of the crystal structure. [16]. The components found in each type of adsorbent are presented in Table 3.

Table 3. Analysis results of XRD

Component	**% Weight**			
	TK:FA of 100:0	**TK:FA of 70:30**	**TK:FA of 30:70**	**TK:FA of 0:100**
SiO_2	12.6	73.5	81.3	36.4
C	26.2	1.1	-	63.6
Al_2O_3	39.2	10.2	-	-
Fe_2O_3	3.6	6.6	8.9	-
TiO_2	18.3	8.6	9.8	-

The TK: FA of 70:30 and TK: FA of 30:70 adsorbents had more SiO_2 content due to the addition of fly ash. However, at the same time, some components like carbon and Al_2O_3 actually decreased even disappeared in the TK:FA of 30:70 sample.

3.4. Effect of Sorbent Types on Efficiency and Capacity of Methylene Blue Adsorption

The adsorption efficiency of the adsorbent is influenced by several factors, such as contact time. Based on Figure 4, it can be seen clearly that adsorption efficiency increases with increasing contact time between the adsorbate and the adsorbent. The longer the contact time, the concentration of methylene blue decreases until the equilibrium is reached. Based on the results, it was found that the adsorption efficiency for the initial concentration of 100 mg/L with a contact time of up to 140 minutes with sampling every 10 minutes increased. Activated charcoal from coconut shell (TK:FA equal to 100:0) adsorbent has the highest adsorption efficiency up to 97.14%, whereas, adsorbent with combination ratio of TK:FA equal to 30:70 showed the lowest adsorption efficiency of 5.58%. The adsorption rate of the adsorbate is very big during the initial contact time, namely in the 30th minute to the 120th minute. Meanwhile, from the 120th to the 140th minute of adsorption time, the equilibrium was achieved.

The adsorption capacity represents the ratio of the amount of adsorbate absorbed by the adsorbent [17]. The correlation between combination ratio of prepared materials to produce adsorbents to the adsorption capacity can be seen in Figure 5. Figure 5 showed that the longer the contact time between the adsorbent and methylene blue resulted in increasing the adsorption capacity. The highest adsorption capacity was obtained when using only activated charcoal from coconut shell adsorbent that is equal to 45.36 mg/g. Meanwhile, the when using adsorbent with TK: FA ratio of 30:70 showed adsorption capacity of only 2.27 mg/g.

The results showed that the adsorption capacity of adsorbents TK: FA 30:70 and TK: FA 0:100 had the lowest adsorption capacity. This is consistent with the results of the XRD test that showed that no Al_2O_3 compound was found in the sample. Al-Si-O ion bonds form a crystal structure, which has a cavity structure filled with interchangeable water and cations and has a pore of a certain size that can expand the surface of the adsorbent [18].

IOP Conf. Series: Materials Science and Engineering **796** (2020) 012049 doi:10.1088/1757-899X/796/1/012049

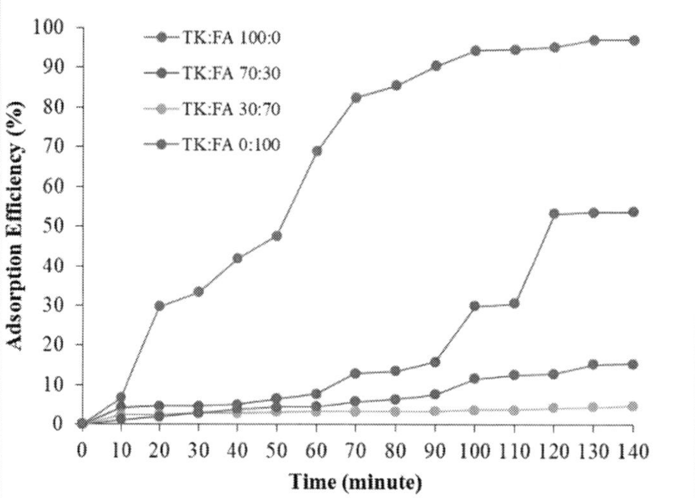

Figure 4. The effect of combination ratio of adsorbents on the adsorption efficiency

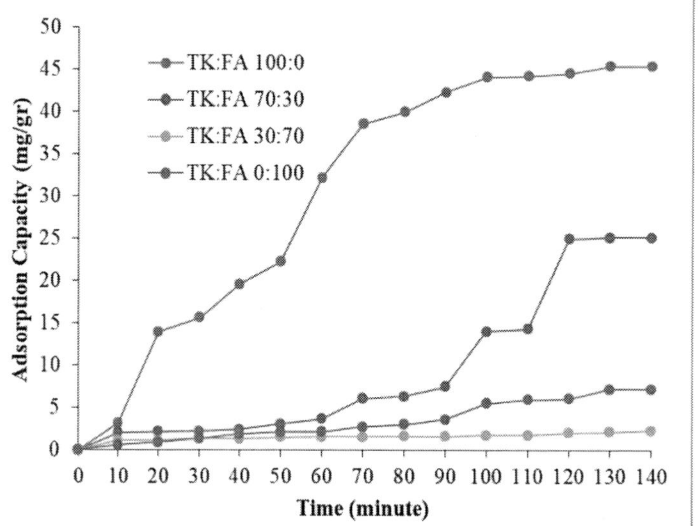

Figure 5. Effect of combination ratio of adsorbents to the adsorption capacity

4. Conclusion

The combination of activated charcoal from coconut shell with coal fly ash has been successfully carried out with the emergence of a new functional group containing P element in FTIR results. The XRD results revealed that the highest SiO_2 content of 81.3% was obtained at ratio sample TK:FA of 30:70. The equilibrium time of methylene blue adsorption using combination of activated charcoal from coconut shell and coal fly ash was reached in minutes 120th. The highest removal efficiency of methylene blue was obtained by using activated charcoal from coconut shell. Similarly, the highest methylene blue adsorption capacity was 45.36 mg/g.

References

[1] Gratuito M K B, Panyathanmaporn T, Chumnanklang R A, Sirinuntawittaya N and Dutta A 2008 *Bioresource Technology* **99** 4887-95
[2] N. Mohd Iqbaldin M, Ismail K, I. Mohd Azlan M, Ghani Z, Surani B and Zubri Z 2013 *Journal of Tropical Forest Science* **25(4)** 497-503
[3] Akhir M F M, Saad N A and Zakaria N A 2017 Potential of coconut shell activated carbon (CSAC) in removing contaminants for water quality improvement: A critical review **1892** 070005
[4] Laine J, Simoni S and Calles R 1991 *Chemical Engineering Communications* **99** 15-23
[5] Roesiani L 2015 *Keefektifan Lama Kontak Karbon Aktif Terhadap Penurunan Kadar Amonia Limbah Cair Industri Tahu di Desa Teguhan Sragen* (Wetan Sragen: Universitas Muhammadiyah Surakarta)
[6] Qi L, Teng F, Deng X, Zhang Y and Zhong X 2019 *Powder Technology* **351** 153-8
[7] Zhao Y-l, Yang W-j, Zhou J-h, Wang Z-h, Liu J-z and Cen K-f 2015 *Journal of Fuel Chemistry and Technology* **43** 266-72
[8] Slamet A A and Imas K 2017 Pemanfaatan Limbah Fly Ash Untuk Penanganan Limbah Cair Amonia **39** 69-78
[9] Agusta D J 2012 *Uji Adsorpsi Gas CO pada Asap Kebakaran dengan Menggunakan Karbon Aktif dari Arang Tempurung Kelapa yang Terimpregnasi TiO2* (Jakarta: Universitas Indonesia)
[10] Idrus R, Lapanporo B P and Putra Y S 2013 *Prisma Fisika* **1(1)** 50-55
[11] Rahayu L H, Purnavita S and Sriyana H Y 2014 *Momentum* **10(1)** 47-53
[12] Oobuchi M Y, Maezawa A and Uchida S J 2004 $Ca(OH)_2$/Diatomaceous Earth Sorbents for HCl Removal in a Bag Filter Reactor **35** 491-4
[13] Ma Z, Chen D, Gu J, Bao B and Zhang Q 2015 *Energy Conversion and Management* **89** 251-9
[14] Hanum F, Gultom R J and Simanjuntak M J 2017 *Jurnal Teknik Kimia* **6(1)** 49–55
[15] Khalili, S. Asghar, A. Ghoreyshi, Jhanshasi, M. 2013 *Chemical industry and chemical engineering Quarterly* **19** 153-154
[16] Das, D, Samal, D P, Meikap, B C 2015 *Journal of chemical Engineering &Process Technology* **6(5)** 1-7
[17] Hokkanen S, Bhatnagar A and Sillanpää M 2016 *Water Research* **91** 156-73
[18] Amsiri 2010 *Penyerapan Merkuri Dalam Limbah Simulasi Menggunakan Zeolit Klinoptilolit* (Jakarta: Universitas Islam Negeri Syarif Hidayatullah)

Preparation and Characterization of Activated Carbon from Gayo Coffee Shell as an Adsorbent for Removal of Lead (Pb) in Liquid Waste

L Juniar[1], Mariana[2*], S Mulyati[2], D Fathira[2], R Safitri[2]
[1]Graduate School of Chemical Engineering, Universitas Syiah Kuala, Banda Aceh, 23111, Indonesia
[2]Department of Chemical Engineering, Syiah Kuala University, Banda Aceh, Indonesia, 23111

E-mail: Mariana_hasyim@yahoo.com

Abstract. Adsorption been a popular low-cost and environmentally friendly method that can be used to reduce heavy metals in wastewater especially with the use of biomass-sourced sorbents. In this study, the adsorbent was prepared from the coffee shell. The Performance of the prepared adsorbent was studied on the adsorption of Pb(II). The initial concentration of Pb(II) solution used was 300 mg/L and contact time was varied to 0, 30, 60, 90, 120 and 150 minutes. The characteristics of the sorbent were studied in terms of moisture content, ash content, and adsorption of I_2. Further characterizations of the coffee shell sorbent were also carried out using Scanning Electron Microscopy (SEM), Fourier Transform Infra-Red Spectrophotometer (FTIR), X-Ray Diffraction (XRD), X-Ray Fluorescence (XRF) and Surface Area Analyzer (SAA). The characterization results showed that the prepared sorbent meets the quality requirements according to SNI 06-3730-1995. The results from the adsorption process revealed that the maximum absorption capacity that can be performed by sorbent was 170.26 mg/g at the equilibrium time of 90 minutes. In conclusion, the coffee shell can be a potential biomass material to prepare adsorbent with high removal performance of heavy metal.

1. Introduction

Indonesia is one of developing countries with high pollution issues particularly in terms of water pollution. Water pollution is defined as a condition which living things, substances, energy or other components presence in the water, causing the quality of the water to be polluted. These inputs are often referred to as pollutant elements, which practically sourced from routine discharges, such as liquid waste discharges. One of common water pollutants is heavy metal such as lead (Pb) [1-3].

In the environment, Pb is usually resulted from mining activities especially during the processing of metal ore with concentrations ranging from 200-500 mg/L [3, 4]. This concentration is considerably very high to be discharged into the environment. According to class I water quality standards, the safe Pb content in water is not more than 0.03 m /L [5].

Adsorption is a separation process that has been rapidly adopted for remediation of contaminants in water and wastewater [6, 7]. Absorbents such as activated carbon plays an

* To whom any correspondence should be addressed

Content from this work may be used under the terms of the Creative Commons Attribution 3.0 licence. Any further distribution of this work must maintain attribution to the author(s) and the title of the work, journal citation and DOI.
Published under licence by IOP Publishing Ltd

important role as an efficient material in this process. However, the producing cost of activated carbon is still expensive; for that reason, many researchers have begun to look for alternative adsorbents made from local and natural sources. Natural-source such as those from plant-based materials can be used as a source to fabricate sorbent for removal of heavy metals from water and wastewater systems [8, 9].
Aceh Province precisely the Central Aceh Regency is one of big producer of highest quality coffee in Indonesia. The farming and production activities of coffee generate a huge load of waste. In most case, those generated wastes especially in the form of coffee shell are directly loaded into the environment as waste without further treatment or use. These wastes continue to grow with the growing production of coffee in Indonesia. Based on this, in this study, coffee shells are utilized to prepare activated carbon with the aim to add more economical and environmental values to the coffee-sourced waste. To the best of our knowledge and based on literature study, the utilization of coffee shells especially those from Gayo arabica coffee is for the first time reported in this work. In earlier work [10], we have performed a similar study by utilizing coffee ground as a bio-sorbent. The coffee shell can be used as a sorbent source due to its organic components such as cellulose, hemicellulose, and lignin. According to Pope [11], materials containing lignin, hemicellulose, and cellulose can be used as raw material for sorbent production due to their effectiveness in absorbing metals in liquid waste. The detailed content or characteristics of coffee endocarp (shell) can be seen in Table 1.

Table 1. Characteristics of coffee shell [10]

Component	**Percentage (%)**
Cellulose	53,19
Hemicellulose	20,17
Lignin	28,58

In detail, this paper reports the preparation of low-cost adsorbent from the shell of Gayo Arabica coffee. The basic characteristics of the prepared adsorbent are evaluated through several analysis. The performance of adsorbent is also tested in adsorption of Pb-containing artificial wastewater solution.

2. Experimental

2.1 Materials
The shell of Gayo Arabica coffee was obtained from Aceh Tengah regency, Indonesia, and used as the main material in sorbent preparation. The other supporting materials were $Pb(NO_3)_2$ (Pudak Scientific, Indonesia) as an artificial contaminant sample. In addition, Iodine solution (Sigma Aldrich, USA) (0.1 N), the starch indicator (1%) (LabChem, Pennsylvania, USA), sodium thiosulfate (The Science Company, Colorado, USA) (0.1 N) and distilled water were also used in this work.

2.2 Preparation of Activated Carbon Sorbent from Coffee Shells
The coffee shell was washed and sun-dried then carbonized at 350°C for 120 minutes using a muffle furnace until the carbon was formed. The resulting carbon was crushed using mortar and pestle, then sifted using a 100-120 mesh sieve.

2.3Characterizations of the Sorbent
The prepared activated carbon sorbent from this research was characterized using Scanning Electron Microscopy (SEM, FEI, Inspect-s50, Japan), Fourier Transform Infra-Red

Spectrophotometer (FTIR, Shimadzu, IRPrestige21, Japan), X-Ray Diffraction (XRD, PanAnalytical, E'xpert Pro, UK), X-Ray Fluorescence (XRF, PanAnalytical, Minipal 4, UK), Surface Area Analyzer (SAA, Nova, Quantachrome, USA) to study the morphological structure, observe the functional groups, investigate the crystalline structure, analyze the chemical composition as well as concentration, and determine the surface area of sorbent, respectively. In addition, some other characterizations were also carried out, as follows:

2.3.1 Water content. 1 gram of sorbent was put into a porcelain dish which has been weighed. The dish was placed into an oven at a temperature of 105°C for 3 hours. Following that, the dried sorbent was cooled down in a desiccator and weighed again.

$$\text{Water content (\%)} = \frac{a-b}{a} x100 \tag{1}$$

a and b represent the initial and dried weights of the sorbent (g), respectively.

2.3.2 Ash content. 2 grams of sorbent was placed into a porcelain dish and weighed. The cup was put into a muffle furnace at 600°C for 6 hours then cooled in a desiccator and weighed.

$$\text{Ash content} = \frac{a}{b} x100\ \% \tag{2}$$

Where a and b are the respective weight of ash and initial dried carbon (g)

2.3.3 Absorption of iodine. 0.15 gram of sorbent was put into a sealed container. Then, 50 mL of 0.1 N iodine solution was added in the container then shaken for 15 minutes then filtered. A total of 10 mL of filtrate was titrated with 0.1 N. sodium thiosulfate solution. If the yellow color of the solution was almost gone, 1% starch indicator was added. Titration continues until the blue color is gone.

$$\text{Absorption of iodine}\left(\frac{mg}{g}\right) = \frac{\left\{10-\left(\frac{N x V}{0,1}\right)\right\}}{S} x12,69\,x\,2,5 \tag{3}$$

Where, V is the volume of used sodium thiosulfate solution, N for sodium thiosulfate concentration, S represents the mass of carbon (g), and 12,69 is the equivalent of iodine to 1ml of 0.1 N sodium thiosulfate solution.

2.4 Adsorption Test on Artificial $Pb(NO_3)_2$ Wastewater. Initially, artificial wastewater solution was prepared in 1000 ppm by dissolving 0,799 gram of $Pb(NO_3)_2$ into 500 ml of deionized water then diluted into 200, 300 and 500 mg/L as varying initial concentrations. A total of 100 ml sample was contacted with 0.1 gram sorbent with a different contact time of 0, 30, 60, 90, 120 and 150 minutes, the process was carried out with a stirring speed of 250 rpm. The concentration of $Pb(NO_3)_2$ in feed solution and after adsorption was measured Pb using Atomic absorption spectrometer (AAS).

3. Results and Discussion

3.1 Water and ash content and

The activated carbon sorbent from the coffee shell was characterized in terms of water content and ash content, as well as adsorption of iodine, the results can be seen in Table 2. The results of analysis which presented in Table 2 indicate that the prepared activated carbon sorbent has the water content, ash content and absorption capacity of iodine that meet the quality standards of activated carbon based on SNI No. 06-3730-1995. These results confirm that the prepared adsorbent in this study can be used for application in the reduction of Pb (II) metal.

Table 2. Characteristics of activated carbon sorbent from the coffee shell

Parameter	SNI Standard No. 06-3730-1995*	Analysis results
Water Content	Max. 15%	4 %
Ash Content	Max. 10%	4 %
Iodine Adsorption	Min. 750 mg/g	994,05 mg/g

3.2 Scanning Electron Microscope (SEM) Analysis
The Scanning Electron Microscope (SEM) analysis was carried out to study the surface morphology of the activated carbon sorbent prepared from the coffee shell. The SEM imaging result is shown in Fig. 1.

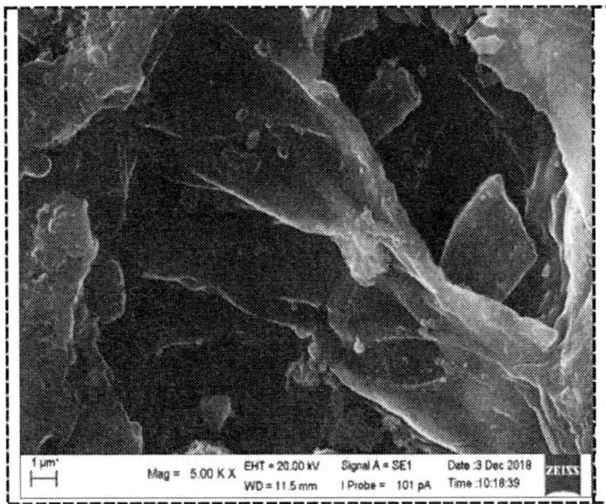

Figure 1. The results of sorbent characterization using SEM with a magnification of 5000 times

Based on Fig.1, it is seen that the sorbent has a porous surface with large size of pores. It is reported that sorbent with a porous surface is advantageous for the adsorption as it provides more contact area [10].

3.3 Surface Area Analyzer (SAA) Analysis
The surface area of the sorbent was determined by means of Surface Area Analyzer (SAA) using the Brunauer, Emmet, and Teller (BET) method. Results of analysis can be seen in Table 3.

Table 3. Analysis results of the surface area of coffee shell activated carbon sorbent

Surface Area (m^2/g)	Pore Volume (cm^3/g)	Pore Size (Å)
3,0438	0,0075	98,5269

Table 3 shows that the sorbent has pores with an average diameter of 98.5269 Å. These results indicate that sorbent is classified as a mesoporous material. As reported by Mariana et al. [10], the mesoporous material has a diameter ranging of 20-500 Å.

3.4. Fourier Transform Infrared (FTIR) Analysis

With the aim to observe the chemical functional groups contained in coffee shell sorbent, the analysis was carried out by means of FTIR instrument. The IR spectra result is provided in Fig. 2.

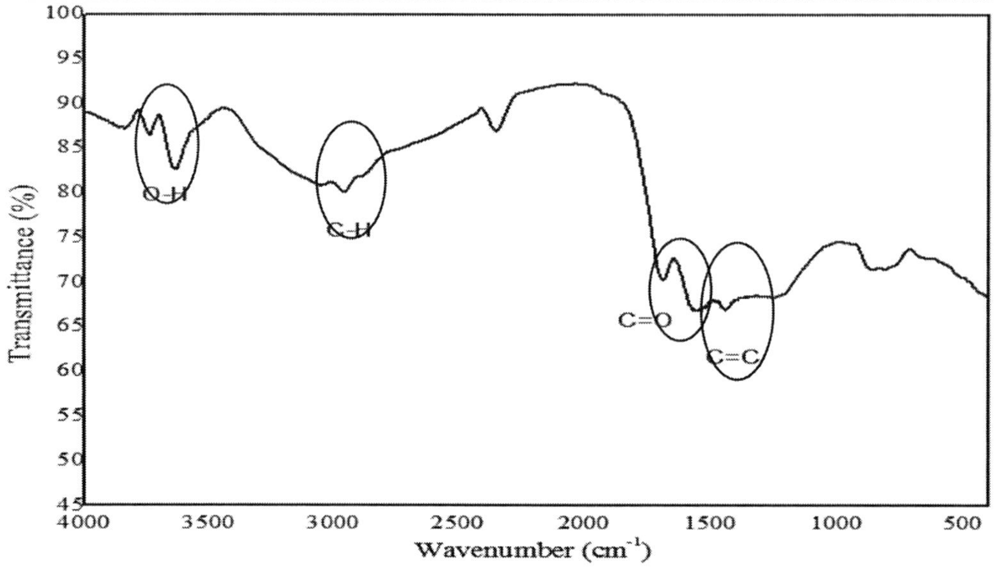

Figure 2. IR Spectra of activated carbon sorbent from the coffee shell

From Fig. 2, it can be observed that there are several peaks that distinguish the spectra of the prepared sorbent. The peaks are representing the presence of –OH group, C-H, C = C and C = O which mostly contributed from the organic components contained in the coffee shell. Detailed information on spectra reading is given in Table 4.

Table 4. IR spectra of activated carbon sorbent from the coffee shell

Functional group	Remarks	*Wavenumbers range* (cm^{-1}) Reference [12]	*Wavenumbers range* (cm^{-1}) This study
H_2O	O–H	4000-3400	3620 and 3743
CH_4	C–H	3000-2700	2964
Aromatic	C=C	1690-1450	1649-1452
Aldehyde, Ketone, Carboxylic acid	C=O	1900-1650	1693

3.5 X-Ray Diffraction (XRD) and X-Ray Fluorescence (XRF) Analysis

The X-Ray Diffraction (XRD) analysis aims to determine the crystalline structure of the coffee shell sorbent which is characterized by peaks at an angle of 2θ. The results of the XRD analysis can be seen in Fig. 3.

Fig. 3 shows the results of XRD pattern of coffee shell sorbent. According to Khalil et al. [13], activated carbon showed an irregular microcrystalline structure in which microcrystals of graphite were randomly oriented. Their result is consistent with the finding in this study, as seen in the sorbent which carbonized at 350°C for 120 minutes, carbon (C) is detected in the 2θ area of 26.28° (d = 3.3887 Å) and 44.27° (d = 2.0443 Å). The small but sharp peak in the XRD pattern indicates an amorphous structure of the sorbent [14].

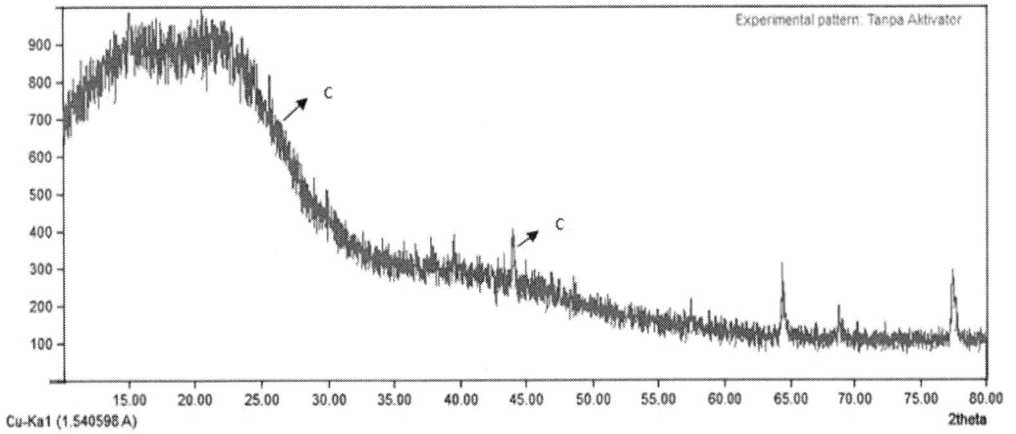

Figure 3. XRD pattern of coffee shell sorbent

Table 5. XRF Analysis Data

Komponen	% Weight
SiO_2	47,32
Al_2O_3	10,46
Fe_2O_3	8,87
MgO	8,84
SO_3	0,16
K_2O	0,45
CaO	11,54
Na_2O	0,07
P_2O_5	0,161
TiO_2	0,377
Mn_2O_3	0,163
CrO_3	0,078

3.6 Effect of Contact Time on Pb (II) Metal Absorption Capacity

Contact time is one of the factors affecting the adsorption process. Determination of the optimum contact time for adsorption aims to determine the time needed by the coffee shell sorbents to absorb the maximum lead (Pb) ion. The relationship between contact time and the adsorption efficiency of Pb (II) metal is shown in Fig. 4.

Based on Fig. 4, it can be seen that the absorption capacity increases with increasing contact time between the adsorbate and sorbent. By prolonging the contact time, the absorbed Pb (II) metal increase until the equilibrium occurs. Based on the results of the experiment, it is found that the absorption capacity for the initial concentration of 302.22 mg/L with contact times of 30, 60, 90, 120 and 150 minutes are 77.29; 105.49; 170.26; 174.44 and 176.97 mg/g, respectively. At the beginning of the process, the absorption rate is very large, especially in the first 30 minutes until minute 90, and from minute 90th until minute 150th (te) in which the absorption equilibrium occurs. Based on this, it can be concluded that the optimum contact time of activated carbon in Pb metal ion absorption is achieved at 90 minutes. Longer contact time resulted in more adsorbed Pb due to the increasing chance of activated carbon to intersect and contact with the Pb particles and the particles adhered to the surface of the sorbent [15].

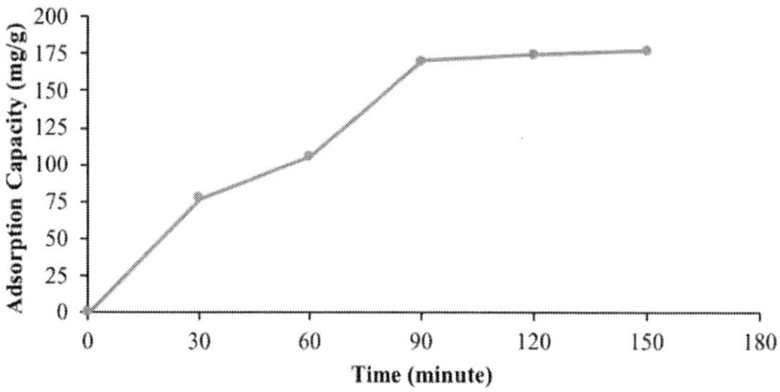

Figure 4. Effect of contact time on adsorption capacity of Pb (II) metal at the initial concentration of adsorbate 302,22 ppm

4. Conclusion

Preparation of activated carbon sorbent from Gayo Coffee Shell has been conducted. Characterization results show that the water content, ash content, and absorption of iodine have met the quality standards of activated carbon based on SNI No. 06-3730-1995. Result of SEM imaging confirms that the sorbent has a structure with a lot of pores formed. BET characterization result indicates that the coffee shell sorbent is classified as mesopore. The FTIR characterization results show that the –OH, C-H, C = C, and C = O functional groups dominate the chemical composition of the activated carbon sorbent from the coffee shell. The XRD characterization reveals a slight sharp peak which is the indication of the amorphous structure. The equilibrium time was achieved at 90 minutes of operation time with an absorption capacity of 170.26 mg/g. From the overall results, it can be concluded that the coffee shell waste has a high potential to be utilized as an adsorbent material.

References

[1] Vareda J P, Valente A J M and Durães L 2019 *J. Environ. Manage.* **246** 101-18
[2] Nour H E, El-Sorogy A S, Abd El-Wahab M, Nouh E S, Mohamaden M and Al-Kahtany K 2019 *Mar. Pollut. Bull.* **144** 167-72
[3] Kasemodel M C, Sakamoto I K, Varesche M B A and Rodrigues V G S 2019 *Sci. Total Environ.* **675** 367-79
[4] Argyraki A, Kelepertzis E, Botsou F, Paraskevopoulou V, Katsikis I and Trigoni M 2018 *J. Geochem. Explor.* **187** 201-13
[5] Ucun H, Bayhana Y K, Kaya Y, Cakici A and Algur O F 2003 *Desalination* **154** 233-8
[6] Sophia A C and Lima E C 2018 *Ecotox. Environ. Safe.* **150** 1-17
[7] Rodriguez-Narvaez O M, Peralta-Hernandez J M, Goonetilleke A and Bandala E R 2017 *Chem. Eng. J.* **323** 361-80
[8] Al-Senani G M and Al-Fawzan F F 2018 *Egypt. J. Aquat. Res.* **44** 187-94
[9] Pyrzynska K 2019 *J. Environ. Chem. Eng.* **7** 102795
[10] Mariana, Marwan, Mulana F, Yunardi, Ismail T A and Hafdiansyah M F 2018 *IOP Conf. Series: Mater. Sci. Eng.* **334** 012029

[11] Pope J P 1999 Activated Carbon and Some Application for Remediation of Soil and Ground Water Pollution. In: *Civil Engineering,* (Virginia: Virginia Tech. USA)
[12] Ma Z, Chen D, Gu J, Bao B and Zhang Q 2015 *Energy Convers. Manag.* **89** 251-9
[13] Khalil H, Jawaid M, Firoozian P, Rashid U, Islam A, and Akil H M 2013 *J. Biobased Mater. Bioenergy* **7** 708-14
[14] Das D, Samal D P, and Meikap P 2015 *J. Chem. Eng. Process Technol.* **6** 1-7
[15] Zikra N R Y, Chairul C and Yenti S 2016 *Jurnal Online Mahasiswa Fakultas Teknik Universitas Riau* **3** 1-8

IOP Conf. Series: Materials Science and Engineering **796** (2020) 012051 doi:10.1088/1757-899X/796/1/012051

Performance of tannic acid/tetraethylenepentamine coated polymeric membrane for the separation of different surfactant-stabilized oil in water emulsions

F Senusi[1,2], S Ismail[1*]
[1] School of Chemical Engineering, Universiti Sains Malaysia, Engineering Campus, Nibong Tebal 14300 Pulau Pinang, Malaysia
[2] Faculty of Chemical Engineering, Universiti Teknologi MARA, Cawangan Pulau Pinang, Permatang Pauh, 13500 Pulau Pinang, Malaysia

E-mail: chsuzy@usm.my

Abstract. The present research reported on the surface coating of PVDF membrane with polyphenolic-amine, which improves the hydrophilicity and underwater oleophobic properties. The separation efficiency and antifouling property of emulsion oil by pristine PVDF and polyphenolic-amine coated PVDF microfiltration membranes were evaluated. The performance of pristine and coated membrane was compared using three different types of surfactant-stabilized oil droplets (T80/Diesel/H_2O, SDS/Diesel/H_2O and CTAB/Diesel/H_2O) in terms of flux recovery ratio (FRR) and removal efficiency. The continuous cycle using cross-flow filtration mode was also investigated. The filtration experiments indicated that modifying the surface properties of the PVDF polymer membrane by polyphenolic-amine coating method can significantly improve the antifouling properties compared to unmodified polymeric membranes.

1. Introduction

Oily wastewater consists of different forms of oils, including free-floating oils, unstable dispersed oils and stable emulsified oils. The dispersed types tend to coalesce into the free floating oils which make both of these types easily separated using conventional methods (gravity, skimming and flotation) [1]. In contrast, emulsified oils with an aqueous phase are rather stable due to the presence of surfactants. The stable emulsified oils which have tiny droplet size (< 10 μm) are effectively treated by membrane filtration methods due to energy efficiency with higher separation quality [2].

Generally, on the membrane market, polymeric membranes such as polyvinylidene fluoride, polysulfone and polyacrylonitrile are most dominant materials used as the commercial membrane that are mainly for the water and wastewater treatment [3]. The focus attention to the improvements of membrane surfaces because of the fouling issues due to its hydrophobic properties resulting in poor flux and rejections as well as cleaning problems [4]. Moreover, the separation of the emulsion was influenced by surfactants depends on the surface charges and types [5]. For examples, Wu et al. reported the effect of membrane surface charges towards cationic and anionic surfactants emulsion [6].

Among the surface modification strategies, surface coating is the most convenient and facile method to be applied on a large scale [7,8]. To address these challenges towards the scalable

[*] To whom any correspondence should be addressed

Content from this work may be used under the terms of the Creative Commons Attribution 3.0 licence. Any further distribution of this work must maintain attribution to the author(s) and the title of the work, journal citation and DOI.
Published under licence by IOP Publishing Ltd

modification techniques of commercial polymeric membrane for oil/water separation, this study report on the improvements of membrane surfaces with hydrophilic and underwater oleophobic properties with low oil adhesion. Inspired by mussel surface chemistry [9], the alternative of polydopamine [10] which showed high similarities in structure and have adhering efficiency with a lower cost are required [11].

In this study, the tannic acid, plant polyphenols with abundant of phenolic hydroxyl groups, will be used. It also tends to be oxidized under alkaline conditions and react with amine groups to form stable hydrophilic coated membrane surfaces. Then, the comparison of fouling ability of pristine and coated membranes was analysed using different types of surfactant-stabilized emulsions. Furthermore, the influences of different surfactant types onto the coated membrane were evaluated.

2. Experimental

2.1. Materials
Tannic acid (TA) and tetraethylenepentamine (TEPA) were supplied by Merck (Darmstadt, Germany). Tris(hydroxymethyl)aminomethane, Tween 80 (T80) and sodium dodecyl sulfate (SDS) were purchased from Friendemann Schmidt Chemical and cetyltrimethyl ammonium bromide (CTAB) supplied by Nacalai Tesque. PVDF microfiltration (flat sheet, 0.22 μm pore size) were purchased from Merck Millipore, which was prewetted in 25% isopropanol for 24 h and soaked overnight with deionized water (DI) prior to the modification process.

2.2. Preparation of Surfactant-Stabilized Emulsion Oil Solution
A diesel oil (Shell Malaysia) emulsion was used as the foulant in crossflow filtration studies. Three types of surfactant-stabilized diesel in water emulsions with the concentration of 1500 ppm (9:1 ratio of oil: surfactant) were prepared including T80/diesel, SDS/diesel and CTAB/diesel in 3L of deionized water. The solutions were emulsified using homogenizer (Ultra Turrax T-50, IKA) operated at 6000 rpm for 30 – 45 min.

2.3. Modification of Membrane
TA (0.2 g) and TEPA (1.2 mL) were dissolved in 50 mM Tris-HCl solution (pH 9, 100 mL). Then, the prewetted pristine PVDF membrane (M1) was stirred into the solution using the open container for 6 h at 30°C. Then, the coated membranes were named as M2 after this procedure was rinsed by isopropanol for 30 min and soaked overnight in deionized water to remove the redundant solution. Finally, M2 was dried in an air-drying oven for 3h at 45°C.

2.4. Characterization
The M2 membranes were characterized by field-emission scanning electron microscope (FESEM, Quanta FEG 650) to ascertain structural morphology and element mapping. The droplet size distribution and zeta potential of each type of emulsion were measured using the Zetasizer Nano Series (Nano-ZS).

2.5. Filtration Studies
Cross-flow filtration mode was used to evaluate the separation performance of the membrane. Prior to the filtration process, the pristine membrane was wetted using 25% IPA, while DI water used for the coated membrane to prevent the surfactant wetting effect during the filtration cycles. Then, the prewetted membrane was compacted at 2 bar for 20 min prior to the filtration process. The pure water flux (J_0) of deionized water was measured at TMP of 1 bar for 60 min, then the feed was replaced by emulsion solution at the same TMP for 60 min. The permeate mass (m) was recorded by a computer-recorded electronic balance (EK3000i, A&D Company) at an interval of 1 min (t) which was used to calculate permeate flux (J_P) according to following formulae:

$$J_P = \frac{\Delta m}{\rho_w A \Delta t} \tag{1}$$

where ρ_w is the density of water, and A is the effective filtration area. Then, the membrane was cleaned using the flushing method of 0.005 M HCl solution for 5 min followed by 10 min of deionized water. Finally, the pure water flux (J_C) were measured again for 60 min. The concentration of permeate flux was then measured using UV-Vis spectrophotometer (DR 6000, Hach). The flux recovery ratio (FRR) was calculated using the following equation:

$$FRR = \frac{J_C}{J_0} \times 100 \tag{2}$$

The continuous cycle of emulsion separation was also evaluated for 30 min and the same cleaning steps was done for each interval. The normalized flux for each cycle (n) was calculated as follows:

$$\text{Normalized flux} = \frac{J_n}{J_0} \tag{3}$$

3. Results and Discussion

3.1. Characterization

The properties of the emulsion solution were evaluated using Zetasizer to determine the charges and mean diameter of droplet sizes for each type. The results were presented in Table 1. The emulsion stabilized by CTAB emulsion was positively charged due to the ammonium groups, while the T80 and SDS were negatively charges. These charges represent the types of surfactant used which CTAB as cationic type, while SDS as anionic surfactant. While, the slightly negative charge of T80 surfactant was attributed to the ionization of polar oil [12]. The droplet size of the emulsion types is in the range of 0.10 - 10 µm with approximately similar mean diameters. The charge of coating solution was -3.96±0.5 mV as illustrated in Figure 1 represented the negative charge of coated surface was dominant by OH groups of catechol units on polyphenolic components [13].

Table 1. Properties of emulsion solution

Emulsion Types	Surfactant Types	Zeta potential (mV)	Mean diameter (µm)
$T80/Diesel/H_20$	Nonionic	- 25.5±0.9	3.67±0.02
$SDS/Diesel/H_20$	Anionic	- 79.8±1.2	3.52±0.23
$CTAB/Diesel/H_20$	Cationic	69.2±3.5	3.74±0.29

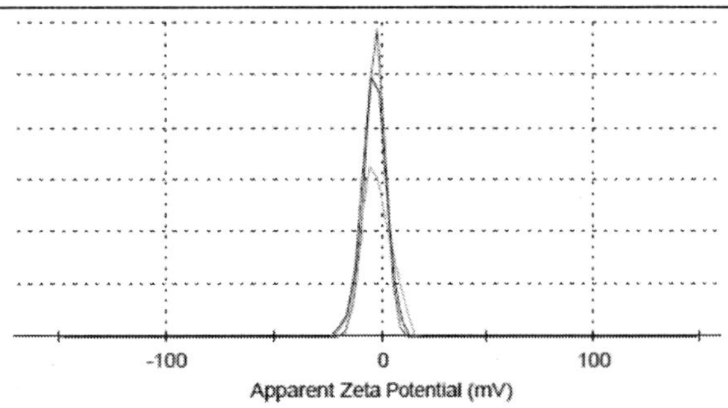

Figure 1. Zeta potential of coating solution (triplicate data)

The surface morphologies of pristine (M1) and coated (M2) membranes are shown in Figure 2 and Figure 3. In comparison on the cross-sectional morphology of both membranes as shown in Figure 2b and 2c, the coated membrane (M2) has denser and compact layer throughout the pore walls. It can be observed that some deposition on the wall of the pores resulting in the increase of roughness on the M2 membrane surface as displays in Figure 3. The EDX elemental mapping on the top surface of M2 indicating that the presences of O and N elements distributed well which were ascribed to the TA/TEPA coating solution. The functional groups between the pristine and coated membrane using FTIR and XPS analysis already reported details in our previous studies [14].

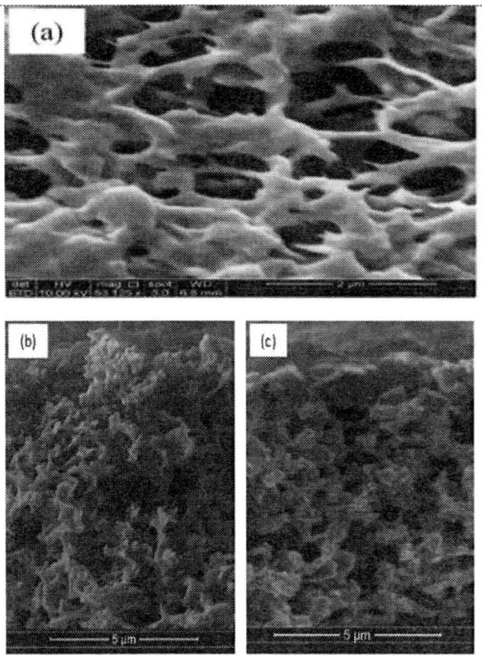

Figure 2. Surface morphology of (a) top surface; (b) cross-section for pristine membrane (M1) and (c) cross-section for coated membrane (M2)

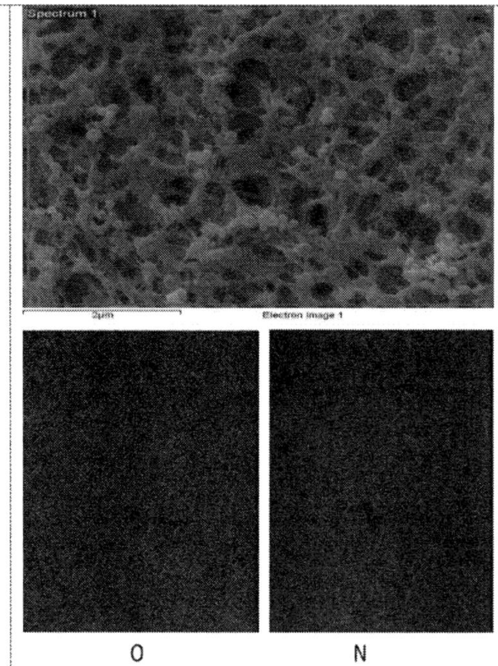

Figure 3. Surface morphology and element mapping for additive of O and N elements on the coated membrane (M2) surface

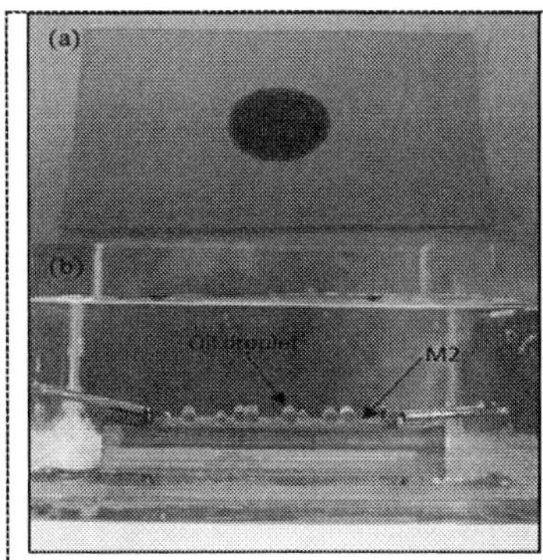

Figure 4. Photograph of water permeation (a) and underwater oil droplets (b) on top of coated membrane (M2) surface

IOP Conf. Series: Materials Science and Engineering **796** (2020) 012051 doi:10.1088/1757-899X/796/1/012051

The effect of hydrophilicity and underwater oleophobicity is shown in Figure 4 by dropping the water and oil droplet (1,2-dichloroethane as model oil), respectively on top of the M2 surfaces. It shows that the coated membrane improved the surface hydrophilicity due to the appearance of the hydroxyl and amine group and successfully coated on the membrane surfaces [15]. Also, the roughness surface with micro/nanostructure created by the coating solution helps the trapped water molecules prevent the permeation of oil onto the membrane surface [16].

3.2. Comparison on Antifouling Properties of Membranes

The performances studies of nonionic (T80), anionic (SDS) and cationic (CTAB) surfactant-stabilized emulsion oil for pristine (M1) and coated (M2) membranes are shown in Figure 5. The enhancement of pure water fluxes of M2 membranes could be observed due to the hydrophilic surface, which makes water easily permeate through this membrane. The water affinity mainly attributed to the hydration effect between hydroxyl groups on coated membrane and water [17].

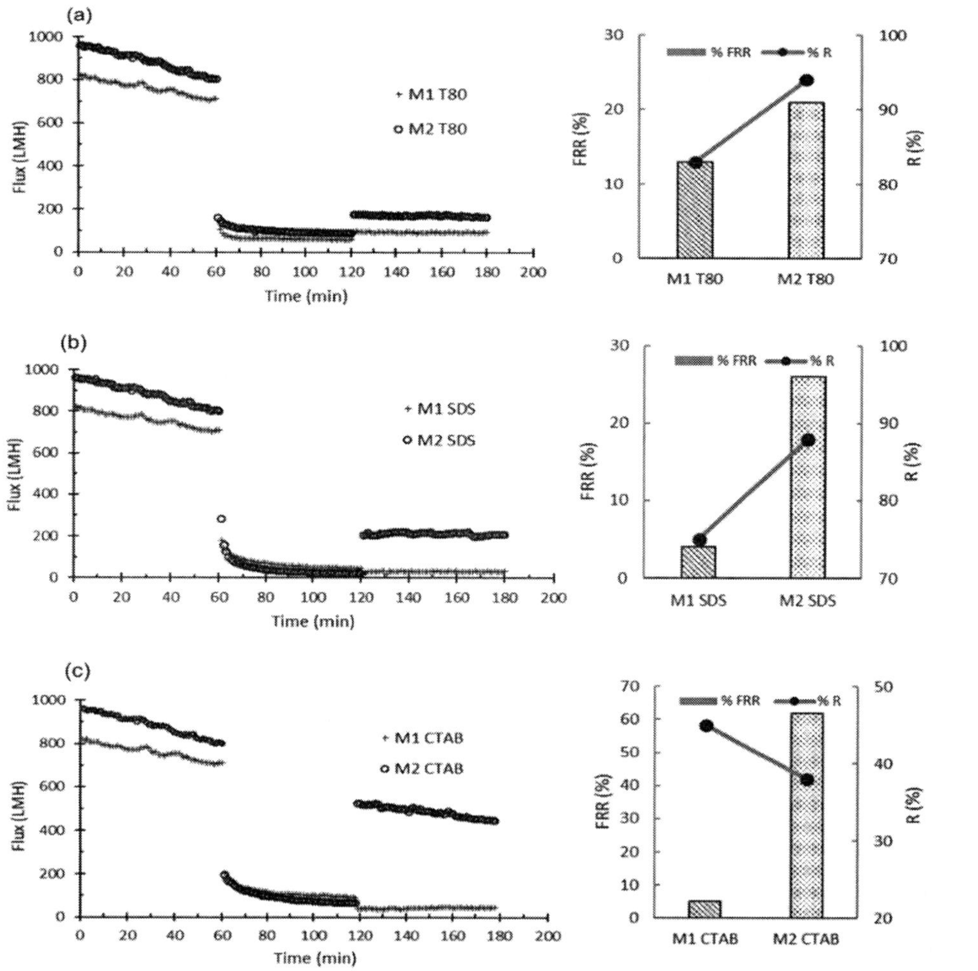

Figure 5. Performances of antifouling studies on different types of surfactant-stabilized emulsion (a) T80/Diesel/H2O (b) SDS/Diesel/H2O (c) CTAB/Diesel/H2O for pristine (M1) and coated (M2) membranes

The antifouling performances of the M1 and M2 membranes were evaluated using time-dependent fluxes. During emulsion filtration separation, the flux sharply reduced for both membranes due to oil deposition and cake formation. After the cleaning process, it could clearly be seen that the flux recovery for the M1 membrane is lower for all types of foulant due to the hydrophobic properties of membranes. Furthermore, for the M2 membranes, the flux recovery ratio (FRR) were significantly improved for all types of emulsion solutions. Among the surfactant types, the CTAB cationic gave higher FRR as 62% followed by anionic and nonionic emulsion types.

However, the removal efficiency for cationic emulsion slightly lower due to the electrostatic interaction between the coated membrane with negative charges and CTAB with opposite charges. The separation efficiencies of T80 nonionic and SDS ionic surfactants were increased using M2 membranes. The coated membrane surface is hydrophilic and rough, which traps water molecules and decreases the contact area between the oil droplet and membrane surface. The improvement of separation efficiency and flux recovery after being subjected to the cleaning process indicates that coated membrane demonstrated low oil-adhesion behavior compared to the pristine membrane. Therefore, the M2 coated membrane exhibits the anti-fouling ability by hydrophilic and underwater oleophobicity properties [18].

3.3. Comparison on Different Separation of Surfactant-Stabilized Emulsion

The separation efficiency and reusability towards oil droplets with different surfactants were further evaluated with a continuous cycle separation by using the same cross-flow system. The normalized flux ratios for each cycle are collected and presented in Figure 6. For the CTAB cationic emulsion, the normalized fluxes after each cycle are 90% higher than SDS and T80 emulsions. It may due to the surfactant barrier effect that promotes the demulsification of oil droplets. Initially, the oil droplets tend to coalesce into larger droplets and stuck on the pore mouth and walls, which could prevent the penetration of tiny oil droplets into membrane pores. Therefore, it can be deduced that electrostatic repulsion reduced the fouling effects [6]. However, the static attraction effects between the cationic foulant and coated surface, resulting in the lower rejection (63%) and higher permeate flux values (~168 LMH) than negatively charge emulsions.

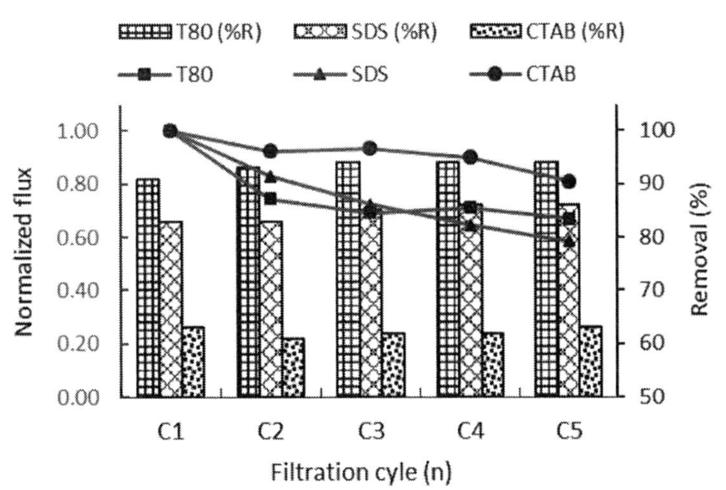

Figure 6. Normalized flux and removal efficiency for five cycle of emulsion filtration process

Besides, for the negatively charged emulsion, the normalized recovery ratio slightly lower for both types. It may due to the low demulsification effects, resulting in the penetration of tiny droplets into the membrane pores which could lead to serious irreversible fouling. The permeation fluxes of anionic and nonionic surfactants are slightly lower about 129 and 82 LMH, respectively. However, the rejection of T80 emulsion is 94% and 85% for SDS emulsion. It may due to the stronger electrostatic repulsion with the negatively charged membrane surface, which caused less surfactant adsorption on top of the membrane surface [19]. After 5-cycles, the coated M2 membrane still displayed the stable removal efficiency with slightly decrease the permeation fluxes.

4. Conclusions

In summary, the hydrophobic PVDF polymer membrane was successfully modified using polyphenolic-amine coating method. The hydrophilic and underwater oleophobic properties of coated membrane surface exhibited better antifouling ability compared to the pristine membrane for nonionic, anionic and cationic surfactant-stabilized emulsion solutions. In addition, the surfactants play the roles towards the demulsification of emulsions depends on the surface charge of the membrane. It was concluded that this coated membrane shows a potential application in oil/water separation.

Acknowledgements

The authors thank the Ministry of Education (MOE) Malaysia for the financial support through FRGS Grant (203/PJKIMIA/6071413). Faraziehan Senusi gratefully acknowledged the Ministry of Education and Universiti Teknologi MARA, Cawangan Pulau Pinang, Malaysia for the financial support throughout her study leave.

References

[1] Chen C, Weng D, Mahmood A, Chen S and Wang J 2019 *ACS Appl. Mater. Interfaces* **11** 11006–27
[2] Munirasu S, Haija M A and Banat F 2016 *Process Saf. Environ. Prot.* **100** 183–202
[3] Wan Ikhsan S N, Yusof N, Aziz F, Misdan N 2017 *Malaysian J. Anal. Sci.* **21** 643–58
[4] Robinson S, Abdullah S Z, Bérubé P and Le-Clech P 2016 *J. Memb. Sci.* **503** 177–87
[5] Chen L, Si Y, Zhu H, Jiang T and Guo Z 2016 *J. Memb. Sci.* **520** 760–8
[6] Wu J, Wei W, Li S, Zhong Q, Liu F, Zheng J and Wang J 2018 *J. Memb. Sci.* **563** 126–33
[7] Liu Q, Li J, Zhou Z, Xie J and Lee J Y 2016 *Sci. Rep.* **6** 1–10
[8] Yang H C, Chen Y F, Ye C, Jin Y N, Li H and Xu Z K 2015 *Chem. Commun.* **51** 12779–82
[9] Lee H, Dellatore S M, Miller W M and Messersmith P B 2007 *Science.* **318** 426–30
[10] Syawaliah, Mulyati S, Muzaitun and Mulyasari R 2018 *IOP Conf. Ser. Mater. Sci. Eng.* **352**
[11] Yang H C, Luo J, Lv Y, Shen P and Xu Z K 2015 *J. Memb. Sci.* **483** 42–59
[12] Liu Y, Su Y, Cao J, Guan J, Zhang R and He M 2017 *J. Memb. Sci.* **542** 254–63
[13] Liebscher J, Mrówczyński R, Scheidt H A, Filip C, Hădade N D, Turcu R, Bende A and Beck S 2013 *Langmuir* **29** 10539–48
[14] Senusi F, Shahadat M and Ismail S 2019 *Int. J. Environ. Sci. Technol.*
[15] Ding L, Wang Y, Zhu P and Bai Y 2019 *Appl. Surf. Sci.* **479** 423–9
[16] Zarghami S, Mohammadi T and Sadrzadeh M 2019 *J. Memb. Sci.* **582** 402–13
[17] Sui Y, Wang Z, Gao X and Gao C 2012 *J. Memb. Sci.* **413–414** 38–47
[18] Zhang G, Deng Y, Zhou X, Shen S, Wyman I and Bai R 2018 *J. Memb. Sci.* **569** 60–70
[19] Lu D, Zhang T and Ma J 2015 *Environ. Sci. Technol.* **49** 4235–44

IOP Conf. Series: Materials Science and Engineering **796** (2020) 012052 doi:10.1088/1757-899X/796/1/012052

Pharmaceutical Emerging Pollutant Removal by Thin Activated Carbon Coating: Equilibrium, Kinetic, and Thermodynamic Studies

Muhamad Sharafee Shamsudin[1], Suzylawati Ismail[1*]
[1]School of Chemical Engineering, Engineering Campus, Universiti Sains Malaysia, 14300, Nibong Tebal, Pulau Pinang, Malaysia.

E-mail: chsuzy@usm.my

Abstract. In Malaysia, issues of pharmaceutical pollutant have been highlighted recently. Acetaminophen (analgesic) also known as paracetamol that usually used to reduce moderate pain such as headaches, menstrual periods, cold/flu aches and fever. There are a large amount of acetaminophen has been detected in sewage treatment plants in Malaysia. Unfortunately, sewage treatment plants are not effective enough to overcome this kind of pollutant. Therefore, in this research study, a new approach applying adsorption concept is introduced. Thin activated carbon coating (TACC) for adsorption of acetaminophen (ACT) was investigated. The TACC is formulated using Epoxidized Neutral Rubber (ENR-50) and poly(vinyl) chloride (PVC) as binders with activated carbon as an adsorbent, then was coated on white cotton fabric via brushing technique. Characterization analysis using SEM-EDX and BET analysis were performed. The pore volume and surface area of the TACC are 0.07 cm^3/g and 64.3 m^2/g, respectively. The TACC was evaluated through varies parameters including different initial concentrations and temperatures. The amount of ACT as adsorbate, 50 mg/L, able to be adsorbed up to 33.3 mg/g into TACC within 5 h. The result of equilibrium and kinetic studies indicated that Freundlich isotherm model and pseudo-second-order, respectively, are best fitted the adsorption of ACT onto TACC. It was decisive that the adsorption process of ACT on TACC is spontenoous ($\Delta G^0 < 0$) and exothermic ($\Delta H^0 < 0$) through thermodynamic studies.

1. Introduction

Pharmaceutical emerging pollutant considered as rare pollutants has been detected nowadays. This kind of pollutant has attracted much attention among the researchers because of growing

[*] To whom any correspondence should be addressed

Content from this work may be used under the terms of the Creative Commons Attribution 3.0 licence. Any further distribution of this work must maintain attribution to the author(s) and the title of the work, journal citation and DOI.
Published under licence by IOP Publishing Ltd

pharmaceutical manufacturing industries and human behavior. In the United States, it was reported detection frequency exceeding 30 % of the water stream from 47 groundwater sites selected [1]. Then, wastewater sample has been collected from five different types of sewage treatment plants in Johor, Malaysia [2]. There are six pharmaceutical compounds (Acetaminophen, Lincomycin, Trimethoprim, Sulfamethazine, Sulfamethoxazole, Clarithromycin, Carbamazepine, Ibuprofen, and Naproxen) are the most frequently consumed by resident living nearby of these sewage treatment plant. Acetaminophen was recorded the highest concentration has been detected able to reach 40.17 mg/L. Consequent of that all of the pharmaceutical compounds are potentially harmful to the aquatic organisms and human life in the future. These substances enter the environment largely from wastewater treatment, aquaculture treatment and leaking landfills [3][4][5]. Other exposure pathways exist, including emission from manufacturing industries, disposal of unused medicine to landfill and flushed down the toilet, irrigation with wastewater from hospitals, veterinary medicine from the hard surface in farmyards and disposal of carcasses of treated animal [6][7][8]. Other than that, this pollutant able to enter the environment is when a person who takes medicine, up to 9 percent of medicine excreted out of the body through urination, then it flushed down the toilet and mixed with wastewater [9].

In Malaysia, pharmaceutical waste from hospital, clinic and any medical store has been regulated under government on disposal of expired, unused and defected medicine through Environmental Quality (Scheduled Waste) Regulation 2005. However, this regulation does not apply to residencies, aquaculture activities and others, where there has an irresponsible citizen that disposed the medicine improper ways to landfills and also mixed together with their daily food waste, then disposed in flushing bowls. Through that pathway, it showed how easy this pollutant may enter the environment. The pollutant concentration stated by a group of researchers in Johor, Malaysia, frequently detected [2]. Moreover, these compounds with long term exposure may lead to mutagenic and genotoxic effects on aquatic life. As a consequence, the compounds drastically decrease the reproductive success in aquatic life. The most critical effect arises from the exposure to these compounds is their ability to cause the liver and reproductive damage in human health, accumulation tissues and inhibition of cell proliferation [10].

The disposal of medicine through flushing in toilet bowls, therefore it directly been through sewage wastewater treatment plant. Unfortunately, the treatment plant did not have any specific treatment for pharmaceutical waste. So, the water will not be treated accordingly before discharged and distributed back to the residents. According to previous research, there are several conventional sewage treatment processes to facilitate in removal of these compounds, such as coagulation, sedimentation, filtration but it inefficient because only able to remove about 10 – 20 % of these compounds [11]. Nowadays, domestic sewage treatment is using biological treatment systems. In Malaysia, commonly included extended aeration, oxidation ditch, oxidation pond, sequencing batch bioreactor. The biological treatment system is a reliable treatment system especially in removing high organic and nutrient content from sewage [12]. However, it is still not completely eliminate the pollutants before distributed to residents [2].

Presently, this research is aimed to develop a novel adsorption technology that will reduce the amount of pharmaceutical pollutant exist in the environment. The existing adsorption process requires several types of equipment to operate such column, pump, valve and need energy consumption to run the system. In this research study, the technology is relatively simple to be applied in the wastewater treatment system. It can be added onto the existing process without affecting their current operation. Therefore, the performance of TACC being investigated in the application of acetaminophen removal from the aqueous solution.

2. Experimental

2.1 Materials

Acetaminophen (ACT) was purchased from BT Science Sdn. Bhd. The chemical structure and physical properties of ACT are shown in Table 1. Epoxidized Neutral Rubber with 50 % mol of epoxidation (ENR-50) was provided by School of Chemistry, Universiti Sains Malaysia. Toluene and dichloromethane were supplied by BT Science Sdn. Bhd. Poly(vinyl) chloride, commercial untreated activated carbon and kaolinite were obtained from Sigma-Aldrich Sdn. Bhd.

Table 1. Properties and chemical structure of acetaminophen [13] [14].

Chemical structure	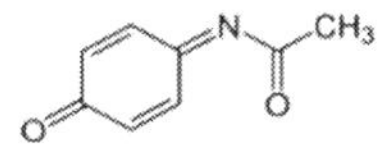
Category	Analgesics
IUPAC name	N – (4-hydroxyphenyl) acetamide
Molecular weight	151.165 g/mol
Odor	Odorless
Color	Large monoclinic prisms from water
Taste	Slightly bitter
Melting point	168 °C
Solubility	> 22.7 µg/mL

2.2 Preparation of adsorbate

The stock solution of adsorbate (ACT) was prepared before the experiment started by weighing 0.4 g of ACT using analytical balance (A&D Company, Limited, model HR-250A) and completely dissolved into 2 L of distilled water. The stock solution was then diluted to several studied concentrations. The dilution is calculated using equation (1). The detection of adsorbate via UV-vis spectroscopy, the calibration curve of ACT have been created with wavelength 287 nm respectively through varies concentration (1, 10, 20, 30, 40, 50 and 100 mg/L).

$$M_1V_1 = M_2V_2 \quad (1)$$

Where:

M_1 = Concentration of stock solution (200 mg/L)
V_1 = Volume of stock solution need to be diluted (L)
M_2 = Concentration required (mg/L)
V_2 = Volume need to be tested (L)

2.3 Preparation of Thin Activate Carbon Coating (TACC)

Epoxidized Neutral Rubber (ENR-50) solid was dissolved using the reflux process, where 24.80 g of solid ENR-50 was mixed with 250 mL of toluene. Then, the mixture was left for reflux under the temperature of 85 – 90 °C for 80 hours [15]. 0.80 g of poly(vinyl) chloride (PVC) in 35 ml of dichloromethane was dissolved via sonication for 30 minutes. Then, 4 g of ENR-50 solution was mixed with 65 mL toluene and the mixture was shaken for 1 minute. Next, 4 g of activated carbon was added and the mixture was sonicated for 6 hours [16]. After being sonicated, the mixture was divided into 6 mL for each coating. The solution was applied onto a cotton fabric (dimension 210 mm x 50 mm) via brushing technique, and dried in the oven at 100 °C for 5 minutes.

2.4 Batch adsorption studies

The acetaminophen solution was prepared from a stock solution (200 mg/L). The adsorption performance was evaluated through various initial concentrations (5 to 100 mg/L) and temperature (30, 40, 50, 60 and 70°C). The TACC strip was weighed prior to the testing. Then, the strip was placed to the interior wall of 250 mL beaker and adsorption process start. The sample solution was collected for every 30 min interval and analyzed it. The experiment was continued for 5h. The adsorption isotherm study was performed applying the recorded experimental results.

The equilibrium adsorption values were calculated using the equations below. The amount of ACT adsorbed, q_t (mg/g) on the coated strip at a particular temperature was determined using Eq. (2):

$$q_t = \frac{(C_o - C_t)V}{w} \quad (2)$$

where C_0 and C_t (mg/L) are initial concentration of ACT and concentration at the time t, respectively. V is the volume (in L) is the volume of ACT solution that has been tested and W is the mass (in gram) of adsorbent used for each coating.

The equilibrium adsorption capacity, (q_e) was determined according to the following equation, Eq. (3):

$$q_e = \frac{(C_o - C_e)V}{w} \quad (3)$$

2.5 Adsorption isotherm studies

The nature of adsorbent coating toward the adsorption of ACT was examined through Langmuir and Freundlich isotherm models.

2.5.1. Langmuir adsorption isotherm

The nature of ACT adsorption (homogeneous monolayer) on the surface of the adsorbent coating was examined using the Langmuir isotherm [17]. The Langmuir adsorption isotherm is shown as:

$$\frac{1}{q_e} = \frac{1}{q_m} + \frac{1}{k_L q_m C_e} \quad (4)$$

The separation factor, (R_L), is used to evaluate if the adsorption if unfavorable ($R_L > 1$), favorable ($0 < R_L < 1$), linear ($R_L = 1$) or irreversible ($R_L = 0$) [18], R_L can be calculated as:

$$R_L = \frac{1}{1 + k_L C_O} \quad (5)$$

2.5.2 Freundlich adsorption isotherm

Heterogeneous and multilayer surface adsorption can be described using the Freundlich model [19]. The equations of Freundlich isotherm are:

$$q_e = k_F (C_e)^{1/n} \quad (6)$$

$$\ln q_e = \ln k_F + \left(\frac{1}{n}\right) \ln C_e \quad (7)$$

Where:

Table 2. Parameters and definition of isotherm models.

Parameter	Definition
C_o	Initial concentration (mg/L)
C_e	Concentration at equilibrium state (mg/L)
	Langmuir model
q_m	Maximum adsorption capacity (mg/g)
k_L	Langmuir isotherm constant (L/mg)
	Freundlich model
k_F	Freundlich isotherm constant (mg/g. (L/mg)$^{1/n}$)
n	Adsorption intensity

2.6 Error analysis

The optimization procedure requires an error function to be defined in order to evaluate the best fit of the isotherm to experimental equilibrium data. Error function was calculated for all isotherm models and those with the lowest value of error analysis was selected as the best model to describe the experimental data. In this study, linear coefficient of determination and non-linear the Chi-square test (χ^2), root mean square error (RMSE) and sum of error squares (SSE) were used.

Chi-square test (χ^2) [20]value was evaluated using the Eq. (8):

$$\chi^2 = \sum_{i=1}^{n} \frac{(q_{e,cal} - q_{e,exp})^2}{q_{e,exp}} \quad (8)$$

Residual root mean square error (RMSE) [21] was determined using the Eq. (9):

$$RMSE = \sqrt{\frac{1}{n-2} \sum_{i=1}^{n} (q_{e,exp} - q_{e,cal})^2} \quad (9)$$

Sum of error squares (SSE) [22] analysis was examined using the Eq. (10):

$$SSE = \sqrt{\frac{\sum_{i=1}^{n} (q_{e,exp} - q_{e,cal})^2}{n}} \quad (10)$$

Where $q_{e,exp}$ (mg/g) and $q_{e,\ cal}$ (mg/g) are the adsorption capacity from experimental data and calculated from the model, respectively. Then, n is the number of data used.

2.7 Kinetic studies

In 1898, Lagergren's kinetics equation was the first one in describing the adsorption of a liquid-solid system based on solid capacity. The first order rate equation is also known as pseudo-first-order (Ho, 2004). The equation is presented as Eq. (11):

$$\log(q_e - q_t) = \log q_e - \frac{K}{2.303} t \quad (11)$$

Where q_e, equilibrium adsorption capacity (mg/g), q_t is the adsorption capacity (mg/g) at time t (min) and K is the constant of pseudo-first-order rate (1/min). The constants of pseudo-first-order can be defined by plotting log ($q_e - q_t$) against t.

In 1999, Ho and McKay proposed a pseudo-second-order equation that explained the amount of adsorbate captured on the adsorbent [24]. The pseudo-second-order is a chemisorption equation and the nonlinear equation was expressed by Eq. (12):

$$\frac{t}{q_t} = \frac{1}{K_2 q_e^2} + \frac{t}{q_e} \quad (12)$$

Where q_t and q_e is the amount of adsorption capacity (mg/g) at time t (min) and at equilibrium, respectively. Then, K_2 is the rate constant of the pseudo-second-order equation (g/mg.min). The constant of pseudo-second-order kinetic can be defined by plotting t/q_t against t.

2.8 Thermodynamic studies

Thermodynamic parameter such as standard enthalpy ΔH^o, entropy change ΔS^o and Gibbs free energy change ΔG^o are studied to evaluate the feasibility of the adsorption process. The experiments are evaluated based on different temperature (303, 313, 323, 333 and 343 K). The thermodynamic value was calculated with the following equations [25]:

$$\Delta G^{o} = -RT\ln K_d \tag{13}$$

$$\Delta G^{o} = \Delta H^{o} - T\Delta S^{o} \tag{14}$$

$$\ln K_d = \frac{-\Delta H^{o}}{R}\frac{1}{T} + \frac{\Delta S^{o}}{R} \tag{15}$$

$$K_d = \frac{q_e}{C_e} \tag{16}$$

Where, R is the universal gas constant (8.314 J/mol K), T is the absolute temperature in Kelvin (K) and K_d is equilibrium constant. The linear graph is plotted ln K_d against 1/T. ΔH° and ΔS° are determined from the slope and intercept of the graph respectively. Then, ΔG° is calculated using Eq. 14 with varies temperatures (303, 313, 323, 333 and 343 K)

2.9 Apparatus and instrumentation

The apparatus used were 250 mL beaker, magnetic stirrer and cuvette (Quartz material). A magnetic stirrer (Multi Hotplate Stirrer WISD Laboratory Instrument SMHS-6) was used for mixing the slurry and adsorbate homogeneously. The UV-vis, spectrophotometer (Shimadzu UV-1800) was employed to measure the concentration of ACT. The coating solution was dissolved by sonication process (model Elmasonic S80H). Drying Oven (Constance Germany) was used to dry the TACC. The surface morphology and content of the elements in TACC was evaluated by using Scanning Electron Microscope – Energy Dispersive X-ray spectroscopy (SEM-EDX) spectroscopy (model Crest System (M) Sdn. Bhd. Quanta Feg 450). For BET analysis, a piece of TACC (dimension 10 mm x 30 mm) was tested. The surface area, total pore volume and pore size of the TACC and commercial AC were determined in nitrogen adsorption isotherm at 77 K using a BET surface analyzer (Micromeritics model ASAP 2000, Nocross, GA).

3. Results and Discussion

3.1. *Characterization*

The Scanning Electron Microscopy analysis captured the images of surface morphology of TACC for before and after adsorption as illustrated in Figure 1. The activated carbon fully coated on white cotton fabric. The adsorbent with varies size was strongly attached through binder (ENR-50 and PVC). There is a huge amount of small particle of activated carbon, therefore, the surface area for adsorption can be increased. No obvious sign of ACT attached to the TACC, but it has been proven through energy dispersive x-ray analysis. The main elements representing the compound of ENR-50, PVC and activated carbon which is carbon, oxygen, and chloride detected in high amounts which weigh percentage 72.13%, 15.46% and 10.27%, respectively, as illustrated in Figure 2(a) and 2(b). The nitrogen (2.14%) has been detected which indicated the element of ACT compound [13] as shown in Figure 2(b). Therefore, it is proven there is an interaction between the ACT and the surface of TACC. The result of BET analysis demonstrated that the surface area of commercial AC decreased drastically when it introduced in TACC (948.1 m^2/g to 64.3 m^2/g). A similar pattern also observed in pore volume analysis, where the decreasing values from 0.94 cm^3/g to 0.07 cm^3/g. However, there is some increment in pore size diameter from 39.6 Å to 45.1 Å. According to the research by Bahrudin & Nawi (2017), similar trends of decreasing surface area and pore volume whilst the pore size diameter is increasing when ENR-50 and PVC were introduced as a coating [16].

(a) (b)

Figure 1. SEM images of TACC (a)before (Mag. x500) and (b) after adsorption process (Mag. x500).

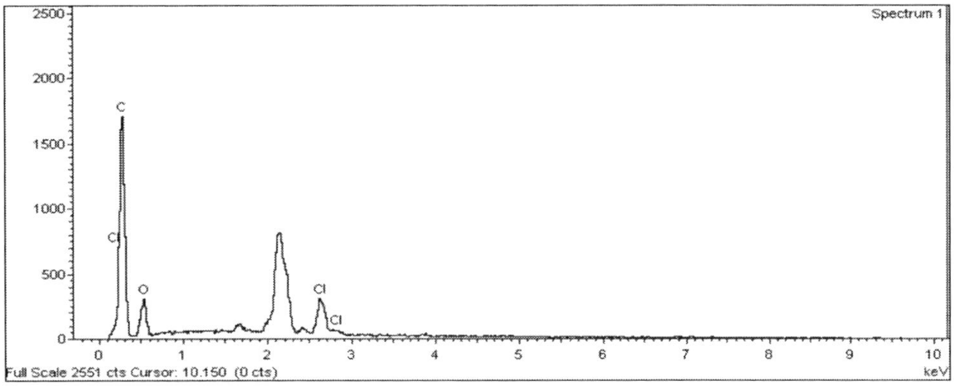

(a)

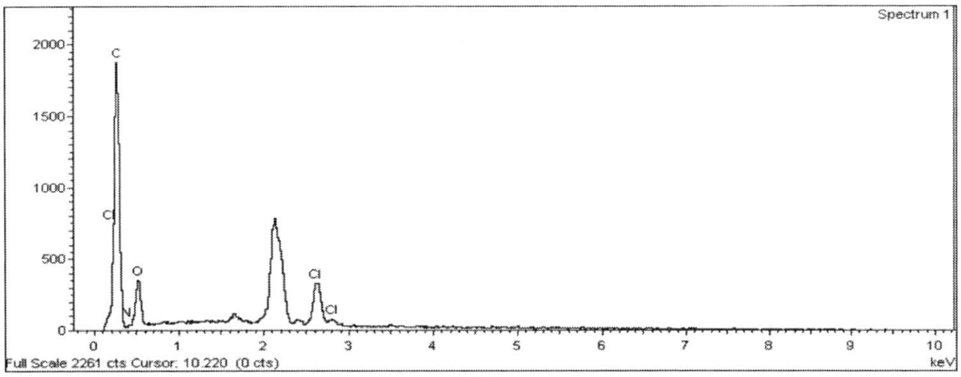

(b)

Figure 2. Energy Dispersive X-ray (EDX) analysis (a) before and (b) after adsorption.

3.2. Adsorption isotherm studies

Table 3 summarising the adsorption isotherm models for Langmuir and Freundlich for the TACC at 30°C. The data were obtained from the experiments, it showed that Freundlich model was the best fitted which determined through correlation coefficient (R^2), where the values for Freundlich model were higher than Langmuir model, 0.9183 and 0.9073 respectively. However, this result was supported by error analysis as showed in Table 4. There was a comparison of R^2 value on non-linear regression with other error function. As the evaluation for Freundlich isotherm is showing the coefficient of determination, R^2 was slightly higher as compared to Langmuir isotherm, then error function obviously the smallest value. The Freundlich model is associated with a heterogeneous and multilayer surface adsorption, which would relate to more selective adsorption [26]. Therefore, for the adsorption process of ACT onto TACC, Freundlich isotherm is the best-fitted model.

Table 3. Summary of parameters from linear adsorption isotherms

Langmuir Isotherm		Freundlich Isotherm	
R^2	0.9073	R^2	0.9183
q_m (mg/g)	103.092	1/n	0.5486
k_L (L/mg)	0.01102	n	1.8228
R_L	0.6447	k_F (mg/g. (L/mg)$^{1/n}$)	3.8126

Table 4. Comparison of the non-linear regression

Langmuir Isotherm		Freundlich Isotherm	
SSE	4.6673	SSE	3.4266
χ^2	3.0808	χ^2	1.9390
RMSE	6.0254	RMSE	4.4237
R^2	0.9808	R^2	0.9835

3.3. Adsorption kinetic studies

The adsorption kinetics of ACT onto TACC were evaluated through pseudo-first-order and pseudo-second-order models on the data that was obtained from the experimental studies. The results were tabulated in Table 5. From the data, it shows that pseudo-second-order has higher correlation coefficient, (R^2 = 0.9950) compared to pseudo-first-order (R^2 = 0.4053). Moreover, the adsorption capacity calculated ($q_{e,\ cal}$) value agree with the experimental data ($q_{e,exp}$). The result approved that adsorption is well described by the pseudo-second-order kinetic model. It means the chemisorption involves in the interaction between ACT and activated carbon [27].

3.4. Thermodynamic studies

Through the effect of temperature adsorption thermodynamic can be determined by varies temperature (303, 313, 323, 333 and 343 K). The Gibb's energy (ΔG^o), enthalpy (ΔH^o) and entropy (ΔS^o) were the parameters for thermodynamic studies were determined through equation (13) to (16). The results were showed (ΔS^o = 137.65 kJ/mol.K) and (ΔH^o = 24759.92 J/mol). The values of ΔG^o (-16947, -18323.5, -19699.9, -21076.4 and -22452.9 kJ/mol) were more towards negativity as temperature rises which were 303, 313, 323, 333 and 343 K respectively. According to the negative values of ΔG^o, ΔS^o and ΔH^o, it is proven that the process is spontaneous and exothermic. The results are comparable with other researher [26].

Table 5. Pseudo-first-order, pseudo-second-order constants and R^2 values for adsorption of ACT on TACC at 30°C

Initial dye concentration (mg/L)	$q_{e\ exp}$ (mg/g)	Pseudo-first-order model			Pseudo-second-order model		
		$q_{e\ cal}$ (mg/g)	K (1/min)	R^2	$q_{e\ cal}$ (mg/g)	K_2 (1/min)	R^2
25	15.1356	11.3370	0.0101	0.7092	17.3910	0.0575	0.9693
50	30.3959	7.9287	0.0106	0.4053	32.5733	0.0017	0.9950
75	35.4314	19.0327	0.0106	0.8698	38.7597	0.0009	0.9896
100	36.6151	18.6251	0.0117	0.7452	39.3701	0.0011	0.9930

4. Conclusion

As a conclusion, it was proven that TACC is adsorbing ACT with 33.3 mg/g of adsorption capacity. The equilibrium studies, adsorption kinetic and thermodynamic studies were examined using vary initial concentrations and temperatures. As a result, it showed that the Freundlich model is best fitted for this adsorption of ACT onto TACC through correlation coefficient ($R^2 = 0.9183$) with supported by smallest error analysis. The Freundlich model is associated with heterogeneous and multilayer surface formation. Moreover, kinetic studies showed pseudo-second-order is well presented the adsorption process. Thermodynamic studies were found that Gibbs energy, ΔG^0 is more negativity as all temperature rises from 303 to 343 K, while both ΔH^0 and ΔS^0 were also negative value. This reflects that process is spontaneous and exothermic. The innovative approach in this research study, by introducing the TACC as a medium for the adsorption process is relevant for research nowadays.

Acknowledgment

The authors would like to acknowledge the Bridging Grant Universiti Sains Malaysia (304/PJKIMIA/6316498) for financial support for this research work.

Reference

[1] Barnes K K, Kolpin D W, Furlong E T, Zaugg S D, Meyer M T and Barber L B 2008 **2**
[2] Yacob H, Ling Y E, Hee-young K, Zainura O, Noor Z, Fadhil M, Din M, Mat S and Hun L T 2017 **173** 165–73
[3] Price O R, Hughes G O, Roche N L and Mason P J 2010 **6** 677–84
[4] Agency E M 2016 **44** 1–77
[5] Agency E M 2018 **44** 1–48
[6] Boxall A B A, Rudd M A, Brooks B W, Caldwell D J, Choi K, Hickmann S, Innes E, Ostapyk K, Staveley J P, Verslycke T, Ankley G T and Beazley K F 2012 **120** 1221–9
[7] Arsson D G J O L 2009 P **28** 2522–7
[8] States U 2018 How PPCPs Enter the Environment 1–6
[9] Division E H 2014 Acetaminophen in Drinking Water
[10] Patneedi C B and Prasadu K D 2015 *Rasayan J.* **8** 2008–11
[11] Lishman L, Smyth S A, Sarafin K, Kleywegt S, Toito J, Peart T, Lee B, Servos M, Beland M and Seto P 2006 *Sci. Total Environ.* **367** 544–58
[12] Verlicchi P, Aukidy M Al and Zambello E 2012 *Sci. Total Environ.* **429** 123–55
[13] PubChem 2018 Acetaminophen 1–116
[14] States U, Medicines E, States U, Paracetamol P, Patent F P, Other A, Pregnancy O, Mechanism C, Chemical P C, Ahfs S R and License M M 2018 Paracetamol **01**
[15] Nawi M A and Zain S M 2012 *Appl. Surf. Sci.* **258** 6148–57
[16] Bahrudin N N and Nawi M A 2017 *React. Kinet. Mech. Catal.* **124** 153–69

[17] Carabineiro S A C, Thavorn-amornsri T, Pereira M F R, Serp P and Figueiredo J L 2012 *Catal. Today* **186** 29–34
[18] Tonucci M C, Gurgel L V A and Aquino S F de 2015 *Ind. Crops Prod.* **74** 111–21
[19] Freundlich Herbert 1907 Über die Adsorption in Lösungen *Zeitschrift für Phys. Chemie* **57U** 385
[20] Khamanur T, Tg A, Zamri M, Sakinah M, Munaim A and Abdul Z 2017 *Int. Res. J. Eng. Technol.* **4** 2–6
[21] Ghaffari H R, Pasalari H, Tajvar A, Dindarloo K, BakGoudarzi B, Alipour V and Ghanbarneajd A 2017 *Int. J. Eng. Sci.* **6** 1–11
[22] Senusi F and Abd. Karim K 2011 *Colloq. Humanit. Sci. Eng. CHUSER 2011* 37–42
[23] Ho Y.S. 2004 Citation Review of Lagergren Kinetic Rate Equation on Adsorption Reactions *Scientometrics* **59** 171–7
[24] Ho Y S and McKay G 1999 Pseudo-second order model for sorption processes *Process Biochem.* **34** 451–65
[25] Mohd N, Fadhil M, Ery A, Fauziah S and Draman S 2015 Isotherm and thermodynamic study of paracetamol removal in aqueous solution by activated carbon **10** 9516–20
[26] Rey-mafull C A, Tacoronte J E, Garcia R, Tobella J, Llópiz J C and Iglesias A 2014 Comparative study of the adsorption of acetaminophen on activated carbons in simulated gastric fluid 1–12
[27] Ferreira R C, Lima H H C De, Cândido A A, Junior O M C, Arroyo P A and Carvalho K Q De 2015 Adsorption of Paracetamol Using Activated Carbon of Dende and Babassu Coconut Mesocarp **9** 717–22

IOP Conf. Series: Materials Science and Engineering **796** (2020) 012053 doi:10.1088/1757-899X/796/1/012053

Effect of Sodium Dodecyl Sulfate (SDS) on Polysulfone/Zeolite membrane for the removal of copper ions

S A Hamid[1] and S Ismail[1*]
[1]School of Chemical Engineering, Universiti Sains Malaysia, Engineering Campus, Nibong Tebal, Pulau Pinang, 14300 Malaysia.

E-mail: chsuzy@usm.my

Abstract. The influence of anionic surfactant, Sodium Dodecyl Sulfate (SDS) on Polysulfone (PSf) polymer impregnated with zeolite adsorbent for removing copper ions from aqueous solution is the main focus of this paper. The flat sheet membrane was prepared via blending and phase inversion methods. Different amount of SDS (1 wt%, 3 wt%, 5 wt%, and 7 wt%) were added into the polymer casting solution. Batch adsorption and continuous filtration studies were carried out to study the effect of SDS dosage on PSf/zeolite membrane for the removal of copper ion. The membrane characterizations such as pure water flux (PWF), water contact angle (WCA) and mean pore size were investigated. From the study, it was found that the presence of a high amount of SDS could enhance adsorption efficiency (99 %) by shortening the equilibrium time with adsorption of 9.4 mg/g. Filtration study indicates that a lower amount of SDS (1 wt%) results in high rejection value (70 %) of copper metal ions from the solution. Besides, as the amount of SDS increase, the PWF increases from 0 L/m^{2}h to 100 L/m^{2}h, WCA decreases from 94.8° to 46.4° and the pore size of membrane increases from 0.082 μm to 2.530 μm.

1. Introduction

The demand for clean water is gradually increased as the world population increases rapidly. Water pollution caused by heavy metal has become the biggest challenge faced by the world as it is one of the most common toxic pollutants detected in water streams especially in industrial effluents (mining operations, paper, batteries, fertilizer and thermoplastics industries) [1][2]. Heavy metals are commonly defined as those with a specific density of more than 5 g/cm^3 such as copper, chromium, mercury, cadmium, lead, nickel etc [3]. Copper is a type of heavy metals that have gained attention due to its toxicity and risky to the surroundings and human health when they are excessively emitted to the environment without proper treatment. Copper is an essential micronutrient that is vital to the body in small amounts [4]. However, human beings may present symptoms of temporary stomach and intestinal disorders to kidney or liver damage in high quantities, depending of the time of exposure [5]. It is therefore necessary to remove copper metal ion from the effluent before it is released into the environment in order to improve the quality of the water.

Conventional treatment techniques for the removal of heavy metal ions from wastewater included chemical precipitation, ion-exchange, coagulation-flocculation, adsorption, membrane filtration, etc [6-9]. Among these processes, adsorption and membrane filtration are widely used for the

[*] To whom any correspondence should be addressed

Content from this work may be used under the terms of the Creative Commons Attribution 3.0 licence. Any further distribution of this work must maintain attribution to the author(s) and the title of the work, journal citation and DOI.
Published under licence by IOP Publishing Ltd

immobilization and separation of metal species. Currently, a major focus has been placed on adsorptive membrane where it has dual function of membrane filtration and adsorption and it is found to be effective to remove traces amount of pollutants such as heavy metals [10]. In recent years, there has been increasing interest in using zeolite as an adsorbent to remove toxic heavy metals from aqueous solutions [11]. Zeolite is a low cost form of clay mineral, abundant availability, high surface area, non-toxic nature, great adsorption and a strong potential for ion exchange [12]. In this work, zeolite adsorbent is stuck on the surface of the Polysulfone (PSf) polymer membrane. PSf is a type of polymer that broadly utilized in membrane fabrication because of its high strength and rigidity [13].

Sodium Dodecyl Sulfate (SDS) is a type of anionic surfactant that can be found in pharmaceutical applications and detergent formulations and usually used in soil remediation [14]. SDS can be used to modify the zeolite surface to enhance the anion exchange capacity thus potentially remove cations, anions and organic compounds [15]. This paper aims to investigate the effects of different SDS amount on PSf/zeolite membrane in terms of the performance of copper removal by batch adsorption and continuous filtration experiment. In current work, PSf /zeolite membrane mix with SDS was prepared by the blending method and phase inversion technique. The pure water flux (PWF), water contact angle (WCA) and membrane pore size were determined.

2. Experimental

2.1. Materilas

All chemical and reagents used were in analytical grade and used as received. Polysulfone (PSf) resin pellets (MW = 35,000 g/mol) and zeolite powder were purchased from Sigma Aldrich. N-methyl-pyrrolidone (NMP) (MW=10,000 g/mol) (Merck) and Sodium docecyl sulfate (SDS) (Friendemann Schmidt) were used as the solvent and surfactant respectively. The copper (II) stock solution was prepared by using copper (II) chloride ($CuCl_2.2H_2O$) (R&M Chemicals). All solutions were prepared using distilled water.

2.2. Membrane preparation

The composition of the casting solutions for the prepared membranes is shown in Table 1. A solution was first produced by dissolving SDS in NMP solvent for 1 h under agitation at 300 rpm. It was followed by adding zeolite into the solution. The solution was then sonicated for 1 h to produce homogeneous casting solution with well-dispersed zeolite particles. Subsequently, PSf polymer was added slowly into the mixture which was stirred at 500 rpm and 70 °C for overnight. The resulting dope solution was once more sonicated and degassed for 1 h to get rid of air bubbles that exist in the solution. Afterward, the polymer solution was casted on a glass plate (250 μm thickness) using an automatic film applicator at room temperature. In order to allow the phase inversion phase to take place, the membrane was air-dried for 30 s and immersed in a water coagulation bath. Water was changed daily for 3 days to remove remaining NMP from the membrane. The detached membrane was finally dried at room temperature until it was used.

Table 1. Composition of the casting solutions for the fabrication of membranes.

Membrane	PSf (wt%)	Zeolite (wt%)	SDS (wt%)	NMP (wt%)
M0 (control)	15	-	-	85
M1	15	5	1	79
M2	15	5	3	77
M3	15	5	5	75
M4	15	5	7	73
M5	15	5	9	71

2.3. Batch adsorption experiment

Batch adsorption studies were performed to assess the behaviour of PSf/zeolite membrane in removing copper ions with and without the addition of SDS. The membrane was cut into small pieces and weighed (approximately 0.2 g). Then, it was added into a beaker containing 200 ml of copper solution with 10 mg/L initial concentration under stirring speed (500 rpm). The pH of the copper solution was set at 5.0 for preventing precipitation of Cu which occurred at pH greater than 6 [16]. The concentration of copper was measured for every time intervals of 1 h for the overall 7 hours experiment. Copper concentration was measured by atomic absorption spectroscopy (AAS) (Model: AA-6800 Shimadzu). Copper ion removal can be calculated using Equation (1)

$$Copper\ removal\ (\%) = \left(1 - \frac{C_f}{C_i}\right) \times 100 \quad (1)$$

where C_f and C_i are the final and initial concentration of copper solution (mg/L). In addition, the amount of copper adsorbed on the membrane (mg/g) was determined by using Equation (2):

$$q_e = \frac{(C_o - C_e)V}{M_m} \quad (2)$$

where q_e is the amount of copper adsorbed per membrane weight at equilibrium (mg/g), C_o and C_e are the initial (mg/L) and equilibrium concentrations (mg/L) of copper in the solution, V is the volume (L) of the copper solution and M_m is the weight (g) of the dry membrane used in the experiment.

2.4. Continuous filtration adsorption experiment

Continuous filtration adsorption experiment was performed using a cross-flow membrane system at 1 bar pressure with 2 L copper feed solution (10 mg/L). The experiment began by placing the membrane to the membrane cell with an effective surface area of 3.154 x $10^{-3}m^2$. The permeate sample was collected after 1 h and the concentration of Cu (II) was analysed by using AAS machine.

2.5. Membrane Characterizations

The pure water flux (PWF) of the membrane was indicated by a cross-flow membrane rig system. Firstly, the membrane sample (3.154 x 10^{-3} m^2) was pressurized at 2 bar for 30 minutes to reduce the compaction effects. The permeation experiment was performed at 1 bar for 1 h. The PWF value of membranes can be calculated using Equation (3):

$$J_w = \frac{Q}{A \times \Delta t} \quad (3)$$

where J_w is the pure water flux (L/m^2h), Q is the volume of collected permeate (L), A is effective membrane area (m^2) and Δt is sampling time (h).

The water contact angle (WCA) was analysed by a DropMeter A-100 contact angle system (Rame-Hart Model 300 Advanced Goniometer, USA). The samples were taped flatly onto a glass slide with a membrane surface facing upwards. The water was dropped on the surface of the membrane using microsyringe. For each membrane, ten points of different spots of the surface were measured to get the averaged water contact angle valued. The liquid-liquid displacement porometer (Porolux 1000, Germany) was used to investigate the pore size of the membrane. The membrane samples were cut and wetted with porefil liquid (surface tension of 16 dynes/cm) for an hour. Next, the nitrogen gas was pressurized on the membrane surface and the pore size of membrane was determined based on wet and dry of nitrogen curves.

IOP Conf. Series: Materials Science and Engineering **796** (2020) 012053 doi:10.1088/1757-899X/796/1/012053

3. Results and discussion

3.1. Batch adsorption experiment

The adsorption behaviour of copper ion on zeolite impregnated with PSf polymer at different SDS mass with respect to time is shown in Figure 1. From the results, no adsorption observed for M0. This is due to no zeolite adsorbent presence in the membrane. The presence of zeolite provides a significant role during adsorption owing to the existence of negatively charged groups, Si-O$^-$ and Al-O$^-$ on its surface that can attract the positive charge of copper metals [17]. It is also observed that all membranes except M0 achieve equilibrium and 99 % removal of copper after 7 hours adsorption period. The maximum adsorption capacity of copper on the zeolite membrane was 9.4 mg/g. At 1 hour experiment, M5 (9 wt% SDS) gives high removal efficiency of copper which is 92 % whereas M1 gives the lowest removal of copper which is 70 % without the addition of SDS surfactant. As can be seen in the results, once the amount of SDS increase, adsorption removal also increase. This is because SDS is an anionic surfactant that consists of a negative head that can also attract positive metal species [18]. Once it mixed with zeolite that has negative charge, strongest negatively charges of the membrane surface was formed. As a result, the modification of the zeolite membrane using SDS could enhance the adsorptive capacity for copper ions.

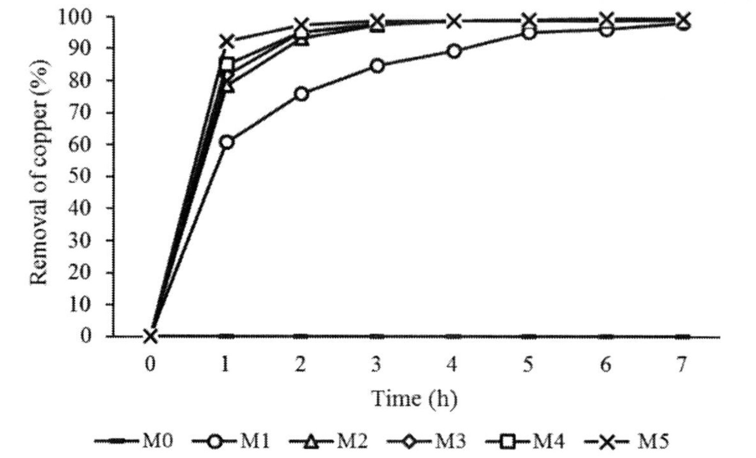

Figure 1. Copper ion removal (%) versus time (h) for the adsorption of copper using PSf/zeolite membranes blend with SDS

3.2. Continuous adsorption filtration experiment

The rejection of copper ions on the zeolite membrane was evaluated by carried out cross flow filtration experiment at low transmembrane pressure (TMP) of 1 bar as plotted in Figure 3. It shows that M2 with the lowest SDS amount (1 wt%) offers the highest percentage of rejection which is 70 % but the membrane with highest SDS (M5) amount (9 wt%) gives only 12 % rejection value. It is apparent that the rejection of copper ions reduced as the amount of SDS in the membrane increase. This is because of a large pore size of the membrane obtained. When the pressure was applied, many copper particles able to pass through the porous membrane then causes the decrement in rejection value of copper ions from the solution. In addition, M0 and M1 give zero rejection value of copper because of zero pure water flux (PWF) as shown in Figure 3.

3.3. Membrane characterizations

The results of pure water flux (PWF) and water contact angle (WCA) through PSf/zeolite membrane with SDS is illustrated in Figure 3. The increment of PWF of the membrane is following the order of M5 (100 L/m^{2}h) > M4 (68 L/m^{2}h) > M3 (40 L/m^{2}h) > M2 (19 L/m^{2}h) > M1 and M0 (0 L/m^{2}h). The water permeability increased when SDS surfactant was blended with PSf/zeolite casting solution. As the amount of SDS increase, the PWF also increases. The existence of SDS makes the membrane more porous as it creates larger macro voids on the structure thus improving the hydrophilicity of the membrane [19]. This is proven by the water contact angle (WCA) analysis shown in Figure 3.

WCA has been measured to estimate the membrane's surface wettability. Generally, the higher value of contact angle represents hydrophobic surface while the lower value represents hydrophilic surface characteristic [20]. The WCA values expected to be dependent on the SDS wt%. The decreasing trend of WCA is illustrated in Figure 3 after zeolite and SDS was added into PSf polymer. M5 membrane which contains the highest amount of SDS demonstrated the lowest WCA value of 46.4°. Apart from that, the PSf membrane, M0 showed the highest WCA value of 94.8° due to the hydrophobic nature of its material [21].

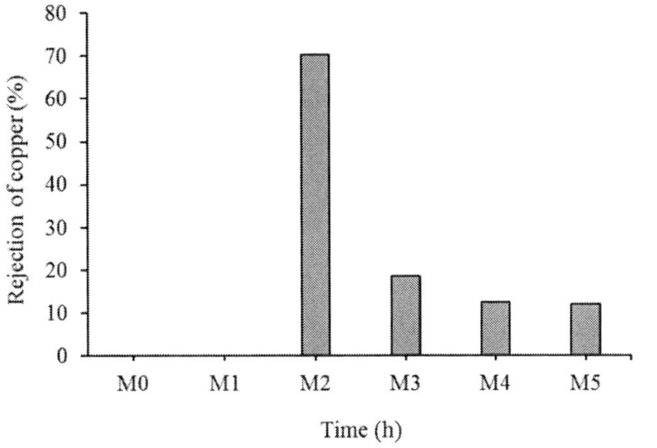

Figure 2. Rejection of copper on PSf/zeolite membrane blend with SDS

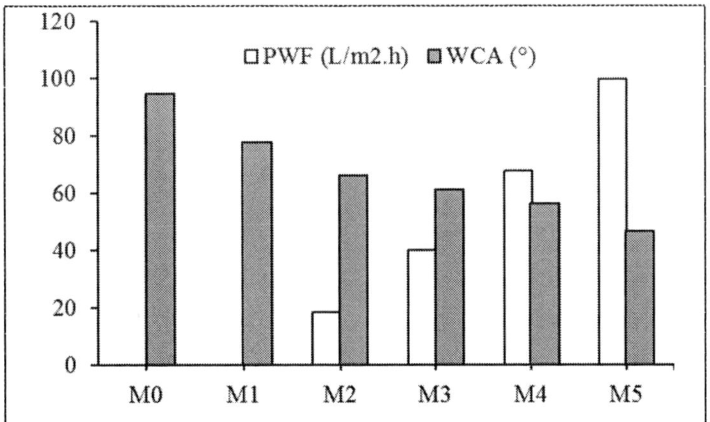

Figure 3. Pure water flux (PWF) and water contact angle (WCA) of PSf/zeolite membrane blend with SDS

The pore size of each membrane was measured to determine the mean pore size distribution on the membrane surfaces. As presented in Table 2, M0 membrane that consists of PSf polymer only possesses the lowest pore size which is 0.082 µm. However, the pore size increased to 0.244 µm when the zeolite was mixed into the polymer solution. Besides, the membrane pore size become larger as the amount of SDS increase. The membrane pore size is shifted from ultrafiltration to microfiltration type with the presence of zeolite and SDS in the mixture. Basically, in the structure of the surfactant comprises of long hydrophobic alkali and hydrophilic ionic group (polar group). As the surfactant enters a watery media, the ionic head tends to water whereas the non-polar tail tends to absorb on a solid surface, air or organic phase. When the alkali chains close to polymer matrix (hydro-carbonic structure), they will join them together meanwhile the hydrophilic and polar portion will remain free. The aggregation of the polar groups in polymer structure causes repulsion between the polymer chains thus forming an open structure [22]. As a result, the porosity of the membrane increases.

Table 2. Mean pore size of different types of membrane (µm).

Membrane	Mean pore size (µm)
M0	0.082
M1	0.244
M2	0.627
M3	1.354
M4	1.597
M5	2.530

4. Conclusions

The flat sheet adsorptive membrane with PSf and zeolite adsorbent with the addition of SDS surfactant was successfully prepared by phase inversion approach for removing copper metals from aqueous solution. The results showed that surfactant plays an important role in influencing membrane performances. This study clearly indicated that the presence of SDS surfactants in the membrane improved the adsorption efficiency of copper metal (99 % removal) by shortening the equilibrium time (1 hour). However, with a high amount of SDS, more copper particles will pass through the membrane resulting in lower rejection value of copper ions (12 %). The finding of this study shows that at a low amount of SDS (1%), high filtration efficiency (70 %) can be achieved. As the amount of SDS increases, the pure water flux (PWF) also increases and the membrane becomes more hydrophilic. Besides, the presence of SDS in the membrane matrix leads to a larger pore size of the membrane.

Acknowledgments

The authors would like to acknowledge the Ministry of Energy, Science, Technology, Environment and Climate Change, Malaysia (MESTECC) in providing the financial support through R & D Fund (305/PJKIMIA/6013394) for this research project.

References

[1] Srivastava N K and Majumder C B 2008 *J. Hazard. Mater.* **151** 1-8
[2] Pawar P R and Bhosale S M 2018 *Int. J.Current Eng. and Technol.* **8** 400-406
[3] Alissa E M and Ferns G A 2011 *J. Toxiol.* **2011** 1-20
[4] Rani K S, Srinivas B, Naidu K G and Ramesh K V 2018 *Mat. Today: Proceedings* **5** 463-469
[5] Azimi A, Rezakazemi M, Azari M and Ansarpour M *ChemBioEng Rev.* **4** 1-24
[6] Barakat M A 2011 *Arabian J. Chem.* **4** 361-377
[7] Gunatilake S K 2015 *J. Multidisciplinary Eng. Sci. Stud. (JMESS)* **1** 12-18

[8] Fu F and Wang Q 2011 *J. Environ. Manag.* **92** 407-418
[9] Shazili N A M, Yunus K, Ahmad A S, Abdullah N, Kamil M and Rashid A 2006 *Aquatic Ecosyst. Health and Manag.* **9** **(2)** 137-145
[10] Khulbe K C and Matsuura T 2018 *Appl. Water Sci.* **8** 1-30
[11] Zanin E, Scapinello J, Oliveira M, Rambo C L, Franscescon F, Freitas L, Mello J M M, Fiori M A, Oliveira J V and Dal J 2017 *Process Safe. Environ. Protec.* **105** 194-200
[12] Uddin M K 2017 *J. Chem. Eng.* **308** 438-462
[13] Moradihamedani P and Abdullah A H 2017 *Water Sci. and Technol.* **75** 2422-2432
[14] Jim M E and Medina D I 2017 *Water 2017* **9** **(235)** 1-12
[15] Alireza N E and Raja G 2013 *J. Chem.* **2013** 1-13
[16] El-Gendi A, Ali Abdalla H and Saied M 2016 *Membr. Water Treat.* **7** 55-70
[17] Ziyath A B, Mahbub P, Goonetilleke A, Adebajo M O, Kokot S and Oloyede A, 2011 *J. Water Res. and Prot.* **3** 758-767
[18] Adeli M and Yamini Y 2017 *Arabian J. Chem.* **10** 514-521
[19] Zainal S H, Hassan A R, Mohd Isa M H 2016 *Malaysian J. Analytical Sci.* **20** 1524- 1529
[20] Winandy J and Shupe T F 2010 *Wood and Fiber Sci.* **42** **(4)** 490-510
[21] Alam J and Arockiasamy L *Appl. Water Sci.* **2** 37-46
[22] Rahimpour A, Madaeni S S and Mansourpanah Y 2007 *J. Membr. Sci.* **296** 110-121

IOP Conf. Series: Materials Science and Engineering **796** (2020) 012054 doi:10.1088/1757-899X/796/1/012054

Stability and durability studies of zwitterionic adsorbent coating for the removal of organic pollutants: Chemical and thermal tolerance

S F Azha and S Ismail*
School of Chemical Engineering, Universiti Sains Malaysia,
Engineering Campus, Nibong Tebal, Penang,
14300 Malaysia.

*E-mail: chsuzy@usm.my

Abstract. The zwitterionic adsorbent coating (ZACC) was used to eliminate both cationic and anionic dyes. The ZACC was developed in a way to improve the current ordinary form of adsorbents (beads, powdery) for flexible working application especially in treating the industrial textile effluents. Since it is in a coating form, a number of factors have to be considered especially in terms of its stability and durability in wastewater. This is to prevent any detected adsorbent leaching and fast loss in treated water. This study was investigated for harsh conditions (chemical and thermal) to prove its strength. The chemical tolerance has established the stability of ZACC strip by showing a highly robust and potent structure and form even after 100 days in acidic, neutral and mix dyes solution. Result also showed the percentage removal of BG dye could attained within 87.4 - 97.1 % as increased the pH from 3 to 11. While higher removal of AR 1 obtained in acidic medium (99.4 %). The percentage weight loss obtained were 0.25 % and 0.11 % respectively for thermal stability test in drying oven and outside building.

1. Introduction

The zwitterionic adsorbent coating (ZACC) is a novel adsorbent used to adsorb both cationic and anionic dyes from aqueous solution. Our previous studies have confirmed the formulation of this adsorbent coating using acrylic polymer emulsion (APE) as binder, epichlorohydrin dimethyl amine (EPIDMA) as surfactant and bentonite as an additive for their functionalized materials involved [1–3]. The performance of ZACC was proven through a series of promising result in adsorbing ionic charges such as Acid Red 1 and Brilliant Green dyes. The adsorbent coating is a new approach of adsorbent where their ordinary form (powder, beads and flakes) has been reformulating and transform into a liquid/slurry form which then coated onto inert surfaces or substrate. After undergoing a drying procedure, a hardened layer of adsorbent was developed. The adsorbent coating modifies the classical adsorption concept, with simpler synthesis procedure and flexible working application afterward. This new innovative solutions aimed at simultaneous increase of performances in terms of dyes removal and decrease the energetic footprint by bringing to cost minimization in operation and time saving [4,5]. Besides, this adsorbent with substrate/support also possesses good mechanical stability that is highly required for different typologies of mechanical stresses including vibrations or collision during adsorption processes [6–8]. The flexibility of the adsorbent coating to be applied at any condition is an

* To whom any correspondence should be addressed

Content from this work may be used under the terms of the Creative Commons Attribution 3.0 licence. Any further distribution of this work must maintain attribution to the author(s) and the title of the work, journal citation and DOI.
Published under licence by IOP Publishing Ltd

advantage since it can be layered or slotted in a narrow, small or bigger spaces within a treatment plant. As long as the support used also flexible.

The sorbent with surface coating requires an optimized chemical and physical stability to strongly adhere to the targetted support materials. This is to prevent peel off of the adsorbent from the substrate during the adsorption process. Otherwise, the consequences will be not favorable if the adsorbent itself creating secondary pollution for the treated water. Harsh condition in effluent and treatment plant especially in terms of thermal situations (high temperature) and pH level were also being investigated. The chemical tolerance towards strong acid/alkali is desired to be deliberated throughout the synthesis and evaluation of the formulation of the adsorbent coating so that it can stay attached at any condition or up to a certain extent. Besides, the durability of adsorbent at various ranges of temperature, pH, in broad types of pollutants and reusability should also be encountered. Therefore, this study evaluated the stability and durability of ZACC in terms of its chemical and thermal tolerance. The experimental test was conducted in order to identify the strength of ZACC so that it can be applied without generating secondary pollution to the environment.

2. Experimental

2.1. Chemical and reagents

The cationic dye, Brilliant Green (BG) and anionic dye, Acid Red 1 (AR1) are analytical grade dyes supplied by Modern Lab Sdn. Bhd. The cationic polyelectrolyte, poly-epichlorohydrin-dimethyl-amine (EPIDMA) (50 wt. % in H_2O) and acrylic polymer emulsion (APE) were obtained from NHA Scientific Resource, Malaysia. The laboratory grade bentonite was purchased from Modern Lab Sdn. Bhd. The preparation method of ZACC has been synthesized as in our previous study [3].

2.2. Adsorbent coating stability studies

2.2.1. Determination of adsorbent coating durability in chemical solutions

The durability of ZACC in chemical solutions was tested under four main conditions; acidic, alkaline, neutral and mixed dyes. The samples (with the dimensions of 10 cm x 10 cm) were immersed in 50 mL solution of hydrochloric acid and sodium hydroxide for 10 days and prolong to 100 days. One mol per liter solution of each solvent was simulated as a harsh environment test under the acidic and alkaline conditions (acidic at pH 1 and alkaline at pH 14). Distilled water and mix dyes solution of BG and AR1 were also simulated for comparison study. The final result was examined qualitatively based on the physical appearance of the samples in terms of coating condition using photographic images and snap using SAMSUNG Tab 7, model SM-T211. The quantitative values obtained from the percentage weight loss of adsorbent before and after experiments were carried out.

2.2.2. Determination of adsorbent coating durability in various pH of effluent solution

The initial pH of Acid Red 1 and Brilliant Green dyes solution was ranging at pH 3, 5, 7, 9 and 11 and adjusted either by adding NaOH (0.1 M) or HCl (0.1 M) solutions. The initial pH was measured using pH meter (model Martini Instrument, Mi 151). The pH of 200 mL dye solution at 50 mg/L concentration was adjusted and pours into a 250 mL beaker containing adsorbent coating clipped in it. The setup was placed on a stirrer at 330 rpm and the temperature was maintained at room temperature. The agitation was carried out for 8 hours to ensure that the equilibrium was reached and dye concentrations were drawn out at 30 minutes time intervals. The concentration of dyes in solution was measured using a spectrophotometer. The experiments were carried out triplicate to assure accuracy and reproducibility and the error bar was shown in the graph.

2.2.3. Determination of adsorbent coating durability under thermal condition

The durability test for ZACC was carried out based on 3 situations and methods; tested on high temperature in the oven, water bath and real weather conditions outside the building. The adsorbent coating with the dimensions of 10 cm x 10 cm in sizes were placed as in condition, duration and temperature stated in Table 1. The final results were employed based on physical appearance and percentage weight loss. The images of adsorbent coating before and after testing was visualized in photographic images and snap using SAMSUNG Tab 7, model SM-T211).

Table 1. Durability test of the adsorbent coating under thermal condition

	Thermal condition	Duration	Temperature
1	Temperature in oven	7 days	100 °C
2	Temperature in water bath	4 hours	30 °C, 50 °C, 70 °C
3	Temperature surrounding (outside the building)	5 days	Ambient temperature (25 °C – 32 °C)

3. Results and discussion

3.1. Chemical stability test

The test for chemical stability of ZACC was evaluated by soaking a piece of the samples in different tests solution. They were acidic (1 M), alkaline (1 M), neutral (pH 7) and colored medium. The physical appearance of strips after 10 days (Figure 1) revealed a highly robust and the texture of ZACC remained intact as the original sample even after undergone different exposure conditions of strong acidic, alkaline, neutral, as well as colored solution. The supernatant liquid was crystal clear and no solid object could be seen in solution by naked eyes. This has established the stability of ZACC in chemical solutions. The stability testing of ZACC was prolonged to 100 days in each of the solutions. For this duration, all of the strips maintained their appearance in square shape except for ZACC in alkaline solution. The coating was about to desquamate from cotton cloth. Therefore, the stability of any adsorbents in a high concentration of hazardous chemicals is essential in a real environment, especially in terms of economic issues.

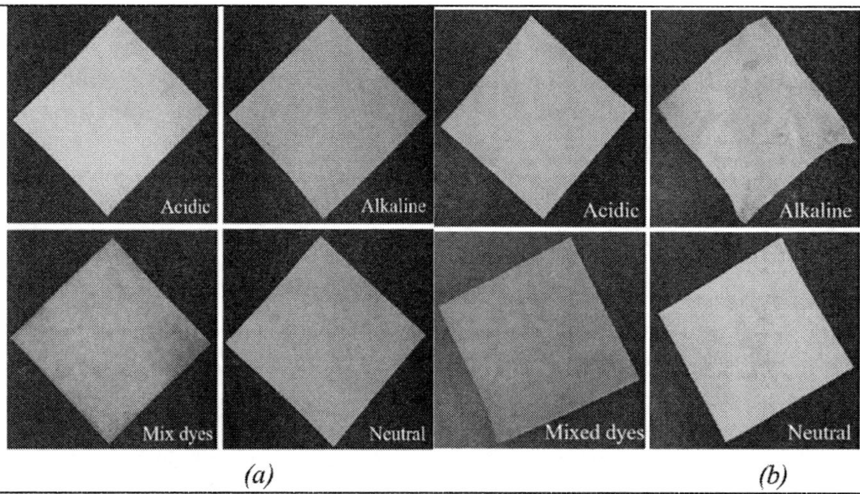

Figure 1. Photographic images of ZACC after 10 days (a) and 100 days (b) of soaking in acidic, alkaline, mixed dyes and distilled water solution

The chemical stability of ZACC was further investigated by pH study. The pH of dye solution can alter the surface charge of the adsorbent, which then can affect the dye sorption efficiency. Besides, another important factor such as the degree of ionization would also impact the solubility of dyes and the dissociation of surface functional groups on the active sites of the adsorbent. AR1 gives anions when dissolved in an aqueous medium while BG gives cations. The dye uptake capacity at different pH (pH 3.14 to pH 11.02) with constant initial concentration (50 mg/L) and 320 min contact time were depicted in Figure 2. Initially, the color of AR 1 and BG solution was changed with the variation of pH values. This was attributed to the change of dye molecular structural stability at the pH range and also due to the resonance among charged canonical structures of dyes [9,10].

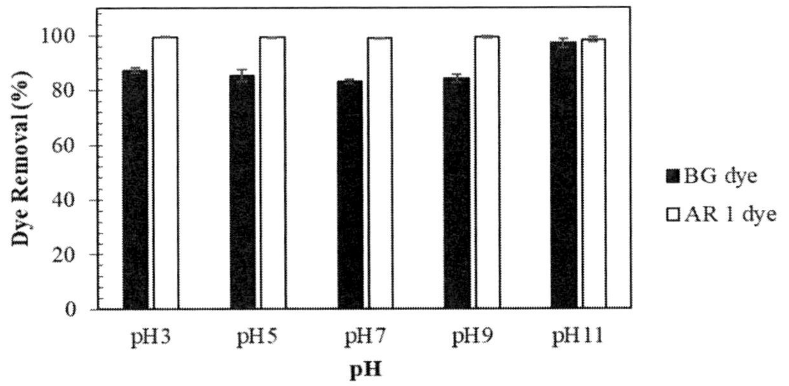

Figure 2. Effect of initial solution pH on adsorption of BG and AR 1 dyes using ZACC.[Constant condition: 50 mg/L of initial dye concentration, atmospheric temperature, adsorbent dosage 0.3 g]

In acidic medium, the adsorbent surface gets protonated due to availability of a large number of H^+ ions and attracted available anions from the medium. AR 1 exists as an anionic dye, where it contains charged sulfonated group $-SO_3Na^+$ in an aqueous medium. Therefore, the electrostatic attraction occurs between protonated ZACC surface and anionic dye, hence higher removal of AR 1 dye could be observed in acidic medium (99.4±0.1 %). Meanwhile, the percentage removal of AR 1 dye was decreased when the pH of AR 1 solution goes in alkaline range (97.4±0.1 %). This is due to the electrostatic repulsion occurred between the anionic dye and deprotonated of adsorbent surface. In addition, the availability of OH^- ions in the alkaline medium will also compete with anionic dye resulting in decreasing in AR 1 percentage removal. Nevertheless, still the adsorption rate of anionic dye shown excellent performance with a very slight variation within 97 % to 99 % in adsorption over a long pH range (3-11). Other possible reasons might be due to the presence of weak van der Waals forces between dyes and ZACC surfaces [11].

In the case of BG dye, dye removal increased once the solution pH increases. BG dye exists in the cationic form in aqueous solution. In acidic medium, electrostatic repulsion occurs between protonated surfaces and BG cations, which decreased adsorption of BG in acidic condition. Besides, there will be a strong competition between BG cations and H^+ ions for the adsorbent surface in acidic medium. While in alkaline medium, deprotonated surface favors cationic dye due to the electrostatic attraction between deprotonated surface of ZACC and BG cations. By increasing pH from 3.14 to 11.02, the percentage removal of BG dye was attained within 87.4 % and 97.1 %. The isoelectric point of ZACC was recorded at pH 8.2. The ZACC surface gets positively charged when the pH of the solution is <8.2 which will favor anionic species while at higher pH >8.2, negatively charged adsorbent surface will favor cationic species. Thus, BG removal was higher in alkaline medium and AR 1 removal was higher in acidic medium.

Theoretically, at low pH, adsorption coating showed lower adsorption for cationic dyes while higher adsorption efficiency was found for anionic dyes. It was due to the presence of a large number of H^+ and H_3O^+ in the aqueous solution which competes with the dye ions for available adsorption sites on the adsorbent surface. Thus, most of the functional groups present in the adsorbent are protonated (present in the form of positive charge). Therefore, electrostatic repulsion between dye molecules and positively charged adsorbent surface resulted in lower adsorption [12]. At higher pH, a number of positively charged decreases and the adsorbent surface appears negatively charged which resulted in higher cationic dye adsorption compared to anionic dye.

Furthermore in Table 2, even initially the pH were set accordingly for both dyes before the adsorption, the pH reading after adsorption were demonstrated a neutral pH (7< pH<8) for each of the solution, except for pH 11 where it was shown the alkalinity of solution (8.29 and 8.57 for AR 1 and BG dyes, respectively). The result provides information that ZACC surfaces have stabilized the AR 1 and BG dyes solution into a normal pH along with the adsorption treatment. Therefore, looking into the situation, by applying ZACC, the water can be neutralized from any acidic and alkaline condition.

As a conclusion, these findings suggested that the adsorption of cationic and anionic dyes on ZACC was not strongly and significantly pH-dependent. The variation in percentage efficiencies was very small and claimed as excellent performance for both types of dyes even in acidic or alkaline solution. Therefore, the amphoteric functionality of ZACC was proved by its accessibility to remove anionic (AR1) and cationic (BG) dyes using a single strip even in various pH. The result is consistent and in lined with earlier literature by [13] that found their cationic and anionic dyes were fairly stable in relatively wide pH range from 4 to 8 [13]. Similar findings were observed in other studies [10,11].

Table 2. Value of pH before and after adsorption of AR 1 and BG dyes.

		pH variation				
AR 1	Before adsorption	3.14	5.275	7.045	9.175	11.017
	After adsorption	7.48	7.38	7.42	7.57	8.29
BG	Before adsorption	3.008	5.087	7.065	9.045	11.087
	After adsorption	7.38	7.49	7.34	7.7	8.56

3.2 Thermal stability test

The test for the harsh environment under thermal stability study was classified into three conditions. There was thermal stability in extreme dry temperature (1), in water (2) and outside building (3). The tests were carried out in such conditions in order to examine the durability and stability of ZACC in various situation.

The thermal stability study was conducted on ZACC strips (5 cm x 5 cm) in order to investigate the stability of an adsorbent coating at high temperature. Figure 3 exhibited the photographic images of four pieces of ZACC strips after 5 days in 100 °C drying oven. All four pieces of ZACC had similar sizes and were used for repetition studied. The physical appearance of ZACC samples (A), (B), (C) and (D) after 5 days in the oven with a high degree of temperature shown no changes in sizes and color. The sizes remained as original, stayed in white color and not get burnt. Table 3 provides a percentage weight loss of ZACC after thermal stability test in a drying oven. The percentage average was about 0.25 % and the value can be claimed as very low. This also indicated that an adsorbent coating was strongly attached/coated on cotton cloth. Higher the temperature will bring to the weight loss of adsorbent coating. As a conclusion, the ZACC has strong thermal stability, especially under dried temperature condition. This characteristic is required for the adsorbent to stand longer under various environmental circumstances in the industry, specifically.

The thermal stability of ZACC was further investigated by immersing the coating strips in water with different temperature condition, at 30 °C, 50 °C and 70 °C. The studied was aimed to examine the strength of adsorbent coating anchored on cotton cloth especially in water circumstances. The thermal stability of ZACC was determined based on changes of spectrum through wavelength measurement using UV-VIS spectrophotometer in 4 hours duration from 800 nm to 300 nm wavelength reading. The result was plotted in the graph as shown in Figure 4. At 30 °C, there were no changes of spectrums from 0 hours of ZACC once exposed to water up to 4 hours of the process. This result showed the stability of ZACC at atmospheric temperature. However, as time increased and temperature increase, the measurement through absorbance value started to increase in a very medium gap.

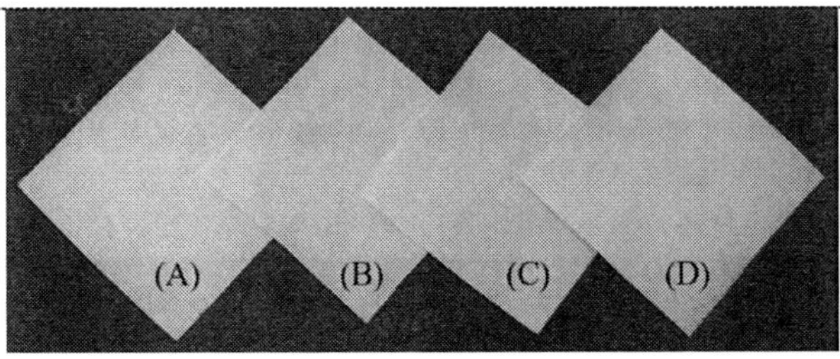

Figure 3. Photographic images of ZACC after 5 days in 100 °C drying oven.

Table 3. Percentage weight loss of ZACC after thermal stability test in drying oven.

Samples	Weight before (g)	Weight after (g)	Percentage weight loss (%)
A	0.3518	0.3512	0.17
B	0.3333	0.3323	0.30
C	0.3226	0.3222	0.12
D	0.3428	0.3414	0.40
		Average	0.25

The water solution at 50 °C showed no changes in absorbance value up to 3 hours of experiment, however, increased in absorbance gap after 4 hours. The graph shows upwards trend especially at wavelength approaching 300 nm. The trend was similar in a water temperature of 50 °C and 70 °C. The changes in absorbance value were due to the disturbance of water content. The water might be disturbed/affected by the presence of adsorbent coating in a solution. In a nutshell, the presence of ZACC in the water at higher temperature showed very minimum changes in absorbance value after 4 hours of experiment especially at temperature 50 °C and 70 °C.

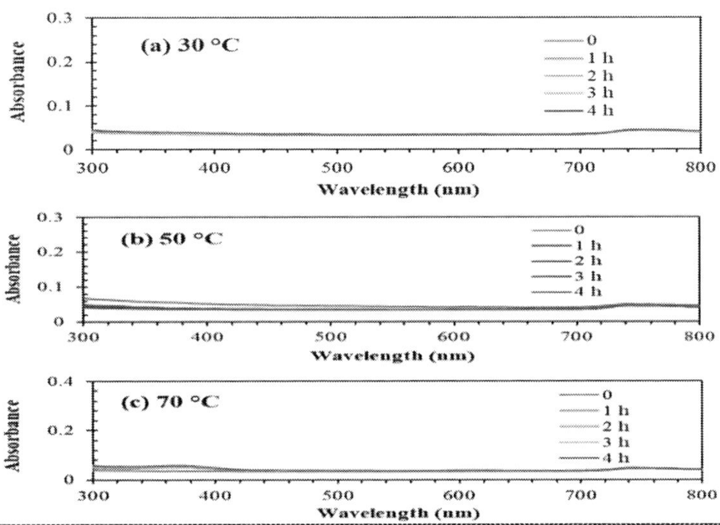

Figure 4. Wavelength changes detected in water at temperature (a) 30 °C, (b) 50 °C, (c) 70 °C

In addition, the other study related to the thermal stability of ZACC samples were tested at the outside of the building. The temperature outside the building was based on the environment condition of the located samples. Four ZACC strips were exposed to the various temperature conditions outside building aimed to get information regarding the stability of samples. Along seven days of testing starting from 3rd March 2019 to 9th March 2019, the environment temperatures were approximately at 30 °C at noon and around 25 °C at night. The samples have also been exposed to the raining day, once. Similar ZACC samples marked as (A), (B), (C) and (D) were cut into 5 cm x 5 cm sizes and used for repetitive studied (Figure 5). The observation based on the physical appearance of all samples after 7 days been exposed to environment temperature outside the building seems no changes in sizes and colors. However, quantitatively, the percentage weight loss of ZACC was calculated and result shown in Table 4. The average weight loss obtained was about 0.109 % and considered very less. The ZACC proved a good strength of adsorbent coated on cotton cloth even been exposed to various weather condition and open to the outside building environment.

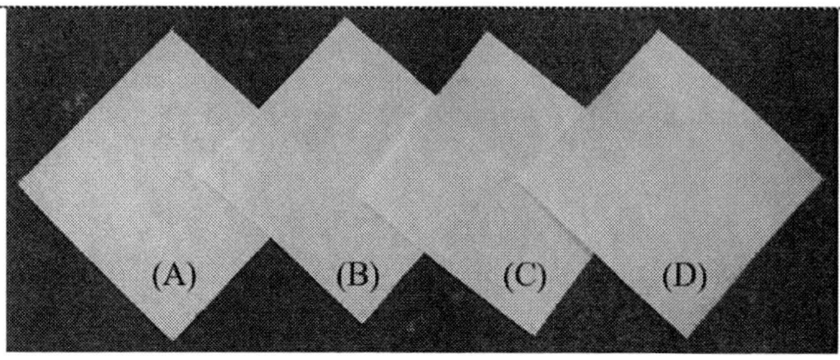

Figure 5. Thermal stability testing outside building

Table 4. Percentage weight loss of ZwitAd after thermal stability test outside building

Samples	Weight before (g)	Weight after (g)	Percentage weight loss (%)
A	0.4107	0.4097	0.243
B	0.4014	0.4013	0.025
C	0.4156	0.4150	0.144
D	0.4000	0.3999	0.025
		Average	0.109

4. Conclusions

The stability and durability of adsorbent materials were considered as an important factor in order to deals with the harsh environment such as at high chemical and thermal condition. The additional procedure after the adsorption process such as centrifugation and filtration should be prevented. This extra process will indirectly increase the operation cost, energy consumption and maintenance and time-consuming. The zwitterionic adsorbent coating revealed their tolerance in both chemical and thermal stability and durability studied. The ZACC was stable in various pH condition (acidic to alkaline) and could maintained the strength of coating without peel off at extreme pH studied. The ZACC was also can stand especially under extreme temperature (100 °C for 5 days), exposed to various weather condition and at outside building environment as proved from a very minimum percentage weight loss tested (within 0.11 % to 0.25 % loss).

Acknowledgments

The authors would like to acknowledge the Bridging Grant (304/PJKIMIA/6316498) and Kementerian Pengajian Tinggi for providing MyPHD as financial support throughout the study.

IOP Conf. Series: Materials Science and Engineering **796** (2020) 012054 doi:10.1088/1757-899X/796/1/012054

References

[1] Azha S F, Shahadat M and Ismail S 2017 *Dye. Pigment.* **145** 550–60
[2] Azha S F, Sellaoui L, Sharafee M, Ismail S, Bonilla-petriciolet A, Ben A and Erto A 2018 *Chem. Eng. J.* **351** 221–9
[3] Azha S F, Shamsudin M S, Shahadat M and Ismail S 2018 *J. Ind. Eng. Chem.* **67** 187–98
[4] Junyi L, Wong L M, Gurudayal G, Wong L H, Chiam S Y, Yau Li S F and Ren Y 2016 *J. Mater. Chem. A* **4** 13280–8
[5] Jawad A H, Azharul Islam M and Hameed B H 2017 *Int. J. Biol. Macromol.* **95** 743–9
[6] Nawi M A, Sabar S, Jawad A H, Sheilatina and Ngah W S W 2010 *Biochem. Eng. J.* **49** 317–25
[7] Tahseen K, Anwar Y, Khan S B, Chani M T S, Asiri A M, Zhang L, Hu P, Wang J and Huang R 2016 *Appl. Surf. Sci.* **369** 558–66
[8] Das I and De G 2015 *Sci. Rep.* **5** 1–11
[9] Jing Z, Lü Q-F and Luo J-J 2017 *J. Clean. Prod.* **167** 739–48
[10] Srivastava V and Sillanpää M 2017 *J. Environ. Sci. (China)* **51** 97–110
[11] Kai L, Chen L, Huang L and Lai Y 2016 *Carbohydr. Polym.* **151** 1115–9
[12] Reddy D H K and Lee S M 2013 *Adv. Colloid Interface Sci.* **201–202** 68–93
[13] Yabo W, Xie Y, Zhang Y, Tang S, Guo C, Wu J and Lau R 2016 *Chem. Eng. Res. Des.* **114** 258–67

IOP Conf. Series: Materials Science and Engineering **796** (2020) 012055 doi:10.1088/1757-899X/796/1/012055

Production Of 5-Hydroxylimethylfurfural From Starch of Durian Seed (*Durio zibhetinus*) Through Dehydration Reaction Using Solvent-Based Choline Chloride: Glycerol

R Manurung[1*], I Zulkarnain[1], and A G Siregar[1]
[1]Department of Chemical Engineering, Faculty of Engineering, Universitas Sumatera Utara, Padang Bulan, Medan 20155, Indonesia.

Email: Renita.manurung@usu.ac.id

Abstract. The energy demand is increasing and the production of fossil fuel is decreasing, which is a challenge for humans to produce eco-friendly renewable energy resources. Biomass is the raw materials of renewable energy. One of the plentiful biomass in Indonesia is the durian seed (Duriozibethinus). Durian production in Indonesia in 2014 reached 859.12 tons, where the weight of durian seeds 20-25% of total fruit weight. Durian seeds have a high content of starch which is 50.40%. 5-HMF (Hydroxymethylfurfural) is one of the most representative monomers of material based on biomass. 5-HMF can be synthesized from carbohydrates, in this case, durian seed is going through two stages of chemical reactions, those are hydrolysis and dehydration reaction. DES (Deep Eutectic Solvents) is a sub-class of IL (Ionic Liquids) that can be used as co-solvents in 5-HMF synthetic from durian seeds. This study applies DES-based Choline Chloride: Glycerol. The variables in this study are sulfuric acid concentration used in hydrolysis reactions and glucose mass ratio: DES, reaction time, glucose concentration in dehydration reactions. The reaction conditions with the highest yield were 61.93% with a glucose mass ratio: DES of 1: 5 and a reaction temperature of 80°C.

1. Introduction

Cellulose Increasing energy needs and the decreasing production of fossil fuel resources is a challenge for humanity to produce environmentally friendly renewable energy sources [1]. Biomass is one of the renewable raw materials on this earth [2]. Conversion of biomass waste into fuel, chemicals and commodity materials has the potential to reduce national dependence on non-renewable raw materials. Utilization of biomass is one of the efforts that have been carried out since a long time ago. One product that can be produced is 5-Hydroxymethylfurfural (5-HMF). 5-HMF is one of the most representative monomers of biomass-based materials. 5-HMF contains the active-aldehyde and hydroxymethyl groups, 5-HMF can be used as an intermediate for many reactions, materials for the synthesis of macrocyclic compounds, polymer synthesis monomers. At present, 5-HMF is produced by dehydration of carbohydrates [3].

Durian (Durio zibethinus) is the most popular seasonal fruit in Southeast Asia, especially Malaysia, Indonesia, Thailand, and the Philippines. Durian seeds have a starch content that is high enough so that it has the potential as an alternative substitute for ingredients that require starch properties. In one study, HMF production was studied using ionic liquid / Ionic Liquids (IL) [4]. DES (Deep Eutectic Solvents)

* To whom any correspondence should be addressed

Content from this work may be used under the terms of the Creative Commons Attribution 3.0 licence. Any further distribution of this work must maintain attribution to the author(s) and the title of the work, journal citation and DOI.
Published under licence by IOP Publishing Ltd

is a subclass of IL (Ionic Liquids), showing characteristics that are comparable, cheaper and easier to produce because the cost of raw materials is lower, less toxic and often bioavailable, making it a valuable alternative solvent for green analytical chemistry.

Assanosi, et al., 2015 reported the use of DES as a catalyst and solvent consisting of a mixture of N, N-diethylethanolammonium chloride as organic salts and p-Toluenesulfonic Acid Monohydrate (p-TSA) as HBD in fructose dehydration reactions. So that the yield of 5-HMF was 84.8% at 80 ° C with a 1 hour reaction time using a molar mixing ratio of DES 1: 0.5, salt: p-TSA and a feed ratio of 5% [1]. Zhang, et al., 2016 reported that DES containing [Emim] Cl and isopropanol was the best in converting fructose to HMF catalyzed by [HNMP] Cl with yields of more than 89% after 3 hours of reaction time at 25°C [5].

Zuo, et al., 2017 reported that DES consisting of MeCN and Choline Chloride with a small concentration of HCl as catalyst produced yield of 90.3% at 100°C operating conditions and a reaction time of 4 hours [6]. In this study, researchers are interested in knowing the effect of DES usage, DES number, and reaction temperature on the 5-HMF yield on processing durian seed waste and are expected to help increase the utilization of durian seed waste in Indonesia.

2. Experimental

2.1 Materials

The materials used in this study are Durian seeds, Choline chloride (ChCl), Glycerol ($C_3H_8O_3$), Hydrochloric Acid (HCl), Aquadest (H_2O), Sulfuric Acid (H_2SO_4).

2.2 Typical process for the conversion of glucose into 5-HMF

In a typical run, the conversion of glucose into 5-HMF was conducted in a glass flask (100 ml) with a condenser in a hot plate. The DES system was formed with ChCl: glycerol (1:2). 1.2 mol% HCl was added as the catalyst. The flask was then placed in hot plate and heated with vigorous stirring. After the reaction, 5- HMF has analyzed UV-vis spectrophotometer.

2.3 Product analysis

In this study, the 5-HMF product was degrading due to UV-vis spectrophotometer, at a wavelength of 284 nm and 336 nm. 5-HMF concentrations are calculated as the following :

$$\text{5-HMF (mg/100mg)} = \frac{(A284-A336)\times 14.97\times 5}{\text{sample weight (g)}} \quad [1]$$

2.4 Synthesis and characterization of DES

DES used in this research was made from a mixture of choline chloride as salt and glycerol as HBD with a molar ratio of 1: 2 for 60 minutes at 80°C with a stirring speed of 300 rpm [8]. Density, viscosity, and pH are important characteristics of DES physical properties [9]

2.5 Characterization of product

The FTIR spectra of 5-HMF analysis of compounds before and after dehydration was carried out to identify differences in groups in the compound. recorded on Fourier Transform Infrared Spectrometer Shimadzu IR-21. It was measured in the wavelength range from 4000 cm^{-1} to 500 cm^{-1}.

3. Results and Discussion

3.1 Characterization of Deep Eutectic Solvent (Des) Based on Chcl: Glycerol

The Characteristic test of ChCl-based DES: glycerol which is carried out is pH test, density, and viscosity. The density value of DES at 30°C was obtained at 1.1892 g / ml, the viscosity was 246.919 mPa.s and the pH was 7.2. This is consistent with the study of Yadav, et al., In 2014 who reported that DES-based ChCl: glycerol with a ratio of 1: 2 at a temperature of 30 °C has a density value of 1.1885 g / ml and viscosity 246.7928 mPa.s [10]

3.2 Effect of Acid Concentration (%) on Glucose Yield (%)
Hydrolysis Process of Durian Seed Starch was carried out using sulfuric acid (H_2SO_4) as a catalyst with various concentrations, namely 1.5%; 3%; 5% and 10% with the ratio of starch of durian seeds: acid (1:10 b/v), temperature of 100°C, time of 90 minutes and stirring speed of 300 rpm.

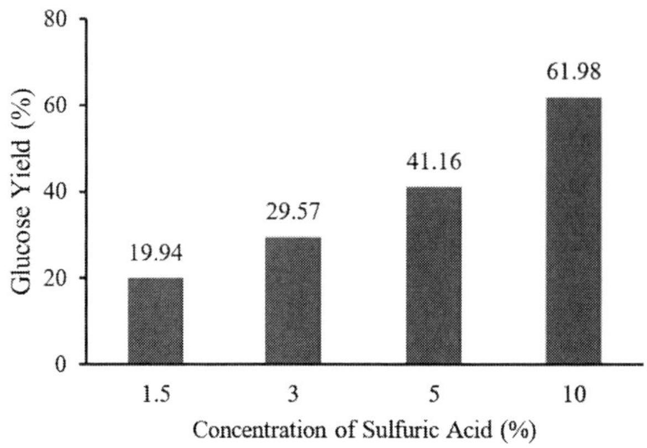

Figure 1. Effect of Sulfuric Acid Concentration (%) on Glucose Yield (%)

In the process of hydrolysis of durian seed starch using sulfuric acid, the catalyst is expected to produce high glucose yield to be used as a reactant in the dehydration process to produce 5-HMF. Figure 1 shows that the yield of glucose increases with increasing concentrations of sulfuric acid used. The highest glucose yield was produced at 10% acid concentration, which was 61.98% and the lowest glucose yield was produced at 1.5% sulfuric acid concentration, which was 19.94%. Addition of sulfuric acid as a catalyst in the hydrolysis reaction of starch in durian seeds is directly proportional to the yield of glucose at the same reaction time because H^+ ions in strong acids can help break the glycoside bonds found in starch to form a glucose chain[11]. The role of H + ions is the main factor causing the increase in glucose.

3.3 FTIR (Fourier Transform Infra-Red) Analysis Results on Compounds and After Dehydration
FT-IR analysis of compounds before and after dehydration was carried out to identify differences in groups in the compound. From the functional group analysis using FT-IR obtained the results of the spectrum in the form of graphs which can be seen in Figure 2.

From Figure 2 it can be seen the difference of the groups in the compound before and after the dehydration reaction, where after the dehydration reaction there is the addition of C = C (Aromatic) and C-H (Aldehyde) groups. According to Pavia, et al., 2001, the presence of C = O groups and weak absorption of C-H showed aldehyde groups. If there is a weak absorption area of the C = C group near 1650 cm^{-1} which is accompanied by the absorption area of the C-H group at an area of 3000 cm^{-1}, then it shows that there is an aromatic ring group in it. The OH group with H bound followed by the appearance of the C-O absorption area shows an alcohol group [12]. 5-Hydroxymethylfurfural consists of a furan ring, aldehyde group and alcohol [13]. Furan is a class of organic compounds with an aromatic heterocyclic ring consisting of 1 oxygen atom and 4 carbon atoms [14]. Based on this theory, the results of FTIR showed that the samples of the process after dehydration contained aldehyde, alcohol, and furan (aromatic) rings, so that it could be concluded that the dehydration results contained 5-HMF.

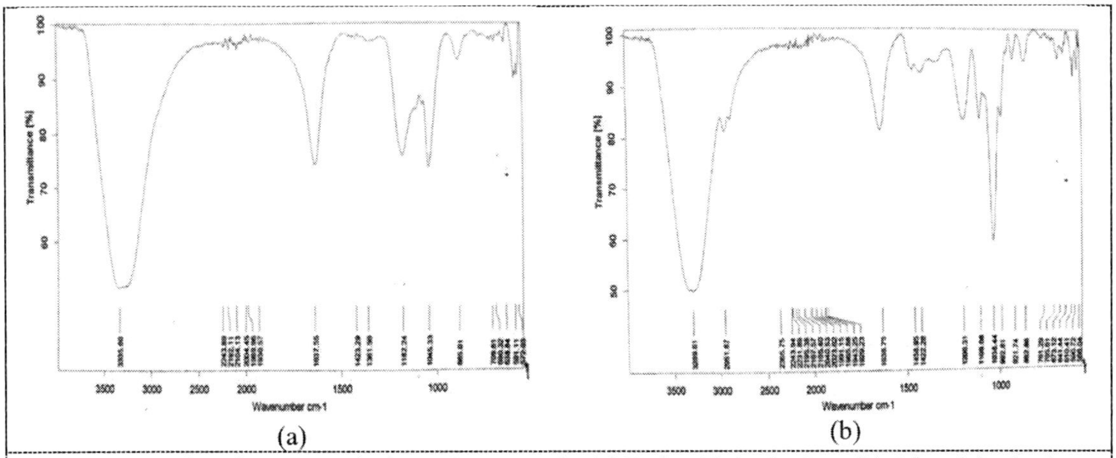

Figure 2. Results of FTIR Analysis on Compounds: (a) Before Dehydration, (b) After dehydration

3.4 Comparison of Dehydration Reactions without DES and with DES against 5-HMF Yields

Figure 3 shows the comparison of dehydration reactions without DES and with DES-based ChCl: glycerol against 5-HMF yield. Figure 3 shows that the highest 5-HMF yield is obtained when adding DES in the reaction. The research conducted provides data that can show that overall the glucose dehydration reaction with the use of DES obtained a higher 5-HMF yield than glucose dehydration reaction without using DES. In this study, the highest 5-HMF yield was obtained when the dehydration reaction using DES with a temperature of 80°C was equal to 22.08%. Meanwhile, the lowest 5-HMF yield was obtained when the dehydration reaction did not use DES with a temperature of 90°C which was 11.65%. According to the previous description, the addition of DES-based ChCl: glycerol into the system can reduce the viscosity of the system so that it can facilitate the transfer of mass and heat in it. Clotting which is contained in DES is also able to interfere with the hydroxyl bond in glucose so that 3 water molecules are released during the dehydration reaction. This result can be attributed to the reaction with the addition of DES.

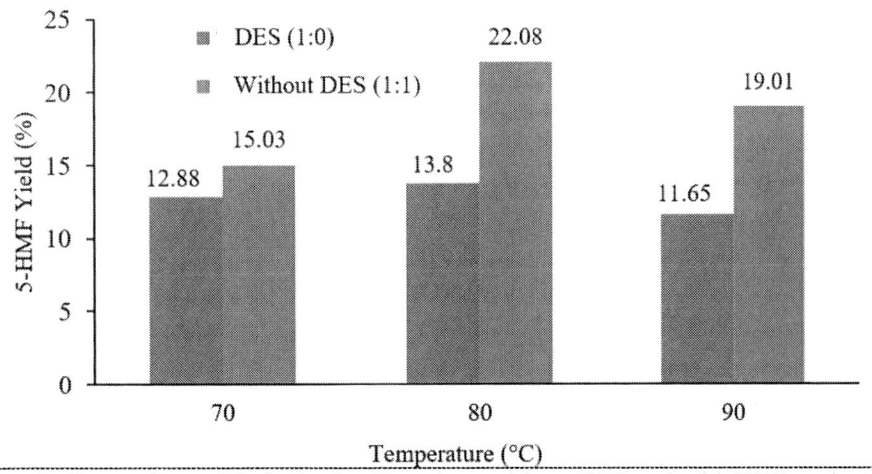

Figure 3. Comparison of Dehydration Reactions without DES and with DES against 5-HMF Yields

3.5 Effect of DES Number and Reaction Temperature on 5-HMF Yield

process of glucose dehydration was carried out using 1.2 mol HCl of glucose as a catalyst and using ChCl-based DES: Glycerol. The effect of glucose mole ratio and reaction temperature on 5-HMF yield was observed in a temperature range of 70°C, 80°C and 90°C and glucose mass ratio versus DES of 1: 1, 1: 3 and 1: 5 with a reaction time of 60 minutes, and stirring 300 rpm. The following is a graph that shows the effect of mass ratio and reaction temperature on the 5-HMF yield.

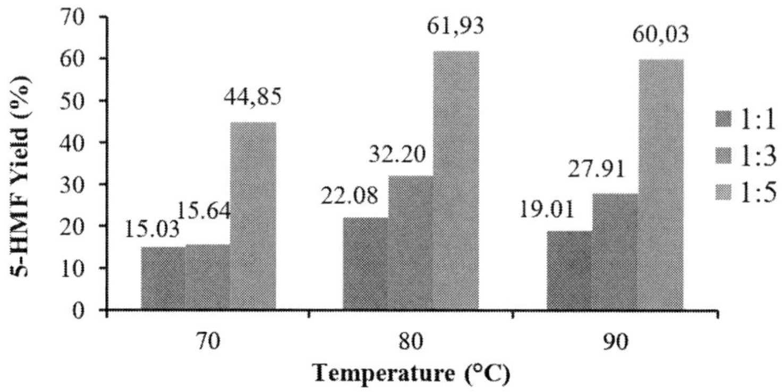

Figure 4. Effect of DES Number and Reaction Temperature on 5-HMF Yield

Figure 4 shows that the 5-HMF yield increases as the number of DES used increases and the 5-HMF yield increases at a temperature increase of 70-80°C, but the 5-HMF yield decreases at a temperature drop of 80-90°C. The highest 5-HMF yield was obtained at the glucose mole ratio versus DES 1:5 and at 80°C, which was 61.93%. The decrease in 5-HMF yield is caused because the 5-HMF formed undergoes a polymerization reaction and forms humin. This is based on the color of the polymer and humin is dark brown and the color of the sample is getting brown as the reaction temperature increases [15]. The 5-HMF yield increase increases as the reaction temperature increases and then decreases when the reaction time is extended. This is due to a dehydration reaction because the water produced from the dehydration reaction is a by-product [16]. By the previous description that the reaction temperature is very dependent on the activation energy that the reaction has. This can indicate the cause of the high yield of 5-HMF obtained at a reaction temperature of 80°C. While at a temperature of 70, 80 and 90°C it is suspected that an unwanted reaction has occurred to form humin or levulinate acid and formic acid.

3.6 Effect of Initial Glucose Concentration on 5-HMF Yield

The process of glucose dehydration was carried out using 1.2 mol HCl of glucose as a catalyst, glucose mass ratio versus DES of 1: 5 with a reaction time of 60 minutes, and stirring 300 rpm. The effect of glucose concentration on 5-HMF yield was observed in variations in glucose concentration of 10.05%; 14.90%; 20.74% and 31.24%. The following is a graph showing the glucose concentration of the 5-HMF yield.

From Figure 5 it can be seen that the 5-HMF yield decreases as the initial glucose concentration increases. The highest 5-HMF yield was 70.75% when the glucose concentration was 10.05% and the lowest 5-HMF yield was 61.93% when glucose concentration was 31.24%. The high initial glucose concentration results in side reactions and produces unwanted products.

Research conducted by McKibbins, et al. (1962) stated that the yield of 5-HMF will increase when the initial concentration of glucose is reduced. The reaction mechanism can be seen as shown Figure 6.

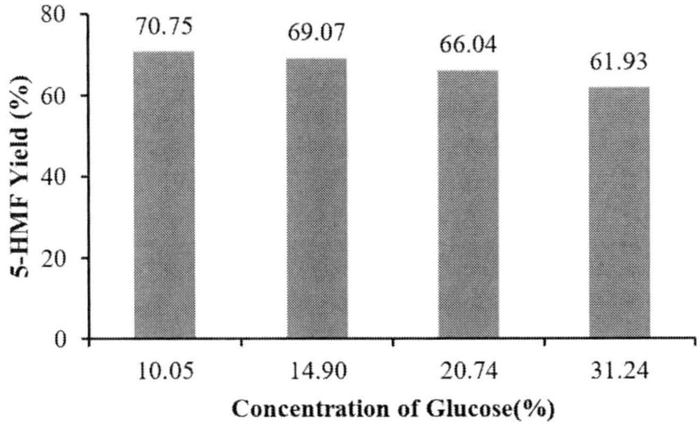

Figure 5. Effect of Glucose Concentration on 5-HMF Yield

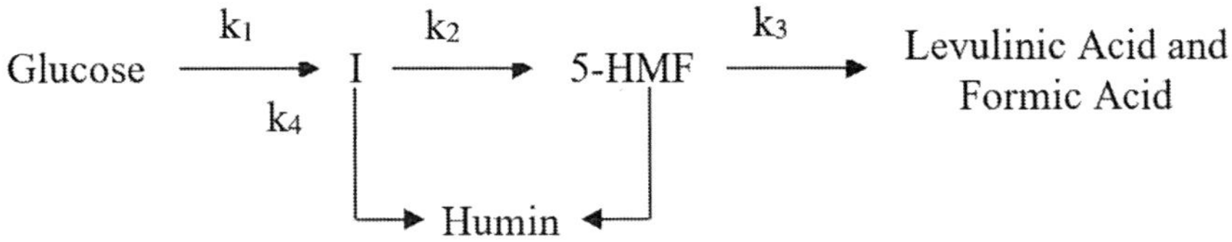

Figure 6. 5-HMF Formation Reaction Mechanism [18]

Through Figure 6 it can be seen that there is I in the reaction, where I am an intermediate; k1, k2, and k3 are reaction speed constants which have first order; and k4 is the reaction speed constant which has an order higher than k1, k2, and k3. [17] For parallel reactions, the concentration of reactants is the key to controlling the product produced in the reaction. A high reaction order requires a high concentration of reactants to increase the reaction product and a low reaction order requires a low concentration of reactants to form the product in the reaction, where the concentration level does not affect the product for reactions with the same order [18]. By this description, it can be indicated that the decrease in 5-HMF yield at high initial glucose concentration is caused by the occurrence of side reactions that form humin and is marked by changes in the color of the sample to dark.

4. Conclusion

In summary, we have shown that 5-HMF can be synthesized from glucose resulting from hydrolysis of durian seed starch. Use of ChCl-based DES: glycerol can increase the 5-HMF yield and reduce side reactions and further reactions in the dehydration process. The maximum yield was obtained in reaction conditions with a glucose mass ratio of DES of 1: 5 and a reaction temperature of 80 ° C with a 5-HMF yield of 61.93%. This study provides a promising alternative to the production of 5-HMF from biomass.

5. Acknowledgments

The authors gratefully acknowledge that the research is supported by Directorate General of Higher Education Republic of Indonesia.

References:

[1] Assanosi, Amhamed A, Mohamed M. F, Joseph W, Bushra Al-D, 2015 *C. R. Chimie* **19** 450-456.
[2] Liu, Bing B, Cong J, Mengmeng, Zhang Z. 2015. *Industrial Crops and Products.***76**. 781–786.
[3] Yang, Liu., Yuhuan Liu., Rongsheng Ruan., Yunpu Wang., Wenweng Zen., Chengmei Liu., jinsheng Zang. 2010. *Scientific Research*. **3(1)**. 332-337

[4] Ariyanti S, Zakaria M, Nawshad M, Amir Sada K, Wan S, Wan H, Asyraf H, Abdul R, Zahoor U, Cecilia Devi W, *Ultrasonics Sonochemistry* **37** 310-319
[5] Zhang J, Yu X, Zou F, Zhong Y, Du N, Huang X, 2015 *ACS Sustainable Chem. Eng.* **3** 3338.
[6] Zuo M, Li Z, Jiang Y, Tang X, Zeng X, Sun Y, Lin L, 2017 *RSC Adv* **6** 27004–27007.
[7] SNI (Standard Nasional Indonesia), "Madu". Badan Standarisasi Nasional. SNI 01-3545-2004.
[8] Ahmad A, Nagoor G, Dachyar A, Enas Muen N, Parveen J. *Bioresource Technology*. S0960-8524(15)00077-2.
[9] Naser J, F. Mjalli, B. Jibril, S. Al-Hatmi dan Z. Gano. 2013 *International Journal of Chemical Engineering and Applications* **4** 114-118.
[10] Yadav A, Shruti T, Rewa R, Siddharth P. 2014 *Fluid Phase Equilibria* **367** 135–142.
[11] Sarip, Harun Sohrab H, Mohamad Azemi M. N, Karim A, 2016 *Bioresources 11* 10625-10653.
[12] Pavia, L. Donald, Gary M. Lampman, George S. Kriz. 2001 Thomson Leraning. Third Edition. Washington. page. 26
[13] Ghaderi, Faranak, Mohammad Reza I, Shadbad, Molood H, 2015 *Pharmaceutical Sciences* **21**. 1-5
[14] Amupam, Verma, Pandeya S.N, Shinha S. 2011 *IJRAP* **2** 1110-1116.
[15] Wang F, Shi A W, Qin X X, Liu C.-L, Dong W S, 2011 *Carbohydrate research* **346** 982-985
[16] Li C P, Li D, Zou S S, Li Z, Yin J M, Wang A L, Cui Y N, Yao Z. L, Zhao Q, 2013 *Green Chemistry* **15**. 2793-2799.
[17] McKibbins, Samuel W, John F. H, Jerome F.S, 1962 *Chemical Conversion of Wood Residues*. **V** 17-23.
[18] Levenspiel, Octave. 1999. Chemical Reaction Engineering. Third edition. Replika Press Pvt. Ltd: India. page 153-154 and 235-236.

IOP Conf. Series: Materials Science and Engineering **796** (2020) 012056 doi:10.1088/1757-899X/796/1/012056

Transesterification of refined bleached deodorized palm oil (rbdpo) using novozym® 435 to produce biodiesel

R Manurung[1], R Hasibuan[1], B I Sari[1], R Wulandari[1], and A G Siregar[1]**
[1]Department of Chemical Engineering, Faculty of Engineering, Universitas Sumatera Utara, Padang Bulan, Medan 20155, Indonesia.

Email: Renita.manurung@usu.ac.id

Abstract. As a renewable energy source, biodiesel is an alternative fuel that is environmentally friendly, and non-toxic. Conventionally, biodiesel is mostly produced by a transesterification reaction of triglycerides with short-chain alcohol in the presence of an acidic or basic catalyst, but it requires high energy, difficult in the purification of glycerol and there is the possibility of soap formation. Currently, the process production of biodiesel has been directed to the enzymatic process. The enzymatic process can produce high-purity products because it is easy in removing by-products such as glycerol. Novozym® 435 have been reported as a biocatalyst which indicates a high conversion and able to catalyze a variety of organic chemical reactions. It is biodegradable, no possibility of soap formation, and can be used repeatedly. Ethanol used as an acyl acceptor because it is more renewable than methanol. The reaction parameters that used in this research were the molar ratio of raw material to ethanol, temperature reaction, and reaction time. The best result obtained in 7 hours reaction time at 40 °C was 98,83 % of ester yield with a molar ratio of ethanol and RBDPO was 6:1 and the catalyst dosage was 30 % (w/w). Based on the results, it shows that Novozym® 435 is a good biocatalyst in enzymatic transesterification process for biodiesel synthesis

1. Introduction

World demand for energy is increasing while the exploration of fossil fuels continuously that causing world oil reserves dwindle. To overcome this, we need alternative energy that is renewable and environmentally friendly [1]. In recent years, research on biodiesel is mostly done to find environmentally friendly alternative fuels. In contrast to fossil fuels, biodiesel is biodegradable energy, free from toxic and sulfur [2].

Biodiesel can be made using raw materials of vegetable oils or animal fats that can react with alcohol and then form an ester. However, the manufacture of biodiesel from waste animal fat is hard to produce, because it contains 10-15% fatty acids that will lead to the formation of a higher soap. Palm oil (palm oil) can be used as a raw material in the manufacture of high-quality biodiesel [3,4]. Indonesia is one of the country's largest producer of crude palm oil in the world. Therefore, RBDPO (Refined Bleach Deodorized Palm Oil) is the main ingredient that has great potential in the manufacture of biodiesel which RBDPO a fraction of palm oil derivative that has been purified. Currently, Indonesia is the country with the largest acreage of palm oil in the world with more than 7 million ha. In 2009-2010 Indonesia's palm oil production increased by 1.20402 million tons [5]. So the use of palm oil as a raw

[*] To whom any correspondence should be addressed

Content from this work may be used under the terms of the Creative Commons Attribution 3.0 licence. Any further distribution of this work must maintain attribution to the author(s) and the title of the work, journal citation and DOI.
Published under licence by IOP Publishing Ltd

material in the manufacture of biodiesel can be considered, given the available raw materials have increased each year.

Recently, production of biodiesel chemically using alkali has been investigated. However there are some problems, the amount of energy required, the difficulty in the purification of glycerol and the difficulty of throwing soap may have formed [6]. The production of biodiesel is directed to the enzymatic process. Enzymatic biodiesel production can produce high purity product and easy the separation of by-products such as glycerol [7]. In the study Veny, et al, 2010 using Jatropha oil as raw materials, using enzyme Lipozyme IM with a time of 24 hours the conversion yield of 54% was obtained [8]. While the research Kumari et al, 2009 using Jatropha oil as raw materials and enzymes immobilized lipase from Enterobacter aerogenes with the reaction time of 48 hours was obtained conversion yield of 94% [9]. Compared with the enzyme that has been reported, Novozyme® 435 has the advantage of cheap and efficient. Go, et al, 2013 has been researching the production of biodiesel from soybean oil using Novozyme® 435 with the time of 48 hours resulted in a yield of 96.4% [10]. Novozym® 435 (Candida rugosa) has the advantage can be obtained commercially well than the other sources lipase.

The most commonly used as a biocatalyst to produce biodiesel is lipase produced by some species of fungi, bacteria, and yeast [11]. Novozym® 435 can catalyze a variety of organic chemical reactions [12]. The use of enzymes which are reported by the researchers used purified oil, so it is necessary to study more about RBDPO-based biodiesel production is transesterification using Novozym® 435 catalyst with ethanol as a substrate.

2. Experimental

2.1 Materials

RBDPO raw materials with 0.15% FFA content and 1.2% moisture content obtained from Indonesia. The alcohol used was ethanol Merck with a purity of> 99.5%. Novozym®^{435} (Candida rugosa) as a solid biocatalyst obtained from Sigma-Aldrich Pte Ltd.

2.2 Enzymatic Transesterification Process

The amount of RBDPO, ethanol, catalyst Novozym® 435 was prepared with a specific weight. RBDPO and ethanol with determined molar ratio into Erlenmeyer that heated in a heater. The thermometer was used to measure the temperature of the solution. The constant of dose biocatalyst Novozym® 435 and heated until reached the temperature of reaction and mixed it using a shaker for homogenous mixture during the specified time [13]. After the reaction completed, the solution was put into separating funnel to separate ethyl ester with biocatalyst, glycerol, and some impurities such as water and ethanol that was not converted. Analysis can be done after ethyl esters prepared.

2.3 Product analysis

The produced biodiesel was analyzed for the content of ethyl ester using by the GC-MS Shimadzu-2010 gas chromatograph.

2.4 Experimental Design

This research was conducted by the independent variable that is the mole ratio of a substrate (N1, N2), temperature (T1, T2, T3) and reaction time (t1, t2, t3) in the transesterification reaction using complete random. The specifications of the models and test conditions are shown in Table 1.

Table 1. Test Model Specifications and Test Conditions

Run	Molar Ratio of Substrate	Temperature (C)	Reaction Time (hour)
1			t_1
2	N_1		t_2
3		T_1	t_3
4			t_1
5	N_2		t_2
6			t_3
7			t_1
8	N_1		t_2
9		T_2	t_3
10			t_1
11	N_2		t_2
12			t_3
13			t_1
14	N_1		t_2
15		T_3	t_3
16			t_1
17	N_2		t_2
18			t_3

3. Results and Discussion

3.1 Effect of Ratio Molar on Yield Gains

In general, the lipase is not stable in the short-chain alcohol, such as methanol. This is due to the methanol insoluble and joins mixed with oil [14]. So, in this study, acyl acceptor used is ethanol. Ethanol was chosen as the acyl acceptor because ethanol is not toxic and can be renewed from the biomass [15]. In figure 4 shown the highest yield obtained at the molar ratio of 1:6 give an optimum result than the molar ratio 1:9 in the equal condition

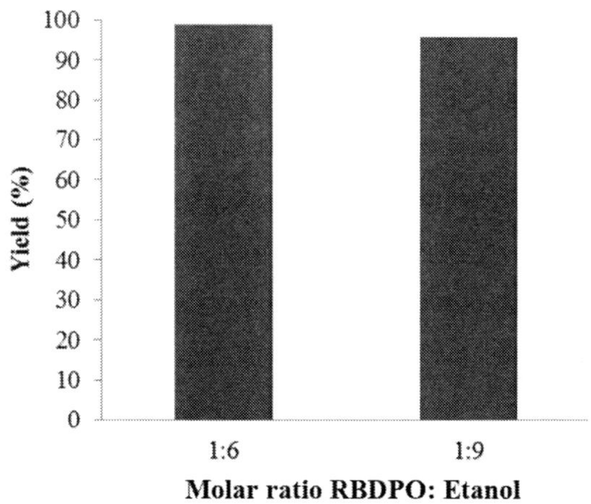

Figure 1. Effect of Ratio Molar on Yield Gains

In the process of the formation of biodiesel chemically, excess alcohol in the triglycerides are always improving the yield of biodiesel. The optimum molar ratio of alcohol used for enzymatic biodiesel production also depends on the actual reaction system. In an organic solvent system, little excess alcohol

IOP Conf. Series: Materials Science and Engineering **796** (2020) 012056 doi:10.1088/1757-899X/796/1/012056

can be used to achieve higher yields. While the system is solvent-free, add a little alcohol is needed to prevent the inactivation of the enzyme. Alcohol should be added to the reaction mixture in small amounts in a row [14,16]. It shows that the optimum molar ratio for this reaction is 1: 6. Addition of excess alcohol will cause the closing of the enzyme active site and make the cost of the enzyme is not optimal. Thus % biodiesel yield will decline, due to the ability of the enzyme has been reduced

3.2 Time and Temperature Effect on Yield Gains

Variable time and temperature simultaneously applied to see its effect on yield gains. From the result of experiments conducted as shown in figure 2 and 3. In figure 2 and figure 3 the highest yield obtained at the molar ratio of 1:6 and a temperature of 40 °C with a conversion yield of 98.83%. While the highest yield obtained at the molar ratio of 1:9 and a temperature of 45°C with a conversion yield of 97.28% at the same time, 7 hours. Time gives an effect on the increase in the percent yield. It is demonstrated that increasing the reaction time give effect on % yield with a constant dose of biocatalysts variable 30%. It is because the longer the reaction time, make the more time for the enzyme to convert alcohol and oils into biodiesel. However, the longer reaction time will not rise the yield of biodiesel produced, even a decline in yield [17].

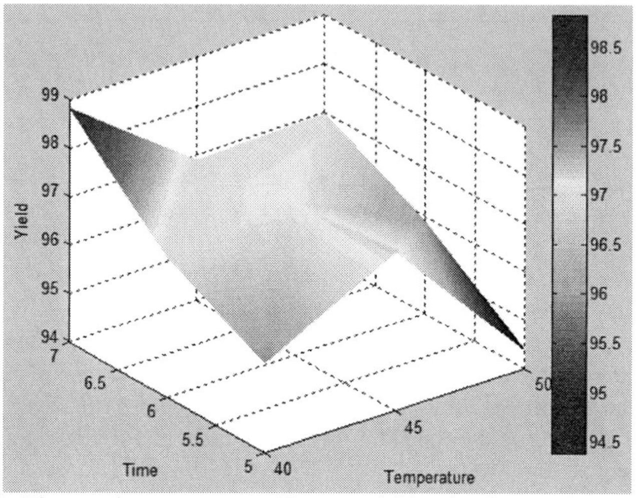

Figure 2. Time and temperature effect on yield gains in a molar ratio of 1:6

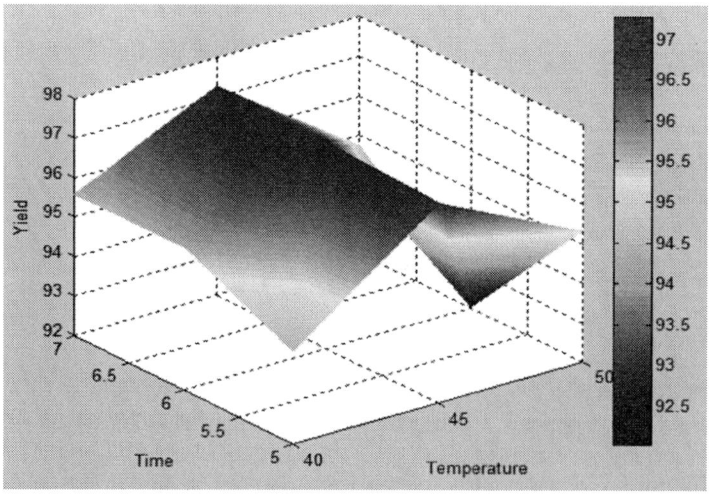

Figure 3. Time and temperature effect on yield gains in a molar ratio of 1:9

Martin and Christina have used soybean oil as feedstock and ethanol as acyl acceptors on the reaction temperature of 25°C for 7 hours and give resulted in the conversion of 97% [18]. While in this study with a time of 7 hours and a temperature of 40°C and 45°C obtained biodiesel yield of 98.83% and 97.28%. This proves that for Novozym® 435 with a time of 7 hours, the desired reaction has been reached. The optimum temperature for the enzymatic transesterification process, depending on the stability of lipase used, the molar ratio between alcohol and oils and types of organic solvents. Normally the temperature on transesterification process for lipase between 30°C - 50°C [14]. Other studies have shown an increase in the synthesis of biodiesel with temperature happens gradually from 30°C to 40 °C, but the temperature rises further at 50°C led to a decrease in enzyme stability [19].

Cervero.et al (2010), reported that the results of biodiesel yield increased at the reaction temperature 37°C and decreased when the reaction temperature has reached above 50°C, when converting soybean oil to the ethyl ester using Novozym 435 [16]. Hung Min Chang. Et al (2005), reported that optimum conditions at a temperature of 38°C with a biodiesel yield of 97.9 % [6]. For raw materials, RBDPO used in this study when seen from Figure 2, the optimum temperature for the actual work Novozym® 435 is $\leq$ 40°C. Theoretically, high temperatures can help increase the rate of reaction. Otherwise, enzyme depends on temperature and easily deactivated at high temperatures. The effect of temperature on the activity of Novozym 435 examined in the range of 30-70°C. Enzyme activity increases as the temperature increases from 30 to 40 °C, followed by a decrease at higher temperatures [20]. After the passing of time and the optimum reaction temperature, lipase activity is inhibited by ethanol and glycerol. This is because the higher reaction temperatures and longer able to stop the activation of the enzyme lipase.

4. Conclusion

RBDPO as raw material to produce biodiesel without going through the pre-treatment (refined) can be used with the aid of a basic catalyst Novozym® 435 and give % yield relatively good. The highest content of ethyl ester is 98.83% at 40°C with a molar ratio of ethanol to refined bleached deodorized palm oil was 6: 1 and the catalyst dosage 30% (w/w). This research also revealed that the appropriate reaction time was 7 hour with a temperature of Novozym® 435 is 40-45

5. Acknowledgments

The authors gratefully acknowledge that the research is supported by Directorate General of Higher Education Republic of Indonesia and facilities support from Chemical Process Laboratory and Microbiological Technic Laboratory Department of Chemical Engineering, Faculty of Engineering, University of Sumatera Utara, North Sumatera, Medan 20155, Indonesia.

References:

[1] Hanifa T, Sulaiman Al-Z, Ali H, Al-M, Yousef H and Mohammed M. F. 2011. *Enzyme Research* 25.
[2] Ghaly. (2010). *American Journal of Biochemistry and Biotechnology* **6** (2): 54-76.
[3] Santhosh K.. 2013. Dalhousie University, Halifax, Nova Scotia.
[4] Alkabbashi A, Alam S, Mirghanil S; and Al-Fusaiel 2010 *Applied Sciences*, **17**: 3166-3170.
[5] Rashid A, Ni Made Suyastiri Y P, and Antik S 2012 SEPA. **9**(1): 125 – 133.
[6] Hong W, Min-Hua Z, Qian L, and Hua-Chang W 2003. Am. *Chem. Soc., Div. Fuel Chem.* **48**(2): 533.
[7] Srivathsan V R, Srinivasan L N, and Karuppan M 2008 *Bioresource Technology*. **99**: 3975–3998.
[8] Harumi V, Mohammed KA, and Nik Meriam N S 2010 *International Conference on Chemical, Biological, and Environmental Engineering*: 251–253.
[9] Kumari, Annapurna, Paramita M, Vijay K G and Rintu B. 2009. *Biotechnology for Biofuels*. **2**.
[10] A-Ra G, Youngrak L, Young H K, Sehkyu P, Joongso C, Jinwon L, Sung Ok H, Seung W K and Chulhwan P 2013. *Technology*. **53**: 154– 158.
[11] Lew P. C, Hemanathan and Vasudeo P Z 2014 *Applied Energy* **119**: 497–520.

[12] Nemanja M, Aleksandra N and Katja L. 2012 *Applications Bioresource Technology*. **115** 126–135.
[13] Renita M, Rosdanelli H, Taslim, Nimpan B, Rahayu W, and Bunga I S 2016 *Journal of Engineering and Applied Sciences* **11**.16
[14] Xiaohu F 2012. *Lipid Technology*. **24**(2), 31-32
[15] Rahmath A and Pogaku R 2013. *Biomass and Bioenergy* **56** 8-13.
[16] Cervero J.M, Alvarez J.R; and Luque S 2013. *Biomass and Bioenergy* **30** 1-7.
[17] Mahabubur R T, Jin C W, Thi Bich V N, Ng M F and Yeo Li S M 2009. *Journal of Molecular Catalysis B: Enzymatic*. **60**:106–112
[18] Estela H M and Cristina O 2008. *Bioresource Technology* **99** 277–286.
[19] Adriana G, Marius R, Monica T, Csaba P, and Florin D I 2012 *Renewable Energy* **30** 10-16.
[20] Dahai Y, Li T, Hao W, Song W, Ye W, Dongxiao M, and Xuexun F 2010 *Process Biochemistry*. **45**:519–525

IOP Conf. Series: Materials Science and Engineering **796** (2020) 012057 doi:10.1088/1757-899X/796/1/012057

Preparing of Carbon Nanodots from Binchotan Carbon by Electrochemically Sonification and Dialysis

Nirwan Syarif [1,2,3*], Dedi Rohendi [1,2,3], Sri Haryati [4], Lee Chew Tin [5]
[1] Department of Chemistry, Universitas Sriwijaya, Indralaya, Sumsel, Indonesia
[2] Research Center of Excellence for Fuel Cell and Hydrogen
[3] National Center for Sustainable Transportation Technology, Bandung, Indonesia
[4] Department of Chemical Engineering, Universitas Sriwijaya, Indralaya, Sumsel
[5] Department of Bioprocess and Polymer Engineering, Universiti Teknologi Malaysia, Johor Bahru, Malaysia

E-mail: nsyarif@unsri.ac.id

Abstract. This paper reports the research of preparing an aqueous solution of binchotan carbon electrochemically and followed by its purification through sonification and dialysis. Binchotan rod and platinum plate were used as anode and cathode. The electrodes were placed on both sides of beaker glass that contained 200 ml of 14 M sulphuric acid, oxalic acid and phosphoric acid solutions as electrolytes, respectively. The potential between anode and cathode was 14 V. The solution was moved from beaker glass after 24 hours, filtered to produce a solid product for further examinations. The solid was washed with distilled water until pH 7, dried in 100 ℃ ovens. Two types of materials were examined, i.e. the product before and after the purification. The solid was then characterized by its morphology, crystallography, and surface functional groups by SEM, XRD, and FT- IR. The morphology of the binchotan after running the electro-oxidation process generally were aggregates and dots. Both shapes were reduced in size after purification. The diffractogram showed a significant change at $2\Theta=14^{o}$- 44° meaning that the purification process changes the crystallography of carbon. The FTIR showed the presence of a functional group of aromatic, carboxylic, alkene, and alcohol on the surface of the carbon.

1. Introduction

Biomass carbonization basically produces carbon material containing a few functional groups because during pyrolysis the availability of oxygen in the reactor atmosphere reacts with carbon is very limited [1]. In the carbon surface modification, various types of functional groups are generated as the oxidation process is applied according to its power [2]. In practice, oxidation of carbon surfaces is done by reacting carbon sources with oxygen or oxygen-containing chemical compounds. Oxygen atoms that are inserted or added on the surface can also come from carbon raw materials that contain oxygen in their structures such as wood, sucrose, or phenol-formaldehyde resins, or oxidizing agents. The oxygen atom is cut off from the structure of its compound and diffuses and reacts with other carbon atoms on the surface of carbon forming a surface oxide or carbon oxide.

Surface oxides affect each other depending on interactions between adjacent functional groups. The overall interaction will affect carbon reactivity. The use of biomass-based carbon or biocarbon as a

* Corresponding author: nsyarif@unsri.ac.id

Content from this work may be used under the terms of the Creative Commons Attribution 3.0 licence. Any further distribution of this work must maintain attribution to the author(s) and the title of the work, journal citation and DOI.
Published under licence by IOP Publishing Ltd

starting material in the oxidation process is more economically beneficial because such carbon already contains more or less functional groups on its surface [3].

One of the biocarbon has the potential to be developed for technical purposes is binchotan. Binchotan has the highest quality charcoal by burning for 14 days at high temperatures and cooling without using water. Binchotan has a high carbon content (> 95% by weight) with ~ 82% carbon hexagonal aromatic ring network arrangement [4], so it has the property to conduct electricity (conductor). Locally, the raw material of binchotan carbon is made from gelam wood, pelawan, eucalyptus, and other hardwoods. The woods are endemic crops of the island of Sumatra and some other islands of Indonesia [5]. Previous studies have shown several applications of binchotan as electrodes, both supercapacitors [6] and batteries. Furthermore, the conductive nature of binchotan is utilized to produce other functional carbon, i.e., carbon dots. The production is an analogy of the processes applied to graphite. Therefore, this work has never been done before, it is a novelty of this research.

The method commonly to produce carbon dot is electrochemical method [7]. The use of electrochemical methods for oxidation the carbon is more stable, safe, simple and can be expanded for mass production [8]. Electrochemical oxidation is carried out in simple equipment to produce uniform dots [9]. The electrochemical oxidation of graphite produces oxidized carbon with a high CO ratio and makes it possible to mass-produce the oxidized carbon (Ambrosi et al, 2015). The potential for electrochemical oxidation is 10 volts. At a voltage of 10 volts, the peeling process is slow and inefficient while a voltage greater than 10 volts makes the peeling rate take place so that the carbon particles will become thicker [10].

2. Materials and Methods

2.1. Oxidation of Binchotan

Binchotan rod and platinum plate were used as anode and cathode. Both electrodes were placed on both sides of the beaker glass that contained 200 mL of 14 M sulfuric acid solutions as electrolytes. The electrode was connected to the two poles of the power supply with two pieces of 2 mm diameter cable. The voltage applied to the anode and cathode was 14 volts for 24 hours operation. The same procedure was carried out for a solution of phosphoric acid and oxalic. At this stage in an aqueous solution of binchotan was obtained.

Purification of binchotan aqueous solution is carried out by sonification and dialysis. The solution was placed into the vial and rinsed in water. Sonification was done in an ultrasonic bath and operated for 2 hours. The ultrasonicated solution was removed from the vial and put into a dialysis tube. The solution and dialysis tube were into a beaker glass containing demineralized water. The water in the tube was stirred with a magnetic stirrer for 10 hours so that a high concentration of carbon dot solution in the tube was obtained and stored in the vial.

2.2. Material Characterizations

Binchotan aqueous solution before and after purification were moved from a vial, filtered to produce solid for further character examinations. The solid was washed with distilled water until pH 7 and was dried in 100℃ ovens. The solid was then characterized by its morphology, crystallography, and surface functional groups by SEM, XRD, and FT- IR. SEM measurement was in JEOL JSM 9390A. XRD patterns were produced from Shimadzu 7000 equipment, and FTIR spectroscopy was carried out by Shimadzu Prestige 21.

Acrylic glass plate 3 mm thick cut with a size of 4 cm x 1 cm, then washed with alcohol for 10 minutes in 50 mL beaker glass, then cleaned again with warm water. The acrylic glass was dried in the oven for 10 minutes at 100℃. Binchotan aqueous solution was taken 5 mL with 1 ml of permanent ink in a 50 ml beaker glass. Carbon was coated on the acrylic glass using a brush and dried in 100 °C oven for an hour. At this stage, a transparent electrode is produced which is ready to determine the electrical conductivity and band gap values. The same procedure is also applied for carbon dot to an obtained electrode.

2.3. Performance Test

The transparent electrode is connected to the multimeter to measure the electrical resistivity (R) of carbon. The thickness of the glass before and after coated with the pasta was measured by using a digital caliper. The difference is the thickness of carbon to have d, and s is the measurement distance of the multimeter probes in the electrode. Therefore, K represents the conductivity. The resistivity of each glass is determined using the following mathematical formula:

$$\rho = R.\, s / d \tag{1}$$

$$K = 1 / \rho \tag{2}$$

The determination of the optical properties of biocarbon provides a transparent electrode. Spectrogram of the transparent electrode was obtained from UV-Vis spectrophotometer Shimadzu UV 7000 measurements. Binchotan transparent electrodes are inserted in the sample compartment to measure the transmittance at wavelengths of 400-800 nm in every the increasing of 20 nm. The optical band gap value is determined from the calculation of bandgap energy (bandgap) in a semiconductor material. Measurement with UV-Vis equipment will produce transmittance values in the maximum wavelength range. Electronic properties are determined based on the results of the bandgap energy calculation. Gap energy values can be obtained from the Tauc Plot method.

3. Results and Discussions

3.1. Material Characterisation

The SEM image presents a morphological interpretation of the prepared material using acids as electrolytes (figures 1). It represents the morphology of binchotan aqueous solutions produced by using 14 V of electrical potential. It can be shown that a part of the oxidized carbon is aggregated and the other is in the form of dots. In the foreground, the morphologies are dominated with aggregates while in the background are dots. A clear image of the dots that fill the solution can be seen in figure 2. The aggregates have a size range from 0.5 to 80 micron, and the dots have a size range from 40 to 150 nm. The dots tend to have regular morphology, inversely in two other electrolyte solutions, i.e. sulphate and phosphate (figure 1A and 1C). The same dot can be found in the oxidation results of polystyrene-graphite sintering [11]. The surface of binchotan aqueous oxalate (figure 1B) solution is rougher than the graphite oxidation results by pyrolysis which shows the surface of a circular hole formed. The same results can be found in binchotan aqueous sulphate solution.

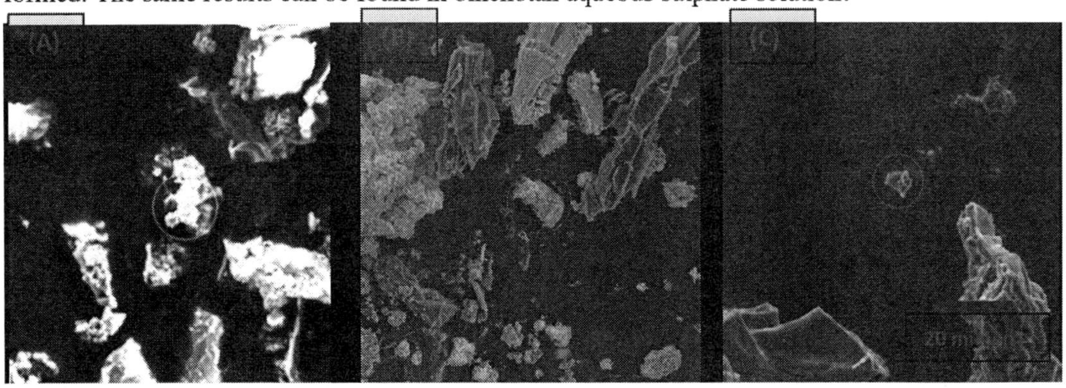

Figure 1. SEM images of binchotan aqueous solution contain electrolytes (A) sulphate (B) oxalate (c) phosphate.

Figure 2. SEM image of binchotan aqueous solution contain (A) phosphate and (B) sulphate 14M, to show the dots in the carbon matrix.

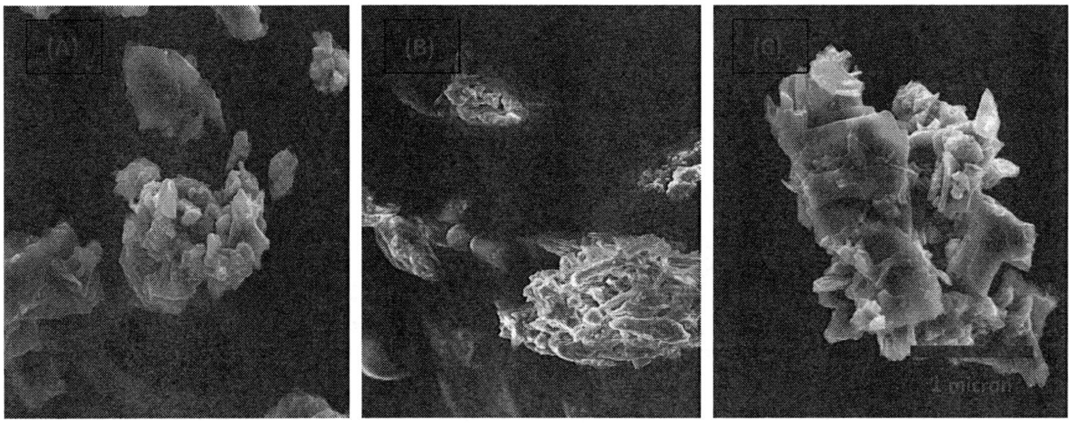

Figure 3. SEM images of binchotan aqueous solution after purification with electrolytes (A) sulphate (B) oxalate (c) phosphate.

Figure 1 (A) and (B) show the most eroded carbon surface is binchotan carbon which is oxidized electrochemically in sulphate and oxalate. The changes occur in the binchotan carbon surface after oxidation in sulphate solution, where the carbon surface becomes rougher (figure 2 B). This is because when oxidized using strong acid causes the reaction runs faster. The faster redox reaction the more oxidized binchotan carbon will also be rougher on its surface as explained in other papers [12]. The results of reduced carbon binchotan oxide show thinner aggregates on the surface of carbon that has been verified and has been observed with magnification 1000 times there is a form of carbon dot.

SEM image of binchotan aqueous with sulphate electrolyte after purification (Figure 3) shows aggregate sizes range from 80 to 2000 nm, with dots sizes range from 10 to 40 nm. Aggregates and dots size from the sulphate are the smallest compared to the other two solutions but relatively smooth (figure 3A). Figure 3B shows the aggregates and dots size of binchotan aqueous with sulphate electrolyte after purification that relatively the same as the sulphate with a more abrasive surface. Bigger aggregates dan dots can be found in binchotan – phosphate aqueous solution. Generally, the aggregates and dots become smaller after purification. The size reductions are caused by the sonification process that induces flaking of the carbon layer in binchotan and reducing the surface of the carbon. The use of sonification also affects increasing temperature. The increase in temperature causes the bonds of oxygen and carbon to become weak so that hydrogen gas can easily push oxygen away from carbon, called the reduction process. The reduction in oxygen in the carbon causes loss of carbon mass [13].

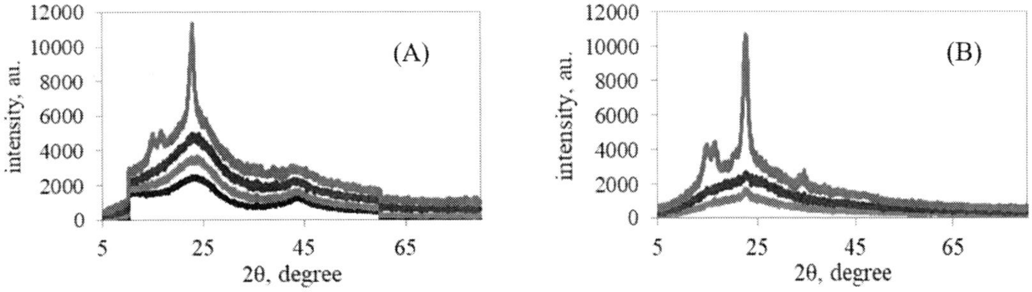

Figure 4. Diffractogram of binchotan (black line) and binchotan aqueous solution contain electrolytes sulphate (blue line), oxalate (red line), phosphate (green line); (A) before and (B) after purification.

The results using X-ray measurement can be seen in Figure 4. The diffractograms show that there is no significant change in carbon crystallography binchotan after oxidation using weak electrolytes (oxalate and phosphate). Inversely, crystallography of carbon is sharper after oxidation in sulphate, it can be shown that the peak in $2\Theta = 23$ is sharper with addition to peak in $2\Theta = 14^{o}$ that belong to graphene-like binchotan. The addition of acid as electrolytes is more likely to carry out the oxalate reaction with hydroxide from carbon so that more gas is formed, where the reaction occurs as follows (xxxx, 0000):

$$X^{-} + C\text{-}OH \longrightarrow HOX + C^{+} + e^{-}$$

$$(COOH)_2 + XOH \longrightarrow 2CO_2 + H_2O + HX$$

Where X is an anion that is in an electrolyte solution. Therefore, the addition of HX will increase the production of oxidative products, in this case, HX is carbon. At the same time, some of the acids decrease with an increase in the initial concentration of acid. This causes an increase in the amount of oxalic acid needed for the amount of OX^{-} which is higher for oxidation. Conversely, higher concentrations of acid can reduce the diffusivity of the ions from the mass of the solution to the surface of the electrode and consequently reduce the rate of oxalic acid removal (Alhamed et al, 2016). In electrochemical oxidation, three types of acids, hydroxyl and carboxylic functional groups are formed. Both functional groups have the ability to form bonds between layers of carbon in addition to the electron cloud from C = C conjugated to the carbon layer. This is because the existence of functional groups in sulphate is relatively lower than weak electrolyte (oxalate and phosphate). The decrease in the number of functional groups increases the mobility of the carbon layer. This increase in carbon mobility provides greater opportunities for formation in carbon crystallites. The increasing size in carbon crystallites was detected by the presence of sharp peaks in the diffractogram [14].

The results of the purification of carbon binchotan oxide carried out using ultrasonic waves which are shows in the diffractograms Figure 4B. In addition, the reduction of shows the presence of a new diffraction peak, which is around the region of $2\Theta = 35^{o}$. This peak comes from diffraction by silica/alumina [15]. This is because, in addition to reduction, the carbon surface treated by the ultrasonic method also exfoliates the carbon so that the presence of smaller particles begins to appear. This causes no significant changes in binchotan aqueous solution after purification as shown in figure 4B.

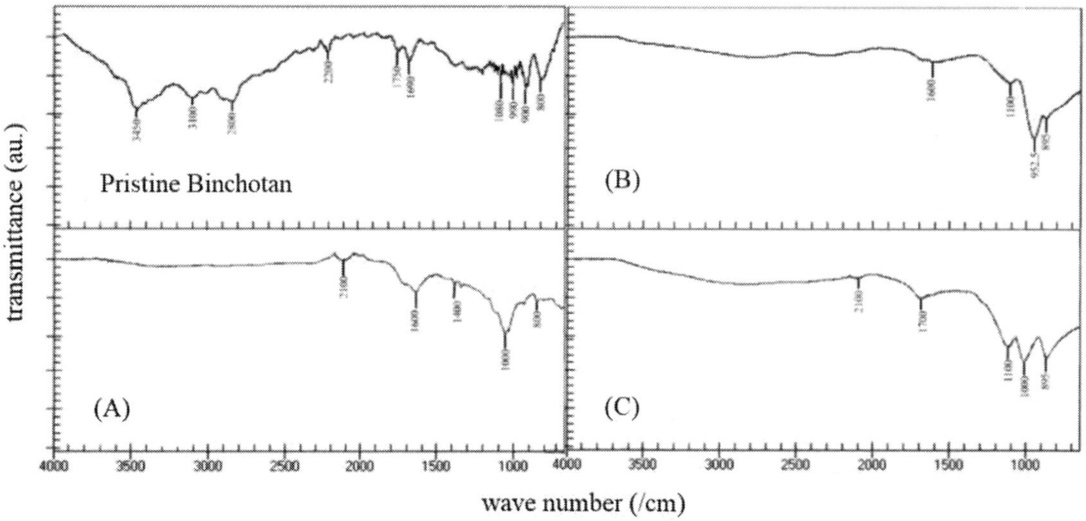

Figure 5. FTIR Spectrogram of prinstine binchotan and binchotan aqueous solution contain electrolytes (A) sulphate, (B) oxalate, (C) phosphate before purification.

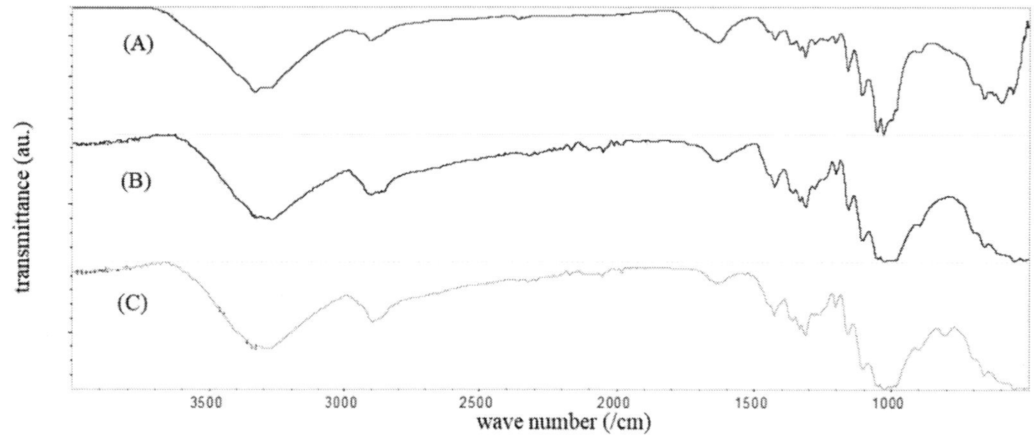

Figure 6. FTIR Spectrogram of pristine binchotan and binchotan aqueous solution contain electrolytes (A) sulphate, (B) oxalate, (C) phosphate after purification.

Analyses of FTIR provide more insight on the functional group of binchotan aqueous solution in electrolytes before purification (Figure 5). In general, FTIR spectras show that all binchotan aqueous solution have absorption at 3450 cm $^{-1}$ for CO in alcohol, 3100 cm^{-1} for CH in aromatic, 2800 cm^{-1} for CH in alkane, 2200 cm^{-1} for C≡C, 1750 cm^{-1} - 1690 cm^{-1} for C = O in carboxylic acid, and 1080 cm $^{-1}$ - 800 cm $^{-1}$ for C = C. FTIR sulphate (Figure 5A) spectra showed several absorption areas including at 2100 cm $^{-1}$ for C≡C alkaline functional groups, 1600 cm -1 for C = C aromatic functional groups, 1400 cm $^{-1}$ for CO alcohol functional groups, and 1000 cm^{-1} – 800 cm^{-1} for the alkene C = C. The FTIR spectra of the oxalate (Figure 5B) shows the absorption at 1600 cm -1 for the aromatic functional group C = C and 1100 cm^{-1} - 952.5 cm^{-1} for the alkene C = C functional group. The FTIR (Figure 5C) for the phosphate shows the peak at 2100 cm^{-1} for the C≡C alkaline functional group, 1700 cm $^{-1}$ for the carboxylic acid C = O function group, and 1100 cm^{-1} – 895 cm^{-1} for the C = C alkene group.

Analyses of FTIR provide more insight into the functional group of binchotan aqueous solution in electrolytes after purification (figure 6). The spectra for sulphate after purification in figure 6A shows the peak, i.e. 3333.36 cm^{-1} for the OH alcohol function group, 2900.94 cm^{-1} for the alkane CH group, 1633. 61 cm^{-1} for the alkene functional group C = C, 1427, 75 cm^{-1} for the alkane functional group, 1315.31 cm^{-1} - 1030.56 cm^{-1} for the CO of alcohol functional group. FTIR spectra for oxalate after purification in figure 6B shows the absorption at an area of about 3400 cm -1 for OH functional groups, 2928.69 cm^{-1} for alkane CH functional groups, 1700 cm^{-1} for carboxylic acid functional groups C = O, 1137.06 cm $^{-1}$ -1022.83 cm $^{-1}$ for the CO alcohol functional group, and 867.65 cm $^{-1}$ for the C = C alkene functional group. FTIR spectra for phosphate after purification in Figure 6C shows the presence of 2847.72 cm^{-1} alkanes CH functional groups, 2319.40 cm $^{-1}$ for OH alcohol functional groups, 1621.05 cm^{-1} there were alkene functional groups C = C, 1104.45 cm^{-1} there is an alcohol functional group CO, and 971.98 cm^{-1} there is an alkene functional group C = C.

According to Quezada-Renteria, et.al [16], the change in functional groups is caused by the carbon oxidation occurred at 1.2 Volts forming hydroxyl groups on carbon. The oxidation strength is affected by the strength of the electrolyte. Where in this study the electrolyte strength series is $H_2SO_4 > H_3PO_4 > H_2C_2O_4$. The strength of oxidation helps the change in the hydroxyl group to the carbonyl group further to a carboxylate [17]. So that there is a large hydroxyl content in binchotan aqueous sulphate solution. In the reduction process, the ultrasonic treatment occurs in the formation of C-H bonds in the reduced carbon binchotan. This is due to the insertion of hydrogen atoms from water molecules into the carbon layer aided by ultrasonic waves. Besides, the O-H group from water is also bound to carbon as evidenced by the absorption peak in the FTIR spectra which tends to increase compared to the absorption peak in the O-H vibrational region.

4. Conclusions

Morphological interpretation of SEM images of the binchotan aqueous solution using acids as electrolytes shows that a part of the oxidized carbon is aggregated and the other is in the form of dots. In the foreground, the morphologies are dominated with aggregates while in the background are dots. SEM images show that the aggregates have a size range from 0.5 to 80 micron, and the dots have a size range from 40 to 150 nm for binchotan aqueous solution before purification and range from 80 to 2000 nm, with dots sizes range from 10 to 40 nm for binchotan aqueous solution after purification.

The results using X-ray show that there is no significant change in carbon crystallography binchotan after oxidation using weak electrolytes (oxalate and phosphate). Inversely, crystallography of carbon is more sharp after oxidation in sulphate, it can be shown that the peak in $2\Theta = 23$ is sharper with addition to peak in $2\Theta = 14°$ that belong to graphene like binchotan. FTIR spectras show that all binchotan aqueous solution have absorption at 3450 cm $^{-1}$ for CO in alcohol, 3100 cm^{-1} for CH in aromatic, 2800 cm^{-1} for CH in alkane, 2200 cm^{-1} for C≡C, 1750 cm^{-1} - 1690 cm^{-1} for C = O in carboxylic acid, and 1080 cm $^{-1}$ - 800 cm $^{-1}$ for C = C. The oxidation strength is affected by the strength of the electrolyte. Where in this study the electrolyte strength series is $H_2SO_4 > H_3PO_4 > H_2C_2O_4$.

Acknowledgments

The financial support for this research was provided by Universitas Sriwijaya. LPPM Universitas Sriwijaya is gratefully acknowledged for the award of 'hibah kompetitif' and 'hibah profesi' grant and supported by USAID through Sustainable Higher Education Research Alliances (SHERA) Program.

References

[1] Ertl P 2017 *J. Cheminformatics* **9**
[2] Chen J, Zhang Y, Zhang M, Yao B, Li Y, Huang L, Li C and Shi G 2016 *Chem. Sci.* **7** 1874–81
[3] Lim S, Yoon S-H, Mochida I and Chi J 2004 *J. Phys. Chem. B* **108** 1533–1536
[4] Chia C H, Joseph S D, Rawal A, Linser R, Hook J M and Munroe P 2014 *J. Anal. Appl. Pyrolysis* **109** 215–21
[5] Henri H, Hakim L and Batoro J 2017 *Biosaintifika J. Biol. Biol. Educ.* **9**
[6] Syarif N 2014 *Energy Procedia* **52** 18–25

[7] Ahirwar S, Mallick S and Bahadur D 2017 *ACS Omega* **2** 8343–53
[8] Zhang J and Yu S-H 2016 Mater. Today 19 382–93
[9] Fu Y, Gao G and Zhi J 2019 J. Mater. Chem. B 7 1494–502
[10] Yu Z, Tetard L, Zhai L and Thomas J 2015 Energy Env. Sci 8 702–30
[11] Rassaei L, Sillanpää M, Bonné M J and Marken F 2007 *Electroanalysis* **19** 1461–6
[12] Yi Y, Weinberg G, Prenzel M, Greiner M, Heumann S, Becker S and Schlögl R 2017 *Catal. Today* **295** 32–40
[13] Zhang G, Wen M, Wang S, Chen J and Wang J 2018 RSC Adv. 8 567–79
[14] Borah D, Satokawa S, Kato S and Kojima T 2008 Appl. Surf. Sci. 254 3049–56
[15] Syarif N, Ivandini Tribidasari and Wibowo W 2012 Int. Trans. J. Eng. Manag. Appl. Sci. Technol. 3 21 – 34
[16] Quezada-Renteria J A, Ania C O, Chazaro-Ruiz L F and Rangel-Mendez J R 2019 Carbon 149 722–32
[17] Bottani E J and Tascón J M D 2008 *Adsorption by carbons* (Amsterdam ; Boston ; London: Elsevier)

IOP Conf. Series: Materials Science and Engineering **796** (2020) 012058 doi:10.1088/1757-899X/796/1/012058

Degradation of Nitrate, Ammonium and Phosphate in Domestic Wastewater by Aquatic Plants, *Actinoscirpus grossus* in Floating Treatment Wetland System (FTWs)

Syafrizal [1], Suhendrayatna[1], Muhammad Zaki[1], Elvitriana[2]
[1]Chemical Engineering Department Faculty of Engineering, Syiah Kuala University, Darussalam, Banda Aceh, 23111, Indonesia.
[2]Environmental Engineering Department, University of Serambi Mekkah, Banda Aceh, 23246, Indonesia.

E-mail: suhendrayatna@unsyiah.ac.id

Abstract. Ecological technologies such as wetlands constructed for wastewater treatment are innovative solutions for environmental protection and restoration. This study examines a Floating Treatment Wetlands System (FTWs), which is a new treating concept using macrophytes rooted in modified growing aquatic plants with floating systems. An aquatic plant, *Actinoscirpus grossus* obtained from rice fields in Banda Aceh. FCWs filled with five plants per shoot with five compartments were fed with domestic wastewater with a flow rate of 7 L/hour. Plant height was varied by 90cm - 150cm in Pond 1 and 50cm - 90cm in Pond 2, and Pond 3 was prepared as a control (without plants). Water quality in influent and effluent was analyzed every two weeks with a duration of 18 weeks and nine times sampling. Results showed a decrease in the concentration of nitrate, ammonium, and phosphate in the effluent flow with degradation efficiency (% DE) on average, NO_3-N: 76.34%; NH_4-N: 97.75%; and PO_4^{3-}: 89.45%; respectively. The degradation of domestic wastewater showed very significant results. The periodic harvest management process becomes an important part of aquatic plants, *Actinoscirpus grossus* to achieve optimum results in treating waste, i.e., for 112 days with a maximum plant growth height of 165cm and 173cm for both variations of experimental ponds.

1. Introduction

Nowadays, the most dominant pollutants in water bodies derived from domestic wastewater, which reaches 60 - 70%[1]. The technology of treating wastewater with energy and centralized costs has proven to be ineffective in solving problems related to complex water discharge resulting from rapid urbanization [2], especially the City of Banda Aceh [3]. The idea for artificial wetlands (CWs) has emerged and is currently a viable option for wastewater treatment, which is recognized as an attractive alternative to conventional wastewater treatment methods. This system offers the overall need for low and sustainable operational costs [4][5]. Various types of CWs include (i) surface free water flow (FWS); (ii) subsurface flow (SSF); (iii) hybrid system; and (iv) floating wetlands (FTWs). The selection of all types of artificial wetland constructions above is very dependent on the target needs, especially concerning the maintenance system, geographic, cost, available area, and processing purpose [6]. Today's rapid development in processing domestic waste, one of which is the bioremediation system [7][8] using phytoremediation techniques (Phyto which means plants and

Content from this work may be used under the terms of the Creative Commons Attribution 3.0 licence. Any further distribution of this work must maintain attribution to the author(s) and the title of the work, journal citation and DOI.
Published under licence by IOP Publishing Ltd

IOP Conf. Series: Materials Science and Engineering **796** (2020) 012058 doi:10.1088/1757-899X/796/1/012058

remediation which means cleaning) by utilizing local natural green plants directly to absorb/remove pollutants through roots and translocate to the surface plant parts, various plant species are used [9]. As shown by [10], the plant species used for phytoremediation must be native plants and have a rapid growth rate, many root systems, high yields of biomass, can adapt to various habitats, have a high tolerance and can accumulate pollutants on the top of the land. Some environmental factors such as temperature, pH, solar radiation, and water salinity can affect plant growth and its performance in phytoremediation. From several previous studies conducted by researchers [2], [11], and [12], aquatic plant *Actinoscirpus grossus* is very suitable for phytoremediation systems that superior compared to some other similar plants, both decreasing levels of BOD, COD, NH_4^+-N, NO_3-N, Total Phosphate, coliform fecal, and ammonium. Besides, as a local plant, this plant has resistance to the season, weather, many root systems and strong, and also has a large biomass yield which very suitable in the tropics [13]. This study assessed the effectiveness of domestic waste treatment by the aquatic plant *Actinoscirpus grossus* using the FTWs.

2. Materials and Methods

2.1 Materials

Constructed Wetland (CWs) pond was modified from wood and vinyl materials available on the market. Stereoform and wood were used between plants to support plants do not fall and spread in the CWs. Influent was accommodated in a storage tank equipped with a water discharge controller before being distributed to the treatment pond (3 ponds) which were pumped from drainage with a peristaltic pump. The effluent flow was collected in a 600mL bottle container.

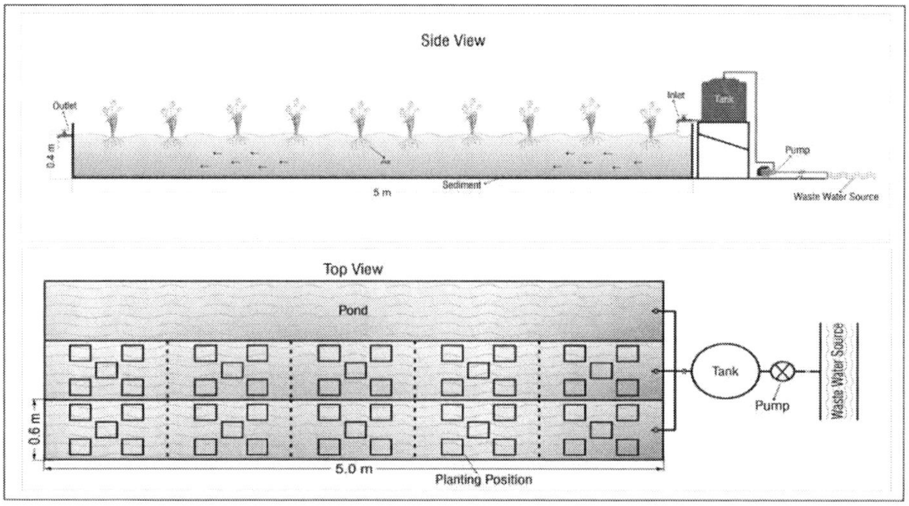

Figure 1 View of *Floating Treatment Wetland System (FTWs)*

2.2 CWs Design

Three CWs ponds were prepared with a size of 5.0m x 0.6m x 0.4m (long x wide x height) respectively, two ponds for treatment test and one pond as a control (without plants). *Actinoscirpus grossus* plant was taken from the rice fields in Ilie Ulee Kareng Village, Banda Aceh using a manual procedure. The size of plants was grouped into two groups according to the research variable; the first group has an average height of 50-90cm while the second group was 100-150cm high. CWs were filled with domestic wastewater, then each CWs was filled with five plants per shoot with 25 shoots per CWs.

2.3 Wastewater Degradations Process

The plants were acclimated for 1-2 weeks by ensuring the availability of sufficient wastewater in the continuously FTWs. The degradation process occurred when plants have been able to grow and adapt which was indicated by the growth of new roots. This condition proved that plants have been adapted to the new environment and already used to promote phytoremediation in FTWs. The FTWs system could be carried out for the treatment process that flowed continuously with new domestic wastewater with a flow rate of 7 L/hour. Water quality in influent and effluent was observed and analyzed every two weeks with a duration of 18 weeks with nine times sampling analysis. Samples were analyzed using the analysis method of the Indonesian National Standard.

3. Analysis Results

The average temperature and pH of influent were 28±4°C and 6.69±0.7, respectively. The effect of plants on water quality was observed as a significant reduction in the parameters of NH_4-N, NO_3-N, and PO_4^{-3} (Table 1). The concentration of the test parameters above, which were significantly degraded indicates that the FTWs system was functioning properly. This condition occurred because of the low Hydraulic Loading Rate (HLR) of 7 L/hour which results in a high Hydraulic Retention Time (HRT) so that it can make good contact between roots with wastewater.

Table 1. Degradation Efficiency (% DE) of domestic wastewater laboratory results on FTWs system with various plant heights (T).

No.	Parameter		pH	NH_4-N (mg/L)	NO_3-N (mg/L)	PO_4^{-3} (mg/L)
		Indonesian Standard	6.5 – 8.5	0.06	10	400
1.	Control (without plant)	Average DE (%)	7.07	63.22%	34.27%	45.84%
2.	Plants T 50-90cm		6.65	97.14%	72.74%	88.03%
3.	Plants T 90-150cm		7.04	98.36%	79.94%	90.87%
	% DE Average		6.92	97.75%	76.34%	89.45%

The most significant degradation in domestic wastewater with the FTWs system results from the three experimental ponds occurred in ponds with plant heights T90-150cm, NH_4-N 98.36%; NO_3-N 79.94% and PO_4^{-3} 90.87%, respectively.

3.1 The Relationship Between Degradation of Nitrogen Compound Content (NH_4^+ and NO_3^-) with Phosphate Compounds (PO^{4-}) on the Growth of Aquatic Plants Actinoscirpus grossus

Nitrogen (N) and Phosphorus (P) are nutrients that are needed by plants in large quantities. Nitrogen was important in the formation of chlorophyll, protoplasm, proteins, and nucleic acids. This element has an important role in the growth and development of all living tissues [14]. Phosphorus (P) was an important component of compounds for energy transfer (ATP and other nucleoproteins), for genetic information systems (DNA and RNA), for cell membranes (phospholipids), and phosphoprotein, P in CWs systems being deposited as microorganism nutrients and absorbed directly by plants and is needed for growth, roots and flowering [15]. Neutral pH conditions between 6.5-7.5 cause the absorption of NH_4^+ to be dominant compared to NO_3^-, the plant itself could reduce the need for nitrate absorption because ammonium compound could be immediately transported to the cell through translocation to other plant organs, starting from stems, leaves, canopy and other parts without the need to undergo the process of nitrification first. Nitrogen compounds (NH_4^+ and NO_3^-) were directly translocated so that they have a significant influence in providing protein for plants; chlorophyll synthesis also increases, which results in a photosynthetic activity that produces assimilates. The asymylate results are then translocated to the more dominant parts, namely shoots or apical. If the results of assimilating have been fulfilled in the shoots, then it is translocated to other organs, especially around the canopy so that the fresh weight of the plant can increase its growth. *Actinoscirpus grossus* adsorb phosphorus in the form of PO_4^{-3} through roots, which will be reduced through the harvest process, and the presence of microorganisms in the roots can reduce pollutants in the waste [16]. Therefore, the presence of phosphorus will increase plant growth that will also have an impact on the optimal absorption of nitrogen, causing the absorption of phosphor (as nutrients) to be more effective.

Figure 2 reports that the average reduction efficiency of NH_4^+ for treatment plants (T 50-90cm and T 90-150cm) was found to be 96.57% which was more dominant than the average decreasing of NO_3^- (77.04%). Optimal absorption of nitrogen nutrients (NH_4^+ absorption is more dominant than NO_3^-) results in maximum nutrient P being degraded by aquatic plants with an average phosphate absorption (PO4-3) of 89.92% for treatment plants and 45.92 % for control plants, respectively.

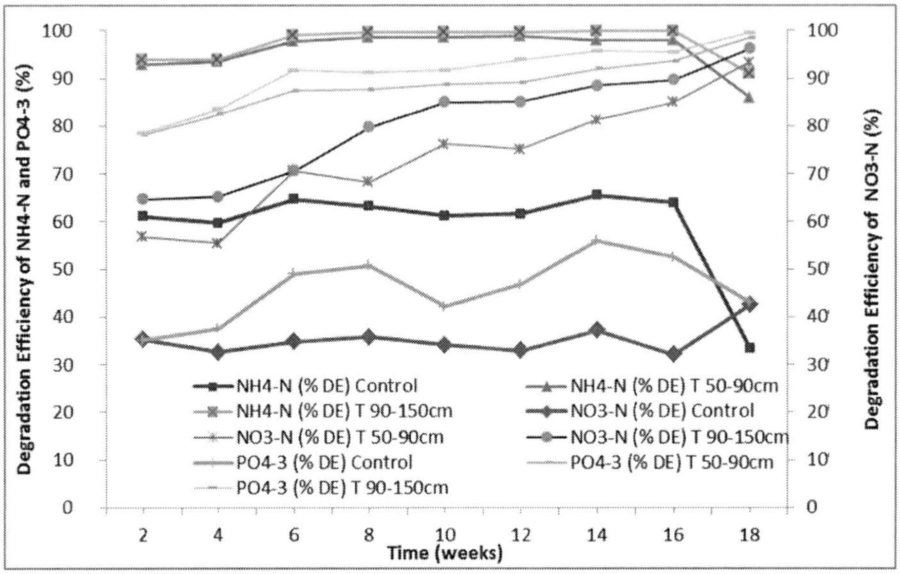

Figure 2 Degradation efficiency of NH_4^+ and NO_3^-

3.2 Periodic Impact of System Sustainability on Harvesting of Aquatic Plants Actinoscirpus grossus

Figure 3 shows the growth and development of the Actinoscirpus grossus plant when the study begins which is observed regularly every week. It takes about 60 days since the first plant was exposure into the experimental pond to stabilize the first growth until the plant is ready for the first harvest (initial cutting). Based on observations observed, it occurred that the average strand length had been stable with growth. Although macrophyte productivity influenced by the growth rate after the first harvest, however, the growth and development process at this initial stage shows a stable process, this is because the water plant *Actinoscirpus grossus* has undergone a stabilization process before the experiment begins so that every process of research *Actinoscirpus grossus* has gone through its recovery process. From the data, it could be found that the maximum height of this plant was 165 cm for plants with an average height of 50-90cm and 171cm for plant height 90-150cm. When the plant has experienced a maximum height, the first harvesting process begins (Figure 3). When the plant has experienced a maximum height, the first harvesting process begins.

Research data shows that planting and harvesting *Actinoscirpus grossus* could be done periodically and sustainably. Throughout the research, the process of domestic wastewater supply was carried out continuously with a flow rate of around 7 L/hr; this maximum water velocity caused contact between wastewater and plant roots became very maximal. FTWSs systems in floating aquatic plants were planted with *Actinoscirpus grossus* on October 28, 2018, and continue to be given domestic waste until the end of June 25, 2019. Plant growth was observed every week. Variations in shoot height were shown in Figure 3. The average maximum shoot height was 165cm in the initial plants T 50-90cm and 173cm for the initial plants 90-150cm in the first harvest. The growth curve showed that shoots reach a maximum growth of 120 days during the first harvest. Subsequent growth was expected to be faster, possibly only needing under 120 days to reach maximum shoot height after each harvest [13].

Figure 4 compares the average plant height above the water level in the FTWs that appeared one month to 4 months after the first harvest. [17] reported that harvest can affect nutrient storage in annual plant tissues, depending on harvest time and the type of biomass (leaves, rhizomes, or both) harvested and that it can also reduce carbohydrates available for early season growth and strand strength. Harvesting on land has been reported harm on biomass production, although total productivity was much higher if shoots were moved during the period of growth. It can also be seen the results in Figure 4, which showed that the maximum height achieved in sustainable activities in the harvest of *Actinoscirpus grossus.*

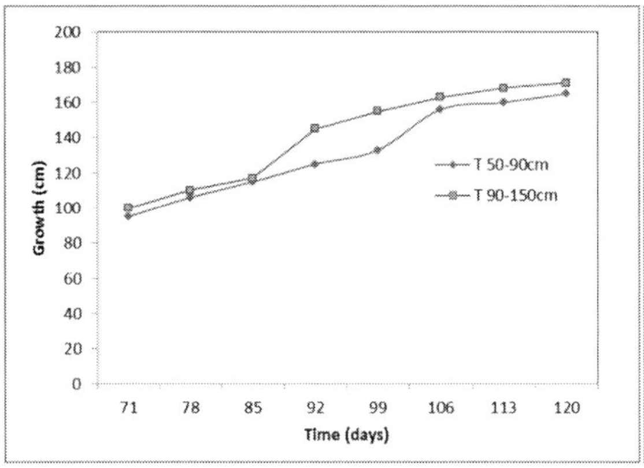

Figure 3. Growth of *Actinoscirpus grossus* before harvesting.

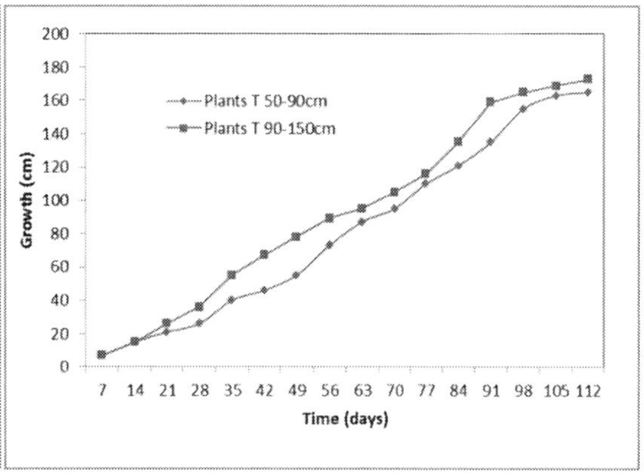

Figure 4. Growth of *Actinoscirpus grossus* after first harvesting.

Harvesting can be considered as an effective management tool for achieving proper vegetation balance (balance between adequate vegetation for recovery while controlling excess detritus formation if needed for optimal processing performance. If excess vegetation is not harvested, most nutrients that have been absorbed into plant tissues will be returned to water through decomposition. The built-up wetland system can act as a reservoir for nutrients during active growth [18], and sustainable harvesting in tropical conditions does not directly encourage active growth throughout the year — high above-ground biomass in wetlands built on situations of general hydraulic and nutritional loading [19].

4. Conclusion

This study concludes that the aquatic plants, *Actinoscirpus grossus* can degrade NH_4-N, NO_3-N, and PO_4^{-3} from the domestic wastewater with degradation efficiency (%DE) on average, NO_3-N: 76.34%; NH_4-N: 97.75%; and PO_4^{3-}: 89.45%; respectively. The system run with the HLR 7 L/hour and HRT 6 days with a pond volume of 1 m^3, pond size (5.0m x 0.6m x 0.4m); planting plants; 5 shoot and 25 shoots per ponds, and all of these characteristics can degrade significantly.

References

[1] Filliazati M Apriani I and Zahara T A 2011 Pengolahan limbah cair domestik dengan biofilter aerob menggunakan media bioball dan tanaman kiambang *Tek. Lingkung.* 1.
[2] Liu Y Zhang D Q Ng W J Gersberg R M Tan S K and Jinadasa K B S N 2014 *J. Environ. Manage.* **141** 116.
[3] Bappeda, 2011 *Laporan Evaluasi Rencana Pembangunan Jangka Menengah (RPJM) 4 Tahun 2007-2010* Banda Aceh.
[4] Kivaisi A K 2001 *Ecol. Eng.* **16** (4) 545.

[5] Haberl R 1999 *Water Sci. Technol.* **40** (3) 11.
[6] Horner J E Castle J W Rodgers J H Gulde C M and Myers J E 2012 *Water. Air. Soil Pollut.* **223** (5) 1945.
[7] Vijayaraghavan K and Balasubramanian R 2015 *J. Environ. Manage.* **160** 283.
[8] Suhendrayatna M Andriani R and Fajriana Y, Elvitriana E 2012 *Int. J. Eng. Innov. Technol.* **2** 6.
[9] Sharma S Singh B and Manchanda V K 2014 *Environ. Sci. Pollut. Res.* **22** (2) 946.
[10] Valipour A and Ahn Y H 2016 *Environ. Sci. Pollut. Res.* **23** (1) 180.
[11] Tanaka N Jinadasa K B S N Werellagama D R I B Mowjood M I M and Ng W J 2006 C *J. Environ. Sci. Heal. - Part A Toxic/Hazardous Subst. Environ. Eng.* **41** (10) 2221.
[12] Kantawanichkul S Somprasert S Aekasin U and Shutes R B E 2003 *Water Sci. Technol.* **48** (5) 199.
[13] Jinadasa K B S N Tanaka N Mowjood M I M and Werellagama D R I B 2006 *Ecol.* **21** (4) 603.
[14] Tanner C C Kadlec R H Gibbs M M Sukias J P S and Nguyen M L 2002 *Ecol. Eng.* **18** 499.
[15] Grant C A Flaten D N Tomasiewicz D J and Sheppard S C, 2000 The importance of early-season phosphorus nutrition *Can. J. PLANT Sci.*
[16] Chen C Zhao T and Liu R, 2017 Performance of five plant species in the removal of nitrogen and phosphorus from an experimental phytoremediation system in the Ningxia irrigation area.
[17] Dubbe D R Garver E G and Pratt D C, 1988 Production of Cattail (Typha) **17** p. 79–104.
[18] Hoagland C R Gentry L E David M B and David A, 2014 Plant Nutrient Uptake and Biomass Accumulation in a Constructed Wetland October 2014 p. 37–41.
[19] Tanner C C and Headley T R 2011 *Ecol. Eng. 37 474–486* **37** p. 474–486.

IOP Conf. Series: Materials Science and Engineering **796** (2020) 012059 doi:10.1088/1757-899X/796/1/012059

Performance of *Terminalia Catappa* Leaves Extract as Bio-Corrosion Inhibitor for Mild Steel in H_2SO_4 Solution

Marta Pramudita[1,2], Sukirno[1], M Nasikin[1*]

[1]Chemical Engineering Department, Faculty of Engineering, Universitas Indonesia, Kampus Depok Baru, 16640 Indonesia.
[2] Chemical Engineering Department, Faculty of Engineering, University of Sultan Ageng Tirtayasa, Banten 42124 Indonesia

E-mail: mnasikin@che.ui.ac.id

Abstract. *Terminalia catappa* leaves extract has a high content of polyphenol compounds. Polyphenol compounds inhibit the corrosion rate for mild steel. This study aimed to assess the ability of *terminalia catappa* leaves extract to inhibit bio-corrosion on the surface of mild steel placed in a corrosive medium of 1M sulphuric acid, and to compare this with that of tannin and sodium phosphate. The method used was the loss of weight at temperatures of 313K, 333K and 353K. It found that the inhibition efficiency was susceptible to increased temperature. The efficiency of bio-corrosion inhibition of sodium phosphate> tannin>*Terminalia catappa* leaves extract; that is, the ability of *Terminalia catappa* leaves extract to inhibit bio-corrosion was relatively lower than that of tannin and sodium phosphate. *Terminalia catappa* leaves extract's inhibiting corrosion rate of mild steel was very good at 333K and 1250 ppm concentration, with an efficiency inhibition of 64%. The adsorption mechanism is a Langmuir isotherm, and the process is endothermic and spontaneous.

1. Introduction

An inhibitor is defined as a substance that can decrease the rate of corrosion when added to an environment in small concentrations [1]. The addition of inhibitors is common in liquid-containing systems, such as boilers, condensers, heat exchanger, pipelines, and industrial petroleum and other chemical processing industries [2]. The use of chemical inhibitors such as sodium phosphate is considered to have sufficient ability to lower corrosion rates, but its use has an impact on the environment. So, it is necessary to obtain data on inhibitors that would be safe in the environment, and which are abundantly available and renewable. These are known as bio-inhibitors. Bio-inhibitors derived from natural materials ideally would not contain harmful metals, so that they can be used in the food, beverage and pharmaceutical industries. They generally come from natural material extracts that contain several organic compounds, and elements including nitrogen, oxygen, phosphorus and sulphur. These atoms are coordinated, and react with metal ions to form layers on the metal's surface to prevent corrosion [3].

Several studies on the ability of plant extracts to inhibit corrosion rates have been published, including those on *Sida acuta* [4], *Jasminum nudiflorum* [5] and *Rothmania longi flora* [6]. *Terminalia catappa* a tropical plant, widely found in Indonesia, and belonging to the *Combretaceae* family.

[*] To whom any correspondence should be addressed

Content from this work may be used under the terms of the Creative Commons Attribution 3.0 licence. Any further distribution of this work must maintain attribution to the author(s) and the title of the work, journal citation and DOI.
Published under licence by IOP Publishing Ltd

Terminalia is *catappa* leaves extract (TLE) contains polyphenol compounds that can inhibit corrosion. Tannin is a widespread polyphenolic compound found in plants, and TLE contains high concentrations of tannins [7], along with *saponin, phlobatin, antraquinone*, flavonoids, terpene and alkaloids. Tannin is a compound that can form complex compounds with iron ions. Tannins can be used in the environment because the molecules tend to be stable at low pH [8]. The chemical properties of TLE provide it with an excellent potential for development as a corrosion inhibitor, along with its proven ability to survive for 168 days, with an inhibition efficiency of up to 87% at 298K in 1M sulphuric acid solution [9]. Its abilities at higher temperatures have not been tested, however; this is necessary so that its use can be extended.

In this study, a TLE corrosion test was performed at various temperatures (313K, 333K, 353K). As an indicator of TLE capability, corrosion tests were also performed using sodium phosphate, representing a chemical inhibitor, and tannin, representing effective compound in TLE.The results of this study provide information on the thermodynamic and kinetic properties of TLE and, whether its ability can be improved so that it can find wider application.

2. Experimental

2.1. Materials

The mild steel used had a composition (wt%) of 0.54Mn, 0.05Si, 0.01S, 0.01P, 0.16C and the remaining Fe was used. The sample size was 3 x 2 x 0.5 cm. The sample was cleaned using grit 250–1000 abrasion paper, then washed with distilled water and cleaned again with acetone and ethanol.

2.2. Preparation of the Corrosion Inhibitor

The *T.catappa* leaves used were collected at the Universitas Indonesia Campus. The samples were cleaned, and then dried using sunlight for one week. Following this, a blender was used to reduce the leaves to powder. TLE was obtained through a maceration process for four days, using ethanol as a solvent at a ratio of 1:4[10]. The TLE solution was then filtered, and separated using a rotary evaporator.

2.3. The Corrosive Medium

The corrosive medium used was 1M sulphuric acid, with the concentration of the inhibitor being 0, 250, 500, 750, 1000 and 1250 ppm. The temperatures used were 313K, 333K and 353K, and the immersion time was 6 h.

2.4. Weight Loss Method and Analysis

Weight loss refers to the difference in mass of a sample before and after being immersed in a corrosive medium. This was measured for various concentrations, temperatures and times. From the data obtained,a corrosion rate was calculated, based on the weight loss, using the formula (ASTM G31):

$$CR\ (mmpy) = \frac{(87500\ x\ \Delta W)}{A\ \rho\ t} \tag{1}$$

where CR is corrosion rate, ΔW is weight loss (g), A is the surface area in cm^2, ρ is the density of mild steel (g/cm^3), and t is immersion time (hr).
The efficiency of inhibition was calculated using the formula:

$$I.E\ \% = \left(1 - \frac{CR_{inh}}{CR_{blank}}\right) x\ 100 \tag{2}$$

where CR_{inh} and CR_{blank} are corrosion rates in the presence and absence of inhibitor, respectively.
Finally, surface coverage was calculated using the equation:

$$\theta = \left(1 - \frac{CR_{inh}}{CR_{blank}}\right) \tag{3}$$

2. Results and Discussion

2.1. Effect of temperature on the corrosion inhibition efficiency of TLE, sodium phosphate and tannin

Table 1 shows the reaction rates derived from the weight loss calculations for the corrosion inhibitors at different concentrations and temperatures. From the data in Table 1, it can be seen that, for all types of inhibitors (sodium phosphate, tannin and TLE), a temperature increase causes an increase in corrosion rate. The higher the temperature, the more reaction occurs. This relates to the theory of collision, whereby, as the temperature rises, the collision between reactants will be faster. Also, in an acidic medium – in this case, sulphuric acid – the corrosion that occurs on the metal (mild steel) will be accompanied by the evolution of hydrogen gas. Higher temperatures cause a faster corrosion reaction, and the solubility of the metal increases as more and more hydrogen gas is produced [13].

Table 1. Corrosion rates for various concentration at different temperature

T K	Bio Inhibitor	Concentration ppm					
		0	250	500	750	1000	1250
313	Na_3PO_4	575	100	91	77	71	60
	Tannin	575	238	227	209	189	175
	TLE	575	245	233	237	238	226
333	Na_3PO_4	706	228	215	187	184	175
	Tannin	706	338	297	282	275	254
	TLE	706	337	320	315	308	324
353	Na_3PO_4	813	328	333	317	290	288
	Tannin	813	492	484	459	449	420
	TLE	813	582	560	508	554	456

Table 1 shows that the TLE corrosion rate is higher than that of sodium phosphate and tannin (sodium phosphate < tannin < TLE); that is, its ability to inhibit corrosion is less than that of sodium phosphate and tannin. Table 1 also shows the effect of temperature on the corrosion rate of all three inhibitors. At the same temperature, with the highest inhibitor concentration, the rate of corrosion decreases. Table 1 shows that the higher the concentration of the inhibitor, the greater is its ability to inhibit the rate of corrosion. This is because more complex compounds are formed, and these function to inhibit the corrosion rate of mild steel. To obtain the value of the activation energy, Arrhenius' equation is used[6, 14]:

$$Log\ CR = \log A - \left(\frac{E_a}{2{,}303\ RT}\right) \tag{4}$$

where *CR* is the mild steel corrosion rate, *A* is the Arrhenius pre-exponential factor, *Ea* is the activation energy, *R* is the gas constant and *T* is the temperature. Table 2 shows the activation energy values for TLE, tannin and sodium phosphate, which are TLE < tannin < sodium phosphate. The activation energy is the minimum energy required for a reaction to proceed. The higher the value, the higher the energy needed to react. When connected to the corrosion rate, the higher the activation energy, the lesser the corrosion rate, and vice versa. Surface coverage is assumed to be an inhibitor molecule that covers a protected steel surface. The Langmuir adsorption kinetics model is used to provide a characteristic description of the adsorption of a metal-ion monolayer on the surface of the adsorbent. The Langmuir isotherm illustrates that, on the surface of the adsorbent, there are a certain number of active sites that are proportional to the surface area, each active site being only one molecule that can be adsorbed.
The efficiency of inhibition indicates the ability of an inhibitor to inhibit the corrosion rate. The higher the efficiency value of inhibition, the better the inhibitory ability. The results on Figure 4 show that the

efficiency of inhibition is ranked sodium phosphate >tannin> TLE. The temperature rises also caused the inhibition efficiency to decrease, since the inhibitory efficiency is related to the corrosion rate. When associated with time, the corrosion rate decreased the longer the mild steel was immersed in the 1M sulphuric acid solution. Although all of the inhibitors saw this decrease, the sodium phosphate inhibitor decrease was lower than that of tannin and TLE.

Table 2. Activation energy for various concentration

Concentration	TLE	Tannin	Sodium Phosphate
ppm	Ea kJ/mol	Ea kJ/mol	Ea kJ/mol
0	7,97	7,97	7,97
250	21,72	20,05	27,33
500	21,91	19,62	29,89
750	22,86	24,05	32,40
1000	23,37	22,67	32,46
1250	23,25	23,10	36,29

2.2. Kinetics of Reaction

The natural interaction between a molecular inhibitor and a metal surface is approached through isotherm adsorption. The preferred adsorption method is Langmuir isotherm adsorption. The equivalent isotherm equation is[15]:

$$\frac{C}{\theta} = \frac{1}{K_{ads}} + C \tag{5}$$

where C is the concentration of the inhibitor, θ is the surface coverage and K_{ads} is the adsorption equilibrium constant. Its adsorption properties can explain the mechanism of corrosion protection. Isotherm adsorption is important for explaining the mechanism of the organo-electrochemical reaction. From this, there are known interactions among metal inhibitors, molecular inhibitors and a metal surface. The presence of metal interfaces/solutions is due to the formation of an electrolyte, or covalent bonds between the adsorbate and the atoms on the metal surface [16]. From Table 3, it can be seen that TLE shows that the adsorption of Langmuir isotherm is suitable for use at 313K and 333K.

Table 3. Langmuir isotherm parameters for the adsorption of Terminalia catappa leaves extract on the surface of mild steel

Temperature K	Kads	R^2
313	15,48	0,99
333	74,43	0,99
353	0,99	0,85

In Table 3 the R^2 value for 313K and 333K can be seen to be 0.99, while at 353K, $R^2<0.9$. At 313K and 333K, the K_{ads} value increased, while at 333K and 353K, the K_{ads} value decreased. Based on the tendency of the K_{ads} value to rise and fall when associated with temperature, K_{ads} values obtained using Langmuir isotherms, where by the values decreased with increasing temperature, indicated that the adsorption of TLE on mild steel surfaces was excellent when used at low temperatures[17]. Similarly, if the K_{ads} value increases with rising temperature, then high temperatures can be used instead.

2.3. Thermodynamic Activity Parameters ΔG, ΔH and ΔS

The thermodynamic activity parameters used are ΔG, ΔH and ΔS, and the equations are as follows[18]:

$$\Delta G^{o}_{ads} = -2.303\ RT \log(55.5\ K_{ads}) \tag{6}$$

$$K_{ads} = \frac{\theta}{(1-\theta)}\ [C] \tag{7}$$

$$\Delta S = \frac{\Delta H_{ads} \Delta G_{ads}}{T} \tag{8}$$

By making a linear graph between log (CR/T) and(1/T), then ΔH_{ads} is derived from the slop (-ΔH_{ads}/2.303R). The thermodynamic parameter shows the ability of the adsorption of the inhibitor on a mild steel surface submerged in a 1M sulphuric acid solution, the data can be seen in Table 4. From the results obtained from the calculations, ΔG TLE is negative, which indicates that the adsorption of molecule inhibitors on the surface of mild steel takes place spontaneously; in general, the reaction takes place by physisorption. The value of ΔH_{ads} shows the amount of heat adsorbed during the adsorption process. ΔH_{ads} on a positive-value table show that the process is endothermic. The higher the temperature, the more heat is adsorbed in the inhibition process. The entropy value indicates whether the reaction is reversible or non-reversible, and the stability of the inhibitory layer that formed on the mild steel surface. From Table 4, ΔS can be seen to be negative, indicating the instability of the inhibitory layer formed, and that the reaction takes place non-reversibly.

Table 4. Thermodynamic adsorption parameters at different temperature

	Na3PO4			Tannin			TLE		
T K	ΔHads kJ/mol	ΔG kJ/mol	ΔS kJ/mol.K	ΔHads kJ/mol	ΔG kJ/mol	ΔS kJ/mol.K	ΔHads kJ/mol	ΔG kJ/mol	ΔS kJ/mol
313	6,15	-14,25	-0,28	8,24	-15,10	-0,39	8,24	-17,58	-0,46
333	6,55	-12,88	-0,25	8,77	-15,87	-0,42	8,76	-23,05	-0,61
353	6,94	-11,58	-0,23	9,29	-14,37	-0,38	9,29	-11,76	-0,31

3. Conclusions

1. The ability of TLE to inhibit the corrosion rate of mild steel at 333K and at a concentration of 1250 ppm is a 64% inhibition efficiency. At this temperature, the performance of TLE is optimal. The adsorption mechanism is the Langmuir isotherm, the process occurring endothermically, and the inhibition layers tending to be unstable.
2. The inhibition efficiency of TLE is still less than that of sodium phosphate and tannin, so it will be necessary to undertake further research in order to improve this ability in TLE.
3. TLE's sensitivity to temperature is quite significant, characterised by higher temperatures causing lowered efficiency of corrosion inhibition.

Acknowledgement

This work is supported by Hibah PITTA 2018 by DRPM Universitas Indonesia No. 5000/UN2.R3.1/HKP.05.00/2018.

References

[1] N. International. NACE International Glossary of Corrosion-Related Terms [Online].
[2] M. A. Amin, K. F. Khaled, Q. Mohsen, and H. A. Arida 2010 *Corros. Sci* **52** 1684-1695
[3] Y. M. A. K. Shalabi, Hala M. Hassan, A.S. Fouda 2014 *Int. J. Electrochem. Sci* **9** 1468 - 1487
[4] S. A. Umoren, U. M. Eduok, M. M. Solomon, and A. P. Udoh 2016 *Arab. J. Chem* **9** S209-S224 2016.

[5] S. Deng and X. Li 2012 *Corros. Sci* **64** 253-262

[6] C. O. Akalezi, E. E. Oguzie, C. E. Ogukwe, and E. A. Ejele 2015 *Int. J. Ind. Chem* **6** 273-284
[7] A. Howell 2004, "Hydrolyzable Tannin Extract Plants Effective at Inhibiting Bacterial Adherence to surface," *United States Patent Application Publication*
[8] M. F. Montemor 2016 Active Protective Coatings, Springer Series in Materials Science **233** 107-137
[9] N. O. Eddy, P. A. Ekwumemgbo, and P. A. P. Mamza 2009 *Green Chem. Lett. Rev* **2** 223-231
[10] M. Pramudita, Sukirno, and M. Nasikin 2018 *IOP Conf. Ser.: Mater. Sci.Eng.* **345** 012023
[11] Y. Hao, L. A. Sani, T. Ge, and Q. Fang 2017 *Corros. Sci* **123** 158-169
[12] G. Socrates 2001 Infrared and Raman Characteristic Group Frequencies, Third ed. Baffins Lane, Chichester: John Willey & Sons Ltd
[13] S. Rekkab, H. Zarrok, R.Salghi, A.Zarrouk, Lh.Bazzi, and B.Hammotuti 2012 *J. Mater. Environ. Sci* **3** 613-627
[14] O. M. Onukwuli O. Dominic 2016 *J. Chem. Technol. Metall* **51** 302-314, 2016.
[15] S. Deng and X. Li 2012 *Corros. Sci* **5** 407-415
[16] R. T. Loto, C.A. Loto, and A. P. I. Popoola 2016 *J. Mater. Environ. Sci* **3** 613-627
[17] K. K. Alaneme, S. J. Olusegun, and O. T. Adelowo 2016 *Alexandria Eng. J* **55** 673-681
[18] F. Bentiss, C. Jama, B. Mernari, H. E. Attari, L. E. Kadi, M. Lebrini 2009 *Corros.Sci* **51** 1628-1635

Photocatalytic Degradation of Organic Waste in Visible light using TiO_2 Nanotubes Array

Indar Kustiningsih[1*], Cecep Saripudin[1], Siti Suwansih[1], Denni Kartika Sari, Jayanudin[1], Slamet[2]
[1]Department of Chemical Engineering, Universitas Sultan Ageng Tirtayasa, Serang, Indonesia.
[2] Department of Chemical Engineering, Universitas Indonesia, Depok, Indonesia

E-mail: indar.kustiningsih@untirta.ac.id

Abstract. Photocatalytic degradation of organic waste using TiO_2 nanotubes array has been investigated. The phenol was used as organic waste in this research. The Titania nanotubes were obtained by anodization method with 98% ethylene glycol and 0.5%wt NH_4F as electrolyte solution at 50 V for 2 hours. The prepared catalysts were characterized by different techniques such as electron microscopy (SEM), X-Ray diffraction (XRD), Fourier transform infrared (FTIR) and UV-vis reflectance spectroscopy (UV-DRS). The effect of pH and initial concentration of phenol has been evaluated. The results showed that the best pH for degradation phenol was obtained at pH 5. In this condition, the phenol has been reduced from 40 ppm to 6.9 ppm for 180 min.

1. Introduction

Phenol is a representative contaminant in industrial effluents such as wastewaters from steel industries, pharmaceutical industries, plastic and varnish industries, textile units, pesticide plants, textile and pulp, and paper manufacturing plants [1-2]. The presence of Phenol in the environment represents a serious ecological problem due to its toxicity for human life, aquatic life and others [3]. The conventional technologies are used to eliminate phenol from wastewater such as oxidation, adsorption, coagulation and enzyme oxidation [4]. However, these conventional methods could not completely degrade organic pollutants and could produce secondary pollutants [5-6].

The advanced oxidation process (AOPs) may become an interesting alternative to traditional technologies. Among AOPs, heterogeneous photocatalysis with nanoparticles has shown high efficiency in the removal of a wide range of organic contaminants in wastewater [7- 11]. Titania oxide is a well-known effective semiconductor photocatalyst due to generates hydroxyl (OH) radicals, which are more powerful than ozone and other advanced oxidation processes (AOPs) used to degrade organic pollutants. Moreover, TiO_2 has high photocatalyst activity, chemical stability relatively low-cost and especially non-toxicity [12]. Furthermore, TiO_2 has been applied in photocatalysis such as water purification and decomposition of pollutants, hydrogen generation, air purification and also medical applications, such as drug delivery, orthopaedic and dental implants [9,13-15].

* Corresponding author: indar.kustiningsih@untirta.ac.id

Content from this work may be used under the terms of the Creative Commons Attribution 3.0 licence. Any further distribution of this work must maintain attribution to the author(s) and the title of the work, journal citation and DOI.
Published under licence by IOP Publishing Ltd

However, photocatalysts in water purification and decomposition of pollutants are often used in a slurry system which needs separation and recovery of catalyst, moreover, it causes damages to the recirculation pumps. This disadvantage is overcome realizing the photocatalysis thin film. One of the photocatalysts in the form of thin films is TiO_2 nanotubes arrays (TNTs) which fabricated by anodization method. In recent years TNTs have received considerable attention due to their large surface area, high photocatalytic activity, and convenient recycling when compared with conventional TiO_2 powder [16].

The anodization method is a relatively simple process for the synthesis of the TiO_2 nanotube arrays due to the easily controlled synthesis parameters. The different structures of the TiO_2 nanotube arrays will be synthesized by adjusting anodization time, the composition of electrolyte and anodization voltage [17-18]. The highly ordered TiO_2 nanotube array has synthesized by the anodization method in this study. Anodization voltage was adjusted to control the structure of the TiO2 nanotube. The photocatalytic activity of the TiO_2 nanotube arrays was examined for phenol degradation.

2. Materials and Method

2.1. Materials.
All chemicals, used in the synthesis were of analytical grade and used as received. Ethylene glycol, HF, HNO_3, and phenol were purchased from Sigma-Aldrich.

2.2. Preparation of Titania nanotubes arrays.
The nanotube arrays were synthesized by anodic oxidation of titanium foils (3 cm x 2 cm) with 99.6% purity and 0,3 mm in thickness. Before anodization, the Ti foils were cleaned by water. Afterward, the samples were chemically etched by immersing in a mixture of 3.2 HF and 5.4 HNO_3 for 40 s and immediately rinsed with deionized water in a sonicator for 10 s, dried under air and used immediately. The anodization was performed in a two-electrode configuration bath with the Ti foil as the anode and the platinum (Pt) foil as the cathode. We used an electrolyte solution of ethylene glycol with consisting of NH_4F (0.5 wt.%) and H_2O (2 wt.%) and placed in a beaker glass for 2 h at different potential (20-60 V) and at room temperature using DC power supply (ESCORT 6030SD). After anodization, the obtained nanotube electrodes were rinsed immediately with deionized water and then dried in air. The prepared samples were annealed at 500°C for 3 h.

2.3. Sample Characterization.
The morphology of the synthesized catalyst was examined using scanning electron microscopy (SEM, JEOL 6390 A). The crystal structure was analyzed using X-ray diffraction (XRD, Analytical type EMPYREAN). Element analysis and the chemical bond of the catalyst were examined using Fourier transform infrared spectrophotometer (FT-IR, Shimadzu IR Prestige-21). The bandgap value was analyzed using UV-Vis diffuse reflectance spectra (UV-Vis DRS, Shimadzu type 2450). The amount of phenol degradation was determined using a UV-Vis spectrophotometer (Genesys 10).

2.4. Degradation of Phenol
The degradation of phenol was performed in a photocatalytic reactor with a mercury lamp as a visible light source in atmospheric pressure for 3 h. PHILIPS HPL-N 250 W mercury lamp is used as a photon source placed in a test box with a distance of 3 cm from the catalyst. This lamp is a mixture of 17.25% UV light and 82.75% visible light. The phenol solution as much 250 ml with a different initial concentration (10 ppm, 20 ppm, and 40 ppm) and different pH values (ph 5, no adjust, ph 8) was used in the degradation process. During the irradiation, samples were taken from the irradiated solutions at different time intervals 30 s.

3. Results and Discussion

3.1. Characterization

In order to investigate the effect of potential anodization on the formation of TiO_2 nanotubes, samples were prepared using different potential anodization. The structures of the prepared samples were characterized by SEM. The SEM image of TiO_2 TNTs at different voltage are shown in figure 1. Figure 1 shows the potential anodization affecting the formation of the nanotubes in the sample. Anodization with a voltage of 30 V has not produced a perfect nanotubes morphology. TiO_2 TNTs have been successfully fabricated on the surface of Ti by potential anodization at 50 V as shown in figure 2(c). The diameter and length of TNTs for various potential anodization could be seen in table 1.

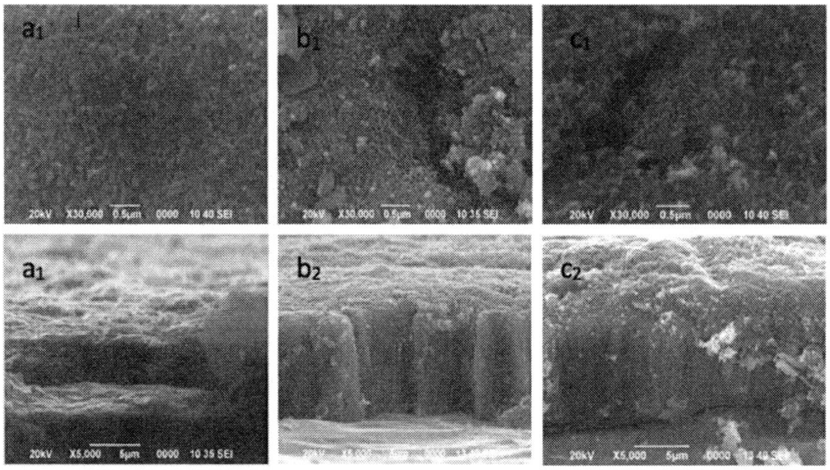

Figure 1. SEM image of top view (1) and crossed section view (2) of TNTs prepared for (a)30V (b)40V (c)50V

Table 1. indicated the greater potential anodization produced TiO_2 nanotubes which are the longer and larger diameter. This result indicated that the formation of TiO_2 nanotubes can be influenced by a voltage that was used during the anodization process. Lai C. W. and Sreekantan S. (2011) reported the greater the potential of anodization, produced the longer the tube. It is due to the greater the voltage then the driving force of ions to move through the barrier layer on the bottom of the tube becomes larger, resulting in deeper pores and longer tubes [16].

3.1.1. XRD

The XRD patterns of catalyst TNTs uncalcined and calcined at 500°C and with potential anodization showed in figure 2. It shows that without calcination exhibited an amorphous structure. The phase of TNTs at various voltage was identified to be anatase and had crystallinity were almost the same. Anatase peaks were observed at 25.3° and 48°. The peaks at 38°, 40°, 53° and 63° were identified as the main peak of Ti [19].

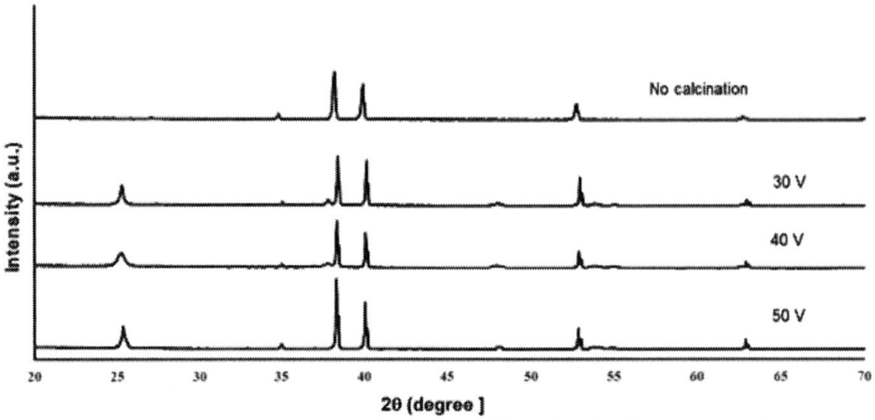

Figure 2. XRD patterns of TNT uncalcined and Bi-TNT samples.

3.1.2. FT-IR and UV-Vis DRS

The FTIR images and UV-Visible diffuse reflectance spectra for samples are shown in figure 3. Figure 3 (a) shows visible valleys in the FTIR transmittance spectrum at wavenumbers of about 3800 cm^{-1} and 1500 cm^{-1} show the stretching OH bond of the hydroxyl structure attached to Ti (Ti-OH) and OH bending (TNTAs absorb water molecules) [12]. The valleys observed in the wavenumbers 840 and 520 cm^{-1} are Ti-O-Ti stretching bonds of TiO_2 crystals.

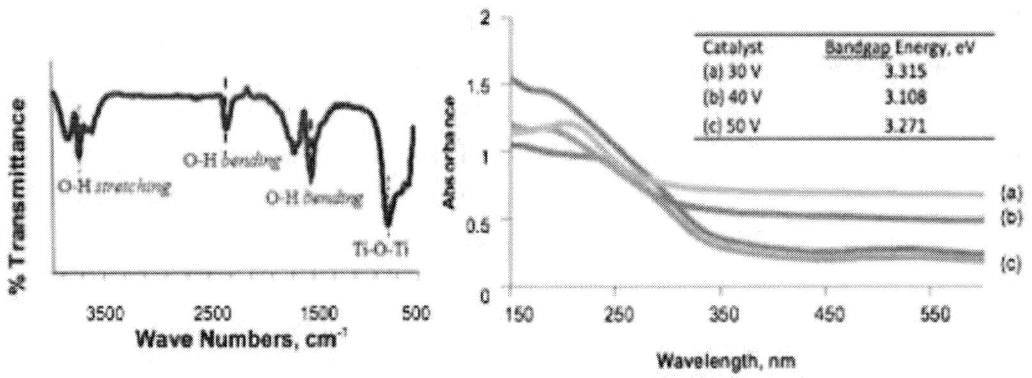

Figure 3. FTIR image of catalyst 50 Volt (a) and UV-Vis DRS (b)

The UV-Visible diffuse reflectance spectra for samples are shown in figure 3(b). It shows the value of bandgap energy. The TiO_2 catalyst with the anatase crystalline phase has a bandgap energy value of 3.2 eV [12]. The test results performed, the average bandgap energy value for the four catalysts of 3.21 eV. With this value, the catalyst will be active at the wavelength of 387 nm which is included in the visible light spectrum (380-700 nm). Sunlight contains 5% UV rays and 45% visible rays.

3.2. Photocatalytic activity.

The photodegradation of phenol was conducted to evaluate the TNTs catalyst. The relationship between potential anodization and degradation of phenol was also investigated, as shown in figure 4.

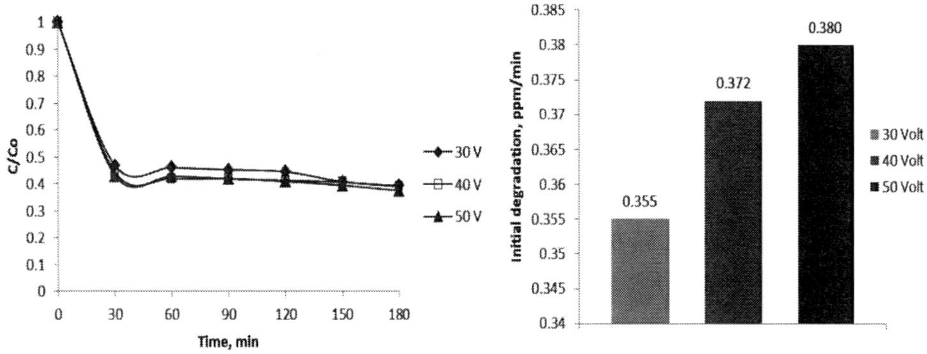

Figure 4. Effect of differential potential anodization on (a) phenol degradation and (b) initial degradation rate of phenol.

Figure 4 shows the results of phenol degradation using TNTs. It was found that TNTs 50V exhibited higher activities than other samples. The initial degradation rate of this TNTs was 0.380 ppm/min. The crystallinity of all samples was almost the same, therefore the result indicated that the morphology of TiO_2 plays an important role in phenol degradation. It was primarily attributed to the morphology of the nanotubes. Khan and Yang (2009) reported tubular structure facilitated fast electrons transfer due to reduced grain boundary and was expected to improve charge separation in photocatalytic reactions [20]. Ghicov et al (2009) reported longer nanotube larger specific area and more active sites which are favorable for the oxygen evolution reaction. Therefore it caused light absorption more efficiently [21].

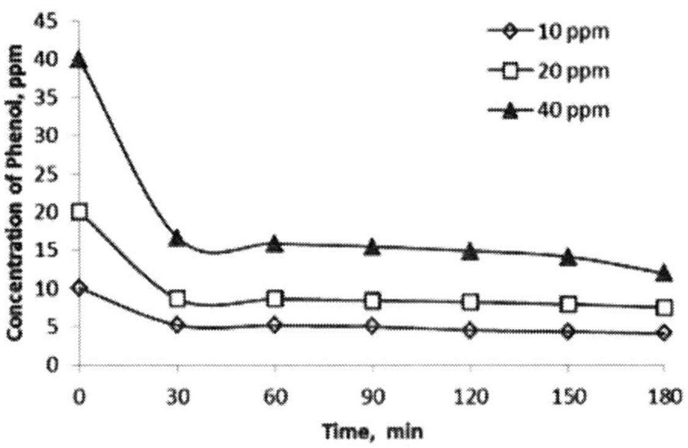

Figure 5. Effect of the initial concentration of phenol on the performance of TNTAs 50V photocatalyst.

The effect of initial concentration at phenol degradation was also observed and can be seen in figure 5. The pH of this examination was 6. The initial concentration of 40 ppm shows the highest percent removal. It shows the increase in the initial concentration of phenol caused a greater percentage of phenol degradation. It was probably due to the higher initial concentration caused by the increasing phenol adsorption on the surface of TiO_2. Therefore the phenol was easily degraded.

The effect of pH on the degradation of phenol can be seen in figure 6. This experiment was carried out at an initial concentration of 40 ppm of waste using a 50-volt catalyst. Figure 7 shows the best phenol degradation percentage at pH 5 with a final concentration of 3 hours to 6.9 ppm. Lian Yu (2016) reported that slightly acidic pH a phenol compound can easily dissolve by releasing H^+ ions from its hydroxyl group to form C_6H_5O-phenoxide. This condition will facilitate radical species to attack phenoxide to produce the final product in the form of CO_2 and H_2O [6].

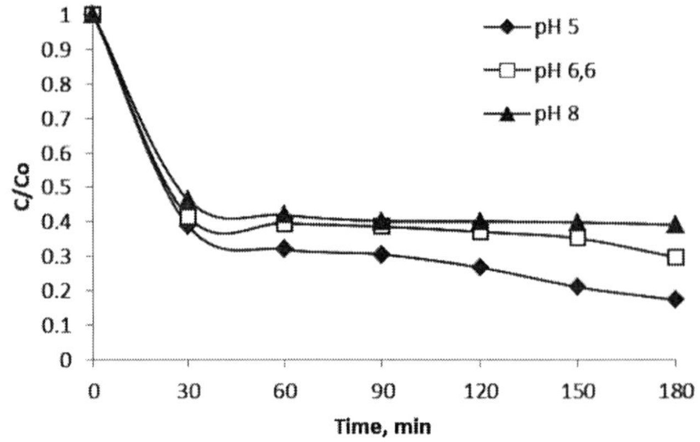

Figure 6. Effect of pH conditions on the degradation of phenol waste 40 ppm by using TNTAs 50V photocatalyst.

The results of this research could potentially be applied to the industrial scale that produces phenol waste, so that waste products can be safely discharged into the environment. The catalyst synthesis can also produce hydrogen gas which is useful for industrial needs such as the ammonia-making industry and urea fertilizer. The resulting catalyst has a great potential for repeated use because of its easy separation process with degraded wastes.

4. Conclusions

The effect degradation of phenol was influenced by TiO_2 morphology, pH and initial concentration of phenol. The longer nanotubes provide better degradation of phenol. The best pH for degradation phenol was obtained at pH 5. In this condition, the phenol has been reduced from 40 ppm to 6.9 ppm for 180 min.

Acknowledgement

The main financial support from the Directorate General of Higher Education, Ministry of National Education Indonesia by HIBAH PENELITIAN 2019 is gratefully acknowledged.

References

[1] Kumar, A., Kumar, S., Kumar, S., 2005, Biodegradation kinetics of phenol and catechol using Pseudomonas putida MTCC 1194, Biochemical Engineering Journal, 22, 151–159
[2] Edalatmanesh, M., Mehrvar, M., Dhib, R., 2008, Optimization of phenol degradation in a combined photochemical–biological wastewater treatment system, Chemical Engineering Research and Design. 86, 1243–1252
[3] Agarwal, G. K., Ghoshal, A. K., 2008, Packed bed dynamics during microbial treatment of wastewater: Modelling and simulation, Bioresource Technology, 99, 3765–3773
[4] Shahrezaei, F., Mansouri, Y., Zinatizadeh, A. A. L., Akhbari, A., 2012, Process modeling and kinetic evaluation of petroleum refinery wastewater treatment in a photocatalytic reactor using TiO_2 nanoparticles, Powder Technology, 221, 203–212, 2012
[5] Gómez, J., Bódalo, A., Gómez, E., Hidalgo, A., Gómez, M., Murcia, M., 2008, A transient design model of a continuous tank reactor for removing phenol with immobilized soybean peroxidase and hydrogen and hydrogen peroxide, Chemical Engineering Journal, 145, 142–148
[6] H. Ling, K. Kim, Z. Liu, J. Shi, X. Zhu, J. Huang, 2015, Catal, Today 258, 96–102
[7] Bavasso I., Vilardi G., Stoller M., Chianese A., Di Palma L., 2016, Perspectives in nanotechnology based innovative applications for the environment, Chemical Engineering Transactions, 47, 55-60

[8] Stoller, M., Vuppalab, S., Matarangoloc, M., Vaianoc, V., Sanninoc, D., Chianesea, A., Cianfrinib, C., 2017, About a Novel Production Method for N-Doped Magnetic Nanocore Nanoparticles of Titania by Means of a Spinning Disk Reactor, 60, 43-38

[9] Sulaiman, F., Sari, D.K., Kustiningsih, I., 2017, The influence of ozone on the photocatalytic degradation of phenol using TiO2 photocatalyst supported by Bayah natural zeolite, AIP Conference Proceedings, 1840, 110014

[10] Vaiani, V., Saccoa, O., Sanninoa, D., Capuab, DG., Femiab, N., 2017, Enhanced Performances of a Photocatalytic Reactor for Wastewater Treatment Using Controlled Modulation of LEDs Light, Chemical Engineering Transactions, 553-558. et al., 2017

[11] Mammadov, G. Sh., Ramanzanov, M.A., Kanaev, A., Hasanova, U.A., Huseynov, K., 2017, Photocatalitic Degradation of Organic Pollutans in Air by Application of Titanium Dioxide Nanaoparticles, Chemical Engineering Transactions, 60,241-246. ,G.Sh. et al., 2017

[12] Pang, Y. L., Abdullah, A. Z., 2013. Effect of Carbon and Nitrogen Co-Doping on Characteristics and Catalyst for Degradation of Rhodamine B in Water. Chemical Engineering Journal, 214, 129-138

[13] Shokuhfar, T., S. Sinha-Ray, C. Sukotjo, A. L. Yarin, 2013, Intercalation of anti-inflammatory drug molecules within nanotubes, RSC Advances 3 (38), 17380

[14] Kustiningsih, I., Slamet, Purwanto, W.W., 2014, Synthesis of tittania nanotubes and titania nanowires by combination sonication-hydrothermal treatment and their photocatalytic activity for hydrogen production. International Journal of Technology, 5(2), 133-141

[15] Kustiningsih, I., Sutinah, Stefirizky, M., Slamet, Purwanto, W.W., 2018, Optimization of TiO2 NanowiresSynthesis Using Hydrothermal Method for Hydrogen Production, Journal of Mechanical Engineering & Sciences, 12(3), 3876-3887

[16] Zhang, S., Peng, F., Wang, H., Yu, H., Zhang, S, Yang, J., Zhao, H., 2011, Electrodeposition preparation of Ag loaded N-doped TiO2 nanotube arrays with enhanced visible light photocatalytic performance, Catalysis Communications 12, 689–693

[17] Khudhair, D., A. Bhatti, Y. Li, H. Hamedani, H. Garmestani, P. Hodg- son, S. Nahavandi., 2016, Anodization parameters influencing the morphology and electrical properties of TiO2 nanotubes for living cell interfacing and investigation, Materials Science & Engineering C, 59, 1125-1142

[18] Teodorescu-Soare, C.T., Catrinescu, C., Dobromir, M, Stian G, Arvinte, A., Luca, D., 2018, Growth and characterization of TiO2 nanotube arrays under dynamic anodization. Photocatalytic activity, Journal of Electroanalytical Chemistry, 823, 388-396

[19] Rico-Oller, Beatriz, Amel Boudjemaa, Hasliza Bahruji, Mohammed Kebir, Sanjiv Prashar, Khaldoun Bachari, Mariano Fajardo, Santiago Gomez-Ruiz., 2015, Photodegradation of Organic Pollutants in Water and Green Hydrogen Production Via Methanol Photoreforming of Doped Titanium Oxide Nanoparticles, Science od The Total Environment,563,921-932

[20] Khan, M.A., Yang, O.B., 2009, Photocatalytic water splitting for hydrogen production under visible light on Ir and Co ionized titania nanotubes, Catalysis Today 146, 177-182

[21] Ghicov, A.,S.P. Albu, R. Hahn, D. Kim, T. Stergiopoulos, J. Kunze, C.-A. Schiller, P. Falaras, P. Schmuki, 2009, TiO2 nanotubes in dye-sensitized solar cells: critical factors for the conversion efficiency, Chemistry-An Asian Journal, 4, 520–525

Hydrothermal Synthesis of Activated Carbon/α-Fe_2O_3 Nanocomposite and Its Application for Removing Tartrazine Dye

S Lubis[1], M Ramli, Sheilatina and Y Ermanda**
[1]Department of Chemistry, Faculty of Mathematics and Natural Sciences, Universitas Syiah Kuala, Darussalam 23111, Banda Aceh, Indonesia

E-mail: suryalubis@unsyiah.ac.id

Abstract. Hematite containing activated carbon (AC/α-Fe_2O_3) nanocomposite has been prepared by hydrothermal method and used as photocatalyst for removing tartrazine dye. Hematite was extracted from iron sand while activated carbon was derived from coffee waste which was activated using hydrochloric acid. The structure and morphology of photocatalyst was characterized by means of X-ray diffraction (XRD), nitrogen adsorption–desorption (BET) and Scanning Electron Microscopy-Energy-Dispersive X-Ray Spectroscopy (SEM-EDS). These techniques indicate that the AC/α-Fe_2O_3 nanocomposite has an average crystallite size of 35.47 nm and specific surface area of 41.52 m^2/g. The photocatalytic activity evaluation showed that various factors influencing the removal of tartrazine dye including pH of dye solution, photocatalyst dosage, dye concentration and irradiation time. The results showed that AC/α-Fe_2O_3 nanocomposite can remove tartrazine up to 91.01% under UV light and 93.69% under solar light which were obtained at pH value of 2, photocatalyst dosage of 250 mg and initial concentration of tartrazine is 15 ppm with irradiation time for 210 min under UV light and 180 min under solar light. The AC/α-Fe_2O_3 nanocomposite also can remove tartrazine dye contained in noodle waste water up to 94.67% and 97.86 under UV light irradiation for 150 min and 120 min under solar light irradiation, respectively.

1. Intoduction

Photocatalysis has been widely used in the degradation of organic pollutants such as dye effluents, phenols and pesticides in waste water. Semiconductor such as TiO_2, ZnO, ZnS and α-Fe_2O_3 could be used as photocatalyst under ultra violet or solar light. These processes are based on the generation of electron in the conduction band (CB) and hole in the valence band (VB) by light irradiation. The electrons and holes initiate reduction and oxidation reactions with species adsorbed on photocatalyst surface. Electron as reducing agent while hole as oxidizing agent. On the other hand, the electrons and holes can recombine, releasing the absorbed light energy as heat, without any chemical reaction occuring. On the semiconductor surface, the excited electron and the hole can participate in redox reactions with water, hydroxide ion (OH^-), oxygen or organic compounds leading to mineralization of the pollutant [1,2].

Heterogeneous photocatalysis utilizing hematite (α-Fe_2O_3) is the most promising water treatment technology due to its low cost, chemical stability, nontoxicity, and high oxidation ability [3,4]. Hematite (α-Fe_2O_3) is one of the most stable iron oxides with the narrow band gap energy ($E_G \approx$

[*] To whom any correspondence should be addressed

Content from this work may be used under the terms of the Creative Commons Attribution 3.0 licence. Any further distribution of this work must maintain attribution to the author(s) and the title of the work, journal citation and DOI.
Published under licence by IOP Publishing Ltd

2.2 eV) which can be used with visible and solar light. However, there are some weakness of α-Fe_2O_3 including low adsorption ability and flash recombination of photogenerated electron and hole which led to decrease its photocatalytic activity. It has been reported that activated carbon (AC) in combination with various iron oxide related materials is an excellent adsorbent for a various dagerous water pollutants [5,6]. The combination of activated carbon and iron oxide can inrease adsorption and photodegradation and eliminate organic pollutants under UV and/or solar illuminations. Therefore, iron oxide and AC are able to effectively remove organic and inorganic substances by simultaneous adsorption-desorption in the water treatment process [7].

The aims of the present work were to synthesis the AC/α-Fe_2O_3 nanocomposite and to investigate its photocatalytic activity on the removal of tartrazine dye (CI. No.19140). α-Fe_2O_3 can be provided naturally from iron sand and iron ore. The high content of iron oxide in iron sand taken from Syiah Kuala Beach, Banda Aceh Indonesia has been reported [8]. Tartrazine is a synthetic dye with the formula $C_{16}H_9N_4Na_3O_9S_2$ has a wide application in food, drug, pharmaceuticals, cosmetics and textil industry [9]. The full chemical name of tartrazine is trisodium 5-hydroxy-1-(4-sulfonatophenyl)-4-(4-sulfonatophenylazo)-H-pyrazol-3-carboxylate, with the structural formula was described in Fig. 1. Based on the national agency of Drug and Food Control of Republic of Indonesia the acceptable daily intake of tartrazine for humans is 0-7.5 mg/kg, while in food is 70 mg/kg. Tartrazine is very toxic in high concentration and is known to cause allergic reactions such as asthma, migraines, eczema, thyroid cancer, lupus, blurred vision, itching, hyperactive and fertility for human being [10, 11]. Thus, waste water containing tartrazine must be treated before discharge to environment. The degradation of tartrazine in TiO_2 fixed bed has been reported. The removal of tartrazine was only 30% under UV light irradiation for 200 min and 99% under solar light irradiation [12].

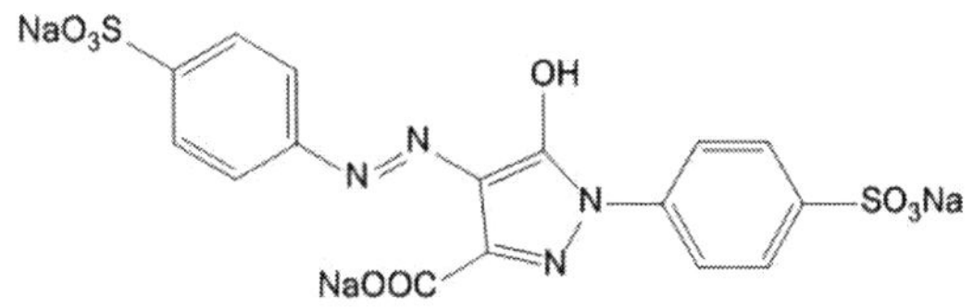

Figure 1. Chemical structure of tartrazine

2. Experimental

2.1. Materials

Tartrazine, a water soluble dye, was purchased from PT. Karya Muda Indochem, Indonesia while waste water containing tartrazine was taken from the noodle industry in Banda aceh, Indonesia. Hematite (α-Fe_2O_3) was extracted from iron sand taken from Syiah Kuala Beach, Banda Aceh, Indonesia. Coffee waste was taken from some coffee shops in Banda Aceh, Indonesia. Other chemicals such as HCl, NH_4OH and NaOH were purchased from Merck.

2.2. Preparation of Hematite (α-Fe_2O_3) and activated carbon

The hematite was extracted from iron sand according to previously report [13]. Hydrochloric acid was used to dissolve the iron oxide, following by adding NH_4OH as precipitation agent to the solution. The precipipate obtained was filtered and rinsed until it reached pH 7 (neutral), dried in an oven at 100 °C for 12 h and calcined in air at 700 °C for 5 h to produce α-Fe_2O_3. Activated carbon was obtained from coffee waste by hydrothermal carbonization process on a Teflon-lined stainless steel autoclave at 170 °C for 5 h. The resulting product was dried at 100 °C for 5 h. In order to activate the carbon material obtained, the sample was immersed in 1 M HCl solution for 24 h. The resulting solid product was washed with water to neutralize its acidity. The sample was dried and calcined at 250 °C for 3 h.

2.3. Production of AC/α-Fe_2O_3 nanocomposites

The hematite containing activated carbon (AC/α-Fe_2O_3) nanocomposite was prepared by hydrothermal method. Briefly, 2.5 g of activated carbon and 7.5 g of α-Fe_2O_3 were mixed in 75 ml of distilled water under magnetic stirring. The mixture was transferred into a Teflon-lined stainless steel autoclave and then hydrothermally treated at 170 °C for 5 h. The resulting solid was filtered, washed with distilled water, dried at 100 °C for 5 h and calcined at 250 °C for 3 h.

2.4. Characterization of materials

X-ray diffraction (XRD) patterns were recorded with a Shimadzu X-ray diffractometer (XD-7000) equipped with Cu-Kα radiation source (λ = 1.54056 Å) operating at 40 kV and 30 mA. The scan rate is 5° min^{-1} with Bragg angle 2θ was ranged from 10 to 80°. The crystallite sizes of as-synthesized samples (D) were examined from broadening of diffraction peaks according to Scherrer's formula [14], where k is shape factor (the value is 0.9), λ is the X-ray wavelength, β is the full width at half maximum (FWHM) measured in radians of the reflection peak and θ is the Bragg's angle.

$$D = \frac{k\lambda}{\beta \cos\theta}$$

The morphologies of the samples surface were examined by Scanning Electron Microscope (SEM) (JEOL JSM6510LV) operated at an acceleration voltage of 15 kV. The nitrogen adsorption-desorption isotherms were carried out by Quantachrome instruments (Nova touch 4LX). The Brunauer–Emmett–Teller (BET) surface area was measured at 77 K. Prior to measurement, the samples were out-gassed at 100 °C under a N_2 flow for 2 h.

2.5. Photocatalytic experiment

The study of photocatalytic degradation of tartrazine was carried out under ultra violet (using a 6-W UV lamp, λ = 365 nm) or solar light irradiation. The effect of various parameters, such as pH of tartrazine solution, photocatalyst dosage and tartrazine concentration were investigated. The entire arrangement was placed in a box to avoid the passage of other light from outside. The pH of tartrazine solution was measured on HI98107 pHep pH tester and was adjusted to desired pH value by adding small amount of 0.5 M HCl or 0.5 M NaOH solution. Typically, 250 mg of AC/α-Fe_2O_3 nanocomposite was added to 25 mL of 15 ppm tartrazine solution on the Pyrex glass vessel. The mixture was placed about 10 cm from the UV lamp. The experiments were carried out for 120 min consisting of 30 min in darkness and 90-180 min under UV light irradiation. After regular time interval of 30 min, the suspensions were collected and centrifuged for 10-15 min to completely remove the photocatalyst. The concentrations of tartrazine were determined with a UV–vis spectrophotometer (Shimadzu UV mini 1240 spectrophotometer) at λ_{max} = 427 nm. The removal (%) of tartrazine was estimated using the following equation (eq. 1), where C_0 and C_t were represented the initial and final dye concentration (ppm) [15].

$$\text{Removal (\%)} = (C_0-C_t)/C_0 \times 100 \qquad (1)$$

The experiment was also performed using AC/α-Fe_2O_3 nanocomposite and 25 mL of tartrazine solution in the absence of UV-light.

3. Result and Discussion

3.1 Characterization of Materials

3.1.1 XRD analysis

The XRD patterns of iron sand, α-Fe_2O_3 samples before and after addition of activated carbon are shown in figure 2. The iron sand mainly composed of magnetite (Fe_3O_4) (JCPDS card no. 96-900-6243) and a small amount of hematite (α-Fe_2O_3) (JCPDS card no. 96-900-9783). After being extracted and calcined at high temperature almost all of magnetite converted into hematite (α-Fe_2O_3) phase. The activated carbon (Fig. 2c) showed a broad peak at 2θ = 10 - 20°, corresponding to the diffraction of amorphous carbon. The AC/α-Fe_2O_3 nanocomposite (Fig 2d) exhibited XRD patterns almost similar to those of α-Fe_2O_3. There is no clear refflection corresponding to amorphous carbon was observed. This is might be due to the low amount of carbon because of the decomposition of some carbon after calcination at high temperature (250 °C). Thus, only few of carbon remain on the surface of α-Fe_2O_3.

The average crystallite sizes of samples was estimated from XRD peaks broadening using Scherrer formula. The α-Fe_2O_3 exhibited the smallest crystallite size (27.54 nm), which size increased after addition of activated carbon (35.47 nm). This observation was confirmed that activated carbon has been added to the surface of α-Fe_2O_3. Larger crystallite size were belong to iron sand (47.09 nm) and activated carbon (44.57 nm).

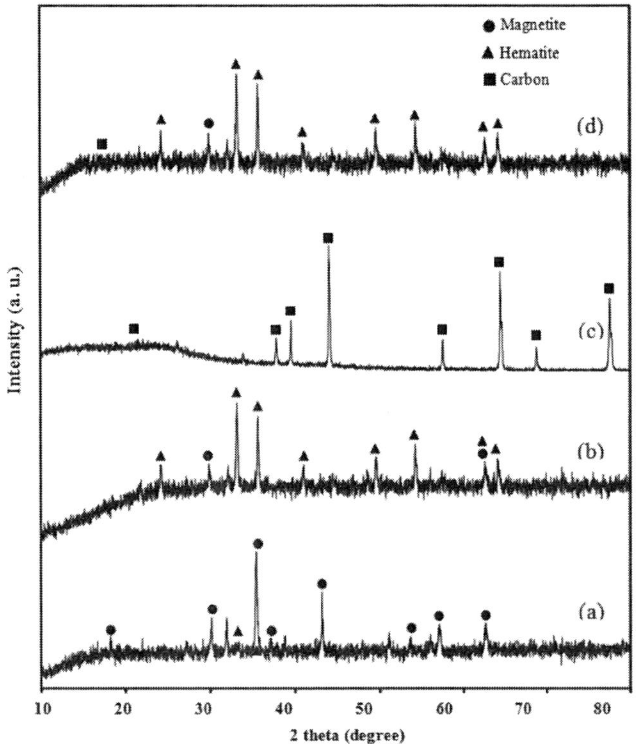

Figure 2. XRD patterns of a) Iron sand, b) α-Fe_2O_3, c) Activated carbon and d) AC/α-Fe_2O_3

3.1.2 SEM-EDS analysis

The surface morphology and elemental composition of the samples were investigated employing SEM and EDS. SEM images of α-Fe_2O_3, carbon from coffee waste, activated carbon and AC/α-Fe_2O_3 at 20,000 x magnification are shown in figure 3 which is show a clear change of the carbon and α-Fe_2O_3 morphologies. α-Fe_2O_3 has the smallest particle size which was accordance with XRD data. The small particle size leads to the agglomeration of particles [16] which can be prevented by adding the activated carbon. It can be seen from SEM image of AC/α-Fe_2O_3 nanocomposites (Fig 3d.) that the agglomeration of α-Fe_2O_3 particles was decreased. The results of the elemental analysis of α-Fe_2O_3 and AC/α-Fe_2O_3 indicated that Fe is the main constituents of the materials, which is estimated to be 87.96 and 69.98%, respectively. Furthermore, carbon content in AC/α-Fe_2O_3 is 4.19% and no carbon content was observed in α-Fe_2O_3. EDS analysis also reveals that activated carbon contain only C and O element.

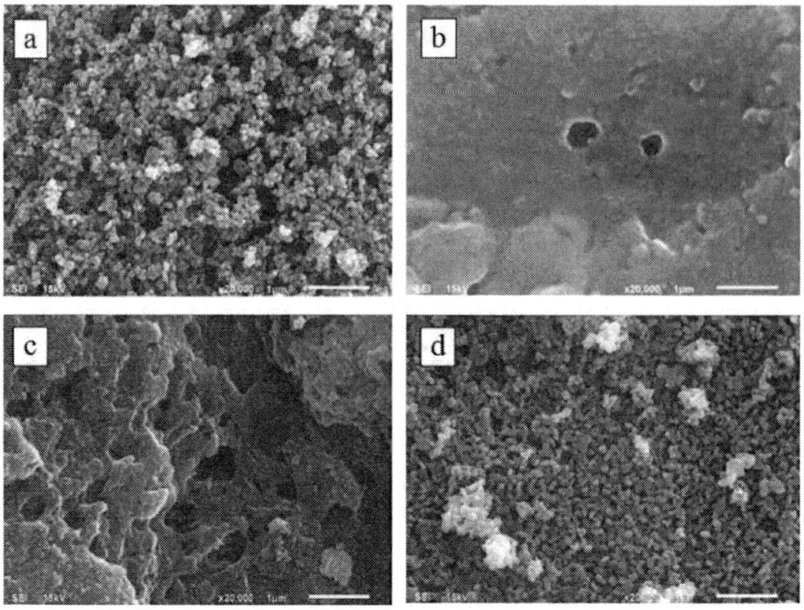

Figure 3. SEM images of a. α-Fe_2O_3 b. Coffee waste, c. Activated carbon (AC) from coffee waste and d. AC/α-Fe_2O_3

3.1.3 Nitrogen adsorption-desorption analysis

The nitrogen adsorption–desorption was applied for determination of BET surface area of activated carbon derived from coffee waste, α-Fe_2O_3 and AC/α-Fe_2O_3. It can be observed from Table 1that the addition of carbon to the α-Fe_2O_3 changes the value of surface area, pore radius and pore volume of α-Fe_2O_3. The results show an increasing of specific surface area of α-Fe_2O_3 due to the combination with activated carbon which has large specific surface area. The result also show that the addition of activated carbon cause a decreasing of average pore diameter of α-Fe_2O_3. Based on IUPAC classification all the materials prepared are categorized as mesopores materials with the pore diameter 2 – 50 nm [17] .

Table 1. Result of adsorption desorption measurements of sample

Samples	Specific surface area (m^2/g)	Pore volume (cc/g)	Pore diameter (nm)
AC	160.463	0.110	2.754
α-Fe_2O_3	22.005	0.057	10.450
AC/α-Fe_2O_3	41.528	0.037	3.596

3.2 Photocatalytic Activity

3.2.1 Effect of initial pH of solution

pH has an important role in dye photodegradation process because it will influence the ionic state of dye molecule or surface of photocatalyst in the aqueous media. The photodegradation experiments were conducted at pH value of tartrazine solution 1, 2, 3, 4 and 5. The results showed that the photocatalytic activity of AC/α-Fe_2O_3 increased from pH 1 to 2 and then decreased from 3 to 5 (Fig.

4). Tartrazine is anionic in nature and at the low pH value the positive charge site of AC/α-Fe_2O_3 surface increase, which contributed to the electrostatic interactions between the dye molecules and the AC/α-Fe_2O_3 surface. The increasing of tartrazine molecule on the photocatalyst surface leads to the greater amount of tartrazine molecule which reacts with photogenerated electron, hole, $^{\bullet}OH$, or $O_2^{\bullet-}$ in photocatalysis. Thus, the photocatalytic degradation increases at low pH value. At higher pH value the positive charge site of AC/α-Fe_2O_3 surface decreases and the contribution to the photodegradation process will decrease. The photocatalytic activity was also evaluated at different time points (30–120 min) and the data show that tartrazine photodegradation increased with the increase of time. At the pH value of 2 and irradiation time 120 min, the maximum dye removal was 50.14% (Fig. 4), so the further experiments were conducted at pH value of 2.

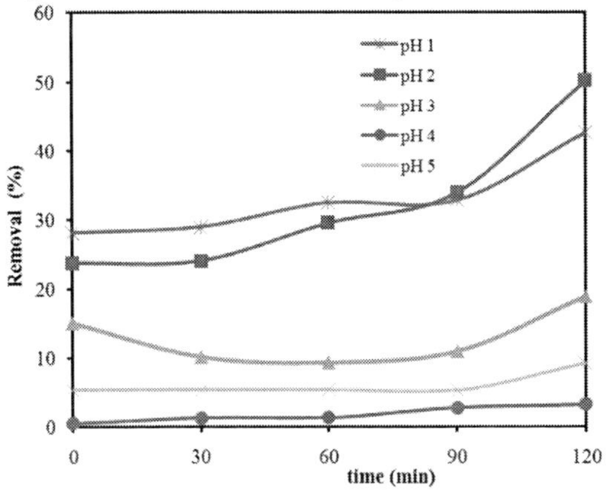

Figure 4. Removal of tartrazine vs time for different pH of dye solution, using 250 mg AC/α-Fe_2O_3 nanocomposite, C_0 = 15 ppm, UV light

3.2.2 Effect of photocatalyst dosage

The amount of the photocatalyst is also an important parameter which influences the photodegradation process. In this study, the photocatalyst dosage used are 150 to 350 mg and the experiments were conducted at pH value 2. The results illustrated that, with an increase in photocatalyst dosage, the removal of tartrazine increased. This can be explained by more active sites being available for dye photodegradation at higher photocatalyst dosages [18]. The maximum dye removal (%) was achieved at a photocatalyst dosage of 250 mg. A further increase in the photocatalyst dosage resulted in a decrease in photocatalytic activity, which can be correlated to the limited light penetration through the tartrazine solution decreasing the photocatalytic activity [19].

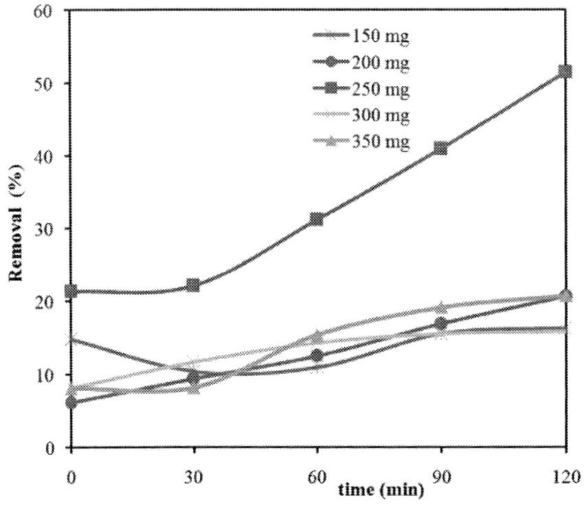

Figure 5. Removal of tartrazine vs time for different photocatalysts dosage, pH of dye solution = 1, C_0 = 15 ppm, UV light.

3.2.3 Effect of dye concentration

To evaluate the effect of the concentration of tartrazine, experiments were conducted for dye concentration from 5 to 15 ppm. The experiments were performed under pH value of 2 and photocatalyst dosage of 250 mg (Figure 6). The results showed that the dye photodegradation increased with an increase in dye concentration. The photodegradation was proportional to the concentration, in agreement with an increase in the molecules adsorbing on the surface of the photocatalyst [20].

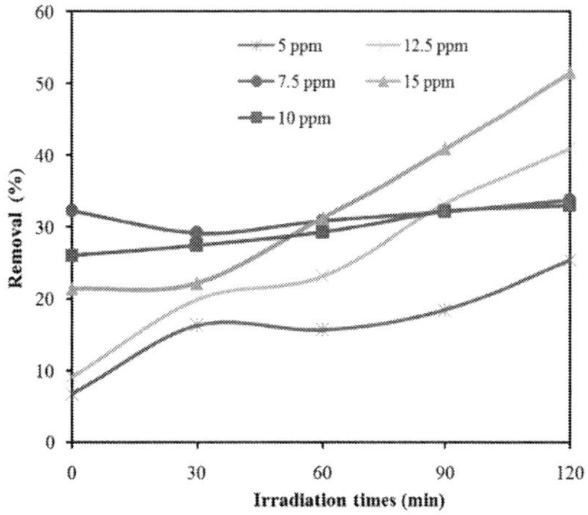

Figure 6. Removal of tartrazine vs time for different concentration of dye solution, using 250 mg $AC/\alpha\text{-}Fe_2O_3$ nanocomposite, pH of dye solution = 1, UV light.

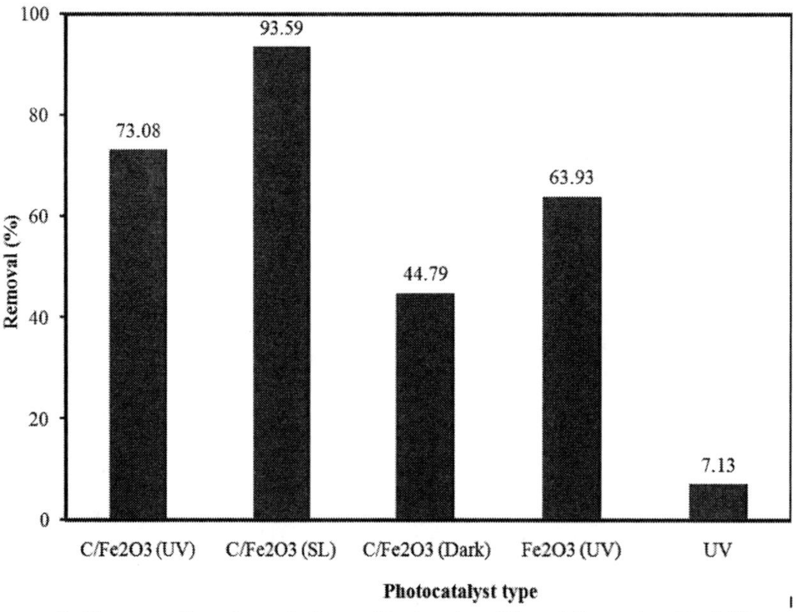

Figure 7. Comparative degradation of tartrazine (C_0 = 15 ppm) with 250 mg of photocatalyst under 180 min of ultra violet (UV) or solar light (SL) irradiation

3.2.4 Effect of photocatalyst type and light source

Figure 7 shows the photodegradation of tartrazine dye over different photocatalysts for 180 min at optimum conditions (pH = 2, photocatalyst dosage = 250 mg and dye concentration = 15 ppm). It is obvious that AC /α-Fe_2O_3 nanocomposite has higher photocatalytic activity than that of bare α-Fe_2O_3. This confirms that addition of activated carbon can enhance the photocatalytic activity of α-Fe_2O_3. Activated carbon has the important role to enhance the adsorption of tartrazine molecule to the photocatalyst surface thus enhance the photodegradation. It can be seen from experiment which was conducted in the dark condition. In addition, the photogenerated electrons can easily transfer from the α-Fe_2O_3 conduction band to carbon and thus inhibited the recombination of photogenerated electrons and holes [21]. Figure 7 also exhibits that AC/α-Fe_2O_3 nanocomposite has higher photocatalytic activity under solar light irradiation than that of under UV light. This observation was accordance with the previously report that tartrazine was more efficiently degraded by TiO_2 under solar radiation than with UV lamp radiation [12]. The photolysis of tartrazine dye under UV light in the absence of photocatalyst were very low, and which can be neglected.

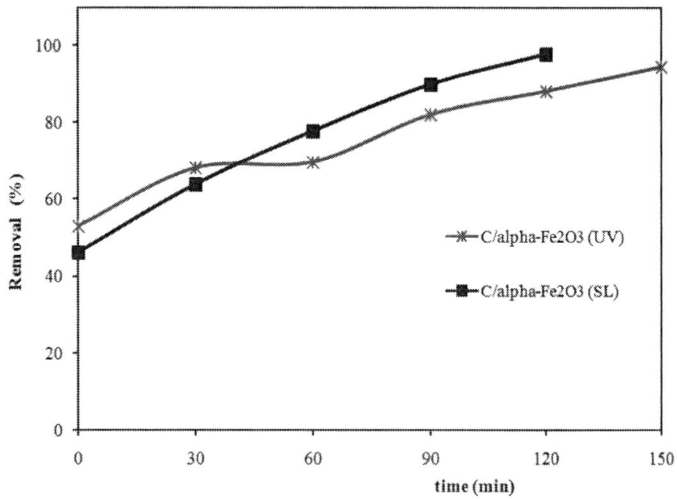

Figure 8. Comparative degradation of tartrazine contained in noodle waste water with 250 mg of AC/α-Fe_2O_3 nanocomposite under ultra violet (UV) or solar light (SL) irradiation

The photocatalytic activity of AC/α-Fe_2O_3 was also evaluated on photodegradation of tartrazine contained in noodle waste water. The experiments were conducted at pH value of noodle waste water using 250 mg of AC/α-Fe_2O_3 (Fig. 8). It was observed that AC/α-Fe_2O_3 can remove tartrazine up to 94.67% under UV light irradiation for 150 min and 97.86% under solar light irradiation for 120 min.

4. Conclusions

In the present study, AC/α-Fe_2O_3 nanocomposite was prepared through a hydrothermal method. XRD analysis confirms that the prepared AC/α-Fe_2O_3 was nano structure and size of the particle was calculated as 35.47 nm. SEM-EDS analysis also confirms the carbon content in AC/α-Fe_2O_3 nanocomposite which was supported by the increasing of specific surface area of α-Fe_2O_3 obtained from nitrogen-desorption isotherm analysis. Photocatalytic acitivity of prepared sample was tested on degradation of tartrazine dye. The results showed that AC/α-Fe_2O_3 has the high photocatalytic activity compare to bare α-Fe_2O_3. The highest photocatalytic activity of AC/α-Fe_2O_3 was achieved at pH value of tartrazine solution is 1, photocatalyst dosage is 250 mg and tartrazine concentration is 15 ppm. The maximum removal of tartrazine for the optimal conditions is 91.01% and 93.69% after being irradiated for 210 min under ultra violet light and 180 min under solar light, respectively. AC/α-Fe_2O_3 nanocomposite also showed high photocatalytic activity on degradation of tartrazine dye contained in noodle waste water.

Acknowledgment

This work was funded by the PD Research Grant from Ministry of Research, Technology and Higher Education by number: 59/UN11.2/PP/SP3/2019 from Universitas Syiah Kuala.

References

[1] Ibhadon A O and Fitzpatrick P 2013 *Catalysts* **3** 189-218.
[2] Saien J and Soleymani A R 2007 *J. Hazard. Mater.* **144**(1-2) 506-512.
[3] Yu K, Huang LY and Lou L L 2015 *Environ. Technol.* **36** 1008–1015.
[4] Li C Q, Sun Z M, Xue Y L, Yao G Y and Zheng S L 2016 *Adv. Powder Technol.* **27** 330–337.
[5] Gupta V K and Suhas 2009 *J. Environ. Manage.* **90** 2313–2342.
[6] Sharma Y C, Srivastava V, Singh V K, Kaul S N and Weng C H 2009 *Environ. Technol.* **30** 583–609.
[7] Kadirova Z C, Katsumata K, Isobe T, Matsushit N, Nakajima A and Okada K 2014 *J. Environ. Chem. Eng.* **2**, 2026–2036.
[8] Lubis S, Sheilatina and Sitompul D W 2018 *IOP Conf. Ser.: Mater. Sci. Eng.* **509** 012142.
[9] Tanaka T 2006 *Food Chem.Toxicol.* **44**(2) 179-187.
[10] Moutinho I L D, Bertges L C and Assis RVC 2007 *Braz. J. Biol.,* **67**(1): 141-145.
[11] Wawrzkiewicz M and Hubicki Z 2009 *J. Hazard. Mater.* **164** 502–509
[12] Chekir N, Tassalit D, Benhabiles O, Merzouk N K, Ghenna M, Abdessemed A and Issaadi R 2017 *Int. J. Hydrogen Energy* **42**(13) 8948-8954.
[13] Lubis S, Sheilatina and Murisna 2018 *J. Phys.: Conf. Ser.* **1116** 042016.
[14] Gnanaprakasam A, Sivakumar V M, Sivayogavalli P L and Thirumarimurugan M 2015 *Ecotox. Environ. Safe.* **121** 121–125.
[15] Shojaie A, Fattahi M, Jorfi S and Ghasemia B 2017 *J. Environ. Chem. Eng.* **5**(5) 4564–4572.
[16] Yousefi A, Allahverdi A and Hejazi P 2013 *Constr. Build. Mater.* **41** 224–230.
[17] Zdravkov B D, Cerm´ak J J, Sefara M and Janku J 2007 *Cen. Eur. J. Chem.* **5**(2) 385–395.
[18] Sin J C, Lam S M, Mohamed A R and Lee K T 2012 *Int.J. Photoenergy* **185159** 1-23.
[19]. Sherly E, Vijaya J J, Selvam N C S and Kennedy L J 2014 *Ceram. Int.* **40** 5681-5691.
[20] Ahmed S, Rasul M, Brown R and Hashib M 2011 *J. Environ. Manag.* **92** 311-330.
[21] Cao Z, Qin M, Gu Y and Jia B 2017 *J. Phys. Chem. Solids.* **107** 42-49.

Optimization adsorption of lead (II) using Aceh natural-bentonite by response surface methodology

J. Jakfar[1], H. Husin[1,2*], A. Muslim[2], D. Darmadi[2], F. Nasution[1], A.R. Thursina[2], E. Rahmawati[2]
[1]Doctoral School of Engineering Sciences
[2] Department of Chemical Engineering, Universitas Syiah Kuala, Banda Aceh, 23111

Email: husni_husin@che.unsyiah.ac.id

Abstract. Response Surface Methodology was employed for the removal of lead on synthetic water. The process parameters including bentonite mass (0.5; 1.25; 2.0 g), lead concentration (10; 55; 100 mg/L), and agitation time (5; 17.5; 30 minutes) were optimized using statistical Box-Behnken Design. The experimental data were obtained and analyzed by analysis of variance (ANOVA) and fit to a quadratic polynomial equation (Regression Analysis, R^2 = 0.9996) for 17 runs experiment and 5 runs of the central-point (confidence level at 95%). The numerical optimation was used to identify the optimum condition for maximum removal of lead. The maximum adsorption efficiency expressed as the adsorption capacity, which was found to be 7.42 mg/L, at agitation time 17.5 minutes and adsorbent dosage 2.0 g.

1. Introduction

Heavy metal contamination of water is a relevant issue in the context of water resources protection. The industrial activities such as paint, pesticide, mining, textile, lead batteries and the manufacturing release a large number of heavy metal ion to aquatic ecosystem. Among the heavy metal ion, lead is one of most dangerous pollutant in industrial wastewater even at low concentration. Lead are non-biodegradable ions and known as a toxin with effect on several organ systems. According to the World Health Organization (WHO), the safe level for lead ions is 0.1 gm^{-3} [1-8].

In order to handle the heavy metal ion on waste water, several methods has been developed to eliminate it from waste water before disposal, such as chemical precipitation, membrane filtration, solvent extraction and adsorption. Among these methods, adsorption is a common and effective method for heavy metal ions removal, due to its features of easy operation, eco-friendly, strong practicability, high efficiency, and low cost [9-13]

In recent years, the used of bentonite as adsorbent have been carried out in handling heavy metal waste on waste water, but the used of bentonite (especially Aceh natural bentonite) as adsorbent has not been utilized optimally. It has a different composition with other natural bentonite. So in this study, Aceh natural bentonite was used as adsorbent to remove lead ions from waste water (lead aqueous solution). However, the adsorption capacity of natural bentonite is low because it still contains impurities [14-16]. Therefore, before being used in industry, it must be purified. The natural bentonite is modified by physical process. The modification results are to remove the impurities (CO_2) and to enhance the adsorption capability [17-22].

* To whom any correspondence should be addressed

Content from this work may be used under the terms of the Creative Commons Attribution 3.0 licence. Any further distribution of this work must maintain attribution to the author(s) and the title of the work, journal citation and DOI.
Published under licence by IOP Publishing Ltd

The most important stages in an environmental process are modeling and optimization to improve the system and increase the efficiency of the process [23]. The main objective in this study is to evaluate of Aceh natural bentonite adsorption capacity as adsorbent in lead (Pb) removal through the application of Response Surface Methodology (RSM). In order to have a better understanding of the optimization of adsorption at various operating parameters were studied.

2. Experimental

2.1. Materials
Bentonite was obtained from Blang Karing, Aceh. The $Pb(NO_3)_2$ and acetate-buffer solution (Merck, Germany) and distilled water were purchased from Rudang Jaya Company, Medan Indonesia.

2.2. Adsorbent Preparation
Firstly, the bentonite were grinded and sieved to work particle size (200-mesh) using ball mill grinder. After that, the bentonite was heated at 105°C for 2 h and kept in desiccator. The sifted and heated bentonite was added in 250 mL acetate-buffer solution (pH = 4.8) and stirred for 5 h until no CO_2(g) formed. Furthermore, the mixture was separated by centrifugal for 10 minutes.

2.3 Adsorption Experiment
The adsorption of lead onto bentonite was investigated in stirred-batch experiment [10]. Purified natural bentonite of 25 g was put into a flask, and added by 500 mL water. It was stirred by magnetic stirrer for 24 h at 70°C. The mixed solution was separated by decantation. The precipitated was dried at 105°C. After that, 100 mL of lead aqueous solution with different concentration and bentonite dose was placed in a flask. Then, the mixture was agitated at 200 rpm and filtered to analyze the residual of lead. The lead adsorption onto natural bentonite was calculated according to the following equation [24]:

$$qt = \frac{(C0-Ct)V}{m} \quad (1)$$

Where, C_0 is the initial lead concentration ; C_t is the remaining lead concentration at any time ; V is the volume of the lead solution ; and *m* is the weight of bentonite.

Table 1. Experimental and predicted results

Run	Bentonite (g)	Time (minutes)	Concentration of Pb (mg/l)	Ion Pb adsorption (mg/l)
1	1.25	17.50	55.00	3.94
2	0.50	17.50	10.00	1.12
3	2.00	5.00	55.00	4.33
4	1.25	5.00	100.00	3.05
5	1.25	17.50	55.00	3.94
6	0.50	5.00	55.00	1.08
7	1.25	30.00	100.00	5.22
8	1.25	5.00	10.00	1.92
9	2.00	30.00	55.00	7.42
10	1.25	30.00	10.00	3.30
11	0.50	30.00	55.00	1.85
12	1.25	17.50	55.00	3.94
13	1.25	17.50	55.00	3.94
14	2.00	17.00	100.00	7.11
15	1.25	17.50	55.00	3.94
16	0.50	17.50	100.00	1.78
17	2.00	17.50	10.00	4.49

IOP Conf. Series: Materials Science and Engineering **796** (2020) 012062 doi:10.1088/1757-899X/796/1/012062

2.4. Box Behnken Experimental Design and Optimation by Response Surface Methodology
Response surface methodology is a statistical method for optimizing the independent factors to achieve the desirable response by carrying out a set of experiment [9].

In this study, a three factor - Box-Behnken design (Design Expert Vers. 6.0.8) as sociating with response surface methodology to determine and to predict the optimum condition of adsorption. In this study obtained 17 run-experiments in a factorial design and 5 run of the central-point.

3. Results and Discussions

3.1 Box-Behnken Design and Model Evaluation.
The Box-Benhken method was analyzed and the results of the analysis of variance (ANOVA) from regression coefficient are presented in Table 2. and Table 3.

Table 2. Analysis of variance (ANOVA) for Response Surface Quadratic Model

Source	Sum of Square	DF	Mean Square	F Value	Prob>F	Characteristic
Model	53.43	9	5.94	174862	<0.0001	Significant
A - Natural Bentonite (g)	38.66	1	38.37	13149.65	<0.0001	Significant
B - Times (minutes)	6.86	1	6.86	2352.24	<0.0001	Significant
C - Concentration of Pb (mg/L)	5.01	1	5.01	1716.54	<0.0001	Significant
A2	0.0069	1	0.0069	2.35	0.1688	Non-Significant
B2	0.32	1	0.32	111.32	<0.0001	Significant
C2	0.39	1	0.39	135.34	<0.0001	Significant
AB	1.35	1	1.35	461.16	<0.0001	Significant
AC	0.96	1	0.96	329.15	<0.0001	Significant
BC	0.16	1	0.16	53.47	0.002	Significant
Residual	0.02	7	0.00292			

From Table 2, it can be seen that the analysis interaction of variables and the effect of individual quadratic were significant. Respectively, it was deduced that the adsorption of lead (Pb) were significant at 95% of confidence level which is an indication of the validity of the model. The significance of each term (variables) were evaluated using second order model and determined by P-value (Prob>F) which is listed in Table 2 and Table 3 (Design Expert Software 6.0.8). The statistic models are feasible and valid for the present work. The resulting of Response Surface Methodology model equation is following:

$$Y = 3.94 + 2.19 X_1 + 0.93 X_2 + 0.79 X_3 - 0.00875X_1^2 - 0.26 X_2^2 - 0.31X_3^2 \ldots$$
$$+ 0.58 X_1X_2 + 0.49 X_1X_3 + 0.20 X_2X_3 \ldots \quad (2)$$

From Equation (2), It can be seen that the factor of each variables (X_1 = Bentonite (g), X_2 = Agitation time (minutes), X_3 = Concentration of Lead (mg/L), X_1X_2 = interaction of X_1 and X_2 as well as $X_1 X_3$, and X_2X_3 is the effect of the interaction of these two variables) have a positive effect on the adsorption, the positive value represents an effect of optimization. As seen in Table 3, this models is very appropriate because the value of R^2 close to 1.00 (R^2 = 0.9996 ; Adj-R^2 = 0.9990; Pre-R^2 =

0.9929). So that, the correlation between actual and prediction is almost same and the suggested mathematical model in this study is quadratic model equation .

Table 3. Statistical Result Model

Response	Source	Std Dev	R-Square	Adj-R^2	Pre-R^2
Concentration of Pb (mg/l)	Linier	0.50	0.9391	0.9251	0.8793
	2 FI	0.27	0.9859	0.9774	0.9546
	Quadratic	0.058	0.9996	0.9990	0.9929

3.2 Effect of interactive variable

The effect of interactive variables can be seen at Figure 1. Figure 1 present the plots of experimental and predicted (responses surface). The predicted value that is proposed by design response surface equation has high accusation and low standard deviation. It can be seen from the distribution point spread above the prediction line. To obtain the decent model, that needs to check the accusation of normal probability value (%). As seen in Figure 2, the proposed equation model is suitable for adsorption (lead) by using bentonite. So that, the suitability indicates the analysis of variance were used for this study.

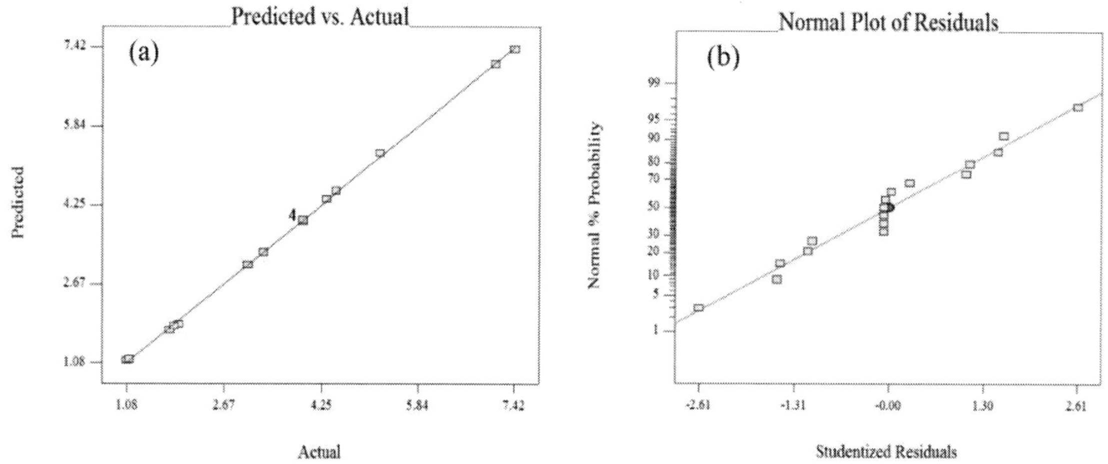

Figure 1. (a) Plot of experimental and predicted responses surface ; (b) plot of studentized residual and normal % probability

3.3 Response Surface Analysis.

The interactive effect of any two of three independent variables on the dependent variable was shows graphically by contour and 3-D response surface. 3-D plot describes the relation between dependent and independent variable as well as to analyze the interaction effect (response surface). Based on the polynomial regression, the suggested equation is Quadratic Model, Box-Behnken (Design Expert Software). The relation between adsorbent (Bentonite mass), agitation time, and adsorbent concentration were shown in (Figure 2 to 4). (Figure 2 to 4) indicated that with an extended of agitation time, the adsorption uptake was increased and then remained to the saturated value. This phenomenon indicated the adsorption reached equilibrium at a certain time [25].

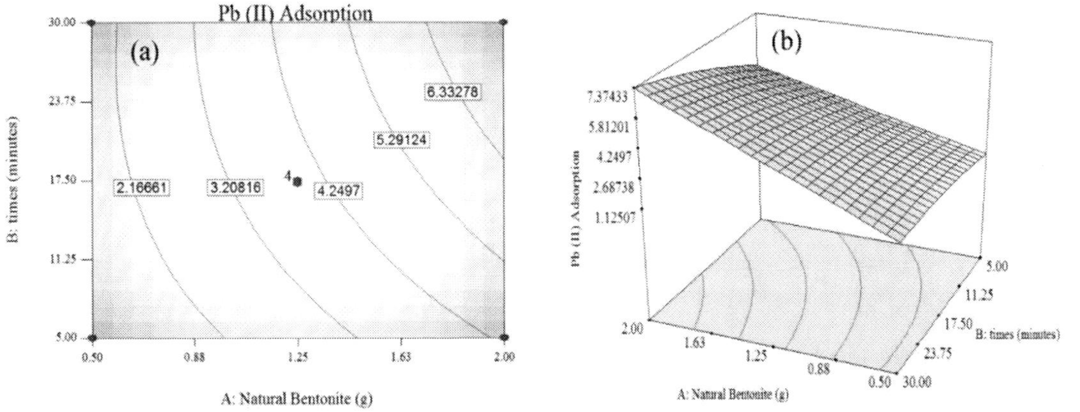

Figure 2. Response Surface – The effect of bentonite mass and agitation time on Pb(II) ion adsorption capacity (a) Contour (b) 3D Graphics

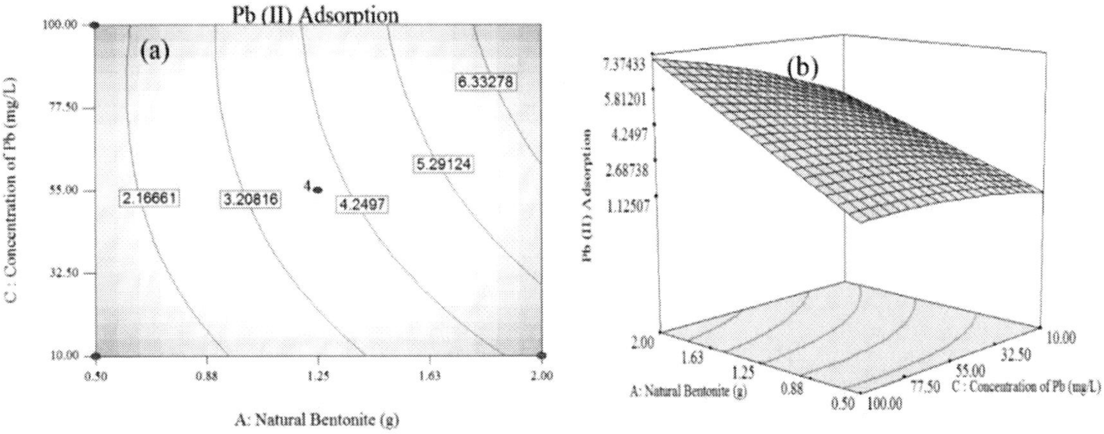

Figure 3. Response Surface – The effect of bentonite mass and adsorbate concentration on Pb(II) ion adsorption capacity (a) Contour (b) 3D Graphics

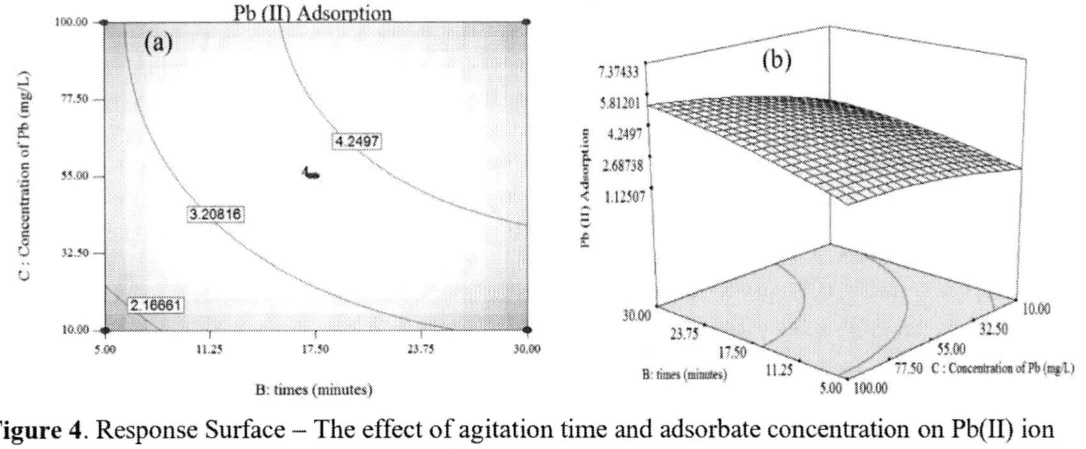

Figure 4. Response Surface – The effect of agitation time and adsorbate concentration on Pb(II) ion adsorption capacity (a) Contour (b) 3D Graphics

4. Conclusion

The statistical methodology, Box-Behnken Response Surface Method is demonstrated to be effective and reliable in finding the optimal conditions for the adsorption of Lead (Pb) onto bentonite. The response surface plots were used for estimating the interactive effect of variables (bentonite mass, agitation time, adsorbate concentration) on the response (adsorption capacity). The results showed that, the quadratic equation mathematical model was developed by regression analysis ($R^2 = 0.9996$) of the experimental data obtained for 17 runs. Optimization of adsorbent (bentonite) dose 2.0 g and agitation times 17.5 minutes gave a maximum uptake (ion) of 7.42 mg/L.

Acknowledgement

The authors acknowledge the financial support from the Ministry of Research, Technology and Higher Education, Republic of Indonesia, through the Research Grant of Doctoral Dissertation.

References

[1] Astuti W, Handayani A D and Wulandari D A 2018 *Jurnal Rekayasa Kimia dan Lingkungan* **13** 189-200
[2] WHO, "Lead Poisoning and Health. Tech. Rep. World Health Organization." [Online]. Available: [Accessed: 11-Aug-2019].
[3] Da'na E 2017 *Microporous and Mesoporous Materials* **247** 145-57
[4] Futalan C M, Kan C-C, Dalida M L, Pascua C and Wan M-W 2011 *Carbohydrate Polymers* **83** 697-704
[5] Han Y, Boateng A A, Qi P X, Lima I M and Chang J 2013 *Journal of Environmental Management* **118** 196-204
[6] Peng W, Li H, Liu Y and Song S 2017 *Journal of Molecular Liquids* **230** 496-504
[7] Husin H and Rosnelly C M 2006 *Jurnal Hasil Penelitian Industri (HPI)* **19**
[8] Husin H and Rosnelly C M 2007 *Jurnal Hasil Penelitian Industri (HPI)* **20**
[9] Gupta A and Garg A 2019 *Journal of environmental management* **250** 109474
[10] Zaki M, Maulana A, Tirtayani F, Alam P N and Husin H 2019 *Journal of Physics: Conference Series* **1402** 055010
[11] Belbachir I and Makhoukhi B 2017 *Journal of the Taiwan Institute of Chemical Engineers* **75** 105-11
[12] Dardir F M, Mohamed A S, Abukhadra M R, Ahmed E A and Soliman M F 2018 *Journal of Pharmaceutical Sciences* **115** 320-9
[13] Derakhshani E and Naghizadeh A 2018 *Journal of Molecular Liquids* **259** 76-81
[14] Misran E, Panjaitan F and Yanuar F M 2016 *Jurnal Rekayasa Kimia dan Lingkungan* **11** 92-8
[15] Mirzaei S M J, Heidarpour M, Tabatabaei S H, Najafi P and Hashemi S E 2013 *International Journal Of Recycling of Organic Waste in Agriculture* **2** 20
[16] Maliki S, Rosnelly C M, Adisalamun A, Husin H and Bilqis N 2019 *Journal of Physics: Conference Series* **1402** 055007
[17] Ahmad T, Guria C and Mandal A 2018 *Process Safety and Environmental Protection* **116** 703-17
[18] Arus V A, Nousir S, Sennour R, Shiao T C, Nistor I D, Roy R and Azzouz A 2018 *International Journal of Hydrogen Energy* **43** 7964-72
[19] Bertagnolli C, Kleinübing S J and da Silva M G C 2011 *Applied Clay Science* **53** 73-9
[20] Kobayashi I, Owada H, Ishii T and Iizuka A 2017 *Soils and Foundations* **57** 683-97
[21] Li M, Wang L, Chen J-y, Jiang Y-l and Wang W-j 2014 *Journal of Fuel Chemistry and Technology* **42** 1266-72
[22] Mohammed A A and Samaka I a S 2018 *Environmental Technology & Innovation* **10** 162-74
[23] Shojaeimehr T, Rahimpour F, Khadivi M A and Sadeghi M 2014 *Journal of Industrial and Engineering Chemistry* **20** 870-80

[24] Asnawi T M, Husin H, Adisalamun A, Rinaldi W, Zaki M and Hasfita F 2019 A *IOP Conference Series: Materials Science and Engineering* **543** 012096
[25] Hasan R and Setiabudi H D 2019 *Journal of King Saud University - Science* **31** 1182-8

IOP Conf. Series: Materials Science and Engineering **796** (2020) 012063 doi:10.1088/1757-899X/796/1/012063

Application of Clustering and VARIMA for Rainfall Prediction

Asep Rusyana[1], **Nidya Tatsara**[1*], **Rozatul Balqis**[1], **Syafinatur Rahmi**[1],
Department of Statistics, Faculty of Mathematics and Natural Sciences, Syiah Kuala University, Darussalam, Banda Aceh, Aceh, 23111, Indonesia

E-mail: nidya.t@mhs.unsyiah.ac.id

Abstract. Global warming causes climate change throughout the world which has an impact on changes in erratic rainfall. For fishermen, the estimated rainfall is an important thing to know before sailing. VARIMA is divided into VAR, VMA, VARMA, and VARIMA. This study uses a rainfall dataset consisting of rainfall (mm), temperature (°C), humidity (%) and wind speed (km/hour) with observations per day for one year in 2018. Training data is data that starts from 1 January 2018 until 30 November 2018 and the testing data starts on 1 December 2018 until 31 December 2018. Clustering was carried out in the area around Lampulo Pelabuhan Perikanan Samudera (PPS) which are Bandar Raya, Lampineung, Kuta Radja, and Kuta Alam. The best model obtained is VARIMA (1,1,2), which is based on the smallest AIC and BIC values, which are 8.5104 and 9.0581. The VARIMA (1,1,2) model also fulfills the white noise test, so it can be concluded that the data does not have autocorrelation and model can be used for predictions. Estimated values of rainfall, temperature, humidity and wind speed are not much different from actual values, with MAPE of all variables are relatively low.

1. Introduction

Climate change is the impact of global warming. One of the effects of climate change is the irregular pattern of rainfall. Rainfall is the amount of water falling on a flat ground surface during a certain period measured with a height of a millimeter (mm) above a horizontal surface. Each millimeter of rainfall means that in the area of one square meter in the flat is a millimeter high water [1]. The factors that affect high low rainfall are temperature (°C), humidity (%) and wind speed (km/h) [2]. One sector that perceives the effects of irregular rainfall patterns is the maritime sector, especially for fishermen.

Aceh is the westernmost region in Indonesia and surrounded by 295,370 km^2 of the ocean. Because the position makes Aceh is the most strategic area especially in the maritime sector. Banda Aceh as the capital of Aceh has an international ocean fishing port called Pelabuhan Perikanan Samudera (PPS) Lampulo. Lampulo lies in the Kuta Alam area, see Figure 1. Therefore, some inhabitants in Lampulo in Banda Aceh generally work as fishermen. Information about rainfall prediction is crucial for fishermen before sailing. It is useful to avoid high waves and storms in the ocean so that it can reduce the risk of accidents.

*Corresponding Author: nidya.t@mhs.unsyiah.ac.id

Content from this work may be used under the terms of the Creative Commons Attribution 3.0 licence. Any further distribution of this work must maintain attribution to the author(s) and the title of the work, journal citation and DOI.
Published under licence by IOP Publishing Ltd

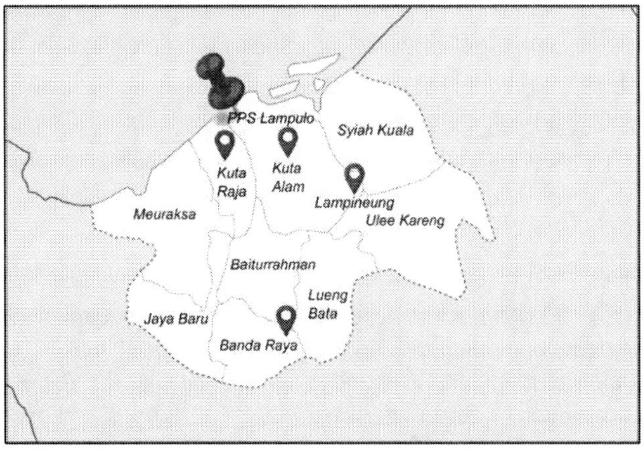

Figure 1. Location map of PPS Lampulo.

Clustering is a data grouping that has similarities or nearby territories. Information about future conditions can not be determined but can be predicted. One of the models for the multivariate time series prediction is Vector Autoregressive Integrated Moving Average (VARIMA). VARIMA is the generalization of Autoregressive Integrated Moving Average (ARIMA) which is a method of analysis of univariate time series. A univariate time series model sometimes results in less representative predictions. This is due to the influence of other variables that do not count towards the model. Therefore, this issue can be resolved by using a multivariate time series analysis such as VARIMA.

VARIMA model is a vector form of the ARIMA model that works by taking into account data and error data from self-changing and other reviewers simultaneously. So that the assessed changes should have a significant interchanges correlation. VARIMA models consist of 4 parts, namely Vector Autoregressive (VAR), Vector Moving Average (VMA), Vector Autoregressive Moving Average (VARMA) and Vector Autoregressive Integrated Moving Average (VARIMA). ARIMA is not only developed to VARIMA but also the other methods such as ARIMA-Autoregressive Conditionals Heteroscedastic (ARCH) and Seasonal ARIMA (SARIMA). Research has been conducted to predict nutmeg prices in South Aceh District by using ARCH[3].SARIMA has been applied to estimate the number of tourists who land at the Kualanamu International Airport in Medan [4].

Vector Autoregressive (VAR) is the generalization of AR to multiple parallel time series. VAR is a system of dynamic equations, with the estimation of a modified period of time depending on the movement of the variable and other variables involved in the system in the previous period [5]. VAR is used to project time-series variables by analyzing the dynamic impact of the interference factor contained in the variable [6]. The formulation for VAR order p or VAR(p) is as follows [7].

$$Y_t = \Phi_0 + \sum_{i=1}^{p} \Phi_i Y_{t-i} + a_t \qquad (1)$$

Where,

Y_t : (n x 1), n variables at t time

Φ_0 : vector averages(n x 1)

Φ_p : coefficientmatrix (n x n), with $i = 1, 2, 3 \ldots p$

a_t : error vector(n x 1)

Vector Moving Average (VMA) is the generalization of MA. VMA is a system of dynamic equations, with the estimation of a modified period of time depending on the error of variable and other error of variables involved in the system in the previous period. The formulation for VMA order q or VMA(q)is as follows [7].

$$Y_t = \Theta_0 + \sum_{i=1}^{q} \Theta_i a_{t-i} + a_t \quad (2)$$

Where,

Y_t : (n x 1), n variables at t time

Θ_0 : vector averages(n x 1)

Θ_q : coefficient matrix (n x n), with $i = 1, 2, 3... q$

a_t : error vector (n x 1)

Vector Autoregressive Moving Average (VARMA) is the generalization of ARMA. VMA is a system of dynamic equations, with the estimation of a modified period of time depending on the movement of the variable, error of variable and other variables and error of variables involved in the system in the previous period. The formulation for VARMA(p,q)is as follows[7].

$$Y_t = \Phi_0 + \Phi_1 Y_{t-1} + ...\Phi_p Y_{t-p} + a_t - \Theta_1 a_{t-1} - ... - \Theta_q a_{t-q} \quad (3)$$

Where,

Y_t : (n x 1), n variables at t time

Φ_0 : vector averages(n x 1)

Φ_p : coefficient matrix of VAR (n x n)

Θ_q : coefficient matrix of VMA (n x n)

a_t : error vector (n x 1)

VARIMA is a VARMA model that is experiencing the differencing process. The formulation of the VARIMA model is as follows [7].

$$\Phi_p(B)(I-B)^d Y_t = \Phi_0 + \Theta_q(B) a_t \quad (4)$$

Where,

Y_t : (n x 1), n variables at t time

Φ_p : coefficient matrix of VAR (n x n)

Θ_q : coefficient matrix of VMA (n x n)

$(I-B)^d$: differencing components

a_t : error vector (n x 1)

B : backshift operator

Or it can also be written as follows:

$$\begin{aligned} Y_t = (I+\Phi_1)Y_{t-1} + (\Phi_2 - \Phi_1)Y_{t-2} + ... + (\Phi_p - \Phi_{p-1})Y_{t-p} - \Phi_p Y_{t-p-1} \\ + a_t - \Theta_1 a_{t-1} - .... - \Theta_q a_{t-q} \end{aligned} \quad (5)$$

The problem of this research is how the rainfall, temperature, humidity, and wind speed overview in the area around of PPS Lampulo, Banda Aceh and what are the results of rainfall, temperature, humidity, and wind speed prediction in Banda Aceh for an upcoming period using the Vector Autoregressive Model Integrated Moving Average (VARIMA). The purpose of this research is to know the overview of rainfall, temperature, humidity and wind speed around the PPS Lampulo area in Banda Aceh and find out the prediction of rainfall, temperature, humidity and wind speed in Banda Aceh for an upcoming period.

2. Methodology

2.1. Data and Variable

The data used in this research is secondary data obtained from Meteorological, Climatological, and Geophysical Agency (BMKG) meteorology station class 1 Sultan Iskandar Muda at Blang Bintang, Aceh Besar and Meteorological, Climatological, and Geophysical Agency (BMKG) climatology station at Indrapuri, Aceh Besar.

The data used in this research for descriptive analysis (clustering) is the rainfall data in the area surrounding Lampulo, Banda Aceh which has a rain post that is Kuta Alam, Lampineung, Kuta Radja, and Bandar Raya. While data for the analysis of inference using rainfall data, temperature, humidity and wind speed in Banda Aceh. The period of data on rainfall, temperature, humidity and wind speed used in this research is daily data of 2018. The variables used in this research are presented in Table 1.

Table 1. Research variables description.

Variable	Description	Unit	Type of Data
Y_1	Rainfall	mm	Ratio
Y_2	Temperature	°C	Interval
Y_3	Humidity	%	Ratio
Y_4	Wind speed	km/hour	Ratio

The data used in this research is divided into training and testing data. Training data is used to create models for predictions i.e. data from 1 January 2018 until 30 November 2018 while testing data is used to measure the accuracy of the model in conducting predictions i.e. data from 1 December 2018 Until 31 December 2018.

2.2. Data Analysis Procedure

The procedure of data analysis in this research has the following stages:

1. A descriptive analysis is the clustering of rainfall, temperature, humidity and wind speed in the area around PPS Lampulo. There are Bandar Raya, Lampineung, Kuta Radja, and Kuta Alam.
2. Time series plotting and analyzing of rainfall, temperature, humidity, and wind speed
3. Multiple correlation testing. The hypothesis for multiple correlation test is:
   H_0: $R = I$ (there is no correlation between rainfall, temperature, humidity and wind speed)
   H_1: $R \neq I$ (there is a correlation between rainfall, temperature, humidity and wind speed)
4. Stationaritytestfor rainfall data, temperature, humidity and wind speed. The stationarity test consists of:
   a. Stationarity of variance test using Lambda Box-Cox. If nonstationary then do the transformation
   b. Stationarity of mean test using an Augmented Dickey-Fuller (ADF) test. If nonstationary then differencing. The ADF test hypothesis is:
      H_0: $\rho = 1$ (nonstationary data)
      H_1: $\rho \neq 1$ (stationary data)
5. Modeling using VARIMA method. This stage consists of:
   a. Determine the model order using the Extended Cross-Correlation Matrix (ECCM). The hypothesis for this test is:
      H_0: $\rho_{j+1}\left(w_{it}^{(j)}\right) = \ldots = \rho_l\left(w_{it}^{(j)}\right) = 0$ (Order model is nonsignificant)
      H_1: $\rho_v\left(w_{it}^{(j)}\right) \neq 0$ for $v > j$ (Order model significant)
      The best models are models with the smallest Akaike's Information Criterion (AIC) and Bayesian Information Criteria (BIC) values.
   b. Make a parameter estimate to know the significance of a parameter in a model
   c. Perform a diagnostic model of white noise test. The following is a hypothesis for the test:

H_0: $\Gamma i = 0$ (Errors data together have white noise)
H_1: $\Gamma i \neq 0$ (Errors data together not white noise)

   d. If the model does not qualify for point c then the process is repeated from point a.

6. Calculated prediction error using Mean Absolute Percentage Error (MAPE). MAPE is used to determine the percentage of deviation of prediction results.
7. Predicted rainfall, temperature, humidity, and wind speed.
8. Concluded the analysis of the results.

3. Result and Discussion

A descriptive or clustering analysis carried out only the rainfall of four regions around Ocean Fishing Port (PPS)Lampulo, Banda Aceh with a rain post. There are Banda Raya, Lampineung, Kuta Radja, and Kuta Alam. Table 2 is the result of a summary of the rainfall in the region.

Table 2. Rainfall summary in Banda Raya and Lampineung.

Region	Banda Raya						Lampineung					
Months	Rainfall (mm)						Rainfall (mm)					
	0-20	21-50	51-100	>100	-	x	0-20	20-50	50-100	>100	-	x
Jan	4	2	1	-	24	-	3	2	-	-	26	-
Feb	3	0	-	-	25	3	4	-	-	-	24	3
Mar	4	0	-	-	27	-	2	-	-	-	18	11
Apr	3	4	-	-	23	1	6	1	-	-	17	7
May	6	2	1	-	22	-	14	-	-	-	17	-
June	2	3	-	-	25	1	1	-	-	-	17	13
July	8	1	-	-	22	-	10	-	-	-	21	-
Aug	7	1	-	-	23	-	6	-	-	-	25	-
Sept	10	3	-	-	18	-	11	2	-	-	18	1
Oct	14	2	-	-	16	-	13	-	-	-	19	-
Nov	17	3	-	-	10	1	14	-	-	-	16	1
Dec	8	4	-	-	19	-	8	2	1	-	20	-
Total	86	25	2	0	254	6	92	7	1	0	238	36

Table 3. Rainfall summary in Kuta Radja and Kuta Alam.

Region	Kuta Radja						Kuta Alam					
Months	Rainfall (mm)						Rainfall (mm)					
	0-20	20-50	50-100	>100	-	x	0-20	20-50	50-100	>100	-	x
Jan	6	2	-	-	23	-	6	1	-	-	24	-
Feb	6	-	-	-	22	3	3	-	-	-	15	13
Mar	4	1	-	-	26	-	6	1	-	-	24	-
Apr	11	1	-	-	18	1	9	-	-	-	21	1
May	14	-	-	1	16	-	11	-	-	1	19	-
June	7	1	-	-	22	1	10	1	2	-	17	1
July	12	-	-	-	19	-	13	-	-	-	18	-
Aug	10	1	-	-	19	2	11	-	-	-	20	-
Sept	13	1	-	-	16	1	7	1	-	-	22	1
Oct	12	3	-	-	16	-	8	1	-	-	1	21
Nov	13	3	1	1	12	1	13	3	1	-	13	1
Dec	9	2	3	-	17	-	10	2	1	-	18	-
Total	117	15	4	2	226	9	107	10	4	1	212	38

IOP Conf. Series: Materials Science and Engineering **796** (2020) 012063 doi:10.1088/1757-899X/796/1/012063

Description:

0 - 20 mm	= Mild	> 100 mm	= Very dense
21- 50 mm	= Moderate	-	= No rain
51 - 100 mm	= Dense	x	= Absence of observation

According to Table 2 and Table 3 shows that the occurrence of mild rain most occurred in Kuta Radja, the occurrence of moderate rain occurs most in Banda Raya while the rain dense category rarely occurs in all areas. No rain occurs most often in a year that is more than 200 days a year. The absence of most observations occurred in Kuta Alam for38 days. Based on Figure 2 explaining the average rainfall in the areas of Banda Raya, Lampineung, Kuta Radja, and Kuta Alam are located below the average rainfall of Banda Aceh.

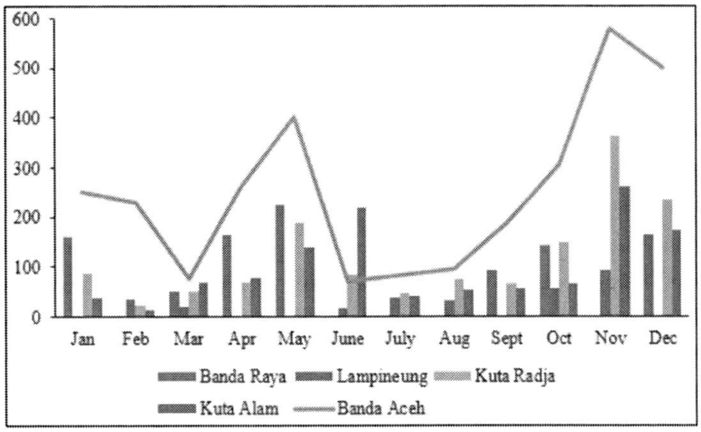

Figure 2. Rainfall comparison in 2018.

The inference analysis done is the application of VARIMA model. Here are the stages of getting a VARIMA model.

- Multiple correlation test
  This correlation test uses a simultaneous correlation test namely Bartlett Sphericity Test. Based on the test was obtained P-value is less than 0,000. The value is smaller than α (0.05) so H_0 is rejected. This indicates that there is an analyzed correlation between the variables so that these variables are appropriate for analysis using the VARIMA model.
- Stationarity test
  The stationarity test divided into two tests, variance and mean. The stationarity of variance resulted that the four variables must be transformed because it is nonstationary. The stationarity of the mean test resulted that only Y_1 and Y_4 stationary because P-value is $< \alpha$ (0.05). While Y_2 and Y_3 have P-value $> \alpha$ (0.05). So, it can be concluded that the two variables are nonstationary and should be done a differencing process.
- VARIMA modeling
  Before modeling, VARIMA performed vector model identification using ECCM. A significant order is an order with the P-value $< \alpha$ (0.05). Seven models can be tested at the modeling stage. Models with the smallest AIC and BIC values are VARIMA models (1, 1, 2) with AIC value is 8.5104 and the BIC value is 9.0581. Then it can be concluded that the best model is VARIMA (1, 2, 2).
- Parameter estimation and diagnostic model
  Parameter estimation is done on the best model of VARIMA (1, 2, 2). The significant parameter is a parameter that has a P-value$< \alpha$ (0.05) so there is a result that two significant parameters are Y_4*4 and Y_4*11. The following are VARIMA (1, 1, 2) formulation that is formed.

$$\hat{Y}^*_{1,t} = 0{,}5977Y^*_{1,t-1} - 4.497Y^*_{2,t-1} + 0.1310Y^*_{3,t-1} + 0.6328Y^*_{4,t-1} - 0.0054a_{1,t-1} - 1.7746a_{2,t-1} + 0.2704a_{3,t-1} - 0.1576a_{4,t-1} - 0.0328a_{1,t-2} - 2.265a_{2,t-2} - 1.1540a_{3,t-2} + 0.2503a_{4,t-2}$$

$$\hat{Y}^*_{2,t} = -0.0054Y^*_{1,t-1} - 0.563Y^*_{2,t-1} - 0.1600Y^*_{3,t-1} + 0.0086Y^*_{4,t-1} - 0.0080a_{1,t-1} + 0.0469a_{2,t-1} - 0.1513a_{3,t-1} - 0.00426a_{4,t-1} - 0.0012a_{1,t-2} + 0.443a_{2,t-2} + 0.1014a_{3,t-2} + 0.0505a_{4,t-2}$$

$$\hat{Y}^*_{3,t} = -0.0678Y^*_{1,t-1} + 1.750Y^*_{2,t-1} + 0.1000Y^*_{3,t-1} + 0.1091Y^*_{4,t-1} - 0.0053a_{1,t-1} + 1.5861a_{2,t-1} + 0.6953a_{3,t-1} + 0.1731a_{4,t-1} - 0.0328a_{1,t-2} - 1.442a_{2,t-2} - 0.0348a_{3,t-2} - 0.4284a_{4,t-2}$$

$$\hat{Y}^*_{4,t} = -0.0046Y^*_{1,t-1} + 0.9510Y^*_{2,t-1} - 0.0672Y^*_{3,t-1} + 1.0066Y^*_{4,t-1} - 0.0047a_{1,t-1} + 0.6970a_{2,t-1} - 0.0087a_{3,t-1} + 0.6175a_{4,t-1} + 0.0019a_{1,t-2} - 0.5520a_{2,t-2} + 0.0004a_{3,t-2} + 0.1241a_{4,t-2}$$

The diagnostic model obtained P-value is 0.999. The decision is H_0 cannot rejected, because P-value > α (0.05). It can be concluded that the data has no autocorrelation, and the model has qualified white noise. Thus VARIMA (1, 1, 2) models can be used to predict rainfall, temperature, humidity and wind speed.

- Measuring error prediction
  Error prediction measured using MAPE. The value of MAPE obtained are Y_1 = 38%, Y_2 = 10%, Y_3 = 6.2% and Y_4 = 19%. The value indicates that the predicted error is relatively low.
- Predictions
  Results comparison of the predictions data and actual data (testing data) of rainfall, temperature, humidity and wind speed are shown in Figure 3.

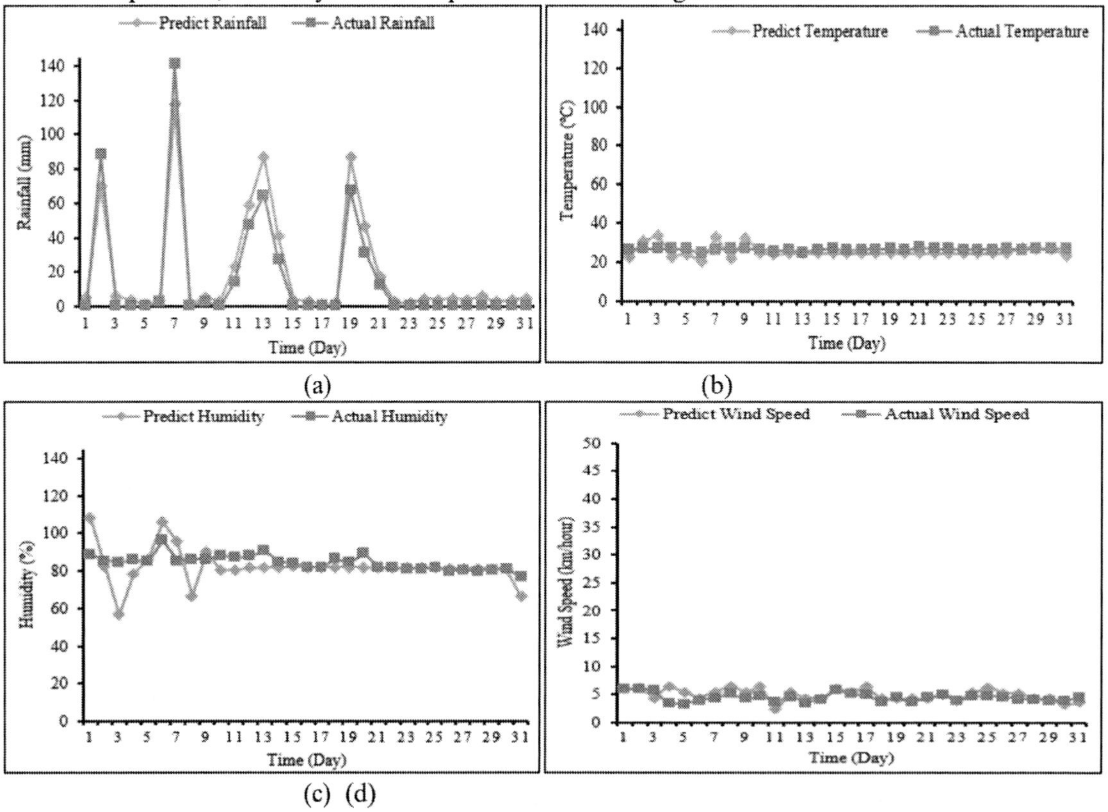

Figure 3. Comparison of predicted data and actual data (a) rainfall, (b) temperature, (c) humidity, (d) wind speed.

Based on Figure 3shows that the predicted data of rainfall, temperature, humidity, and wind speed are close to the actual data so that the approach with the VARIMA model is appropriate. The prediction results of rainfall, temperature, humidity and wind speed in January 2019 using the VARIMA model (1,1,2) are presented in Table 4.

Table 4. Predicted results of rainfall, temperature, humidity and wind speed in January 2019.

Date	Rainfall	Temperature	Humidity	Wind Speed	Date	Rainfall	Temperature	Humidity	Wind Speed
	mm	°C	%	km/h		mm	°C	%	km/h
1/1/2019	5.9	21	108	6	17/1/2019	1.8	25	82	6
2/1/2019	65	28	83	6	18/1/2019	2.9	25	82	4
3/1/2019	5.9	35	56	4	19/1/2019	70.3	25	82	4
4/1/2019	5.9	20	79	6	20/1/2019	46.8	25	82	4
5/1/2019	2.9	20	86	3	21/1/2019	11.7	25	82	4
6/1/2019	5.9	23	106	4	22/1/2019	5.8	25	82	4
7/1/2019	100.2	31	95	5	23/1/2019	5.8	25	82	4
8/1/2019	2.9	25	67	7	24/1/2019	5.8	25	82	4
9/1/2019	5.9	33	90	5	25/1/2019	5.8	25	81	4
10/1/2019	4.9	25	80	7	26/1/2019	5.8	24	81	4
11/1/2019	29.4	24	80	2	27/1/2019	5.8	24	81	4
12/1/2019	64.7	25	82	5	28/1/2019	5.8	27	81	4
13/1/2019	82.3	25	82	4	29/1/2019	5.8	27	81	4
14/1/2019	35.2	25	82	4	30/1/2019	5.8	27	81	4
15/1/2019	5.9	25	82	6	31/1/2019	5.8	24	66	4
16/1/2019	5.9	25	82	5					

4. Conclusion

Based on the research that has been done can be concluded the occurrence of mild rain category occurred in the Kuta Radja, the most moderate category occurrence of rain occurred in Banda Raya while the rainy category frequently occurs in all areas. No rain occurs most often in a year that is more than 200 days a year. The absence of most observations occurred in Kuta Alam for 38 days. Besides, the average rainfall in the area around PPS Lampulo is below the average rainfall of Banda Aceh.

The best model for rainfall predictions, temperatures, humidity and wind speeds is VARIMA (1, 1, 2) with the AIC value equal to 8.5104 and BIC value equal to 9.058. The results of rainfall prediction, temperature, humidity, and wind speed are close to the actual data, so it can be concluded that the VARIMA model is appropriate and the predicted error is relatively low and.

Information on the results of rainfall prediction rainfall that will be received by fishermen will be more accurate so that the number of marine accident events caused by extreme weather be decreased.

Acknowledgments

The author's acknowledgements to Kemristekdikti for providing funds for the implementation of this research, BMKG meteorology station class 1 Sultan Iskandar Muda at Blang Bintang, Aceh Besar and BMKG climatology station at Indrapuri, Aceh Besar for permission to use this research data.

References

[1] Swarinoto YS and Sugiyono 2011 *Utilization of Air Temperature and Air Humidity in Regression Equations for Prediction Simulation of Total Rainy Monthly in Bandar Lampung BMKG Bandar Lampung*

IOP Conf. Series: Materials Science and Engineering **796** (2020) 012063 doi:10.1088/1757-899X/796/1/012063

[2] Tjasyono B 2004 *Climatology* (Bandung: ITB)

[3] Rusyana A, Ferdhiana R and Putri ME 2019 *Application of ARCH model on nutmeg price forecasting in South Aceh DistrictIOP Conf. Ser. Mater. Sci. Eng.* **523** 012003 https://iopscience.iop.org/article/10.1088/1757-899X/523/1/012003/pdf

[4] Rusyana A, Nurhasanah, Marzuki and Flancia M 2016 *SARIMA model for forecasting foreign tourists at the Kualanamu International Airport* In: The 12th International Conference Mathematics, Statistics, and their Applications (ICMSA). (Banda Aceh, Indonesia: IEEE) p. 153–158 https://ieeexplore.ieee.org/document/7954329/

[5] Enders W 1995 *Applied Econometric Time* (New York: Series Willey and Sons, Inc.)

[6] Nahrowi ND 2006 *A Popular and Practical Approach to Econometrics for Economic and Financial Analysis* (Jakarta: Lembaga)

[7] Wei WWS 2006 *Time Series Analysis Univariate and Multivariate Methods* 2nd ed. (Boston: Pearson Education, Inc,)

Grouping of Retention Index on Gas Chromatography using Cluster Analysis

I Husna[1], A Rusyana[5], Muslem[1], G M Idroes[3], R Suhendra[4]and R Idroes[1,2,*]

[1] Department of Chemistry, Faculty of Mathematics and Natural Sciences, Universitas Syiah Kuala, Kopelma Darussalam, Banda Aceh 23111, Indonesia
[2] Department of Pharmacy, Faculty of Mathematics and Natural Sciences, Universitas Syiah Kuala, Kopelma Darussalam, Banda Aceh 23111, Indonesia
[3] Department of Chemical Engineering, Faculty of Engineering, Universitas Syiah Kuala, Kopelma Darussalam, Banda Aceh 23111, Indonesia
[4] Department of Informatics, Faculty of Mathematics and Natural Sciences, Universitas Syiah Kuala, Kopelma Darussalam, Banda Aceh 23111, Indonesia
[5] Department of Statistics, Faculty of Mathematics and Natural Sciences, Universitas Syiah Kuala, Kopelma Darussalam, Banda Aceh 23111, Indonesia

Email: rinaldi.idroes@unsyiah.ac.id

Abstract. Retention index data consisting of 146 compounds commonly found in fuels are used to analyze the level of a psycho-chemical property similarity with cluster analysis methods. Cluster analysis is applied based on the temperature and column dependencies. The results show that the retention index was divided into four groups. The first group consists of benzene and cycloalkanes. The retention index of the compound shows high-temperature dependence and dependence on small column polarity. The second group consists of aromatic compounds and compounds that contain N, S, O. The retention index of these compounds shows dependence on temperature and polarity of large columns. The third group consists of aromatic compounds and hydrocarbons. The retention index of hydrocarbon compounds shows temperature dependence smaller than N, S, O. compounds and also the dependence on small column polarity. And the fourth group consists of only two compounds namely n-Hentriacontane and chrysene. The retention index of the two compounds shows a significant temperature dependence.

1. Introduction

Retention index was first introduced by Erwin Kovats in 1958 which became known as the Kovats retention index. The Kovats retention index uses the n-alkane homologous series as the standard series to change the non-fixed retention time into a system that is relatively stable against the change of analysis conditions [1]. The non-fixed t_R behavior makes this parameter just use as a separation marker [2] of a compound that is subsequently identified by mass spectrometry [3].On another hand, the stable value of the retention index makes it possible to obtain psycho-chemical [4] and thermodynamic [1] information of a compound.

[1]*Coressponding author: rinaldi.idroes@unsyiah.ac.id

Content from this work may be used under the terms of the Creative Commons Attribution 3.0 licence. Any further distribution of this work must maintain attribution to the author(s) and the title of the work, journal citation and DOI.
Published under licence by IOP Publishing Ltd

Besides stability, the retention index also has the lowest systematic error than other retention parameters such as retention time, adjusted retention time, capacity factor, etc [5]. These conditions make it possible to have a relationship between psycho-chemical and thermodynamic properties and retention index. To obtain this relationship, the statistical method is required to analyze the retention index data from the various compounds and classified into the group.

Idroes et al., (2018) have grouped flavor and fragrance retention indexes obtained from some column using the biplot method. The results show that retention indexes are divided into three groups. Each group consists of the similarity of the psycho-chemical property of the compounds. The first group consists of n-alkanes with lower carbon numbers and lower retention indexes. The second group consists of n-alkanes with higher carbon numbers and terpenoid compounds with higher Kow log values. The third group consists of alcohol, aldehydes, esters, and carboxylic acid compounds [6].

In this study, the retention index data of pure compounds that are commonly found in fuels derived from bricks and petroleum are analyzed for similarity levels based on groups formed using cluster methods. The psycho-chemical properties from the compounds in each group are identified and compared to other groups.

2. Literature Review

Cluster analysis is a statistical method for grouping or combining n objects into k groups [7]. Cluster analysis is one of the multivariate techniques for grouping objects into suitable groups based on their resemblance. Objects formed in the same group will be more similar than objects between groups. The similarity between the objects is measured by the Euclidian distance. The equation from the Euclidian distance is:

$$d(x,y) = [(x_1 - y_1)^2 + (x_2 - y_2)^2 + \cdots + (x_p - y_p)^2]^{1/2} \quad (1)$$

Where: d (x,y) is the distance between object i to object j, x_p is the middle value in the group i and yp is the middle value in the group j [8]. Cluster analysis aims to determine the diversity of the object under study and to find out the similarities between objects in a group formed. The greater the similarity in the group, the better the grouping will be.

Group analysis is divided into two, namely hierarchical and non-hierarchical. There are two approaches in the analysis of hierarchical groups, one of which is the agglomerative approach. Agglomerative cluster analysis method grouping objects starting from n objects into one group and it is also called a down to top. To measure the dissimilarity between groups formed can be using the average linkage. Average Linkage measures or looks for similarities between groups by looking at the average distance of observation data. The formula for Average Linkage is as follows:

$$d_{G,A}(G,H) = \frac{1}{N_G N_H} \sum_{i \in G} \sum_{i' \in H} d_{ii'} \quad (2)$$

Analysis of hierarchical groups initially consisted of separate groups so that there were as many groups as observational data. Then similar data will be entered into the same group, repeated until formed into a group that contains all observational data [9]. The results of the cluster analysis are dendrogram based on distances between objects. The hierarchical cluster analysis method requires measurement of inequality (distance) for each data to be grouped to facilitate the development of the dendrogram.

3. Data and Methods

3.1. *Data collection*

This research takes retention index data from Lai's 1995 literary article [10]. The sample used was 146 pure compounds that are generally found in liquid fuels derived from coal and petroleum. The columns used are two capillary columns with different stationary phases of polarity. First is the Rtx-50 column with a 50% phenyl-50% methyl polysiloxane stationary phase and the second is a DB-5 column with a 5% phenyl-95% methyl polysiloxane stationary phase. The column temperature is programmed linearly from levels 40 to 310 at a heating rate of 2, 4, 6 °C min^{-1}.

In this research, the n-alkane series is used as a standard to calculate the retention index of several compounds because it has a robust retention index.

3.2. *Data analysis*

Cluster analysis is performed using XLSTAT software. After the data is entered into the software, the analysis is carried out using the cluster analysis.

4. Result and discussion

The retention index is included in the secondary retention parameter because it is generated through mathematical calculations using primary retention parameter data which is dead time and retention time. The retention index is a parameter that is often used as a standard for gas chromatography using the n-alkane series as an internal standard under isothermal conditions. The retention index is also defined as 100 times n_c (carbon number) [11]. In this study, the retention index data is used to see the level of similarity based on groups formed using the cluster analysis. The compounds analyzed are common compounds found in fuels. Alkanes, cycloalkanes, and aromatics are the three main components that are often found in liquid fuels.

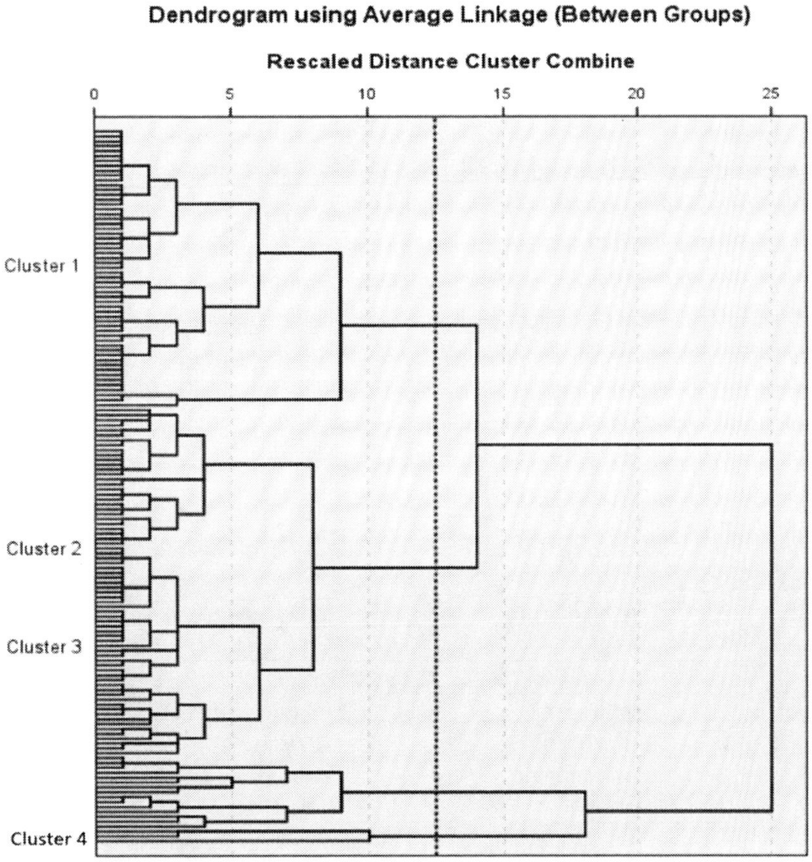

Figure 1. Dendrogram of cluster analysis.

IOP Conf. Series: Materials Science and Engineering **796** (2020) 012064 doi:10.1088/1757-899X/796/1/012064

Table 1. Compounds in their cluster.

Cluster 1			
1-Ethyl-3-Methylbenzene	1,3,5-Trimethylbenzene	*cis*-Octahydro-1H-Indene	*sec*-Butylbenzene
1,2,4-Trimethylbenzene	n-Butylcyclohexane	1-Decane	tert-Butylcyclohexane
n-Decane	Isopropylbenzene	2-Methylindan	1-Methylindan
n-Pentylcyclohexane	*cis*-Decalin	n-Undecane	1-Undecane
trans-Decalin	n-Butylbenzene	Indan	Phenyl Ethyl Ether
1,3-Dimethylbenzene	1,2-Dimethylbenzene	n-Nonane	1-Nonane
n-Propylcyclohexane	Cyclohexene	Benzene	n-Heptane
1-Heptane	Cyclohexane	Methylcyclohexane	Toluene
Pyridine	*cis*-1,2-Dimethylcyclohexane	*cis*-1,4-Dimethylcyclohexane	*trans*-1,2-Dimethylcyclohexane
cis-1,3-Dimethylcyclohexane	n-Octane	1-Octane	1-Methylcyclohexene
n-Hexene	n-Pentane		

Cluster 2			
1,3,5-Triethylbenzene	n-Tridecane	1-Tridecane	Hexylbenzene
Bicyclohexyl	n-Dodecane	1-Dodecane	1,2,3,4-Tetramethylbenzene
1,4-Diisoprppylbenzene	Tetralin	n-Tetradecane	1-Tetradecane
5,6,7,8-Tetrahydro-3-Methylquinoline	2-Methylnaphthalene	Cyclohexylbeneze	1-Indanol
Quinoline	Naphthalene	Benzothiophene	1,4-Dimethylnaphthalene
1,2-Dimethylnaphthalene	1,3-Dimethylnaphthalene	2,7-Dimethylnaphthalene	2-Methylbihenyl
1-Ethylnaphthalene	2-Ethylnaphthalene	2,6-Di-tert-Butylphenol	1,2-Diphenylethane
3-Methylbiphenyl	1,8-Dimethylnaphthalene	5,6,7,8-Tetrahydro-1-Naphthol	n-Hexadecane
1-Hexadecene	n-Octylbenzene	n-Pentadecane	n-Octylcyclohexane
4,4`-Dimethylbiphenyl	Flourene	2-Naphthol	9,10-Dihydroanthracene
1,2,3,4,5,6,7,8-Octahydroacridine	Dibenzyl Ether	n-Decylcyclohexane	n-Heptadecane
n-Octadecane	1-Octadecane	n-Nonadecane	

Cluster 3			
Phenanthrene	Anthracene	1-Phenylnaphthalene	n-Eicosane
n-Henicosane	n-docosane	n-Tricosane	Pyrine
p-Terphenyl	Fluoranthene	n-Tetracosane	n-Pentacosane
n-Hexacosane			

Cluster 4			
	n-Hentriacontane	Chrysene	

IOP Conf. Series: Materials Science and Engineering **796** (2020) 012064 doi:10.1088/1757-899X/796/1/012064

The results obtained from the cluster analysis in the form of a dendrogram showing grouping based on similarities in column polarity and the heating rate between the compounds studied. From the results of the analysis of these compounds are grouped into four main groups, see Figure 1 and Table 1. The first group consists of benzene and cycloalkanes. The retention index of the compound shows high-temperature dependence and dependence on small column polarity. The second group consists of aromatic compounds and compounds that contain N, S, O such as Quinoline, Indanol, and Benzothiophene. The retention index of the compound shows greater temperature dependence and is very dependent on the polarity of the column used. The third group consists of aromatic compounds and hydrocarbons. The retention index of hydrocarbon compounds shows a smaller temperature dependence than N, S, O, and dependence on small column polarity for hydrocarbon compounds, whereas N, S, O compounds are very dependent on column polarity. The fourth group consists of only two compounds namely n-Hentriacontane and Chrysene. These compounds show significant temperature dependence. The retention index of all the compounds studied was reduced by decreasing column polarity (from Rtx-50 to DB-5).

5. Conclusion

Based on the analysis of the retention index data of 146 pure compounds commonly found in liquid fuels can be grouped into four groups. The first group consists of benzene and cycloalkanes. The second group consists of aromatic compounds and compounds that contain N, S, O. The third group consists of aromatic compounds and hydrocarbons. And the fourth group consists of only two compounds namely n-Hentriacontane and Chrysene. Compounds in each group have similarities between one compound with other compounds.

Reference

[1] Didaoui, L., Touabet, A., Meklati, B. Y., Alia, E., Ezzouar, B. P. B., Lewin-kretzschmar, U., & Engewald, W.,1999, Comparison of Retention Indices of Some Explosives and Related Compounds Calculated by Different Mathematical Methods in Reversed-Phase Liquid Chromatography,*Journal of HRC-High Resolution Chromatography*,**11** 613–618

[2] Earlia, N., Rahmad, Amin, M., Prakoeswa, C. R. S., Khairan., and Idroes, R., The Potential Effect of Fatty Acids from Pliek U on Epidermal Fatty Acid Binding Protein: Chromatography and Bioinformatic Studies, Sains Malaysiana, 48, 1019–1024.

[3] Pratiwi, S. U. T., Lagendijk, E. L., Weert, S. V., Idroes, R., Hertiani, T., and Hondel, C. V. D., 2015, Effect of Cinnamomum burmannii Nees ex Bl. and Massoia aromatica Becc. Essential Oils on Planktonic Growth and Biofilm formation of Pseudomonas aeruginosa and Staphylococcus aureus In Vitro, International Journal of Applied Research in Natural Products Vol. 8 (2), pp. 1-13.

[4] Nowotnik, D.P. and Narra, R. K., 1993. A Comparison of methods for the determination of dead time in a reversed-phase high-performance liquid chromatography system used for the measurement of lipophilicity, J. of Liquid Chromatography 16(18), 3919 – 3932.

[5] Idroes, R., 2011, Dead Time Determination and Its Influence to Retention Index in GC System Using Polar Capilarry Column *Maj. Farm. Indones***3** 257-264

[6] Japnur, A. F., Idroes R., Suhendra, R., Rusyana, A., 2018, Kovats Retention Index Analysis of Flavor and Fragrance Compound using Biplot Statistical Method in Gas Chromatography Systems *AIC Conf*

[7] Afifi, A. A., & Clark, V 1999 *Computer-Aided Multivariate Analysis* (3rd ed.). New York: Chapman & Hall/ CRC.

IOP Conf. Series: Materials Science and Engineering **796** (2020) 012064 doi:10.1088/1757-899X/796/1/012064

[8] Johnson, R. A., & Wichern, D. W 2007 *Applied Multivariate Statistical Analysis* (6th ed.). New Jersey: Pearson

[9] Everitt, B. S., Landau, S., Leese, M., & Stahl, D. (2011). *Cluster Analysis* (5th ed.). United Kingdom: John Wiley & Sons, Ldt.

[10] Lai, Wei-Chuan, Chunshan Song 1995 Temperature-Programmed Retention Indices for G.C. And G.C.-M.S. Analysis of Coal- And Petroleum-Derived Liquid Fuels. *Fuel.* **10** 1436-1451

[11] Idroes, R., 2005, Determination of Absolute Index System in High Performance Liquid Chomatography (RP-HPLC) *Malaysian J. Anal. Sci. 3* 224-232.

AUTHOR INDEX